Food Engineering Series

More information about this series at http://www.springer.com/series/5996

Abdul Malik • Zerrin Erginkaya
Saghir Ahmad • Hüseyin Erten
Editors

Food Processing: Strategies for Quality Assessment

Springer

Editors
Abdul Malik
Department of Agricultural Microbiology
Aligarh Muslim University
Aligarh, India

Saghir Ahmad
Department of Post Harvest Engineering and Technology
Aligarh Muslim University
Aligarh, India

Zerrin Erginkaya
Department of Food Engineering
Faculty of Agriculture
Cukurova University
Adana, Turkey

Hüseyin Erten
Department of Food Engineering
Cukurova University
Adana, Turkey

ISSN 1571-0297
ISBN 978-1-4939-1377-0 ISBN 978-1-4939-1378-7 (eBook)
DOI 10.1007/978-1-4939-1378-7
Springer New York Heidelberg Dordrecht London

Library of Congress Control Number: 2014950398

Printed on acid-free paper

Springer is part of Springer Science+Business Media (www.springer.com)

Preface

Food processing is very important in many economies of the world. Food processing may be carried out in the home or by community groups or as cottage industries or more formal commercial formulations with respective sites having increasing levels of sophistication and capital requirements. The aims of the food processing are to ensure microbiological and chemical safety of foods, adequate nutrient content and bioavailability, and acceptability to the consumer and caregiver with regard to sensory properties and ease of preparation. Processing may have either beneficial or detrimental effects on these different properties of food, so each of these factors must be taken into account in the design and preparation of complementary foods.

Food quality is frequently associated with food safety. A good quality food is one that is safe to consume, nutritious, and organoleptically acceptable. Food safety encompasses a whole series of processes and activities both within and outside the food processing plant that will ensure that the food is free of potential chemical, physical, and biological hazards. Among these hazards (which may not be intentional) are naturally occurring toxins, pathogenic microorganisms, and harmful chemicals (pesticides for example).

Quality within a food processing plant may also be related to the notion of quality control. In this regard, quality control has many objectives within a food processing plants, mainly being to maintain the nutritional value of the processed product, to protect customers from the dangers of contaminated food and associated foodborne diseases, to ensure that all food laws and regulations (whether local, national or international) are met, to facilitate international business and commerce, etc. Food and Agriculture Organization (FAO), World Health Organization (WHO) and several national level regulatory organizations such as Fruit products order (FPO), Prevention of Food Adulteration (PFA), Agricultural Produce (grading and marking) (AGMARK), and Meat Food Products Order (MFPO) have been designated to safeguard the health of consumers and to prevent fraudulent practices in food trade.

HACCP (Hazard Analysis and Critical Control Points) is a system used by many food processing plants to ensure the safety and quality of the food products.

The idea behind HACCP is that food processing has critical points at which food contamination (physical, chemical, and/or biological) may occur. By controlling quality tightly at those critical points, it is possible to control the whole food processing.

This book offers a unique dealing with the subject and provides not only an update of state-of-the-art techniques in many critical areas of food processing and quality assessment but also the development of value added products from food waste, Food safety, and nanotechnology in Food and Agriculture Industry and looks into the future by defining current bottlenecks and future research goals. This work will facilitate a ready reference of the current subject matter to students and researchers alike.

This book is not intended to serve as an encyclopedic review of the subject. However, the various chapters incorporate both theoretical and practical aspects and may serve as baseline information for future research through which significant development is possible.

The book has nineteen chapters, with each focusing on a specific topic to cover diverse perspectives. An introductory chapter on Food Processing is included. Other chapters give insights on Processing of fruits, Nutritional Quality Assessment in Dairy Products, Food quality and safety, various Foodborne Microbial Diseases and their Control, use of Microbial Metabolites as Biological Control Agents in Food Safety, Recent Approaches in Risk Assessment of Foods, Microbiological Quality Systems and Microbial Risk Analysis. Furthermore the book also includes chapters on Value Addition and Preservation by Fermentation Technology, Importance of Yeasts and Lactic Acid Bacteria in Food Processing, Application of membrane technology in food processing, Post Harvest Management and Value Addition of Horticultural Crops, Development of Value Added Products from Food Wastes, and Nanotechnology in Food and Agriculture Industry.

With great pleasure, we extend our sincere thanks to all our well-qualified and internationally renowned contributors for providing the important, authoritative, and cutting edge scientific information/technology to make this book a reality. All chapters are well illustrated with appropriately placed tables and figures and enriched with up-to-date information. We are also thankful to the reviewers who carefully and timely reviewed the manuscript.

We are extremely thankful to Springer, New York, USA for completing the review process expeditiously to grant acceptance for publication. We appreciate the great efforts of book publishing team especially Dr. Susan Safren, Senior Publishing Editor, Food Science, in responding to all queries very promptly.

We express sincere thanks to our family members for all the support they provided, and regret the neglect and loss they suffered during the preparation of this book.

Aligarh, India — Abdul Malik
Adana, Turkey — Zerrin Erginkaya
Aligarh, India — Saghir Ahmad
Adana, Turkey — Hüseyin Erten

Contents

Contributors

Erdal Ağçam Department of Food Engineering, Cukurova University, Adana, Turkey

Bilal Ağırman Faculty of Agriculture, Department of Food Engineering, Cukurova University, Adana, Turkey

Saghir Ahmad Faculty of Agricultural Sciences, Department of Post Harvest Engineering and Technology, Aligarh Muslim University, Aligarh, UP, India

Sk. Musheer Ahmad Faculty of Agricultural Sciences, Department of Post Harvest Engineering and Technology, Aligarh Muslim University, Aligarh, India

Samira Umme Aiman Faculty of Agricultural Sciences, Department of Agricultural Microbiology, Aligarh Muslim University, Aligarh, India

Asiye Akyildiz Department of Food Engineering, Cukurova University, Adana, Turkey

Sait Aykut Aytac Department of Food Engineering, Hacettepe University, Beytepe Ankara, Turkey

Abdol Ghafour Badpa Faculty of Agricultural Sciences, Department of Post Harvest Engineering and Technology, Aligarh Muslim University, Aligarh, UP, India

Hakan Benli Department of Food Engineering, Cukurova University, Adana, Turkey

Sezgi Bircan Faculty of Agriculture, Department of Food Engineering, Cukurova University, Adana, Turkey

Erdem Çarşanba Altınozu Vocational School of Higher Education, Department of Food Technology, Mustafa Kemal University, Hatay, Turkey

Ömer Utku Çopur Faculty of Agriculture, Department of Food Engineering, Uludag University, Bursa, Turkey

Mustafa Didin Faculty of Agriculture, Department of Food Engineering, Mustafa Kemal University, Hatay, Turkey

Zerrin Erginkaya Department of Food Engineering, Faculty of Agriculture, Cukurova University, Adana, Turkey

Hüseyin Erten Faculty of Agriculture, Department of Food Engineering, Cukurova University, Adana, Turkey

Cennet Pelin Boyacı Gündüz Faculty of Agriculture, Department of Food Engineering, Cukurova University, Adana, Turkey

Ram Kishor Gupta Director, Central Institute of Post Harvest Engineering and Technology, Ludhiana, India

Mehmet Güven Faculty of Agriculture, Department of Food Engineering, Cukurova University, Adana, Turkey

A. Adnan Hayaloglu Department of Food Engineering, Engineering Faculty, Inonu University, Malatya, Turkey

Selin Kalkan Bahce Vocational School of Higher Education, Department of Food Technology, Osmaniye Korkut Ata University, Osmaniye, Turkey

Sana Khan Faculty of Agricultural Sciences, Department of Agricultural Microbiology, Aligarh Muslim University, Aligarh, India

Showkat Ahmad Lone Faculty of Agricultural Sciences, Department of Agricultural Microbiology, Aligarh Muslim University, Aligarh, India

Abdul Malik Faculty of Agricultural Sciences, Department of Agricultural Microbiology, Aligarh Muslim University, Aligarh, India

Farhana Masood Faculty of Agricultural Sciences, Department of Agricultural Microbiology, Aligarh Muslim University, Aligarh, India

Melike Özcan Technology, Research and Development Center, Mustafa Kemal University, Hatay, Turkey

Emir Ayşe Özer Faculty of Agriculture, Department of Food Engineering, Mustafa Kemal University, Hatay, Turkey

Selvihan Sert Faculty of Agriculture, Department of Food Engineering, Cukurova University, Adana, Turkey

Prabhat Kumar Srivastava College of Agricultural Engineering and Post Harvest Technology, Central Agricultural University, Ranipool, Gangtok, Sikkim, India

Birce Mercanoglu Taban Department of Dairy Technology, Ankara University, Ankara, Turkey

Canan Ece Tamer Faculty of Agriculture, Department of Food Engineering, Uludag University, Bursa, Turkey

Hasan Tangüler Faculty of Engineering, Department of Food Engineering, Nigde University, Nigde, Turkey

Emel Ünal Faculty of Agriculture, Department of Food Engineering, Cukurova University, Adana, Turkey

Food Processing: Strategies for Quality Assessment, A Broad Perspective

Abdul Malik, Farhana Masood, and Saghir Ahmad

1 Introduction

Quality assessment of processed food has become an emerging issue in the present era. The quality factor has broadened and covers all the aspects which satisfy consumer expectations. Subject of quality covers different sections like quality analysis, sensory quality, quality safety quality assurance, and quality standards and regulations. Quality analysis is an approach assisting the functioning of industries, establishing the standards and bringing the manufacturing process for establishment and successful. Every industry, grades the raw materials and implements prevention of Prevention Food Adulteration (PFA) legislation. The analysis of food and food products requires proper sampling as the plant and animal tissues not only having variation in composition of different varieties/brands rather observe variations in individuals of same variety. Sometimes there is variation in various parts of same fruits/vegetables/animal tissues. Food analysis is divided in two groups, namely, proximate and ultimate analysis. The former gives the facts of nutritional/biochemical aspect, while the latter covers the information of particular element or organic compound. Proximate analysis covers the determination of percentage of moisture, ash, crude fiber acidity, proteins, lipid, sugars, and carbohydrate. The ultimate analysis provide information of a particular element like calcium, sodium, iron, magnesium and zinc, vitamins, pigments and antioxidant. Before analysis sampling is very important so as to get the accurate information of analysis. The samples should be taken in sufficient quantity to compensate the variability.

A. Malik • F. Masood (✉)
Faculty of Agricultural Sciences, Department of Agricultural Microbiology,
Aligarh Muslim University, Aligarh 202002, India
e-mail: farhanamasud4@gmail.com

S. Ahmad
Faculty of Agricultural Sciences, Department of Post Harvest Engineering and Technology,
Aligarh Muslim University, Aligarh, UP 202002, India

A. Malik et al. (eds.), *Food Processing: Strategies for Quality Assessment*,
Food Engineering Series, DOI 10.1007/978-1-4939-1378-7_1

A pioneer approach has been adopted by the Government of India to safeguard the health of consumers by establishing the food safety and standard act 2006. This act shall come into force on such time as the Central Government appoint official gazette, and different dates may be appointed for different provisions act at any reference in any such provision to the commencement of this Act shall be used as reference to the coming into force of that provision.

The provision safeguards the interest of consumer to protect them against adulteration.

2 Food Safety

Nowadays global food industry is in a highly competitive market place. There are numerous persons to bring the cost and to improve margin in order to survive. As a result food safety experts are struggling to maintain adequate control of their firm activities. To safeguard the interest of consumer and to prevent fraudulent practices, the food and beverages companies follow food legislation and regulations and their brand names reputation.

Food processor has to go through all steps of productions from grower to the consumer. Each step involves substantially different types of processes and unit operations which require monitoring for safety, compliance of FDA/HACCP, and also profit optimization. Mandate by FDA such as HACCP procedure and ISO-22000-based food safety management system are the basis quality program in the food and beverage industry. To maintain adequate service, a lot of expenses are incurred mostly by companies that maintain improperly trained employee.

2.1 Quality and Safety in Food: Benefits and Risks

The terms "food quality" and "food safety" mean different things to different people. Quality has a vast number of meanings and can encompass parameters as diverse as organoleptic characteristics, physical and functional properties, nutrient content, and consumer protection from fraud. Furthermore, it can cover political and social issues such as wages paid to farm workers, geographical issues such as controlled appellations, and religious issues such as halal and kosher. Safety is more straightforward, relating to the content of various chemical and microbiological elements in food. Clearly, food quality and safety issues need to be addressed along the entire food chain. FAO has adopted this approach, defined as recognition that the responsibility for the supply of food that is safe, healthy, and nutritious is shared by all involved from primary production to final preparation and consumption. Compositional changes, for better or for worse, can be introduced at each and every link in the food chain.

Adopting a food chain framework goes beyond ensuring the safety of food. It facilitates more generally an approach to quality in agriculture and food safety and quality systems that will comprise government, industry, and consumer involvement. This implies potential future shifts in the agricultural sectors in many countries. For example, plant breeders are using genetic resources to increase the nutrient contents of foods at source. Farmers are also exploring new farming and technology choices to meet demands for a safe and healthy diet in response to new regulations and standards, changing global consumption.

FAO recognizes the need to more fully incorporate a food chain approach in its food quality and safety and nutrition strategies and acknowledges that this revised strategic direction will require an integrated and preventive approach to the management of food safety meeting sustainability concerns and building on aspects of the implementation of international commitments such as Agenda 21 (United Nations Conference on Environment Development (UNCED), 1992/1997). While the developments may be largely beneficial, the composition of the foods needs to be monitored to ensure that no harm results to the consumers.

Since 1963, an international food code has been in place to ensure food safety worldwide. Codex Alimentarius, or food code, jointly administered by FAO and the World Health Organization, sets harmonized standards for food. These include specific food standards, guidelines, codes of practice, and recommendations on hygiene; food labeling; food safety risk assessment; contaminants in foods; sampling, analysis, inspection and certification procedures; maximum limits for pesticide residues; food additive provisions; and maximum limits for veterinary drugs in foods. It serves as the basis for many national food standards. Codex has established such well-known safeguards as the "Best if used before" food label and definitions for low-fat and light food. Codex considers independent scientific advice from such bodies as the Joint FAO/WHO Expert Committee on Food Additives (JECFA), the Joint FAO/WHO Expert Meetings on Microbiological Risk Assessment (JEMRA), the Joint FAO/WHO Meeting on Pesticide Residues (JMPR), the Joint FAO/WHO scientific expert consultations on Foods Derived from Biotechnology, and the expert consultations and technical workshops on human nutrition.

Food safety systems in both developed and developing countries face unprecedented challenges arising from demographic change, the globalization of food trade, shifts in food consumption patterns, more processed foods with increased amounts and numbers of additives, increased urbanization, and more intensified food production systems. To help countries respond to these challenges, FAO is redefining its own approach to food safety and quality issues. In a report to the organization's high-level COAG, a comprehensive new system is proposed that would share the responsibility for providing safe food among all players in the food and agricultural sector, from food producers and processors to retailers and consumer households. This "food chain approach" would be strengthened by development of Good Agricultural Practices (GAP) that would help farmers prevent threats to food safety at source.

Parallel with consumer pressure for safer food, economic globalization has raised significantly the risks and costs of food-borne diseases. The globalization of trade means that it will become increasingly difficult to resolve food safety problems of any one country without collaborative international efforts. Once unsafe or contaminated food has entered the food chain, it is distributed more rapidly and to a greater number of consumers; hence the risks are higher (Burlingame and Pineiro 2007).

FAO's framework for development of a food chain approach to food safety is based on three key elements. First, it calls for universal adoption of a "risk-based" approach to the management of food safety hazards. Food control resources should be directed to hazards posing the most serious threats to public health and where the potential gains from risk reduction are greatest. "Establishing risk-based priorities requires sound scientific knowledge and effective systems for reporting the incidence of food-borne diseases," FAO (2003) says. There also needs to be a net distinction between risk assessment, based on independent scientific research, and risk management which "very often involves a political process—the political nature of government decisions on regulation and control of food safety may partially explain why it is essential that risk assessment and management are separate functions."

FAO also advises shifting the emphasis on regulation and control of end products to preventive measures to control the introduction of food hazards along the entire food chain. That requires adoption of GAP in primary production, postharvest treatment, processing, and handling to reduce the risk of microbiological and chemical contamination and to retain optimal quality. In-plant controls of food processing operations should be based on the Hazard Analysis and Critical Control Point (HACCP) system, which identifies, prevents, controls and then monitors the most vulnerable points in a food production system. In order for standards concerning the safety and quality of various types of additives to be developed and enforced at government level, there must be consistent methodology for authentication and validation.

Food safety regulatory agencies such as FDA in the USA and EFSA in the EU, and similar agencies in other countries, have been charged with ensuring the safety of the food supply (Chassy 2008, 2010). The situation is far different in developing countries where a significant portion of the population can suffer from undernutrition or malnutrition, and micronutrient deficiencies are common (FAO 2008, 2009); see also (http://www.gainhealth.org/about-malnutrition/nutrition-facts). Consumers in developing countries may be more concerned with obtaining adequate food supplies and ensuring food security than they are with food safety, although—paradoxically—their food is frequently contaminated with biological and chemical agents that have adverse effects on health (Wild and Gong 2009). Scarcity of food energy and micronutrients takes a staggering toll of the poor, particularly in underdeveloped countries (FAO 2008, 2009) to optimize the risk governance framework, SAFE FOODS proposes to expand conventional risk assessment by incorporating the evaluation of environmental, ethical, and socioeconomic impacts into assessment in terms of risks and benefits associated with food issues. Specifically, a framework for improved risk analysis of foods has been proposed, systematically incorporates risk–benefit assessment, stakeholder consultation, and public participation at appropriate stages in the risk analysis process. The framework includes risk–benefit assessments relating to non-health aspects of food safety. Two main types of assessment are identified; those relating to the risk–benefit assessment of health and environmental impacts, and the assessment of economic, social, and ethical impacts (Koenig et al. 2010; Cope et al. 2010). The dominant model of risk analysis applied in the agrifood sector is that proposed by FAO/WHO. This model comprises three phases: food risk assessment, food risk management, and food risk communication. Risk assessment focuses on estimating the risk that a hazardous food safety incident will negatively affect human health (FAO/WHO 1998).

3 Approaches in Food Safety

New science-based approaches to food safety provide an effective way for governments to protect consumers against food-borne diseases and plan appropriate response measures when necessary. Risk analysis, in particular, allows data on hazards in food to be systematically linked to food-borne disease epidemiological data, making it easier to determine the risk to human health. Risk analysis has demonstrated its ability to improve food safety decision-making processes and produce improvements in public health. It offers governments a framework to effectively assess, manage, and communicate food safety risks in cooperation with the diverse stakeholders involved. By providing a process to establish realistic, science-based targets to reduce the incidence of food-borne disease, plan and implement tailored interventions, and monitor the outcomes (both successful and unsuccessful) of these interventions, risk analysis contributes to continuous improvements in food safety.

3.1 Traditional Food Safety Systems

Food safety is the responsibility of everyone involved with the food chain from regulators to producers to consumers. However, governments are responsible for providing an enabling institutional and regulatory environment for food control. Most developing countries already have some sort of food control system in place, usually based on hygiene and adulteration/fraud inspection. While these vary considerably, they usually incorporate food laws and regulations, food control management, inspection and laboratory services, and sometimes mechanisms for information, education and communication, and monitoring of the food supply.

The increasing globalization of the food trade, urbanization, changing consumption patterns, the intensification of agriculture, increasing travel and tourism, and new types of production and manufacturing systems are just some of the trends that are having a serious impact on food safety in many countries. At the same time, a number of existing and new food safety hazards are of increasing concern. New pathogens are also frequently emerging, and existing ones evolving or reappearing. For instance, the resistance of food-borne pathogens to antimicrobial agents is of increasing concern.

Although traditional food safety systems were somewhat effective in reducing food hazards in the past, they are unable to detect and resolve many current problems, and to effectively deal with the full range of complex, persistent pervasive, and evolving challenges confronting different parts of the food chain. A modern food safety system, with the new Risk Analysis approach has the ability to much sharper diagnose the problems and also to suggest focused interventions to properly deal with them.

3.2 A Science-Based Approach to Food Safety

A number of developing countries are already taking steps to improve and strengthen their systems for food safety management. Several are moving away from the traditional approach focused on end-product control toward a process and science-based approach. Indeed, food safety regulators in many countries are already implementing different types of science-based actions and decision making in their day-to-day work.

Examples of Science-Based Activities

1. Implementation of Hazard Analysis and Critical Control Point (HACCP) systems
2. Establishment of acceptable daily intakes for chemical additives and residues of pesticides and veterinary drugs in food
3. Establishment of tolerable in takes for chemical contaminants, including natural toxins
4. Use of science to develop labels to warn consumers about potential risks including food allergens
5. Use of risk assessment to support food safety regulations
6. Establishment of product safety standards, performance standards, and specifications for use in international trade
7. Resolution of trade disputes based on the Agreement on the Application of Sanitary and Phytosanitary Measures (the SPS Agreement)
8. Establishment of dose-response relations for pathogenic microorganisms
9. Establishment of a Food Safety Objective to achieve an appropriate level of protection (ALOP)

A science-based approach strengthens the capacity of traditional food safety systems to meet current challenges and improve the availability of safe food for consumers. Scientific evidence can be used to minimize the occurrence of food-borne hazards, to reduce and manage risk, and to improve the outcomes of decision making. A science-based approach enhances the ability of food safety regulators to:

1. Identify hazards
2. Characterize the nature and extent of those hazards
3. Assess exposure to the identified hazards
4. Estimate the likelihood and magnitude of the resulting risks and impact on human health
5. Help set priorities between hazards

As a concept, a science-based approach to food safety is not completely new. It is related to processes such as GAP, good hygienic practices, good manufacturing practices, and Hazard Analysis and Critical Control Point system (HACCP), which are already used in many countries. Scientific assessment of chemicals in general has also a rather long "tradition." What is new is the use of risk analysis as a framework to view and respond to food safety problems in a systematic, structured, and scientific way in order to enhance the quality of decision making throughout the food chain.

A science-based risk analysis framework requires modern food safety and public health institutions and infrastructure, as well as an overall environment that values and supports the risk analysis paradigm. Risk analysis is just one part of an effective food safety system. It will also be essential to develop and improve components of food safety systems including food safety policies, food legislation (encompassing food law, regulations and standards), food inspection, laboratory analysis, epidemiological surveillance of food-borne diseases, monitoring systems for chemical and microbiological contamination in foods, and information, education, and communication.

Increasing public awareness of food safety hazards, concern over threats to health attributable to food hazards, and reduced confidence in the ability of current food supply systems to manage food safety risks are additional factors to be considered in the development of a food chain strategy. Information is rapidly disseminated and the media quickly spreads news of food safety emergencies. Consumer organizations concerned with food safety issues continue to increase their political influence and this trend is of great benefit to the consumer. However, food safety concerns and food scares that are not scientifically substantiated may create unnecessary obstacles and potentially hinder development of potentially useful new technology. Consumers are now equally concerned about the quality of their diet with relation to health and risk of chronic diseases. The need to address their concerns with regard to the nutritional quality of the diet can be easily and closely interwoven with food safety during the development of the food chain strategy. In summary, the use of a science-based approach will enable governments to develop and implement a range of general improvements and interventions tailored to specific high-risk areas, which will ultimately improve food safety and reduce the burden of food-borne disease.

References

Burlingame B, Pineiro M (2007) The essential balance: risks and benefits in food safety and quality. J Food Compos Anal 20:139–146

Chassy B (2008) Global regulations of transgenic crops. In: Kriz AL, Larkins BA (eds) Molecular genetic approaches to maize improvement. Springer, New York

Chassy BM (2010) Food safety risks and consumer health. New Biotechnol 27:534–544

Cope S, Frewer LJ, Renn O, Dreyer M (2010) Potential methods and approaches to assess social impacts associated with food safety issues. Food Control 21:1629–1637

FAO (2003) FAO's strategy for a food chain approach to food safety and quality: a framework document for the development of future strategic direction. Item 5 of the Provisional Agenda, Committee on Agriculture (COAG), 17th session. Rome, 31 March–4 April (Document COAG/2003/5/http://www.fao.org/DOCREP/MEETING/005/Y8274e.HTMS)

FAO (2008) The state of food insecurity in the world 2008. FAO available at http://www.fao.org/docrep/011/i0291e/i0291e00.htm
FAO (2009) The state of food insecurity in the world 2009. FAO available at ftp://ftp.fao.org/docrep/fao/012/i0876e/i0876e.pdf
FAO/WHO (1998) The application of risk communication to food standards and safety matters-report of an FAO/WHO Expert Consultation; FAO Food and Nutrition Paper No. 70. FAO, Rome http://www.fao.org/docrep/005/x1271e/x1271e00.htm
Koenig A, Kuiper HA, Marvin HJ, Boon PE, Busk L, Cnudde F et al (2010) The SAFE FOODS framework for improved risk analysis of foods. Food Control 21:1566–1587
United Nations Conference on Environment & Development (1992/1997). Agenda 21/http://www.un.org/esa/sustdev/documents/agenda21/english/agenda21toc.htmS
Wild CP, Gong YY (2009) Mycotoxins and human disease: a largely ignored global health issue. Carcinogenesis. doi:10.1093/carcin/bgp264

Fruit Processing

Ömer Utku Çopur and Canan Ece Tamer

1 Introduction

The quality of processed fruit products depends on their quality at the start of processing; therefore, it is essential to understand how maturity at harvest, harvesting methods, and postharvest handling procedures influence quality and its maintenance in fresh fruits between harvest and process initiation (Kader and Barrett 2005). The specific qualities required in fruits and vegetables will depend on their end-use and the selection of appropriate cultivars for particular products is of paramount importance (Aked 2002). Quality of fresh produce includes appearance (size, shape, color, gloss, and freedom from defects and decay), texture (firmness, crispness, juiciness and toughness, depending on the commodity), flavor [sweetness, sourness (acidity), astringency, aroma, and off-flavors], and nutritive value. Nutritional quality is determined by a fruit's content of vitamins, minerals, dietary fiber, carbohydrates, proteins, and antioxidant phytochemicals (carotenoids, flavonoids, and other phenolic compounds) (Kader 2001; Kader and Barrett 2005).

Despite the beneficial health effects of fresh produce, there is a growing awareness concerning its microbial and chemical food safety (Lynch et al. 2009; Strawn et al. 2011). There was in general an agreement on the main priorities in food safety of fresh produce. Bacterial pathogens were overall considered to be the most important food safety issue for fresh produce, followed by foodborne viruses, pesticide residues, and mycotoxins. Other food safety issues such as antimicrobial resistance, wax coatings, nanomaterials, and genetically modified organisms are increasingly becoming a concern for the fresh produce supply chain

Ö.U. Çopur (✉) • C.E. Tamer
Faculty of Agriculture, Department of Food Engineering, Uludag University, Bursa, Turkey
e-mail: ucopur@uludag.edu.tr

A. Malik et al. (eds.), *Food Processing: Strategies for Quality Assessment*,
Food Engineering Series, DOI 10.1007/978-1-4939-1378-7_2

(Tait and Bruce 2001; Domingo and Gine Bordonaba 2011; Magnuson et al. 2011). Hence, assuring the safety of fresh produce and alertness to maintain consumer trust in fresh produce as a healthy food is of primary importance for stakeholders. This is a challenging task in an increasingly globalized and more complex fresh produce food supply chain. It implies a shared responsibility of the stakeholders within the farm-to-fork continuum (producers, processors, trading companies, retailers, and consumers) and those closely involved in supporting food safety in the supply chain (competent authorities, industry associations, food scientists). Alert systems such as the European Commission's Rapid Alert System for Food and Feed (RASFF) were considered as the most important source of information of food safety issues, followed by reports of international organizations (e.g., WHO, EFSA), legislative documents (e.g., EU legislation), national reports (e.g., on monitoring hazards, foodborne outbreaks), and exchange of information between people. Concerning the control measures, the application of good agricultural practices (GAP) was identified to be the most important control measure to assure the safety of fresh produce, followed by the application of good hygienic practices (GHP) and the certification of food safety management systems (FSMS) (Van Boxstael et al. 2013). Today's management of food safety is to a great extent based on the application of the HACCP system. Originally, the system was introduced to ensure the microbiological safety of food products. Later on, its use was extended to all types of foodborne hazards, including chemical hazards (Motarjemi et al. 2009). As the primary source of raw ingredients for food production, the agricultural sector is a fundamental component of the most food product and supply chains. Consequently, the development of effective HACCP procedures for this sector is essential to the overall success of HACCP (Ropkins et al. 2003). Current attention in chemical HACCP is mainly focused on residual chemicals from the agricultural sector (e.g., pesticides, growth hormones, fumigants, and some natural toxins) (Ropkins and Beck 2003).

Increasing international trade and globalization were overall expected to have a large impact on food safety in fresh produce. Other contextual factors perceived to be important were the food safety policies by governments and the (lack of) food safety knowledge by consumers and other stakeholders of the fresh produce supply chain (Van Boxstael et al. 2013).

Food processing operations have a major influence on the stability of phytochemicals and often damage antioxidants in fruit and vegetables and their products. Domestic, industrial, thermal, and nonthermal processing are widely reported to degrade the level of phytochemicals in processed food products. In order to retain the nutraceutical and pharmacological properties of phytochemicals in processed fruit and vegetable products, the food processor must optimize relevant processing steps in order to restrict the loss of phytochemicals (Tiwari and Cummins 2013). In this chapter, quality criteria in freshly harvested produce, the principal causes of quality deterioration and maintaining quality of fruit products, the impact of thermal and nonthermal processing on nutrients and antioxidants of fruit products are briefly discussed.

2 Principal Causes of Quality Deterioration

Exposure of a commodity to temperature, relative humidity, and/or concentrations of oxygen, carbon dioxide, and ethylene outside its optimum ranges will accelerate loss of all quality attributes. The loss of flavor and nutritional quality of fresh intact or cut fruits and vegetables occurs at a faster rate than the loss of textural and appearance quality (Kader 2001). Many factors can lead to loss of quality in fresh produce, hence the common description of these products as "perishable." Some of these factors are part of the life cycle of living produce, that is, over-ripening of fruits or sprouting in root and bulb crops. Others are a consequence of the act of harvesting. Once severed from the mother plant, the plant organ is deprived of its source of water, nutrients, and antisenescent hormones. As a consequence normal factors such as transpiration and respiration lead ultimately to water loss and senescence of the product. The growth of pathogens or physical damage will cause direct loss of product quality through their visual impact but both also stimulate senescence. Furthermore, the storage environment (temperature, relative humidity, air movement, atmospheric composition, ethylene) will play a highly significant role in determining the speed of all quality changes (Aked 2002).

Fruits and vegetables are naturally contaminated with microorganisms, and many of these microorganisms possess pectin-degrading enzymes, enabling them to produce colonization by using fruit nutrients. Moreover, tissue damage originated by cutting or wounding leads to cell damage, releasing nutrients, and favoring growth of most types of microorganisms. They may also cause spoilage and affect the economic value of produce, not only by decreasing the organoleptic and nutritional quality and shelf-life of produce but also by causing foodborne disease. Therefore, it is important to prevent contamination and growth of microorganisms, in order to reduce degradation of nutrients and maintain fruit safety and sensory attributes. Some of these problems can be solved by improving preharvesting practices, and others need to be addressed through appropriate postharvest handling and processing (Ruiz-Cruz and Arvizu-Medrano 2010).

3 Storage and Packaging Techniques for Maintaining Quality of Fruits

Maintaining quality requires action to be taken to limit factors causing deterioration in fresh fruits (Aked 2002). Fruit storage technology deals essentially with the inhibition of natural, physical, pest-induced, and pathogen-induced decay and damage without going to extremes such as drying or freezing. The object is to maintain fresh quality as long as possible or as long as necessary, depending on market conditions. The two broad categories of obstacles in achieving these objectives are: "the biochemical and physiological activities that proceed within the fruit itself after harvest" and "the introduction and proliferation of microbial pathogens and insects in the storage environment" (Raghavan et al. 2005).

To prolong the storage life of fresh fruits and vegetables, controlled atmosphere (CA) storage is frequently used. The basic CA effect on biochemical reactions can also be used to extend the shelf life of processed and ready-to-use fruit products. These products are often peeled and sliced and they are preferred as fruit dishes by consumers. The technique that provides CA condition for this ready-to-use fruit dishes is usually modified atmosphere packaging (MAP) (Balla and Farkas 2006). MAP involves the modification of the internal atmosphere composition of a package by reducing the amount of oxygen (O_2) and replacing it by carbon dioxide (CO_2) and/or nitrogen (N_2). This process aimed to extend the postharvest life of whole and fresh-cut commodities by reducing their respiration rate and the production of ethylene, minimizing metabolic activity, delaying enzymatic browning, and retaining visual appearance. The gas re-balancing can be achieved either using active or passive techniques inside a package made of various types and/or combinations of films (Saxena et al. 2008; Cui et al. 2009; Ramos et al. 2013). Several studies have reported that modified/controlled atmosphere packaging delayed senescence and microbial growth in fruits and vegetables. On the other hand, it has been observed that the antioxidant content and bioactivity could vary depending on the kind of treated fruit and treatment (Ayala-Zavala et al. 2005, 2007; González-Aguilar et al. 2010).

4 Impact of Processing on Nutrients and Antioxidants of Fruits

Intact fruits and vegetables obviously are prone to deleterious changes induced by respiratory, metabolic, and enzymatic activities, as well as by transpiration, pest and microbial spoilage, and temperature-induced injury. Most such changes may impact adversely on the antioxidant status of these products (Lindley 1998).

In fruits and vegetables, phytochemicals can be bound in the plant cell membranes or exist as free compounds. Food processing such as heating or freezing can disrupt the cell membrane leading to the release of membrane-bound phytochemicals, which implies higher bioaccessibility (Lemmens et al. 2009). Moreover, the amount of phytochemicals retained in fruits and vegetables depends on their stability during food preparation and processing before consumption, which is mostly related to their sensitivity toward oxidation and the environmental conditions (Leong and Oey 2012).

Food processing operations have a major influence on the stability of phytochemicals and often damage antioxidants in fruit and vegetables and their products. Conventional thermal: (blanching, pasteurization, frying, steaming, baking, stewing, roasting), nonthermal (high pressure processing, pulsed electric field, ultrasound, ultraviolet), domestic (washing, peeling, cutting), and industrial (canning, drying, extraction, concentrating by evaporation, extrusion) processing are widely reported to degrade the level of phytochemicals in processed food products (Tiwari and Cummins 2013).

Heat treatment may lead to a decrease in essential nutrients and consequently reduces the nutritional value of some foods. In this context, some water-soluble vitamins (vitamins C, B1, B2, B6, and folic acid) are heat sensitive, while lipid soluble vitamins are relatively stable to heat. The Maillard reaction itself may also lead to the loss of vitamins and proteins due to the transformations involved in the reaction. However, a first and often very important loss of vitamins and minerals already occurs prior to any heat treatment, when the raw materials are physically prepared. This may occur by the practice of peeling fruits or vegetables. The amount of minerals in foods is not much affected by processing, except when this includes discarding certain constituents. Unit operations such as cooking, drying, extrusion, and so on have little effect on the bioavailability of minerals (Burri et al. 2009).

Food processing and subsequent storage conditions may have a positive or negative influence on the stability of phytochemicals (Aaby et al. 2007; Volden et al. 2009; Rawson et al. 2010; Tiwari and Cummins 2013). Maceration, heating, and various separation steps can result in oxidation, thermal degradation, leaching, and other events that lead to lower levels of antioxidants in processed food compared with fresh. This is particularly true in the case of vitamin C and phenolic antioxidants. However, in the case of carotenoids, processing can lead to a dissociation of antioxidants from plant matrix components, an increase in carotenoid antioxidants, and improved digestive absorption (Kalt 2005).

During the processing of fruits and vegetables, several types of oxidative reactions may occur in which electrons are removed from atoms/molecules leading to the formation of an oxidized form. These reactions cause browning, loss or changes of flavor or odor, changes in texture, and loss of nutritional value from destruction of vitamins and essential fatty acids (Dziezak 1986). The oxygen also can play a major role in the flavonoids degradation during the different steps of processing and storage. The presence of oxygen can accelerate the degradation either through a direct oxidative mechanism and/or through the action of oxidizing enzymes as polyphenoloxidase (PPO). For this reason, the degradation of flavonoids is a combination of several mechanisms depending on the operating conditions and the food matrix (Ioannou et al. 2012).

While most vegetables are cooked at domestic level prior to consumption, fruits are consumed raw or undergo minimal processing which has been defined as a combination of procedures, such as washing, sorting, trimming, peeling, and slicing or chopping, that do not affect the fresh-like quality of the food (Odriozola-Serrano et al. 2008a; Tiwari and Cummins 2013). "Fresh-cut" is defined as any fruit or vegetable or combination that has been trimmed, peeled, washed, and cut into 100 % useable product that is then bagged or prepackaged and remains in a fresh state (Lamikanra 2002). Fresh-cut fruits and vegetables are highly perishable products because of their intrinsic characteristics and the minimal processing (Ayala-Zavala et al. 2008a). Microbial growth, decay of sensory attributes, and loss of nutrients are among the major causes of compromised safety and quality of fresh-cut produce. These problems are caused by the steps involved in the minimal processing, such as peeling and cutting, which promote an increment in the metabolic rate, enzymatic reactions, and released juice (Ayala-Zavala et al. 2008b).

Peeling, trimming, depitting, and/or leaf selection may cause a partial or total decrease in flavonol levels (Amarowicz et al. 2009).

Vinha et al. (2013) demonstrated that the removal of the skin of tomato caused a loss of 80 % of lycopene, 63 % of phenolic compound, 57 % of β-carotene, and 26 % of ascorbic acid. Removing the seeds caused 63 % loss of total phenolics. Size reduction (dicing and slicing) results in increasing losses through increasing the surface to volume ratio (Ramaswamy and Chen 2002). Robles-Sánchez et al. (2009) evaluated the losses of bioactive compounds that occur after cutting and cold storage and their contribution to the total antioxidant capacity of fresh-cut mangoes. No significant losses of total phenols were found at the end of storage. Mangoes treated with the antioxidants maintained better quality and higher antioxidant potential compared with controls. Although minimally processing of fruit accelerates ripening of fresh-cut tissues, which could promote an increase in β-carotene content, it is possible that low storage temperature used for mangoes retarded its biosynthesis and accumulation. Plaza et al. (2011) investigated the effect of minimal processing on the health-related characteristics of orange. Carotenoids were retained in minimally processed oranges during refrigerated storage. The flavanone content showed a significant increase throughout refrigerated storage as response to cold stress. Although some vitamin C losses were observed, the antioxidant activity remained stable. Overall, the microbiological quality and potentially health-promoting attributes of minimally processed oranges were preserved during 12 days of storage at 4 °C.

4.1 Thermal Processes

Heating results in enzyme inactivation, texture changes of fruits and vegetables, and unavoidable leaching of water-soluble compounds which could alter the entire phytochemical profile and content of fruit and vegetables. Phytochemicals do not exist as an individual compound; they are mostly bound to other compounds or to cell structures. Due to heat, the disruption of cell membranes occurred. Once the cell is damaged due to heat, this creates an opportunity for the bound phytochemical compounds to be released into the medium, hence they are readily extracted. In fact, heating has been reported to increase the chemical extractability of phytochemical compounds, because of the release of phytochemicals from chromotoplasts leading to an increment of concentration. Heating also encourages the diffusion of cellular fluids, containing phytochemicals, from the plant cell to the water medium (Howard et al. 1999; Leong and Oey 2012).

Jiratanan and Liu (2004) concluded that depending on the particular produce and processing parameters and methods, thermal processing may enhance, reduce, or cause no change in total antioxidant activity from that of fresh produce.

High-temperature processing may lead to thermal destruction of antioxidants. Due to this, long cooking times and sterilization are considered antioxidant-destructive (Grajek and Olejnik 2010). Changes connected to mild hydrothermal processing (<100 °C) are usually advantageous. Due to heating, oxygen is removed from solutions, oxidoreductases are denaturated, and heteroglycosides are hydrolyzed to aglycones. In other respects, increased temperature may lead to higher losses because a portion of water-soluble antioxidants are extracted. Losses in water-soluble vitamins are a good indicator of the antioxidant potential decrease of a given food product. Blanching, where solid material is in direct contact with steam or hot water, effectively inactivates oxidative enzymes and due to that loss of antioxidants. For example, enzymatic oxidation of vitamin C can be eliminated due to the inactivation of ascorbic acid oxidase. If the process is performed at too low temperature, though, it may be ineffective and lead to polyphenol oxidation by PPO. During treatment, a portion of antioxidants leach into the water, which decreases the antioxidant potential of plant materials (Lin and Chang 2005; Amin et al. 2006; Wachtel-Galor et al. 2008; Leong and Oey 2012).

Rickman et al. (2007a, b) reported that, depending on the commodity, freezing and canning processes may preserve nutrient value. The initial thermal treatment of processed products can cause loss of water-soluble and oxygen-labile nutrients such as vitamin C and the B vitamins. However, these nutrients are relatively stable during subsequent canned storage owing to the lack of oxygen. Frozen products lose fewer nutrients initially because of the short heating time in blanching, but they lose more nutrients during storage due to oxidation. Phenolic compounds are also water soluble and oxygen labile, but changes during processing, storage, and cooking appear to be highly variable by commodity. The higher levels of carotenoids typically found in canned as compared to fresh products may be attributed to reporting results on a wet rather than dry weight basis, greater extractability, or differences in cultivars. Minerals and fiber are generally stable to processing, storage, and cooking, but may be lost in peeling and other removal steps during processing. Mineral uptake (e.g., calcium) or addition (e.g., sodium) during processing can change the natural mineral composition of a product. Changes in fiber during processing, storage, and cooking appear to be minimal for intact fruits and vegetables. Outer layers removed or peeled products, however, contained lower amounts of fiber than their unprocessed counterparts. The stability of fiber during storage depends on commodity. Generally, fresh, frozen, and canned fruits and vegetables contained similar amounts of fiber.

Processing of fruit or vegetables can result in a significant reduction in phytochemical content. Thermal processes have a large influence in flavonoid availability in foods which depends on their magnitude and duration (Ioannou et al. 2012; Tiwari and Cummins 2013). In general, the level of phytochemicals in vegetable and fruit processing decreases exponentially with a linear increase in blanching and boiling time (Tiwari and Cummins 2013). Most of heat processes lead to a degradation of flavonoids. Thermal pasteurization treatment (90 °C, 60 s) for strawberry juices had no effect on quercetin and kaempferol contents (Odriozola-Serrano et al. 2008b), whereas it reduced naringin, narirutin, quercetin, naringenin content for grapefruit juices (Igual et al. 2011) and procyanidins in canned peach (Asami et al. 2003). Effect of pasteurization has been reported for mulberry fruit extract, pineapple juice, and cashew apple juice leading to a decrease in the levels of bioactive components such as total anthocyanin, ascorbic acid, and carotenoids (Rattanathanalerk et al. 2009; Zepka and Mercadante 2009; Aramwit et al. 2010; Rawson et al. 2011a).

Pasteurization of grape juice increased the concentration of catechins in cold-pressed juices, but it decreased concentrations in hot-pressed juices. The concentration of most procyanidins was also increased by pasteurization (Fuleki and Ricardo-Da-Silva 2003). An increase of temperature during pressing from 40 to 70 °C allows increasing flavonoid content (50 %) in apple juice (Gerard and Roberts 2004), similar results were found by Renard et al. (2011) an increase of pressing temperature from 5 to 24 °C, increase the extraction of proanthocyanidins. Van Der Sluis et al. (2002) noted that antioxidant activity of juice is lower than that of apple. The flavonoid contents were reduced due to pressing in which most flavonoids were retained in the pomace. In processing blueberries into juice, substantial losses of phenolics occurred; the recovery of anthocyanins, procyanidins, and chlorogenic acid were 32, 43, and 53 %, respectively. Heat-labile enzymes (PPO) in blueberry fruit made a large contribution to the loss in anthocyanins. Approximately 20 % of the anthocyanins in blueberries were retained in the press cake after juicing (Skrede et al. 2000). The total flavonoid content of the juices obtained by manual extraction was less than half that obtained by mechanical extraction the percentage of flavones in the juices obtained manually was always lower than in the juices extracted using industrial methods which implies a possible greater contribution of flavones from albedo and flavedo (Amarowicz et al. 2009). In juice production from concentrates, the range of thermal processing is wider and additionally includes concentration of juice. All these processes lead to decomposition of thermolabile compounds, which include antioxidants (Grajek and Olejnik 2010).

Durst and Weaver (2013) analyzed fresh freestone peaches, fresh cling peaches, and canned cling peaches for vitamins (A, C, E, folate), antioxidants, total phenolics, and total carotenoids to assess how these nutrients were affected by the canning and whether storage further changed these components. The nutritional content of canned peaches was comparable to that of fresh peaches. There were no statistically significant decreases in those nutritional parameters measured between fresh freestone peaches and canned cling peaches. Vitamins A and E along with total carotenoids decreased immediately upon processing, but stabilized after the processing step, showing minimal additional changes upon storage for 3 months. After canning of mandarin orange segments, small proportions of phenolic acids and ascorbic acid were reduced, and about half of flavanone glycosides and total antioxidant capacity were lost. However, in view of that considerable portion of phenolic compounds and ascorbic acid existing in the syrup portion, so the loss was not so significant (Fengmei et al. 2011).

During the heat treatment, the antioxidant activities of flavonoids were also slightly decreased but they remain relatively high. This is due to the fact that the degradation products possess also an antioxidant activity (Murakami et al. 2004; Buchner et al. 2006). Jeong et al. (2004) determined an increase of the antioxidant activity of citrus peels during a heat treatment (50, 100, and 150 °C for 60 min). The degradation of flavonoids is not only a function of temperature and magnitude of heating; it may depend also on other parameters such as pH, the presence of oxygen, and the presence of other phytochemicals in the medium (Ioannou et al. 2012). Degradation of rutin and quercetin is higher under weakly alkaline and neutral reaction conditions (Takahama 1986; Buchner et al. 2006). The presence of oxygen highly induces quercetin and rutin degradations, while the absence of oxygen has the opposite effects (Makris and Rossiter 2000; Buchner et al. 2006). Moreover, the presence of other phytochemicals in the medium like chlorogenic acid plays a protective role (Murakami et al. 2004).

According to Turkmen et al. (2005), food processing and domestic cooking led to an increase in phenol concentration when compared to raw samples. This suggested that temperature-related treatments might produce changes in antioxidant extractability, not only for cellular disruption and dissociation of some phenolic compounds from biological structures but also for the alteration in their chemical structure which could make possible the conversion of insoluble phenolics into more soluble forms (Cohen et al. 2001; Bernhardt and Schlich 2006; Dini et al. 2013).

Processing of strawberries into jam may result in a loss of up to 70 % of the initial anthocyanin content (García-Viguera et al. 1999). Jams produced from various strawberry cultivars differed in terms of pigment and antioxidant capacity retention. Temperature proved to be the most important factor during storage (Wicklung et al. 2005). Brownmiller et al. (2008) determined a reduction of about 43 % in total anthocyanins in purees following blanching and pasteurization comparing to the original levels found in fresh blueberries. Losses of about 23 % of flavonoids were reported in the blackberry juice. Especially blanching, drastically reduced anthocyanins, whereas hot-filling degraded ellagitannins (Gancel et al. 2011). In some cases thermally processed fruits are shown to have higher levels of phytochemicals (Tiwari and Cummins 2013). For instance, Zafrilla et al. (2001) noted that a 2.5-fold increase in free ellagic acid content during the processing of raspberry jams. They suggested that it could be due to the hydrolytic breakdown of ellagitannins to ellagic acid during thermal treatment. In some cases, blanching inactivates enzymes such as PPO, which improves the stability of anthocyanins in processed food. Leong and Oey (2012) evaluated the effects of heating (98 °C, 10 min), freezing (−20 °C), and freeze-drying on anthocyanins, carotenoids, and vitamin C content of cherries, nectarines, apricots, peaches, plums, carrots, and red bell peppers. In most cases, heating increased the anthocyanin content in cherries, peaches, and plums but not in nectarines. It was determined that the heated fruits contained more anthocyanins than the fresh fruits. However, heating decreased the content of carotenoids in apricots, nectarines, and carrots while maintaining the carotenoid content in cherries, peaches, plums, and red bell peppers.

The production of tomato paste from fresh tomatoes involves mechanical homogenization and heat treatment. In this process, bioavailability of β-carotene and lycopene is enhanced, but other labile antioxidants are destroyed. The increase in carotenoids is due to enzymatic degradation, weakening of protein-carotenoid aggregates, and concentration of dry matter during evaporation (Van Boekel et al. 2010). However, conflicting data on tomato carotenoid stability during thermal processing of tomato can be found in the literature. For instance, Capanoglu et al. (2008) showed a significant decrease in the content of both lycopene (32 %) and β-carotene (36 %) during preparation of a tomato paste.

Drying processes lead to flavonoids degradation. The proportion lost depends on the drying method. Freeze-drying is the less aggressive method, whereas hot air drying leads to major losses. As intermediate solutions, microwave and vacuum drying can be used (Zainol et al. 2009; Zhang et al. 2009; Dong et al. 2011; Ioannou et al. 2012). Microwaves directly interact with food and heat is generated volumetrically. Short processing time in microwave drying, sterilization, and thawing is advantageous to reduce quality losses especially for perishable food products (Sumnu and Sahin 2005). Microwave treatments produced small modifications of the quantitative and qualitative composition of carotenoids in papaya and anthocyanins in strawberry. Chlorophylls in kiwi fruit showed significant degradation as a consequence of microwave heating (De Ancos et al. 1999). Igual et al. (2012) compared the drying kinetics and the change in the organic acids, phenolic compounds, and antioxidant activity of dried apricot when using hot air drying and microwave energy. The authors noted that the industrial processing of dried apricots may be improved by using microwave energy, as the drying time is considerably reduced, and the obtained fruit had a higher phenolic content, particularly of chlorogenic acid, catequin, and epicatequin. Nevertheless, as the contribution of these phenols to antioxidant capacity was not significant, microwave dried samples maintained the same antioxidant capacity as the air-dried ones. Fast development allowed new hybrid solutions like microwave-hot air-drying, microwave-vacuum drying, microwave-spouted bed drying, and microwave-halogen lamp drying. These methods allow reduced drying time and maintenance of the high nutritive quality of products (Grajek and Olejnik 2010).

Comparative studies on freeze-drying and hot air-drying of tomatoes showed that freeze-drying retained high levels of antioxidant compounds (8–10 % loss), whereas high temperature treatment caused a tremendous decrease in the content of antioxidants (56–61 % loss) (Chang et al. 2006). Interestingly, the total phenolic and flavonoid contents in both freeze and hot-air-dried tomatoes were significantly higher than in fresh material. Different changes appeared in lycopene content. In freeze-dried tomatoes, lycopene content was reduced by 33–48 %; however, the amounts of lycopene in hot-air-dried tomatoes increased 152–197 %, probably due to breaking of cell walls and weakening of the binding forces between lycopene and the tissue matrix (Grajek and Olejnik 2010).

Compared to heating, freezing could maintain or slightly increase the content of phytochemicals for most of the commodities. Freezing induces the formation of ice crystals that favors localized concentration of solutes (including phytochemicals) and reallocation of water molecules in the cell structure. Nevertheless, the common consequences of freezing due to cell damages by the growth of ice crystals from temperature fluctuation and turgor loss lead to softening texture (Szczesniak 1998). It is noted that the rate of freezing influences the ice crystals formation that impact on the food structure by expanding the separation between cells. In other words, when the samples were rapidly frozen, large amounts of smaller ice crystals formed and caused a lesser degree of cell structure disruption than the samples being frozen slowly, which formed large intercellular ice crystals (Leong and Oey 2012). In general, the manner in which the frozen sample is thawed is a key factor that will attribute to the changes in phytochemical contents (Robards 2003). In contrast, freeze-dried samples mostly resulted in a lower amount of phytochemicals, as compared to fresh, heated, and frozen commodities. Basically, freeze-drying is the combination of dehydration and freezing, i.e., dehydrating the samples by freezing the immobilized water into ice and then removing the ice crystals via sublimation into vapor. While freeze-drying is incapable of inactivating all of the enzymes, it is effective in preserving the sensory and nutritional qualities. Usually, a minor loss of vitamin does occur but extensive reduction of water during freeze-drying will form the fragile porous structure in the end product. Sublimation of ice to vapor caused by drying in the sample slices gave an open and porous texture. The heat utilization in freeze-drying may be harsher than the conventional freezing mechanism as the flavor and aroma compounds were evaporated along with water as volatiles. In practice, thinly sliced samples were used to promote larger surface area available for dehydration had increased the water removal rate. Nevertheless, the phytochemicals in freeze-dried samples were more prone to degradation due to the large surface area exposed during processing. Hence, most of the labile phytochemicals were rapidly oxidized, because the water molecules attached on the sample surface that acted as a protecting film were evaporated as well (Gross 1991; Leong and Oey 2012). Georgé et al. (2011) determined the impact of thermal processing and lyophilization on carotenoids, total polyphenols, and vitamin C in red and yellow tomato cultivars. Micronutrients were analyzed in fresh tomatoes, tomato purée, and lyophilized tomatoes. Processing did not affect the carotenoid content in red tomato, but significantly lowered β-carotene in yellow tomato and also the contents of total polyphenol and vitamin C in both cultivars. Lyophilization lowered the carotenoid content in red tomato but not in yellow tomato; in contrast, the total polyphenol content was preserved in red tomato but lowered in yellow tomato, and the vitamin C content was not affected in both cultivars. Arancibia-Avila et al (2012) determined that the antioxidant activity of lyophilized berry samples subjected to thermal processing at 100 °C for 10 and 20 min did not differ from the non-processed berries, showing high correlation between the total polyphenols, flavanols, and the antioxidant activities. It was found that berries subjected to thermal processing not more than 20 min maximally preserved the bioactivity.

Ohmic heating, also called electric resistance heating, is a direct heating method in which the food itself is a conductor of electricity, taken from the mains that are 50 Hz in Europe and 60 Hz in the USA. It provides rapid and uniform heating, resulting in less thermal damage to the product (Ramaswamy and Chen 2002; Icier and Ilicali 2005; Leizerson and Shimoni 2005). Vikram et al. (2005) reported that the smallest losses of vitamin C were observed in the ohmic-heated orange juices. The highest losses of vitamin C were observed during microwave heating due to uncontrolled temperature generated during processing. Lee et al. (2012) evaluated the efficacy of continuous ohmic heating for inactivating *Escherichia coli* O157:H7, *Salmonella typhimurium*, and *Listeria monocytogenes* in orange and tomato juices with various treatment times and electric field strengths (25–40 V cm^{-1}). The concentration of vitamin C in continuous ohmic-heated orange juice was significantly higher than in conventionally heated orange juice. It was suggested that continuous ohmic heating might be effectively used to pasteurize fruit and vegetable juices in a short operating time and that the effect of inactivation depends on applied electric field strengths, treatment time, and electric conductivity. Yildiz et al. (2009) demonstrated that ohmic heating did not cause any different effect in other quality indices and total phenolic contents of pomegranate juice than the conventional heating.

4.2 *Nonthermal Processing*

Nonthermal technologies are effective at sublethal temperatures, thereby minimizing negative thermal effects on phytochemicals. Several nonthermal techniques such as high pressure processing (HPP), pulsed electric field (PEF), ultrasound/sonication, and ultraviolet (UV) techniques have been investigated on fruit and vegetables and their products (Tiwari et al. 2009a). Recent interests in these technologies are not only to obtain high quality food with "fresh-like" characteristics but also to provide food with improved functionalities (Rawson et al. 2011a).

When innovative processes are used instead of thermal treatments, the importance of food matrix is lower because the flavonoid degradations are limited (Ioannou et al. 2012). Several studies reported the capacity of innovative processes (microwave, infra-red, high-pressure processing) to enhance the flavonoid extraction (Périno-Issartier et al. 2010; Srinivas et al. 2011; Zill et al. 2011). Odriozola-Serrano et al. (2008b) studied the effect of high-intensity pulsed electric fields (HIPEF) process on quercetin and kaempferol contents of strawberry juices and reported that such a process caused no damage on these compounds.

4.2.1 High Hydrostatic Pressure Processing (HHP)

HHP entails the transmission of pressures usually ranging from 300 to 700 MPa to foods, which results into a reduction of microbial loads and thus shelf life extension (Patras et al. 2009a). High hydrostatic pressure (HHP) treatment is considered to be an alternative to thermal pasteurization for fruit and vegetable juices. HHP treatment could preserve nutritional value and the sensory properties of fruits and vegetables due to its limited effect on the covalent bonds of low molecular mass compounds such as color, flavor compounds, and vitamins. HHP processing may enhance the antioxidant activity of juices comparing to those untreated. However, inactivation of important foodborne pathogens in low acid foods by HHP is most urgent and critical (Oey et al. 2008; Garcia-Parra et al. 2011; Pilavtepe-Celik 2013; Uckoo et al. 2013).

Huang et al. (2013) investigated the effects of (HHP) at 300–500 MPa for 5–20 min and high temperature short time (HTST) at 110 °C for 8.6 s on enzymes, phenolics, carotenoids, and color of apricot nectars. Micronutrients and phytochemicals of nectar were well preserved by both HHP and HTST. Compared with HHP treatment (500 MPa/20 min), HTST led to complete inactivation of enzymes, higher total phenolics, epicatechin, ferulic acid, and p-coumaric acid and lighter and more intensity color than those of HHP treatment, since HTST treatment gave better impact on the quality of apricot nectar. PPO, peroxidase, and pectinmethylesterase in apricot nectar were found to be highly resistant to high pressure inactivation, thus in order to maintain the quality of apricot nectar, HHP should be accompanied by additional measures.

Sanchez-Moreno et al. (2003) measured vitamin C, provitamin A carotenoids, and other carotenoids in freshly squeezed juices from oranges that were subjected to HHP.

Total carotenoids and vitamin A (expressed as retinol equivalents) showed an increasingly better extraction when the pressure increased from 100 to 400 MPa. Vitamin C content seems to preserve the carotenoid compounds from oxidation in the treated orange juices. Fernández-García et al. (2001) reported that the vitamin C content of orange and orange–carrot–lemon juices processed at 500 and 800 MPa was not, or only insignificantly, reduced compared to that of unprocessed juices. Vitamin B1, B2, B3, and B6 contents were not changed after pressurizing orange juice (Donsi et al. 1996). In orange juice and kiwi puree, folic acid was relatively pressure stable, in contrast to that in carrot juice (Indrawati et al. 2004). Different folate stabilities among orange juice, kiwi puree, carrot juice, and asparagus seemed to coincide with different levels of ascorbic acid content.

Ferrari et al. (2010) studied the effects of high pressures (400–600 MPa) at 25, 45, 50 °C for 5 or 10 min on phytochemical content of pomegranate juice. Their experimental results indicated that the content of anthocyanins was influenced mainly by pressure and temperature level. At room temperature, the concentration of these molecules decreases with the intensity of the treatment in terms of pressure level and processing time. Therefore, the higher pressure levels or longer processing times caused a decrease of the anthocyanin content. High pressure treatments modified the mechanism of anthocyanin degradation by affecting the enzymes involved in the kinetics of reaction. The residual activity of the enzymes along with a small concentration of dissolved oxygen could cause the degradation of the anthocyanins during the storage of the processed juice.

Keenan et al. (2010) assessed the effect of thermal and HHP on the antioxidant activity and phenolic content of fruit smoothies. Since decreases in levels of antioxidants were noted during long-term storage, it would appear that higher pressure treatments (>450 MPa) might be required for better retention of antioxidant compounds in fruit smoothies. HHP processing of smoothies at moderate temperatures may be a suitable alternative to traditional thermal processing (Keenan et al. 2012). Patras et al. (2009b) reported that levels of phenols increased significantly in HHP treated (600 MPa, 20 °C, 15 min) strawberry and blackberry purees (9.8 and 5.0 %, respectively).

Briones-Labarca et al. (2011) investigated the effect of high pressure on the bioaccessibility of specific nutrients (antioxidant, minerals and starch) in apple. They reported that high pressure processed apple had significantly higher antioxidant capacities, mineral, and starch content when compared to untreated samples. It is possible that changes to the tissue matrix induced by HHP, for example, disruption of the plant cell walls, resulted in the release of compounds with antioxidant actions and increased mineral and starch content into the extracellular environment. Consumption of apple under high hydrostatic pressure may supply substantial antioxidants, minerals, and starch which may provide health promoting and disease preventing effects.

Varma et al. (2010) reported that HHP processing causes conformational changes from the all *trans* to *cis* isomer form of lycopene, indicating that high pressure application can induce isomerization, increasing the availability of the carotenoids in the sample.

Nuñez-Mancilla et al. (2013) analyzed the effects of combined osmotic dehydration and high hydrostatic pressure on physicochemical and quality parameters (color, antioxidant capacity, total phenolic content, and vitamin C) on strawberries stored at 5 °C. The results indicated that quality profiles of strawberry osmotically dehydrated under high hydrostatic pressure between 300 and 500 MPa showed minimal differences when compared to untreated samples. For this reason, it was recommended working at 400 MPa/10 min to obtain processed strawberries with high levels of both nutritional and antioxidant characteristics.

Distinct from the application of HHP for preservation purposes, high pressure treatments have been used to extract secondary plant metabolites from fruits and vegetables. For example, De Ancos et al. (2000) successfully applied HHP (50–400 MPa, 25 °C/15 min) processing to extract carotene from persimmon fruit purees. Different pressure levels at constant temperature gave different release of various carotenes depending on their chemical properties and chromoplast location. The use of high pressure enhances mass transfer rates, which increases cell permeability as well as secondary metabolite diffusion (Dornenburg and Knoor 1993). HHP treatment influences the phytochemical stability and the extraction yield of bioactive compounds. As a consequence, changes in antioxidant activity could also occur during HHP treatment (Rawson et al. 2011a).

4.2.2 Pulsed Electric Field (PEF)

PEF is a technology that has been extensively investigated in recent years for its applications in food processing. PEF pasteurization is a technique based on the delivery of pulses at high electric field intensity (5–55 kV cm^{-1}) to a food in the millisecond range (Lado and Yousef 2002). By the mechanism of electropermeabilization, pulsed electrical fields have proved a valid technology for the production of safe beverage products and shown a positive influence in the texture of solid plant foods, leading to enhanced yields of extraction of metabolites, as well as increased juice yields (Rawson et al. 2011a).

Morales-de la Peña et al. (2010a, b) investigated the effect of PEF on vitamin C in orange/kiwi/pineapple, and soymilk-based beverage immediately after treatment and noted that levels were not different from the thermally processed juice. However, the beneficial effect of the PEF treatment was noticeable over a storage period of 31 days, as an 800 μs treatment at 35 kV/cm showed significantly greater retention than both 1,400 μs treatment and thermal treatment. These results showed that the shorter the PEF treatment time, the higher the vitamin C retention, as previously found in other studies focused on individual fruit juices treated by high intensity PEF (HIPEF). In general, longer exposure PEF treatment times may induce reduction in the retention of vitamin C due to product heating. Longer exposure time may also generate free radicals which may speed up vitamin C degradation. Moreover, the antioxidant capacity of this product during storage decreased to a greater degree in thermally treated samples than in PEF treated samples after a storage period of 60 days.

PEF can retain higher levels of phenolic compounds in fruit juices and improve their stability during storage. Odriozola-Serrano et al. (2008b) observed significantly less phenolic degradation by PEF (49 %) than by thermal pasteurization (55 %) after 56 days of storage of strawberry juice.

Studies evaluating the effects of HIPEF processing conditions on watermelon juices have been demonstrated that HIPEF treatments were effective in reducing the population of pathogenic microorganisms and inactivating spoilage enzymes. Watermelon juice exhibited high retention of lycopene and antioxidant capacity when high electric field strengths, frequencies, and pulse widths were applied. However, severe HIPEF treatments reduced vitamin C content. Maximal relative lycopene content (113 %), vitamin C (72 %), and antioxidant capacity retention (100 %) were obtained when HIPEF treatments were set up at 35 kV/cm for 50 μs using 7 μs bipolar pulses at 200 Hz (Aguiló-Aguayo et al. 2008; Oms-Oliu et al. 2009).

Vervoort et al. (2011) compared the impact of thermal, HHP, and PEF processing for mild pasteurization of orange juice, using processing conditions leading to an equivalent degree of microbial inactivation. Their study provided evidence that HHP and PEF pasteurization do not cause any significant differences in the major components regarding public health that were investigated, in comparison to thermal pasteurization, and therefore no changes in the human metabolism after consumption are to be expected.

4.2.3 Ultrasound

Ultrasound is used at frequencies in the range of 20–100 kHz and requires the presence of a liquid medium for power transmission. It causes chemical and physical changes in biological structures (in a liquid medium) due to intracellular cavitation (Alexandre et al. 2012). In last decade power ultrasound has emerged as an alternative processing option to conventional thermal approaches for pasteurization and sterilization of food products. Ultrasound processing on its own or in combination with heat and/or pressure is an effective processing tool for microbial inactivation and phytochemical retention. Advantages of ultrasound include reduced processing time, higher throughput, and lower energy consumption (Zenker et al. 2003; Rawson et al. 2011a).

Ultrasound treatment of fruit juices is reported to have a minimal effect on the ascorbic acid content during processing and results in improved stability during storage when compared to thermal treatment. This positive effect of ultrasound compared with heating is assumed to be due to the effective removal of occluded oxygen from the juice (Knorr et al. 2004). Ascorbic acid content was found to be significantly higher in guava juice samples treated with carbonation and sonication than in the control. It could be due to cavitation effects caused by carbonation and sonication (Cheng et al. 2007). However, degradation of vitamin C in sonicated orange, strawberry, and tomato juices was observed and the degradation level depended on the wave amplitude and treatment time. Ascorbic acid degradation during sonication may be due to free radical formation and production of oxidative products on the surface of bubbles (Tiwari et al. 2009b, c).

Ultrasonication may be considered a potential technology for processing of red juices because of its minimal effect on anthocyanins (Oms-Oliu et al. 2012). Tiwari et al. (2009a) reported a slight increase (1–2 %) in the pelargonidin-3-glucoside content of the juice at lower amplitude levels and treatment times which may be due to the extraction of bound anthocyanins from the suspended pulp.

Ultrasonic extraction is a well-known commercial method to increase mass transfer rate by cavitation forces. Bubbles in the liquid–solid extraction using ultrasonic extraction can explosively collapse and produce localized pressure, improving the interaction between the intracellular substances and the solvent to facilitate the extraction of the phytochemicals (Saldana et al. 2010). The extraction of lycopene from tomato using ultrasonic-assisted extraction and ultrasound/microwave-assisted extraction was reported (Lianfu and Zelong 2008). Rawson et al. (2011b) determined that sonication temperature played a significant role in preservation of bioactive compounds. Freshly squeezed watermelon juice was subjected to thermosonication treatments with processing variables of temperature (25–45 °C), amplitude level (24.1–60 μm), and processing time (2–10 min) at a constant frequency of 20 kHz and pulse durations of 5 s on and 5 s off. They observed a decrease in the phenolic content of sonicated watermelon juice when the temperature was increased from 25 to 45 °C. Temperature effect was more pronounced at higher processing times.

4.2.4 Radiation Processing

Irradiation treatment generally involves the exposure of food products (raw or processed) to ionizing or non ionizing radiation for the purpose of food preservation. The ionizing radiation source could be high-energy electrons, X-rays, or gamma rays, while the non ionizing radiation is electromagnetic radiation that does not carry enough energy/quanta to ionize atoms or molecules, represented mainly by ultraviolet rays (UV-A: 315–400 nm, UV-B: 280–315 nm, and UV-C: 100–280 nm), visible light, microwaves, and infrared (Prakash et al. 2000; Rawson et al. 2011a). Food irradiation is a physical treatment in which food is exposed to ionizing radiation, i.e., radiation of sufficient energy to expel electrons from atoms and to ionize molecules. Foods treated with ionizing radiation have consistently been shown to be wholesome and nutritious. The effect of irradiation on vitamins has been studied extensively. Sugars may be hydrolyzed or oxidized when subjected to gamma radiation. Free amino acids can be deaminated. Free radicals react with polyunsaturated fatty acids, producing unstable hydroperoxides and a range of further degradation products. Certain vitamins (A, B1, B12, C, E, K), particularly those with antioxidant activity, are degraded when irradiation is carried out in the presence of oxygen (Niemira and Deschênes 2005).

Alighourchi et al. (2008) reported a significant reduction in the total and individual anthocyanin content in pomegranate juice after irradiation at higher doses (3.5–10 kGy). Irradiation effects on anthocyanin pigments depend upon the nature of anthocyanin, for example, diglycosides are relatively stable toward irradiation dose compared to monoglycosides. Reyes and Cisneros-Zevallos (2007) investigated the effect of irradiation (1–3.1 kGy) on mango. The authors did not find a significant impact of irradiation dose on the total phenolic content, while there was a significant increase in flavonols after 18 days storage period for the irradiated fruits (at 3.1 kGy). In contrast, ascorbate content of the fruits decreased when the dose exceeded 1.5 kGy. No major changes in the carotenoids content were recorded. In general, the decrease in antioxidant compounds is attributed to the formation of radiation-induced degradation products or the formation of free radicals (Wong and Kitts 2001; Sajilata and Singhal 2006). The effects of harvest date, storage, and low-dose irradiation on flavanones were investigated in grapefruits. In general; flavanone concentrations increased with increasing irradiation dose even in the late season grapefruit, and storage had a positive effect on flavanone levels (Patil et al. 2004).

It has been reported that irradiation treatments can generate free radicals, thus leading to an induction of stress responses in plant foods, which in turn may lead to an increase in the antioxidant synthesis (Oms-Oliu et al. 2012). Song et al. (2006) observed that total phenolic content of carrot and kale juices substantially increased by applying an irradiation treatment. However, reductions in the total phenolic content have been reported for treatments of more than 10 kGy in some irradiated products (Villavicencio et al. 2000; Ahn et al. 2005).

Irradiation of plant tissues with UV has been shown to have positive interactions, indicating an increase in the enzymes responsible for flavonoid biosynthesis, affecting plant phenolic metabolites apart from induction of abiotic stress. UV-A has been reported to induce anthocyanin biosynthesis in fruits encompassing cherries (Kataoka et al. 1996).

UV-C is the most common applied to fresh fruits and vegetables, since it acts directly or indirectly as an antimicrobial agent. UV-C can cause direct bacterial DNA damage or may induce resistance mechanisms against pathogens in different fruits and vegetables. Low doses of UV-C radiation (254 nm) also reduce decay of a wide range of fruits and vegetables when applied after harvest (Ben-Yehoshua and Mercier 2005; Ramos et al. 2013). Erkan et al. (2008) investigated the changes in antioxidant capacity, enzyme activity, and decay development in strawberry fruit illuminated with different UV-C dosages. Three UV-C illumination durations and dosages, 1, 5, and 10 min (0.43, 2.15, and 4.30 kJ m^{-2}) tested promoted the antioxidant capacity and enzyme activities and significantly reduced the severity of decay during storage at 10 °C compared to the control. All UV-C dosages increased the phenolic content of strawberry fruits as well. Total anthocyanin content increased during storage in all treatments.

Like PEF treatment, UV exposure can kill microorganisms with potentially less impact on food quality (Chen et al. 2013). UV irradiation has proved to be effective against *E. coli* O157:H7 in unpasteurized apple cider (Hanes et al. 2002; Basaran et al. 2004). Guerrero-Beltrán et al. (2009) evaluated the UV-C light effect on *Saccharomyces cerevisiae* inactivation in grape, cranberry, and grapefruit juices. The maximum log reduction (cfu/mL) was 0.53, 2.51, and 2.42 for yeast count in grape, cranberry, and grapefruit juices, respectively, after 30 min of UV light treatment at the maximum flow rate (1.02 L/min).

Noci et al. (2008) reported that UV exposure of apple juice caused a 29 % reduction in total phenolic content, which was much lower than that due to thermal processing (48 %). However, UV exposure has its limitations when treating juices. UV light only penetrates a very short depth into the surface of a juice when compared with clear water (Lu et al. 2010). The penetration of UV light into juices is about 1 mm for absorption of 90 % of the light. As a result, a special conduction of the liquid flow is always used in the UV exposure of juices to minimize the absorption. Lu et al. (2010) designed a small thin film UV reactor to process apple juice with the aim of increasing the microbial inactivation rate and reported its excellent performance in the reduction of microorganisms in various apple juices. The apple juice stability and nutritional qualities were also improved by using this method.

In particular, the combination of UV and PEF as a hurdle may overcome the limitations of the individual techniques and has proven to be more effective for microbial inactivation and maintaining nutritional quality of fruit juice (Chen et al. 2013).

4.2.5 Membrane Filtration

Reverse osmosis (RO) and ultrafiltration (UF) are both unit operations in which water and some solutes in a solution are selectively removed through a semipermeable membrane. They are similar in that the driving force for transport across the membrane is the pressure applied to the feed liquid. However, RO is used to separate water from low-molecular-weight solutes (e.g., salts, monosaccharides, and aroma compounds), which have a high osmotic pressure. A high pressure, 5–10 times that is used in UF ($4{,}000–8{,}000 \times 10^3$ Pa), is therefore necessary to overcome this. Microfiltration (MF) is similar to UF in using lower pressures than RO, but is distinguished by the larger range of particle sizes (0.01–2 μm) that are separated (Fellows 2000).

UF and MF are the most commonly used membrane filtration techniques for fruit juice processing. They have been applied commercially for the clarification of fruit juices. Basically, the membranes retain large molecules such as microorganisms, lipids, proteins, and colloids (UF only) and allow small molecules such as vitamins, salts, sugars, and water to flow through them. Therefore, via this process, "cold pasteurized" products (>5 log reduction or removal of microorganisms) can be produced with better flavors than thermally treated products (Cassano et al. 2003; Rektor et al. 2004; Chen et al. 2013). In contrast to concentration by boiling, RO and UF membranes concentrate foods without heat to produce good retention of sensory and nutritional qualities (Fellows 2000).

Pap et al. (2010) applied reverse osmosis process for the concentration of black currant juice. The researchers reported that enzymatic treatment resulted in the increase of anthocyanin and flavonol content of the juices. The centrifugation process decreased the amount of anthocyanins and flavonols to some extent.

The juice clarified by UF had significantly lower concentrations of anthocyanins and flavonols, while enzymatic pretreatment applied juice had the highest levels of these flavonoids. Enzymatic pretreatment improved the permeate flux in RO during the concentration process and resulted in a juice concentrates highest in anthocyanins and flavonols.

A comparative study by Cassano et al. (2003) on the concentration of blood, orange juice demonstrated that the total antioxidant activity of juice concentrated by evaporation was lower than that of the fresh juice. During UF, the total antioxidant activity was maintained in both permeate and retentate. When RO was applied, a small decrease of the total antioxidant activity was determined. Osmotic distillation, applied as subsequent concentration step after RO, did not cause any significant loss in antioxidant activity of the juice. Cassano et al. (2006) proposed integrated membrane process for the production of kiwifruit juice. Losses of total antioxidant activity after UF and osmotic distillation relative to the fresh juice were 4.4 and 11.1 %, respectively, and the reduction of vitamin content in the final concentrate was also very limited.

References

Aaby K, Wrolstad RE, Ekeberg D, Skrede G (2007) Polyphenol composition and antioxidant activity in strawberry purees; impact of achene level and storage. J Agric Food Chem 55(13):5156–5166

Aguiló-Aguayo I, Soliva-Fortuny R, Martín-Belloso O (2008) Effects of thermal and high-intensity pulsed electric field treatments on quality related enzymes of watermelon juices. In IFT annual meeting, paper 135-03, New Orleans, USA

Ahn HJ, Kim JH, Kim JK, Kim DH, Yook HS, Byun MW (2005) Combined effect of irradiation and modified atmosphere packaging on minimally processed Chinese cabbage (*Brassica ropa* L.). Food Chem 89:589–597

Aked J (2002) Maintaining the post-harvest quality of fruits and vegetables. In: Jongen W (ed) Fruit and vegetable processing: improving quality. Woodhead Publishing and CRC Press, Boca Raton

Alexandre EMC, Brandão TRS, Silva CLM (2012) Efficacy of non-thermal technologies and sanitizer solutions on microbial load reduction and quality retention of strawberries. J Food Eng 108(3):417–426

Alighourchi H, Barzegar M, Abbasi S (2008) Effect of gamma irradiation on the stability of anthocyanins and shelf-life of various pomegranate juices. Food Chem 110:1036–1040

Amarowicz R, Carle R, Dongowski G, Durazzo A, Galensa R, Kammerer D, Maiani G, Piskula MK (2009) Influence of postharvest processing and storage on the content of phenolic acids and flavonoids in foods. Mol Nutr Food Res 53:S151–S183

Amin I, Norazaidah Y, Hainida KIE (2006) Antioxidant activity and phenolic content of raw and blanched Amaranthus species. Food Chem 94:47–52

Aramwit P, Bang N, Srichana T (2010) The properties and stability of anthocyanins in mulberry fruits. Food Res Int 43:1093–1097

Arancibia-Avila P, Namiesnik J, Toledo F, Werner E, Martinez-Ayala AL, Rocha-Guzmán NE, Gallegos-Infante JA, Gorinstein S (2012) The influence of different time durations of thermal processing on berries quality. Food Control 26:587–593

Asami DK, Hong YJ, Barrett DM, Mitchell AE (2003) Processing-induced changes in total phenolics and procyanidins in clingstone peaches. J Sci Food Agric 83:56–63

Ayala-Zavala JF, Wang SY, Wang CY, González-Aguilar GA (2005) Methyl jasmonate in conjunction with ethanol treatment increases antioxidant capacity, volatile compounds and postharvest life of strawberry fruit. Eur Food Res Technol 221(6):731–738

Ayala-Zavala JF, Wang SY, Wang CY, González-Aguilar GA (2007) High oxygen treatment increases antioxidant capacity and postharvest life of strawberry fruit. Food Technol Biotechnol 45(2):166–173

Ayala-Zavala JF, Del-Toro- Sánchez L, Alvarez-Parrilla E, González-Aguilar GA (2008a) High relative humidity in-package of fresh-cut fruits and vegetables: advantage or disadvantage considering microbiological problems and antimicrobial delivering systems? J Food Sci 73(4):R41–R47

Ayala-Zavala JF, Del-Toro-Sánchez L, Alvarez-Parrilla E, Soto-Valdez H, Martin-Belloso O, Ruiz-Cruz S, González-Aguilar GA (2008b) Natural antimicrobial agents incorporated in active packaging to preserve the quality of fresh fruits and vegetables. Stewart Postharvest Rev 4:1–9

Balla C, Farkas J (2006) Minimally processed fruits and fruit products and their microbiological safety. In: Hui YH (ed) Handbook of fruits and fruit processing. Blackwell Publishing, Hoboken

Basaran N, Quintero-Ramos A, Moake MM, Churey JJ, Worobo RW (2004) Influence of apple cultivars on inactivation of different strains of *Escherichia coli* O157:H7 in apple cider by UV irradiation. Appl Environ Microbiol 70:6061–6065

Ben-Yehoshua S, Mercier J (2005) UV irradiation, biological agents, and natural compounds for controlling postharvest decay in fresh fruits and vegetables. In: Ben-Yehoshua S (ed) Environmentally friendly technologies for agricultural produce quality. Taylor & Francis, Boca Raton

Bernhardt S, Schlich E (2006) Impact of different cooking methods on food quality: retention of lipophilic vitamins in fresh and frozen vegetables. J Food Eng 77:327–333

Briones-Labarca V, Venegas-Cubillos G, Ortiz-Portilla S, Chacana-Ojeda M, Maureira H (2011) Effects of high hydrostatic pressure (HHP) on bioaccessibility, as well as antioxidant activity, mineral and starch contents in Granny Smith apple. Food Chem 128:520–529

Brownmiller C, Howard LR, Prior RL (2008) Processing and storage effects on monomeric anthocyanins, percent polymeric color, and antioxidant capacity of processed blueberry products. J Food Sci 73:72–79

Buchner N, Krumbein A, Rhon S, Kroh LW (2006) Effect of thermal processing on the flavonols rutin and quercetin. Rapid Commun Mass Spectrom 20:3229–3235

Burri J, Bertoli C, Stadler RH (2009) Food processing and nutritional aspects. In: Stadle RH, Lineback DR (eds) Process-induced food toxicants: occurrence, formation, mitigation, and health risks. Wiley, Hoboken

Capanoglu E, Beekwilder J, Boyacioglu D, Hall R, De Vos R (2008) Changes in antioxidant and metabolite profiles during production of tomato paste. J Agric Food Chem 56(3):964–973

Cassano A, Drioli E, Galaverna G, Marchelli R, Di Silvestro G, Cagnasso P (2003) Clarification and concentration of citrus and carrot juices by integrated membrane processes. J Food Eng 57:153–163

Cassano A, Figoli A, Tagarelli A, Sindona G, Drioli E (2006) Integrated membrane process for the production of highly nutritional kiwifruit juice. Desalination 189:21–30

Chang CH, Lin HY, Chang CY, Liu YC (2006) Comparison on the antioxidant properties of fresh, freeze-dried and hot-air-dried tomatoes. J Food Eng 77:478–485

Chen Y, Yu LJ, Vasantha Rupasinghe HP (2013) Effect of thermal and non-thermal pasteurisation on the microbial inactivation and phenolic degradation in fruit juice: a mini-review. J Sci Food Agric 93:981–986

Cheng LH, Soh CY, Liew SC, Teh FF (2007) Effects of sonication and carbonation on guava juice quality. Food Chem 104(4):1396–1401

Cohen MF, Sakihama Y, Yamasaki H (2001) Roles of plant flavonoids in interactions with microbes: from protection against pathogens to mediation of mutualism. In: Pandalai SG (ed) Recent research developments in plant physiology. Research Signpost, Trivandrum

Cui X, Shang Y, Shi Z, Xin H, Cao W (2009) Physicochemical properties and bactericidal efficiency of neutral and acidic electrolyzed water under different storage conditions. J Food Eng 91(4):582–586

De Ancos B, Cano MP, Hernandez A, Monreal M (1999) Effects of microwave heating on pigment composition and colour of fruit purees. J Sci Food Agric 79:663–670
De Ancos B, Gonzalez E, Pilar Cano M (2000) Effect of high pressure treatment on the carotenoid composition and the radical scavenging activity of persimmon fruit purees. J Agric Food Chem 48:3542–3548
Dini I, Tenore GC, Dini A (2013) Effect of industrial and domestic processing on antioxidant properties of pumpkin pulp. LWT Food Sci Technol 53:382–385
Domingo JL, Gine Bordonaba J (2011) A literature review on the safety assessment of genetically modified plants. Environ Int 37:734–742
Dong J, Ma X, Fu Z, Guo Y (2011) Effects of microwave drying on the contents of functional constituents of *Eucommia ulmoides* flower tea. Ind Crops Prod 34(1):1102–1110
Donsi G, Ferrari G, Di Mateo M (1996) High pressure stabilization of orange juice: evaluation of the effects of process conditions. Italian J Food Sci 2:99–106
Dornenburg H, Knoor D (1993) Cellular permeabilization of cultured plant tissues by high electric field pulses or ultra high pressure for the recovery of secondary metabolites. Food Biotechnol 7:35–48
Durst RW, Weaver GW (2013) Nutritional content of fresh and canned peaches. J Sci Food Agric 93:593–603
Dziezak JD (1986) Preservative systems in foods, antioxidants and antimicrobial agents. Food Technol 40:94–136
Erkan M, Wang SY, Wang CY (2008) Effect of UV treatment on antioxidant capacity, antioxidant enzyme activity and decay in strawberry fruit. Postharvest Biol Technol 48:163–171
Fellows P (2000) Food processing technology principles and practice. Woodhead, Cambridge
Fengmei Z, Liangbin H, Guihua X, Quanxian C (2011) Changes of some chemical substances and antioxidant capacity of mandarin orange segments during can processing. Procedia Environ Sci 11:1260–1266
Fernández-García A, Butz P, Bognár A, Tauscher B (2001) Antioxidative capacity, nutrient content and sensory quality of orange juice and an orange-lemon-carrot juice product after high pressure treatment and storage in different packaging. Eur Food Res Technol 213:290–296
Ferrari G, Maresca P, Ciccarone R (2010) The application of high hydrostatic pressure for the stabilization of functional foods: pomegranate juice. J Food Eng 100:245–253
Fuleki T, Ricardo-Da-Silva JM (2003) Effects of cultivar and processing method on the contents of catechins and procyanidins in grape juice. J Agric Food Chem 51:640–646
Gancel AL, Feneuil A, Acosta O, Pérez AM, Vaillant F (2011) Impact of industrial processing and storage on major polyphenols and the antioxidant capacity of tropical highland blackberry (*Rubus adenotrichus*). Food Res Int 44:2243–2251
Garcia-Parra J, Gonzalez-Cebrino F, Delgado J, Lozano M, Hernandez T, Ramirez R (2011) Effect of thermal and high-pressure processing on the nutritional value and quality attributes of a nectarine puree with industrial origin during the refrigerated storage. J Food Sci 76(4):C618–C625
García-Viguera C, Zafrilla P, Romero F, Abellán P, Artés F, Tomás-Barberán FA (1999) Color stability of strawberry jam as affected by cultivar and storage temperature. J Food Sci 64(2):243–247
Georgé S, Tourniaire F, Gautier H, Goupy P, Rock E, Caris-Veyrat C (2011) Changes in the contents of carotenoids, phenolic compounds and vitamin C during technical processing and lyophilisation of red and yellow tomatoes. Food Chem 124:1603–1611
Gerard KA, Roberts JS (2004) Microwave heating of apple mash to improve juice yield and quality. LWT Food Sci Technol 37:551–557
González-Aguilar GA, Ayala-Zavala JF, De la Rosa LA, Alvarez-Parrilla E (2010) Phytochemical changes in the postharvest and minimal processing of fresh fruits and vegetables. In: De la Rosa LA, Alvarez-Parrilla E, González-Aguilar GA (eds) Fruit and vegetable phytochemicals: chemistry, nutritional value, and stability. Blackwell, Ames
Grajek W, Olejnik A (2010) The influence of food processing and home cooking on the antioxidant stability in foods. In: Smith J, Charter E (eds) Functional food product development. Blackwell, Ames

Gross J (1991) Pigments in vegetables: chlorophylls and carotenoids. Van Nostrand Reinhold, New York

Guerrero-Beltrán JA, Welti-Chanes J, Barbosa-Cánovas GV (2009) Ultraviolet-C light processing of grape, cranberry and grapefruit juices to inactivate *Saccharomyces cerevisiae*. J Food Process Eng 32:916–932

Hanes DE, Worobo RW, Orlandi PA, Burr DH, Miliotis MD, Rob MG et al (2002) Inactivation of *Cryptosporidium parvum* oocysts in fresh apple cider by UV irradiation. Appl Environ Microbiol 68:4168–4172

Howard LA, Wong AD, Perry AK, Klein BP (1999) Beta-carotene and ascorbic acid retention in fresh and processed vegetables. J Food Sci 64(5):929–936

Huang W, Bi X, Zhang X, Liao X, Hu X, Wu J (2013) Comparative study of enzymes, phenolics, carotenoids and color of apricot nectars treated by high hydrostatic pressure and high temperature short time. Innov Food Sci Emerg Technol 18:74–82

Icier F, Ilicali C (2005) The effects of concentration on electrical conductivity of orange juice concentrates during ohmic heating. Eur Food Res Technol 220:406–414

Igual M, García-Martínez E, Camacho MM, Martínez-Navarrete N (2011) Changes in flavonoid content of grapefruit juice caused by thermal treatment and storage. Innov Food Sci Emerg Technol 12:153–162

Igual M, García-Martínez E, Martín-Esparza ME, Martínez-Navarrete N (2012) Effect of processing on the drying kinetics and functional value of dried apricot. Food Res Int 47:284–290

Indrawati AC, Messagie I, Nguyen MT, Van Loey A, Hendrickx M (2004) Comparative study on pressure and temperature stability of 5-mehytltetrahydrofolic acid in model systems and in food products. J Agric Food Chem 52:485–492

Ioannou I, Hafsa I, Hamdi S, Charbonnel C, Ghoul M (2012) Review of the effects of food processing and formulation on flavonol and anthocyanin behaviour. J Food Eng 111:208–217

Jeong SM, Kim SY, Kim DR, Jo SC (2004) Effect of heat treatment on the antioxidant activity of extracts from citrus peels. J Agric Food Chem 52:3389–3393

Jiratanan T, Liu RH (2004) Antioxidant activity of processed table beets (*Beta vulgaris*, var, conditiva) and green beans (*Phaseolus vulgaris* L.). J Agric Food Chem 52(9):3659–3670

Kader A (2001) Quality assurance of harvested horticultural perishables. In: Ben-Arie R, Philosoph-Hadas S (eds) Proceedings of the 4th international conference on postharvest, 553. Acta Horticulture, Jerusalem

Kader A, Barrett DM (2005) Classification, composition of fruits, and postharvest maintenance of quality. In: Barrett DM, Somogyi L, Ramaswamy H (eds) Processing fruits: science and technology. CRC Press, Boca Raton

Kalt W (2005) Effects of production and processing factors on major fruit and vegetable antioxidants. J Food Sci 70:11–19

Kataoka I, Beppu K, Sugiyama A, Taira S (1996) Enhancement of coloration of Satohnishiki sweet cherry fruit by postharvest irradiation with ultraviolet rays. Environ Control Biol 34:313–319

Keenan DF, Brunton NP, Gormley TR, Butler F, Tiwari BK, Patras A (2010) Effect of thermal and high hydrostatic pressure processing on antioxidant activity and colour of fruit smoothies. Innov Food Sci Emerg Technol 11(4):551–556

Keenan DF, Rößle C, Gormley R, Butler F, Brunton NP (2012) Effect of high hydrostatic pressure and thermal processing on the nutritional quality and enzyme activity of fruit smoothies. LWT Food Sci Technol 45:50–57

Knorr D, Zenker M, Heinz V, Lee D (2004) Applications and potential of ultrasonics in food processing. Trends Food Sci Technol 15(5):261–266

Lado B, Yousef A (2002) Alternative food-preservation technologies: efficacy and mechanisms. Microbes Infect 4:433–440

Lamikanra O (2002) Fresh-cut fruits and vegetables: science, technology, and market. CRC Press, Boca Raton

Lee SY, Sagong HG, Ryu S, Kang DH (2012) Effect of continuous ohmic heating to inactivate *Escherichia coli* O157:H7, *Salmonella typhimurium* and *Listeria monocytogenes* in orange juice and tomato juice. J Appl Microbiol 112:723–731

Leizerson S, Shimoni E (2005) Stability and sensory shelf life of orange juice pasteurized by continuous ohmic heating. J Agric Food Chem 53(10):4012–4018

Lemmens L, Van Buggenhout S, Oey I, Van Loey A, Hendrickx M (2009) Towards a better understanding of the relationship between the beta-carotene in vitro bio-accessibility and pectin structural changes: a case study on carrots. Food Res Int 42(9):1323–1330

Leong SY, Oey I (2012) Effects of processing on anthocyanins, carotenoids and vitamin C in summer fruits and vegetables. Food Chem 133:1577–1587

Lianfu Z, Zelong L (2008) Optimization and comparison of ultrasound/microwave assisted extraction (UMAE) and ultrasonic assisted extraction (UAE) of lycopene from tomatoes. Ultrason Sonochem 15(5):731–737

Lin CH, Chang CY (2005) Textural change and antioxidant properties of broccoli under different cooking treatments. Food Chem 90:9–15

Lindley MG (1998) The impact of food processing on antioxidants in vegetable oils, fruits and vegetables. Trends Food Sci Technol 9:336–340

Lu G, Li C, Liu P, Cui H, Xia Y, Wang J (2010) Inactivation of microorganisms in apple juice using an ultraviolet silica-fiber optical device. J Photochem Photobiol B 100:167–172

Lynch M, Tauxe RV, Hedberg C (2009) The growing burden of foodborne outbreaks due to contaminated fresh produce: risks and opportunities. Epidemiol Infect 137:307–315

Magnuson BA, Jonaitis TS, Card JW (2011) A brief review of the occurrence, use, and safety of food-related nanomaterials. J Food Sci 76:R126–R133

Makris DP, Rossiter JT (2000) Heat-induced, metal-catalyzed oxidative degradation of quercetin and rutin (quercetin 3-o-rhamnosylglucoside) in aqueous model. J Agric Food Chem 48:3830–3838

Morales-de la Peña M, Salvia-Trujillo L, Rojas-Graü MA, Martín-Belloso O (2010a) Impact of high intensity pulsed electric field on antioxidant properties and quality parameters of a fruit juice-soymilk beverage in chilled storage. LWT Food Sci Technol 43:872–881

Morales-de la Peña M, Salvia-Trujillo L, Rojas-Graü MA, Martín-Belloso O (2010b) Isoflavone profile of a high intensity pulsed electric field or thermally treated fruit juice–soymilk beverage stored under refrigeration. Innov Food Sci Emerg Technol 11:604–610

Motarjemi Y, Stadler RH, Studer A, Damiano V (2009) Application of the HACCP approach for the management of processing contaminants. In: Stadler RH, Lineback DR (eds) Process-induced food toxicants: occurrence, formation, mitigation, and health risks. Wiley, Hoboken

Murakami M, Yamaguchi T, Takamura H, Matoba T (2004) Effects of thermal treatment on radical-scavenging activity of single and mixed polyphenolic compounds. Food Chem Toxicol 69:FCT7–FCT10

Niemira BA, Deschênes L (2005) Ionizing radiation processing of fruits and fruit products. In: Barrett DM, Somogyi L, Ramaswamy H (eds) Processing fruits: science and technology. CRC Press, Boca Raton

Noci F, Riener J, Walkling-Ribeiro M, Cronin D, Morgan D, Lyng J (2008) Ultraviolet irradiation and pulsed electric fields (PEF) in a hurdle strategy for the preservation of fresh apple juice. J Food Eng 85:141–146

Nuñez-Mancilla Y, Pérez-Won M, Uribe E, Vega-Gálvez A, Di Scala K (2013) Osmotic dehydration under high hydrostatic pressure: effects on antioxidant activity, total phenolics compounds, vitamin C and colour of strawberry (*Fragaria vesca*). LWT Food Sci Technol 52:151–156

Odriozola-Serrano I, Soliva-Fortuny R, Martin-Belloso O (2008a) Effect of minimal processing on bioactive compounds and color attributes of fresh-cut tomatoes. LWT Food Sci Technol 41:217–226

Odriozola-Serrano I, Soliva-Fortuny R, Martin-Belloso O (2008b) Phenolic acids, flavonoids, vitamin C and antioxidant capacity of strawberry juices processed by high-intensity pulsed electric fields or heat treatments. Eur Food Res Technol 228:239–248

Oey I, Plancken I, van der Loey A, Hendrickx M (2008) Does high pressure processing influence nutritional aspects of plant based food systems? Trends Food Sci Technol 19:300–308

Oms-Oliu G, Odriozola-Serrano I, Soliva-Fortuny R, Martín-Belloso O (2009) Effects of high-intensity pulsed electric field processing conditions on lycopene, vitamin C and antioxidant capacity of watermelon juice. Food Chem 115:1312–1319

Oms-Oliu G, Odriozola-Serrano I, Soliva-Fortuny R, Elez Martinez P, Martín-Belloso O (2012) Stability of health related compounds in plant foods through the application of non- thermal processes. Trends Food Sci Technol 23:111–123

Pap N, Pongrácz E, Jaakkola M, Tolonen T, Virtanen V, Turkki A, Horváth-Hovorka Z, Vatai G, Keiski RL (2010) The effect of pre-treatment on the anthocyanin and flavonol content of black currant juice (*Ribes nigrum* L.) in concentration by reverse osmosis. J Food Eng 98:429–436

Patil BS, Vanamalaa J, Hallmanc G (2004) Irradiation and storage influence on bioactive components and quality of early and late season, Rio Red' grapefruit (*Citrus paradise* Macf.). Postharvest Biol Technol 34:53–64

Patras A, Brunton N, Da Pieve S, Butler F, Downey G (2009a) Effect of thermal and high pressure processing on antioxidant activity and instrumental colour of tomato and carrot purees. Innov Food Sci Emerg Technol 10:16–22

Patras A, Brunton NP, Da Pieve S, Butler F (2009b) Impact of high pressure processing on total antioxidant activity, phenolic, ascorbic acid, anthocyanin content and colour of strawberry and blackberry purees. Innov Food Sci Emerg Technol 10:308–313

Périno-Issartier S, Huma Z, Abert-Vian M, Chemat F (2010) Solvent free microwave-assisted extraction of antioxidants from sea buckthorn (*Hippophae rhamnoides*) food by-products. Food Bioprocess Technol 1–9

Pilavtepe-Celik M (2013) High hydrostatic pressure (HHP) inactivation of foodborne pathogens in low-acid juices. Int J Food Sci Technol 48:673–677

Plaza L, Crespo I, de Pascual-Teresa S, de Ancos B, Sánchez-Moreno C, Muñoz M, Cano MP (2011) Impact of minimal processing on orange bioactive compounds during refrigerated storage. Food Chem 124:646–651

Prakash A, Inthajak P, Huibregtse H, Caporaso F, Foley DM (2000) Effects of low-dose gamma irradiation and conventional treatments on shelf life and quality characteristics of diced celery. J Food Sci 65(6):1070–1075

Raghavan GSV, Vigneault C, Gariépy Y, Markarian NR, Alvo P (2005) Refrigerated and controlled/modified atmosphere storage. In: Barrett DM, Somogyi L, Ramaswamy H (eds) Processing fruits: science and technology. CRC Press, Boca Raton

Ramaswamy HS, Chen CR (2002) Maximising the quality of thermally processed fruits and vegetables. In: Jongen W (ed) Fruit and vegetable processing: improving quality. Woodhead Publishing/CRC Press, Boca Raton

Ramos B, Miller FA, Brandão TRS, Teixeira P, Silva CLM (2013) Fresh fruits and vegetables-an overview on applied methodologies to improve its quality and safety. Innov Food Sci Emerg Technol. doi:10.1016/j.ifset.2013.07.002

Rattanathanalerk M, Chiewchan N, Srichumpoung W (2009) Effect of thermal processing on the quality loss of pineapple juice. J Food Eng 66:259–265

Rawson A, Koidis A, Rai DK, Tuohy M, Brunton N (2010) Influence of sous vide and water immersion processing on polyacetylene content and instrumental color of parsnip (*Pastinaca sativa*) disks. J Agric Food Chem 58(13):7740–7747

Rawson A, Patras A, Tiwari BK, Noci F, Koutchma T, Brunton N (2011a) Effect of thermal and non-thermal processing technologies on the bioactive content of exotic fruits and their products: review of recent advances. Food Res Int 44:1875–1887

Rawson A, Tiwari BK, Patras A, Brunton N, Brennan C, Cullen PJ et al (2011b) Effect of thermosonication on bioactive compounds in water-melon juice. Food Res Int 44:1168–1173

Rektor A, Pap N, Kókai Z, Szabó R, Vatai G, Békássylnár E (2004) Application of membrane filtration methods for must processing and preservation. Desalination 162:271–277

Renard CMGC, Le Quéré JM, Bauduin R, Symoneaux R, Le Bourvellec C, Baron A (2011) Modulating polyphenolic composition and organoleptic properties of apple juices by manipulating the pressing conditions. Food Chem 124:117–125

Reyes LF, Cisneros-Zevallos L (2007) Electron-beam ionizing radiation stress effects on mango fruit (*Mangifera indica* L.) antioxidant constituents before and during postharvest storage. J Agric Food Chem 55:6132–6139

Rickman JC, Barrett DM, Bruhn CM (2007a) Nutritional comparison of fresh, frozen and canned fruits and vegetables. Part 1. Vitamins B and C and phenolic compounds. J Sci Food Agric 87:930–944
Rickman JC, Bruhn CM, Barrett DM (2007b) Nutritional comparison of fresh. frozen and canned fruits and vegetables. Part 2. Vitamin A and carotenoids, vitamin E, minerals and fiber. J Sci Food Agric 87:1185–1196
Robards K (2003) Strategies for the determination of bioactive phenols in plants, fruit and vegetables. J Chromatogr A 1000(1–2):657–691
Robles-Sánchez RM, Rojas-Graüb MA, Odriozola-Serrano I, González-Aguilara GA, Martín-Belloso O (2009) Effect of minimal processing on bioactive compounds and antioxidant activity of fresh-cut 'Kent' mango (*Mangifera indica* L.). Postharvest Biol Technol 51:384–390
Ropkins K, Beck AJ (2003) Using HACCP to control organic chemical hazards in food wholesale, distribution, storage and retail. Trends Food Sci Technol 14:374–389
Ropkins K, Ferguson A, Beck AJ (2003) Development of Hazard Analysis by Critical Control Points (HACCP) procedures to control organic chemical hazards in the agricultural production of raw food commodities. Crit Rev Food Sci Nutr 43(3):287–316
Ruiz-Cruz S, Arvizu-Medrano S (2010) Quality loss of fruits and vegetables induced by microbial growth. In: De la Rosa LA, Alvarez-Parrilla E, González-Aguilar GA (eds) Fruit and vegetable phytochemicals: chemistry, nutritional value, and stability. Blackwell, Ames
Sajilata MG, Singhal RS (2006) Effect of irradiation and storage on the antioxidative activity of cashew nuts. Radiat Phys Chem 75:297–300
Saldana MDA, Gamarra FMC, Siloto RMP (2010) Emerging technologies used for the extraction of phytochemicals from fruits, vegetables, and other natural sources. In: De la Rosa LA, Alvarez-Parrilla E, González-Aguilar GA (eds) Fruit and vegetable phytochemicals: chemistry, nutritional value, and stability. Blackwell, Ames
Sanchez-Moreno C, Plaza L, De Ancos B, Cano MP (2003) Vitamin C, provitamin A carotenoids, and other carotenoids in high pressurized orange juice during refrigerated storage. J Agric Food Chem 51:647–653
Saxena A, Bawa AS, Raju PS (2008) Use of modified atmosphere packaging to extend shelf-life of minimally processed jackfruit (*Artocarpus heterophyllus* L.) bulbs. J Food Eng 87(4):455–466
Skrede G, Wrolstad RE, Durst RW (2000) Changes in anthocyanins and polyphenolics during juice processing of highbush blueberries (*Vaccinium corymbosum* L.). J Food Sci 65:357–364
Song HP, Kim DH, Jo C, Lee CH, Kim KS, Byun MW (2006) Effect of gamma irradiation on the microbiological quality and antioxidant activity of fresh vegetable juice. Food Microbiol 23:372–378
Srinivas K, King JW, Monrad JK, Howard LR, Zhang D (2011) Pressurized solvent extraction of flavonoids from grape pomace utilizing organic acid additives. Italian J Food Sci 23(1):90–105
Strawn LK, Schneider KR, Danyluk MD (2011) Microbial safety of tropical fruits. Crit Rev Food Sci Nutr 51:132–145
Sumnu G, Sahin S (2005) Recent developments in microwave heating. In: Sun DW (ed) Emerging technologies for food processing. Elsevier, San Diego
Szczesniak AS (1998) Effect of storage on texture. In: Taub IA, Singh RP (eds) Food storage stability. CRC Press, Boca Raton
Tait J, Bruce A (2001) Globalization and transboundary risk regulation: pesticides and genetically modified crops. Health Risk Soc 3:99–112
Takahama U (1986) Spectrophotometric study on the oxidation of rutin by horseradish peroxidase and characteristics of the oxidized products. BBA Gen Sub 882(3):445–451
Tiwari U, Cummins E (2013) Factors influencing levels of phytochemicals in selected fruit and vegetables during pre- and post-harvest food processing operations. Food Res Int 50:497–506
Tiwari BK, O'Donnell CP, Cullen PJ (2009a) Effect of nonthermal processing technologies on the anthocyanin content of fruit juices. Trends Food Sci Technol 20:137–145

Tiwari BK, O'Donnell CP, Muthukumarappan K, Culllen PJ (2009b) Ascorbic acid degradation kinetics of sonicated orange juice during storage and comparison with thermally pasteurized juice. LWT Food Sci Technol 42:700–704

Tiwari BK, O'Donnell CP, Patras A, Brunton N, Culllen PJ (2009c) Stability of anthocyanins and ascorbic acid in sonicated strawberry juice during storage. Eur Food Res Technol 228: 717–724

Turkmen N, Sari F, Velioglu YS (2005) The effect of cooking methods on total phenolics and antioxidant activity of selected green vegetables. Food Chem 93:713–718

Uckoo RM, Jayaprakasha GK, Somerville JA, Balasubramaniam VM, Pinarte M, Patil BS (2013) High pressure processing controls microbial growth and minimally alters the levels of health promoting compounds in grapefruit (*Citrus paradisi* Macfad) juice. Innov Food Sci Emerg Technol 18:7–14

Van Boekel M, Fogliano V, Pellegrini N, Stanton C, Scholz G, Lalljie S, Somoza V, Knorr D, Jasti PR, Eisenbrand G (2010) A review on the beneficial aspects of food processing. Mol Nutr Food Res 54:1215–1247

Van Boxstael S, Habib I, Jacxsens L, De Vocht M, Baert L, Van De Perre E, Rajkovic A, Lopez-Galvez F, Sampers I, Spanoghe P, De Meulenaer B, Uyttendaele M (2013) Food safety issues in fresh produce: bacterial pathogens, viruses and pesticide residues indicated as major concerns by stakeholders in the fresh produce chain. Food Control 32:190–197

Van Der Sluis AA, Dekker M, Skrede G, Jongen WM (2002) Activity and concentration of polyphenolic antioxidants in apple juice. 1- Effect of existing production methods. J Agric Food Chem 50:7211–7219

Varma S, Karwe MV, Lee TC (2010) Effect of high hydrostatic pressure processing on lycopene isomers. Int J Food Eng 6:1–20

Vervoort L, Van der Plancken I, Grauwet T, Timmermans RAH, Mastwijk HC, Matser AM, Hendrickx ME, Loey AV (2011) Comparing equivalent thermal, high pressure and pulsed electric field processes for mild pasteurization of orange juice. Part II. Impact on specific chemical and biochemical quality parameters. Innov Food Sci Emerg Technol 12:466–477

Vikram VB, Ramesh MN, Prapulla SG (2005) Thermal degradation kinetics of nutrients in orange juice heated by electromagnetic and conventional methods. J Food Eng 69:31–40

Villavicencio ALCH, Mancini-Filho J, Delincee H, Greiner R (2000) Effect of irradiation on antinutrients (total phenolics, tannins and phytate) in Brazilian beans. Radiat Phys Chem 57: 289–293

Vinha AF, Alves RC, Barreira SVP, Castro A, Costa ASG, Oliveira MBPP (2013) Effect of peel and seed removal on the nutritional value and antioxidant activity of tomato (*Lycopersicon esculentum* L.) fruits. LWT Food Sci Technol. doi:10.1016/j.lwt.2013.07.016

Volden J, Bengtsson GB, Wicklund T (2009) Glucosinolates, L-ascorbic acid, total phenols, anthocyanins, antioxidant capacities and color in cauliflower (*Brassica oleracea* L. ssp. botrytis); effects of long-term freezer storage. Food Chem 112(4):967–976

Wachtel-Galor S, Wong KW, Benzie IFF (2008) The effect of cooking on Brassica vegetables. Food Chem 110:706–710

Wicklung T, Rosenfeld HJ, Martinsen BK, Sundfor MW et al (2005) Antioxidant capacity and colour of strawberry jam as influenced by cultivar and storage conditions. LWT Food Sci Technol 38:387–391

Wong PYY, Kitts DD (2001) Factors influencing ultraviolet and electron beam irradiation-induced free radical damage of ascorbic acid. Food Chem 74:75–84

Yildiz H, Bozkurt H, Icier F (2009) Ohmic and conventional heating of pomegranate juice: effects on rheology, color, and total phenolics. Food Sci Technol Int 15:503–511

Zafrilla P, Ferreres F, Tomás-Barberán FA (2001) Effect of processing and storage on the antioxidant ellagic acid derivatives and flavonoids of red raspberry (*Rubus idaeus*) jams. J Agric Food Chem 49(8):3651–3655

Zainol MM, Abdul-Hamid A, Bakar FA, Dek SP (2009) Effect of different drying methods on the degradation of selected flavonoids in *Centella asiatica*. Int Food Res J 4:531–537

Zenker M, Heinz V, Knorr D (2003) Application of ultrasound assisted thermal processing for preservation and quality retention of liquid foods. J Food Prot 66:1642–1649
Zepka LQ, Mercadante AZ (2009) Degradation compounds of carotenoids formed during heating of a simulated cashew apple juice. Food Chem 117:28–34
Zhang M, Hettiarachchy NS, Horax R, Chen P, Over KF (2009) Effect of maturity stages and drying methods on the retention of selected nutrients and phytochemicals in bittermelon (*Momordica charantia*) leaf. J Food Sci 74(6):C441–C446
Zill EH, Abert-Vian M, Elmaataoui M, Chemat F (2011) A novel idea in food extraction field: study of vacuum microwave hydrodiffusion technique for byproducts extraction. J Food Eng 105(2):351–360

Citrus Juices Technology

Asiye Akyildiz and Erdal Ağçam

1 Introduction

The citrus industry took an upturn in the mid-1940s. This was primarily brought about by the introduction of modern juice extraction processes. Since then, the market for fruit juice and the consumption of fruit juice, and thus sales opportunities, are increasing throughout the entire world. The increasing popularity of citrus beverages with the consumer, together with the importance which food scientists attach to the citrus fruit, has led to an impressive development in this market. Two technological developments were among the most important reasons for this rapid growth process. These developments were automatic juice extractors and gentle concentrate production processes. Nowadays, technical experts and engineers are increasingly devoting their attention to recovering new products from citrus fruit, to increasing the amount of valuable components obtained, to improving the quality of these components and to automating and optimizing the traditional processes (Pecoroni et al. 2013).

Citrus fruit varieties are grown for commercial use in many parts of the world. Cultivation of citrus fruits has since spread worldwide to all regions where the climate is not too severe during the winter months and suitable soil conditions are available. In the USA, the notable growing areas are in Florida and California and in South America, Brazil have taken over the largest share of the world market for oranges and orange juice products. Morocco, South Africa, and parts of Australia have shown increased output during recent years, although within the latter two areas yields are frequently affected by variable weather conditions. China is the largest producer after the USA and Brazil, but over 90 % of its output is for the home market. Details about production of citrus fruits were given in Table 1. In the area of citrus juice production, Brazil, USA, China are the major players, with

A. Akyildiz (✉) • E. Ağçam
Department of Food Engineering, Cukurova University, Balcali, Turkey
e-mail: asiye1@cu.edu.tr; eagcam@cu.edu.tr

A. Malik et al. (eds.), *Food Processing: Strategies for Quality Assessment*,
Food Engineering Series, DOI 10.1007/978-1-4939-1378-7_3

Table 1 Distribution of citrus world production as metric ton in 2011 (FAO 2011)

	Countries	Orange	Countries	Tangerine, Mandarin, Clementine	Countries	Grapefruit (pummelo)	Countries	Lemon, lime
1	Brazil	19,811,100	China	12,707,000	China	3,608,000	Mexico	2,147,740
2	USA	8,079,390	Spain	1,584,290	USA	1,146,680	India	2,108,000
3	China	6,213,829	Brazil	1,004,730	S. Africa	415,679	China	1,313,394
4	India	4,571,000	Japon	982,200	Mexico	397,267	Argentina	1,228,660
5	Mexico	4,079,680	Turkey	873,251	Thailand	218,988	Brazil	1,126,740
6	Spain	2,825,390	Italy	852,562	Turkey	196,000	USA	834,610
7	Egypt	2,577,720	Egypt	848,069	India	189,182	Turkey	790,211
8	Italy	2,469,940	Korea	680,507	Israel	183,682	Spain	700,000
9	Indonesia	1,818,950	USA	570,620	Cuba	112,000	Iran	560,052
10	Turkey	1,730,150	Pakistan	515,372	Tunisia	93,263	Italy	483,088
11	S. Africa	1,496,160	Morocco	473,281	Brazil	75,119	Egypt	296,773
12	Pakistan	1,387,540	Mexico	406,397	Belize	55,862	S. Africa	260,002
13	Iran	1,200,900	Argentina	400,651	Bangladesh	53,842	Peru	224,719
14	Argentina	876,851	Thailand	360,016	Jamaica	47,093	Chile	171,049
15	Morocco	850,000	Iran	261,463	Spain	43,000	Syrian A.R	161,594

Spain and Turkey being notable producers of specialized concentrates. The main citrus varieties for juice processing are the orange, lemon, lime and grapefruit (Pecoroni et al. 2013).

Following is the main types of oranges grown in the individual citrus fruit countries: Hamlin and Pera in Brazil; Navel, Valencia, Jaffa and Pineapple in USA; Biondo Comune, Ovale in Italy; Navel, Valencia, Jaffa in Spain/Marocco; Shamuty, Valencia in Israel; Valencia, Kozan, Dortyol in Turkey.

Because citrus fruits grow throughout the year and because the ripening period can last for a total of 5 months, the fruits obtained at one particular harvest are different from fruits obtained at other periods of the harvest, particularly with respect to acid, sugar, and pigment content. The different types of fruit, depending on type and ripening conditions, also differ in other respects (vitamin C, pectin, carotenoids, flavonoids, etc.). The characteristic properties of the raw product exert a great influence on the final product. Today, the assessment of juice quality is simplified by statistical indices obtained from genuine raw product samples (Pecoroni et al. 2013).

2 Varieties Suitable for Juice Production

The processing method and procedure are determined principally by the type of fruit and the structure of the fruit. The fruit types can be roughly classified as follows (Table 2):

The orange (*Citrus sinensis*) is the most important basic product in the citrus industry, followed by grapefruits, lemons and mandarins. The most important of all citrus fruits is the sweet orange (*C. sinensis*), and this is widely grown in those regions of the world suited to citrus. Each region tends to have its own characteristic varieties. Common varieties to be found growing in various parts are Navel, Valencia, Shamouti, Hamlin, and Parson Brown. Another distinctive variety of orange is the bitter orange (*C. aurantium*), chiefly represented by the Seville orange, which is grown commercially in southern Europe mainly for such products as marmalade. Compared with other citrus crops its yield is small and of little use in the juice market. For the maximization of profitability from processing of oranges, the juice yields of 38.0, 30.6, 43.2, and 42.5 % for Washington, Thompson, Chilena, and Valencia cultivars, respectively, are essential (Erazo et al. 1984).

Table 2 Classification according to the peel color of citrus fruit

Orange-fruit types	Yellow-fruit types
• Sweet oranges (*Citrus sinensis* (L.) Osbeck Common orange, Acidles orange, Pigmented or blood orange, Navel orange • Sour oranges (*Citrus aurantium* L.) • Mandarins or Tangerines (*Citrus reticulate* Blanco)	• Lemons (*Citrus limon* (L.) Burnm. f.) • Lime, limetta (*Citrus aurantifola* Swing) • Grapefruit (*Citrus paradisi* Macfadyen) • Pummelos (*Citrus grandis* (L.) Osbeck)

Some technological characteristics of five orange cultivars [Hamlin, Magnum Bonum, Valencia, Dortyol (Turkish cv.) and Kozan (Turkish cv.)] used in the juice industry have been evaluated by Altan (1995). According to researcher, cultivar Kozan was found superior for production of orange juice.

2.1 *Oranges*

2.1.1 Navel Oranges

Navel oranges are the most popular eating oranges in the world. This is primarily because they are less juicy, the flesh has an excellent and rich flavor, and they are crisp, seedless, and easy to peel. However, a delay limonin bitterness develops upon processing. Navel orange juice can reach a limonin concentration of up to 40 μg/g which corresponds to an intense bitter sensation in the back of the tongue (Chandler and Nicol 1975).

2.1.2 Valencia Oranges

The Valencia or Murcia orange is one of the sweet oranges used for juice extraction. The orange is very juicy and has a high juice yield when processed. The Valencia orange is a late-season fruit and believed to provide the best orange juice. Primarily grown for processing and juice production, however, its excellent taste and internal color make it desirable for the fresh markets, too. Worldwide Valencia oranges are prized as the only variety of orange in season during summer. Valencia orange is quantitatively the richest juice in carotenoids, known to have the most complex pigment pattern among sweet oranges (Gross et al. 1972). The improved color was related to the increased levels of cryptoxanthin (152 μg/mL) in juice samples taken in early April, compared to 90 g/mL in normal "Valencia" juice (Sandhu and Minhas 2006). Valencia is also sold unblended as a premium product. Fruit sizes are very close to each other, the peels are thin and fruit juice content is high.

2.1.3 Hamlin

Hamlin has become the most widely grown orange variety in Florida. The fruit is small, smooth, not highly colored, seedless, and juicy, with a pale yellow-colored juice. The tree is high-yielding and cold-tolerant and it produces good quality fruit, which is harvested from October to December. Its thin rind has a very fleshy pulp, making it one of the most productive oranges for processing. It has a sweet flavor lacking in acid and usually with few seeds. This cultivar is suitable for blending with other orange cultivars to obtain higher flavor.

2.1.4 Jaffa

The Jaffa orange, also called Shamouti orange, is a sweet, almost seedless orange variety with a tough skin that makes it particularly suitable for table consumption. Because of its tough skin juice yield is lower.

2.1.5 Parson Brown

Fruit is medium-large, moderately seedy, well-colored under favorable conditions, medium-thick peel, flesh color dull orange; firm, juicy, well-flavored, very early in maturity.

2.1.6 Blood Orange

Blood oranges are a natural mutation of *C. sinensis*, although today the majority of them are hybrids. High concentrations of anthocyanin give the skin, flesh, and juice of the fruit their characteristic dark red color. The blood orange, with its distinct color and flavor, is generally considered the most delicious juice orange. Kırca and Cemeroğlu (2003) reported that blood orange anthocyanins were found to be very susceptible to high temperatures so they recommended that commercial processing of blood orange into juice is not recommended unless stabilization of anthocyanins is provided by copigmentation or another way. Varieties of blood oranges are Maltese, Moro, Sanguinelli, Scarlet Navel, and Tarocco.

*2.2 Lemons and limes (*C. limon, C. aurantifolia*)*

The characteristic oval-shaped, yellow fruits, apart from their culinary use, are an important source of juice and flavoring for the soft drinks industry. Lemons are primarily processed into juice and lemonade as well as pectin and lemon oil. Limes are a small round fruit; they are green or greenish yellow in color, fresh, and characteristic flavor. Nowadays lemon is processed to clear lemon juice. Clarified lemon juice is a suitable acidifying agent that can substitute citric acid and provide more "natural" products (Uçan 2013).

Lemons are consumed fresh and processed, as juices, jam, jellies, molasses, etc. Chemical industry extracts from lemon bioactive compounds like flavonoids, vitamins, minerals, dietary fiber, essential oils, etc. are used in the food, cosmetic, and pharmaceutical industry (Ozaki et al. 2000; González-Molina et al. 2010).

2.3 *Grapefruits (*Citrus paradisi*)*

A large round citrus fruit with a thick yellow skin and somewhat bitter pulp, the grapefruit is generally accepted to be a hybrid between the pummelo and the orange. Today the commercially important grapefruit is grown in many parts of the world. The most predominant cultivar to be seen in the market is the Marsh Seedless, followed by a red, pigmented version, the Star Ruby. The Star Ruby cultivar has high carotenoids content especially β-carotene and lycopene (İçdemir 2012).

Pink grapefruit varieties are characterized by presence of pink to red pigment of carotenoid lycopene. While it makes the fresh fruit attractive, this pigmentation, like anthocyanins in blood oranges, has a tendency to fade during processing and storage.

2.4 *Mandarins* (Citrus reticulate *Blanco)*

Mandarin is a group name for a class of oranges with thin, loose peel. These are treated as members of a distinct species, *Citrus reticulata* Blanco. The name "tangerine" could be applied as an alternate name to the whole group, but in the trade, it is usually confined to the types with red-orange skin. Mandarins include a diverse group of citrus fruits that are characterized by bright colored peel and pulp, excellent flavor, easy-to-peel rind, and segments that separate easily. The flavor of mandarins is also unique and richer than that of most citrus species. Four basic types of mandarins have been assigned their own separate species classification: common, Satsuma (*Citrus unshiu*), Mediterranean, and King. Satsuma is an important crop in Japan, and the clementine, an important cultivar to be found in Mediterranean areas. Other cultivars of note are the; Tangor, a hybrid of mandarin and orange, and the Tangelo, a hybrid of mandarin and grapefruit. Several high-quality mandarins (Ponkan, Tankan, and Dancy types) have come out of China. Dancy, Robinson (hybrid of Clementine × Orlando), Orlando Tangelo, Nova, Minneola Tangelo, Page, and Kinnow juice are suitable for blending with orange juice to improve the color of the finished product. Orlando tangelo juice does not develop postprocessing bitterness, and Minneola tangelo contains no limonin and is suitable for processing. Page and Nova have good processing quality (Scott and Hearn 1966). Dancy, Robinson, Orlando tangelo, and Nova juices do not develop off-flavors and are suitable for blending to the juice of Temple mandarin (Sandhu and Minhas 2006). Mandarins require much gentler handling and usually cannot be stored for any length of time prior to processing.

3 Citrus Fruit Anatomy

3.1 *Peel (pericarp)*

This is made up of two layers, namely the flavedo and the albedo. The flavedo is the outer layer of the peel. It is pigmented and relatively thin. In ripe fruits, the flavedo contains the vesicles and also the pigments which contain carotene.

The vesicles contain essential oils for flavor and aroma. The outer epidermis is covered by natural wax. This layer of natural wax prevents water losses and protects the fruit against fungal infections.

There is a smooth transition between the flavedo and the underlying albedo. The albedo is so called because of its whitish color. Compared to the flavedo, the cells of the albedo are larger, less compact, and very high pectin content. The albedo layer does not contain essential oils and also this layer is edible part of the fruit. Mandarins are characterized by a looser flavedo/albedo layer that makes them easier to peel but, this loose skin makes juicing operations more difficult. Peel and flesh ratio is important for juice yield of citrus fruit (Kimball 1999).

3.2 Flesh (Endocarp)

The flesh of the fruit is divided up into radial segments. These segments consist of a large number of juice cells in vesicles which are held together by high-molecular adhesive substances. These juice cells are also elongated and attached to the center of the fruit and consist primarily of enlarged vacuoles that contain the juice. The nucleus of these cells and the other organelles are located essentially in the membrane of the expanded juice vacuole. It has been shown that the juice in the vacuole is clear or devoid of cloud material (Bennett 1987). The seeds are located towards the center of each segment, near the fruit axis. Short vascular bundles connect the seeds to the main (thread-like) bundles which run along the main axis of the fruit (Kimball 1999).

3.3 Seed

The peel and seed residue is the primary waste fraction. Peels are a source of molasses, pectin, cold pressed oils, and limonene and can be used as cattle feed, mixed with dried pulps. Seeds are rich in unsaturated fatty acids, but the oil is not extracted commercially; however, seeds can be used to recover limonoids, which are typical citrus fruit triterpenoids, having an extremely bitter taste and, probably, anticarcinogenic/chemopreventive activities (Braddock 1995).

Citrus peels and seeds have an interesting antioxidant activity with regard to citronellal. Perhaps their extracts could well be useful to prevent oxidation in fruit juices and essential oils. The methanolic extracts of mandarin and sweet orange seeds have the best antioxidant properties, while bergamot peels are an interesting source of free phenolic compounds (Bocco et al. 1998). However, little is known about the bioactive potential of the seeds. Yusof et al. (1990) analyzing the content of naringin in a variety of Mexican citrus have detected this flavanone in the seeds of Rough lime. Sun et al. (2010) investigated the flavonoids composition of the different parts of the Chinese mandarin fruit and identified naringin, hesperidin, didymin, tangeretin, and nobiletin in the seeds. Moulehi et al. (2012) reported that the seeds of mandarin and bitter orange are considered valuable, as they provide components with potential for industrial and pharmacological applications as antioxidants.

Moreover, the significant changes of the phenolic composition during ripening suggest that mandarin seeds are a promising source for the extractions of gallic acid at the maturity, while bitter orange seeds could be used as a potential source of neohesperidin at immature stage and naringin at maturity.

4 Compositions of Citrus

Citrus fruits are an important source of antioxidants such as ascorbic acid, carotenoids, flavonoids, limonoid, other phenolic compounds and also sugars, organic acids, and amino acid, pectin, fiber, minerals, and volatile components for human nutrition and quality of citrus juice (Bocco et al. 1998; Gorinstein et al. 2001; Topuz et al. 2005; Abeysinghe et al. 2007; Rapisarda et al. 2008; Ghasemi et al. 2009; Bermejo et al. 2011).

It's well known that vitamin C and carotenoids are abundant in some citrus fruits (Dhuique-Mayer et al. 2005), thus they are very beneficial to human health. In recent years, more attentions had been paid on phenolic compounds of citrus fruits, and some publications have suggested they might play an important role on the antioxidant capacity of citrus fruits (Wang et al. 1996; Rapisarda et al. 1999; Gorinstein et al. 2004). Dietary phenolic compounds of citrus fruits include flavonoids and phenolic acids (Balasundram et al. 2006). Furthermore, narirutin, hesperidin, naringin, and neohesperidin are the major flavonoid glycosides (Rouseff et al. 1987). On the other hand, phenolic acids exist largely in citrus fruits as bound forms, which mostly occur as hydroxycinnamics, such as caffeic, p-coumaric, ferulic and sinapic (Robbins 2003). Recently, some studies have investigated the antioxidant capacity of citrus fruits, and it was assumed that total antioxidant capacity of citrus fruits were mainly attributed to ascorbic acid and phenolic compounds, though there were some divergences as to which compound was the major contributor (Rapisarda et al. 1999; Gardner et al. 2000; Arena et al. 2001; Sun et al. 2002; Yoo et al. 2004). Sugars, organic acids and amino acids are major primary metabolites in the juice sacs of citrus fruit and are important components for internal fruit quality. The contents of sugars and organic acids and their ratios (sugar content/acid content) affect the taste of citrus fruit (Matsumoto and Ikoma 2012).

4.1 Phenolic Compounds in Citrus

Phenolic compounds are divided into two groups as flavonoids and phenolic acids. Flavonoid skeleton is composed of two aromatic rings (namely, A and B), which are connected through a pyrone or hydropyrone ring (C), the flavones or flavanones, respectively (Fig. 1) (Gattuso et al. 2007). More than 60 individual flavonoids have been identified in Citrus sp. and most of them can be classified into three groups: flavanones, flavones and flavonols (Benavente-García et al. 1997). In addition, other phenolic compounds (phenolic acids, etc.) are also present in citrus species.

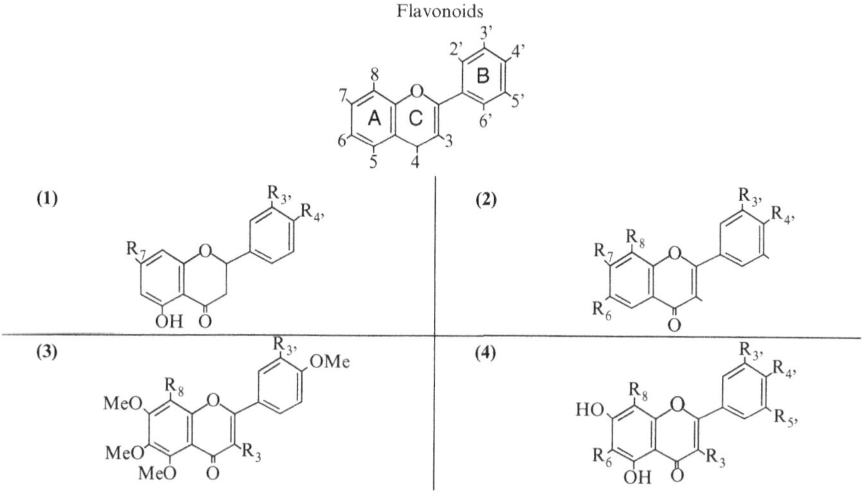

Flavanones (1)	R_7	$R_{3'}$	$R_{4'}$			
Eriocitrin (Eriodictyol 7-*O*-rut)	*O*-rut	OH	OH			
Hesperidin (Hesperitin 7-*O*-rut)	*O*-rut	OH	OMe			
Homoeriodictyol 7-*O*-rut	*O*-rut	OMe	OH			
Naringin (Naringenin 7-*O*-nh)	*O*-nh	H	OH			
Neohesperidin (Hesperetin 7-*O*-nh)	*O*-nh	OH	OMe			
Neoeriocitrin (Eriodictyol 7-*O*-nh)	*O*-nh	OH	OH			
Narirutin (Naringenin 7-*O*-rut)	*O*-rut	H	OH			
Flavones (2)	R_6	R_7	R_8	$R_{3'}$	$R_{4'}$	
Diosmetin 6,8-di-*C*-β-gluc (Lucenin-2 4'-methyl ether)	Gluc	OH	Gluc	OH	OMe	
Diosmin (Diosmetin 7-*O*-rut)	H	*O*-rut	H	OH	OMe	
Vicenin-2 (Apigenin 6,8-di-*C*-gluc)	Gluc	OH	Gluc	H	OH	
Chrysoeriol 6,8-di-*C*-gluc (Stellarin-2)	Gluc	OH	Gluc	OMe	OH	
Lucenin-2 (Luteolin 6,8-di-*C*-gluc)	Gluc	OH	Gluc	OH	OH	
Luteolin 7-*O*-rut	H	*O*-rut	H	OH	OH	
Apigenin 7-(malonylapiosyl)-gluc	H	*O*-Api-Gluc	H	H	OH	
Diosmetin 6-*C*-β-D-gluc	Gluc	OH	H	OH	OMe	
Diosmetin 8-*C*-β-D-gluc (Orientin 4'-methyl ether)	H	OH	Gluc	OH	OMe	
Flavone Aglycon (2)	R_6	R_7	R_8	$R_{3'}$	$R_{4'}$	
Luteolin	H	OH	H	OH	OH	
PMFs (3)	R_3	R_4	$R_{3'}$			
Sinensetin	H	H	OMe			
Nobiletin	H	OMe	OMe			
3,5,6,7,8,3′,4′-Heptamethoxyflavone	OMe	OMe	OMe			
Natsudaidain	OH	OMe	OMe			
Tangeretin	H	OMe	H			
Flavonols (4)	R_3	R_6	R_8	$R_{3'}$	$R_{4'}$	$R_{5'}$
Rutin (Quercetin 3-*O*-rut)	*O*-rut	H	H	OH	OH	H
Quercetin	OH	H	H	OH	OH	H
Kaempferol	OH	H	H	H	OH	H
Myricetin	OH	H	H	OH	OH	OH
Iso-/limocitrol 3-β-D-gluc	*O*-Gluc	OMe	OMe	OH	OMe	H
Limocitrin 3-β-D-gluc	*O*-Gluc	OMe	OMe	OMe	OH	H

Gluc: glucoside; rut: rutinoside; nh: nehesperidoside; PMFs: polymethoxyflavones

Fig. 1 Listing of flavonoids identified in lemon fruit (González-Molina et al. 2010)

Citrus flavonoids are present in the glycoside or aglycone forms. Among the glycoside forms, two types of di-glycosides, L-rhamnosyl D -glucosyl derivatives, are classified as neohesperidosides and rutinosides. Flavonoids can exist as free aglycones but most of them commonly occur as C- or O-glycosides. Neohesperidosides and rutinosides affect the taste of citrus fruits and juices. For example, neohesperidosides, present in grapefruits, are intensively bitter (Garg et al. 2001), while rutinosides, present in lemons, are tasteless (Tomás-Barberán and Clifford 2000; Peterson et al. 2006). Flavanones are the most abundant citrus flavonoids (e.g., 98 % in grapefruit, 96 % in limes, and 90 % in lemons) (Peterson et al. 2006). Their chemical structures are almost specific for every species, which renders them as markers of adulteration in commercial juices (Mouly et al. 1994; Calabrò et al. 2004). Other phenolic compounds such as hydroxycinnamic acids are also known to be present in very low concentrations (caffeic, cholorogenic, ferulic, sinapic, and p-coumaric acids) (Bocco et al. 1998; Manthey and Grohmann 2001; Wang et al. 2007, 2008), in addition to benzoic acids (protocatechuic, p-hydroxybenzoic, and vanillic acids) (Xu et al. 2008) (Fig. 2) (González-Molina et al. 2010). Flavonoids are widely distributed group of polyphenolic compounds with health-related properties due to their antioxidant and radical scavenging activities (Anagnostopoulou et al. 2005). Epidemiological studies have reported that there is a significant positive association between consumption of polyphenols and reduced risk of chronic diseases, such as cancer, cardiovascular disease, diabetes, viral

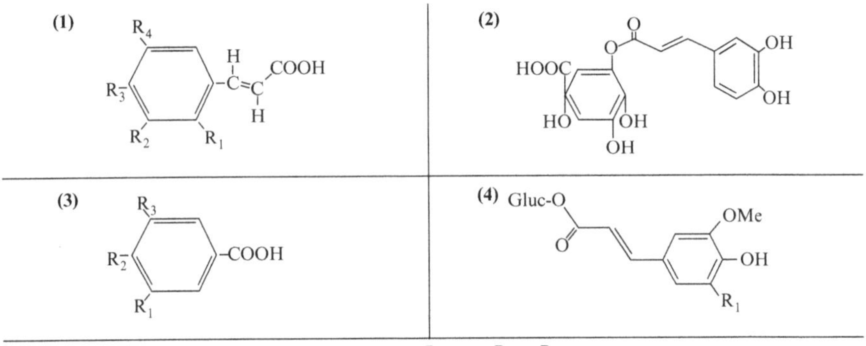

Hydroxycinnamic acids (1)	R_1	R_2	R_3	R_4
Caffeic acid	H	H	OH	OH
Ferulic acid	H	H	OH	OMe
Sinapic acid (or sinapinic acid)	H	OMe	OH	OMe
p-Coumaric acid	H	H	OH	H
Chlorogenic acid (2)				
Benzonic acids (3)	R_1	R_2	R_3	
Protocatechuic acid	H	OH	OH	
p-Hydroxbenzoic acid	H	OH	H	
Vanillic acid	H	OH	OMe	
Others (4)	R_1			
1-Feruloyl-β-D-glucopyranoside	H			
1-Sinapoyl-β-D-glucopyranoside	OMe			

Gluc: glucoside.

Fig. 2 Listing of phenolic compounds identified in lemon fruit (González-Molina et al. 2010)

infection, inflammatory activities, and Alzheimer's disease (Kaul et al. 1985; Kandaswami et al. 1991; Hertog et al. 1993; Galati et al. 1994; Benavente-Garcia and Castillo 2008; Barros et al. 2012). Flavonoid glycoside detection has been used to assess the authenticity of orange juice (Pupin et al. 1998). Analysis of a number of fruit juices such as lemon, grapefruit, and pommerans showed that flavone glycosides fingerprints could be used to differentiate *C. sinensis* from other juices such as grapefruit, mandarin, sour/bitter oranges, and bergamot. Naringin and neohesperidin are flavonoids found only in certain citrus fruits; sweet orange cultivars do not contain these compounds and their presence in orange juice indicates adulteration with juice from certain other citrus fruits such as grapefruit (Widmer 2000; Sandhu and Minhas 2006). The most abundant flavanones in oranges, mandarins, lemons, and limes are mostly hesperidin, whereas in grapefruit the dominant flavanone glycoside is naringin. The sugar neohesperidose (2-O-α-Lrhamnosyl-β-D-glucose), which is high in grapefruits, imparts the tangy or bitter taste to the glycoside naringin. In lemons the flavanone profile is dominated by two specific flavanone glycosides—hesperidin and eriocitrin. In limes, only one flavanone, hesperidin, is dominant. The sugar rutinose (6-O-α-Lrhamnosyl-D-glucose) which is relatively high in lemons and limes, and its flavanone glycosides, hesperidin, and eriocitrin have a neutral taste. Baker and Cameron (1999) have said that flavonoid crystallization is affected by the cultivar of orange, juice extraction pressure, holding time before pasteurization, and pasteurization itself. The authors also found that Shamouti and Satsuma mandarin cultivars both contain high levels of hesperidin, but that most other orange cultivars do not contain much hesperidin. Dhuique-Mayer et al. (2005) measured the hesperidin content in several orange cultivars and found that Shamouti contains about 552 mg/L, Hamlin 317 mg/L, and Valencia 257 mg/L hesperidin in the hand-squeezed juice. Mizrahi and Berk (1970) studied the serum by microscope and found needle-like crystals that were determined to be hesperidin on Shamouti orange juice. Shamouti oranges were determined to be saturated with hesperidin, and this study showed that hesperidin could be a significant factor in juice cloudiness. Rothschild and Karsenty (1974) determined a large increase in cloudiness in fresh Shamouti orange juice that was also attributed to the crystallization of hesperidin. Bennett and Albach (1981) proved that the formation of white spots on freeze damaged Valencia oranges is due to hesperidin crystallization. Hesperidin is unevenly distributed in the cell vacuole of citrus fruit, and crystallization of the hesperidin occurs when the cell membrane is broken. When freezing occurs, the cell membranes are broken, releasing hesperidin, and the uneven distribution of hesperidin in the cells causes the nonuniform white spots on the orange.

4.2 *Carotenoid Compounds in Citrus*

Color is one of the most important attributes of orange juice products and is largely due to the presence of diverse carotenoid pigments. The carotenoid distribution in citrus is extremely complex and subject to considerable variation. A list of more than 115 different carotenoids has been compiled from different citrus cultivars including Valencia, Washington, Navel, and Shamouti oranges (Lee et al. 2001) and some of these compounds were shown in Fig. 3. Carotenoids imparted the yellow, orange or red colors of many foods and in the provitamin activity exhibited by some of them. Recently, interest in these pigments has grown considerably because of their probable relation to the prevention and/or protection against serious human health disorders such as cancer, heart disease, and macular degeneration, among others, which may be somehow linked to their probable antioxidant properties (Krinsky 1989; Ziegler 1989; Olson 1999; Fraser and Bramley 2004; Melendez-Martinez et al. 2004; Melendez-Martinez et al. 2007). Carotenoids are important quality indicators for orange juice. Apart from being responsible for the color of the juice, a number of them have provitamin A activity (e.g., α- carotene, β-carotene, and β-cryptoxanthin) and some are known for their antioxidant capacity (e.g., β-carotene and β-cryptoxanthin, zeaxanthin, and lutein). Oranges are a very complex source of carotenoids, containing the largest number of them among all fruit species (Melendez-Martinez et al. 2007; Meléndez-Martínez et al. 2008). The major carotenoid detected was β-cryptoxanthin; it accounted for about 22.4 % of the total carotenoid content. It is said to be the main contributor to the orange color of the juice since it absorbs light at higher wavelengths, and the main provitamin A carotenoid found in oranges (Mouly et al. 1999; Gama and Sylos 2005; Dias et al. 2009). In addition, the other major carotenoids found were β-carotene, lutein, and zeaxanthin, with a share of 12.6, 11.4 and 11.3 % respectively.

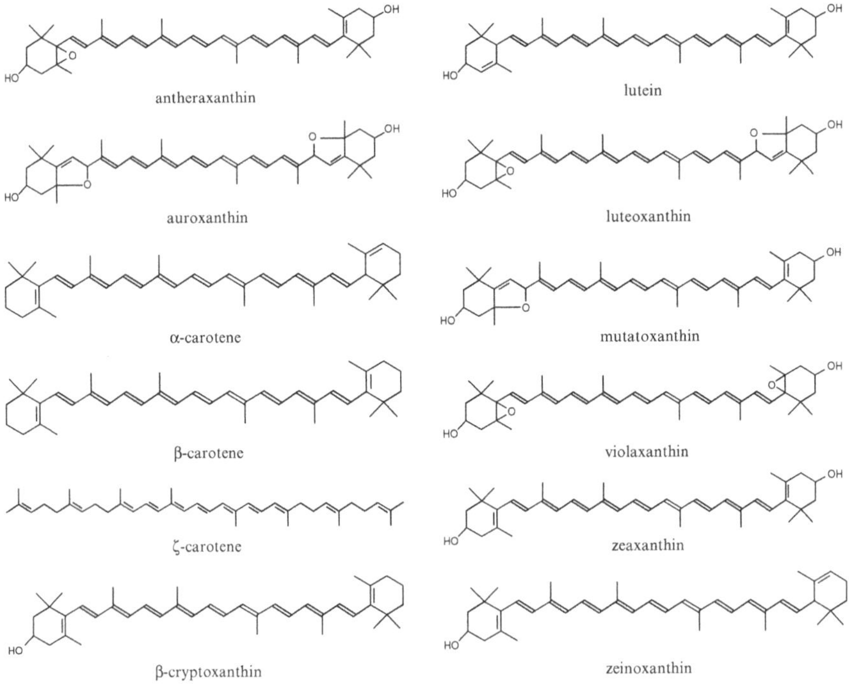

Fig. 3 Chemical structures of some typical orange juice carotenoids (Melendez-Martinez et al. 2007)

4.3 Volatiles Compounds in Citrus

Orange flavor is probably the most widely recognized and accepted flavor in the worldwide food and beverage industry; it is widely used to flavor or aromatise foods and beverages because of its distinctive flavor and aroma (Rouseff et al. 1994; Shaw and Moshonas 1997). Its fresh and unique flavor is the result of a natural combination of volatile compounds in a well-balanced system including sugars, acids, and phenolic compounds. The aroma of orange juice contains more than 200 components that belong to very heterogeneous groups (Table 3). Important contributors to

Table 3 Summary of volatiles reported in fresh orange juices (Perez-Cacho and Rouseff 2008)

Volatiles compounds in orange juices	
Alcohols (27)	*Hydrocarbons (22)*
Methanol	Methane
Ethanol	n-Hexane
1-Propanol	(+)-limonene (d-*p*-Mentha-1,8-diene)
1-butanol	*α*-Terpinene (*p*-mentha-1,3-diene)
3-Methylbutan-1-ol	*β*-Terpinene (*p*-Mentha-1(7),3-diene)
2-Methylbutan-1-ol	*γ* -Terpinene (*p*-Mentha-1,4-diene)
2-methyl-3-buten-2-ol	Terpinolene (*p*-Mentha-1,4(8)-diene)
1-pentanol	Isoterpinolene (*p*-mentha-2,4-diene)
1-Hexanol	*α*-Phellandrene (*p*-Mentha-1,5-diene)
2-Hexanol	*β*-Phellandrene (*p*-Mentha-1(7)-2-diene)
3-Hexen-1-ol	*α*-Thujene (3-Thujene)
1-Heptanol	Sabinene (4(10-thujen-3-ol)
Methyl heptenol	*δ*-Carene (3-Carene)
1-Octanol	*α*-Pinene (2-Pinene)
1-Nonanol	*β*-Pinene (2(10)-Pinene)
2-Nonanol	Valencene
Dodecanol	*α*-Copaene
(Z)-*p*-Mentha-2,8-dien-1-ol	*β*-Copaene
Linalool (3,7-dimethyl-1,6-octadien-3-ol)	*α*-Caryophyllene
p-Mentha-4-en-1-ol	*β*-Caryophyllene
α-Terpineol (*p*-Mentha-1-en-8-ol)	*β*-Myrcene (2-Methyl-6-methylene-2,7-octadiene)
β-Terpineol (*p*-Mentha-8-en-1-ol)	*p*-Cymene (*p*-isopropyltoluene)
Nerol (Z)-3,7-dimethyl-2,6-octadien-3-ol)	*Esters (26)*
Citronellol (3,7-dimethyl-6-octen-1-ol)	Ethyl formate
Geraniol (E)-3,7-dimethyl-2,6-octadien-1-ol)	Ethyl acetate
(E)- carveol (E)-*p*-Mentha-6,8-dien-2-ol	Methyl isovalerate
(Z)- carveol (Z)-*p*-Mentha-1,8-dien-6-ol	Ethyl Butanoate
Aldehydes (19)	Ethyl-n-caprilate
Acetaldehyde	n-Octyl butanoate
Pentanal	Terpinyl acetate
Hexanal	Methyl-n-methyl anthranilate

(continued)

Table 3 (continued)

Volatiles compounds in orange juices	
Alcohols (27)	*Hydrocarbons (22)*
2-Hexanal	n-Octyl isovalerate
2-Hexenal	Ethyl hexanoate
Octanal	Methyl-3-hydroxy-hexanoate
Nonanal	Ethyl-3-hidroxy-hexanoate
Decanal	Octyl acetate
Undecanal	Ethyl octanoate
Dodecanal	Ethyl propionate
Tetradecanal	Ethyl isobutanoate
Neral (Z)- 3,7-dimethyl-2,6-octadienal	Diethyl carbonate
Geranial (E)-3,7-dimethyl-2,6-octadienal	Ethyl-2-methylbutanoate
α-Sinensal (2,6,10-trimethyl-2,6,9,11-dodecatetraenal)	Methyl hexanoate
β-Sinensal (2,6-dimethyl-10-methylene-2,6,11-dodecatrienal)	Methyl butanoate
Perilladehyde (p-mentha-1,8-dien-7-al)	Benzaldehyde Methyl propionate
n-butylaldehyde	Nonyl acetate
Furfural (2-furanal, 2-furaldehyde)	Geranyl acetate
Ketones (7)	Ethyl anthranilate
Acetone	Linalool acetate
Methyl ethyl ketone	Neryl acetate
2-Butanone	*Acids (11)*
2,3-butanedione	Acetic acid
2-Pentanone	Butanoic acid
γ -Decalactone	Hexanoic acid
Carvone (p-menta-1,4(8)-dien-3-one)	Heptanoic acid
Ethers (2)	Octanoic acid
Diethyl ether	Nonanoic acid
Ethyl sec-butyl ether	Decanoic acid
	Undecanoic acid
	Dodecanoic acid
	Tetradecanoic acid
	Hexadecanoic acid

orange juice flavor include esters, aldehydes, ketones, terpenes, and alcohols (Nisperos-Carriedo and Shaw 1990a, b). Investigation of volatile flavor compounds in different orange cultivars has been of interest in numerous studies. Shaw et al. (2005) classified commercial orange juices based on relative amounts of volatile juice constituents; Jordan et al. (2005) investigated the de-aeration and pasteurisation effects on the orange juice aromatic fraction; Selli et al. (2004) determined the volatile aroma components of the orange juice obtained from the Turkish cv. Kozan (*Citrus sinensis* L. Osbeck). They reported that linalool, limonene, β-phellandrene, terpinen-4-ol, and ethyl-3- hydroxy hexanoate were major volatile components of cv. Kozan (Kelebek and Selli 2011). Selli et al. (2004) reported that seven ester components (ethyl butanoate, ethyl hexanoate, ethyl octanoate, dimethyl succinate, ethyl 3-hydroxy-hexanoate, and isoamyl benzoate) were identified. The total amount of esters was 1,256 mg/L and the ethyl 3-hydroxy-hexanoate was the most abundant. Ethyl butanoate is generally the major volatile ester in orange juice and an important contributor to desirable flavor in orange products. Aldehydes have long been considered important to orange flavor (Nisperos-Carriedo and Shaw 1990b).

Two widely accepted orange flavor aldehydes, the *trans*-2-hexenal and the perillaldehyde, were quantified at 113 mg/L in orange juice obtained from the cv. Kozan of Turkey. Five alcohols including 1-hexanol, *trans*-3-hexen-1-ol, *cis*-3-hexen-1-ol, *trans*-2-hexen-1-ol, and octanol were quantified with the total concentration of 330 mg/L. The alcohol composition of Kozan orange juice was in agreement with other orange juices (Moshonas and Shaw 1994; Maccarone et al. 1998). *trans-3-hexen*-1-ol, *cis*-3-hexen-1-ol, and *trans*-2-hexen-1-ol are important contributors to the "fruity-green" flavor in fresh orange juice and in other fruit flavors. Myrcene, limonene, b-phellandrene, g-terpinene, and b-cadinene were determined as terpene hydrocarbons. The total amount of terpene hydrocarbons was 21,568 mg/L. Limonene (18,400 mg/L) and b-phellandrene (2,760 mg/L) were the major components among the terpene hydrocarbons of Kozan orange juice. Limonene has a weak, citrus-like aroma, and is considered one of the major contributors to orange flavor (Nisperos-Carriedo and Shaw 1990a; Shaw 1991). Twelve terpenols were identified with a total concentration of 5,250 mg/L. Linalool makes a positive contribution to orange flavor in combination with several other orange volatiles (Shaw and Moshonas 1997). α-Terpineol is generally considered to make a negative contribution to orange flavor, and its increase is one of the more consistent findings in aged and heat-abused orange juice products (Kefford and Chandler 1970; Shaw 1991). Three ketones were quantified including *cis*-dihydrocarvone, carvone, and nootkatone. The total ketone concentration was found to be 454 mg/L. Kelebek and Selli (2011) reported that total of 58 volatile compounds were identified and quantified in the aromatic extract of Dortyol cultivar. Terpenes and terpenols were found as the main types of volatile components. Selli and Kelebek (2011) expressed that 83 and 78 aroma compounds were identified and quantified in the aromatic extract of Moro and Sanguinello cultivars, respectively. Terpenes, terpenols, and alcohols were found as the main types of volatile components.

4.4 *Pectin and PME in Citrus*

Pectin is found in the flavedo, albedo, membrane, juice vesicles, and core of an orange (Chen et al. 1993). Pectin is a polysaccharide made up of α1-4-linked D-galacturonic acid units and rhamnose inserts in the backbone, with side chains of primarily galactose, arabinose, and xylose (Voragen et al. 1995). PME is an endogenous enzyme to oranges, and it cleaves the methyl esters on pectin molecules, which gives methanol and free pectic acid. The cleavage of the methyl esters turns high methoxyl pectin into calcium sensitive low methoxyl pectin.

4.5 Ascorbic Acid in Citrus

The most important contribution of orange juice to human nutrition is perhaps attributed to its high vitamin C content. Although orange juice is not the only fruit product containing large quantities of vitamin C, it is definitely an important source, because of its relatively high consumption by humans. In addition to its vitamin action, vitamin C is valuable for its antioxidant effect, stimulation of the immune system, and other health-related benefits. An important issue associated with orange juice quality is vitamin C loss during processing and/or storage. Because of its heat-labile properties and instability during storage, ascorbic acid is often used as an indicator for the overall quality of fruits and vegetables, providing information on the loss of other vitamins as well as organoleptic and/or nutritional components (Lee et al. 1976; Polydera et al. 2003; Cortés et al. 2008; Zulueta et al. 2010; Vervoort et al. 2011).

4.6 Minerals in Citrus

Barros et al. (2012) reported that citrus showed high levels of potassium, calcium, and magnesium, and the peels were considered sources of these minerals. Eight minerals were determined, including four major elements (K, Ca, Na, and Mg) and four trace elements (Cu, Fe, Mn, and Zn), in four citrus species grown in Brazil. The citrus fruits showed high potassium content (140 mg in 100 g of orange pulp), while the sodium content is relatively low (2 mg/100 g orange pulp). The ratio of K and Na in oranges plays an important role in maintaining the electrolyte balance of cells in the human body (Ladaniya 2008). However, peels showed 40 times more sodium than pulps. The lowest K content was observed in Sweet lime pulp (101.2 ± 4.9 mg/100 g FW) and the highest value in Pera and Lima orange peels (266.0 ± 12.2 and 258.7 ± 11.0 mg/100 g FW, respectively). The highest Ca content was found in the Tahiti lime peel (214.2 ± 1.8 mg/100 g FW), which can be classified as a "source of" calcium for adults. Therefore, the citric acid in citrus may act as a chelating agent and thus increase the Ca absorption by preventing the formation of insoluble salts (Ladaniya 2008).

4.7 Limonoids in Citrus

The bitter fraction of the orange juice arises from a tetracyclic triterpenoid called limonin. The limonin is produced over time from limonic acid or limonin monolactone, which is found in the seeds and membranes of most citrus fruits. Limonin is the bitter limonoid found in major citrus cultivars such as grapefruit, the Navel orange, and the Shamouti orange (Guadagni et al. 1973; Fayoux et al. 2007; Pichaiyongvongdee and Haruenkit 2009). Limonin is synthesized as a nonbitter form (limonoate A ring lactone) in leaves and transported to fruit and seeds. Thus each part of the fruit contains different amount of limonin (Maier et al. 1977; Fayoux et al. 2007; Pichaiyongvongdee and Haruenkit 2009). Limonoid bitterness occurs gradually after juice processing from some citruses, which are referred to as delayed bitterness (Maier and Beverly 1968) and also in the fruits after freezing or mechanical damage. Many biochemical approaches have been applied to eliminate the delayed bitterness and to produce the acceptable quality of juice (Puri et al. 1996). Normally, intact fruit tissue contains a nonbitter precursor of limonin, limonoate A-ring lactone (Maier and Beverly 1968). The conversion of limonoate A-ring lactone to limonin in the juice proceeds under acidic conditions below pH 6.5 and is enhanced by the action of an enzyme (Fig. 4), limonoid D-ring lactone hydrolase (Maier et al. 1969).

The pattern of accumulation of limonoid glucoside is different between citrus species, but the variety of limonoid compounds is the same. The limonoid content in citrus fruits reportedly decreases during the process of ripening (Hashinaga and Itoo 1981). This decrease is considered to be due to conversion of limonoid to a corresponding glucoside. In citrus juice, the glucoside levels are much higher than the free limonoids (Fong et al. 1990). The free limonoids cause a bitterness problem after juice preparation (Karim and Hashinaga 2002). The taste threshold for limonin is near 6 ppm. Levels of 30 ppm or more are common in early-season Navel juices, which are generally blended to lower the acid levels, diluting the bitterness. However, limonin bitterness usually cannot be reduced below the threshold by blending. Valencia juice also has about 15 ppm limonin early in the season, but by the time it is blended to lower acid levels, the limonin level is lowered to below the taste threshold (Kimball 1991a, b).

The limonin content of the fresh mature Valencia oranges fruit is 2–3 ppm after extraction, rising to 11–13 ppm after pasteurization, but undergoes little subsequent change during storage (Tariq et al. 1974). Limonin concentration (5–6 ppm) in the juices from Italian Tarocco and Sanguinello oranges is not affected by maturity and is lower in juices from first than from second pressing; it could be reduced by removing the peel immediately after juice extraction and by low temperature processing (Trifiro et al. 1984). In addition to the bitter property, limonoids have potential biological functions, nonbitter citrus limonoid glucosides, which are tasteless and water soluble, have also been shown to exhibit anticancer activity in oral carcinogenesis in hamsters and human breast cancer cells in culture (Miller et al. 1994).

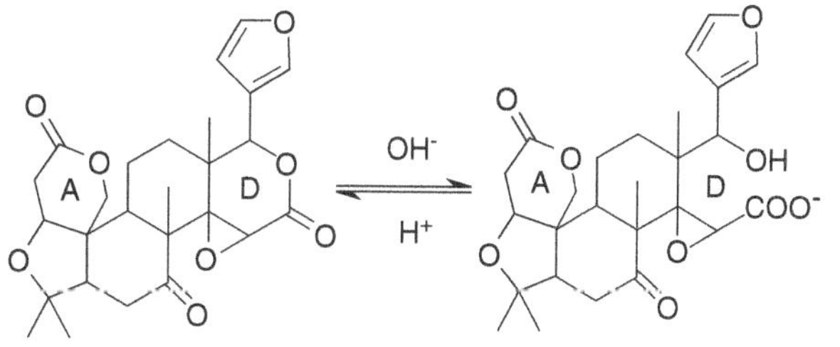

Fig. 4 Formation of limonin in citrus juice

The biological activity of nonbitter limonoid glucosides is equal to the value of bitter limonoids and the water solubility is an important factor for medical application. Moreover, since a large amount of limonoid glucosides (more than 100 ppm) are contained in many kinds of citrus juice and fresh citrus fruits (Berhow et al. 2000). Thus, not only for the processing industry, but also for the consumer, these limonoid glucosides are important compounds (Zaare-Nahandi et al. 2008).

The high content of bitter substances can be linked to several factors: the variety, climatic factors (Hofsommer et al. 1991) and also the rootstock, independently from variety, can concur to determine a high content of limonin in oranges (Di Giacomo et al. 1977). A hard squeezing in juice extractors and in finishers increases the juice yield but also increases the quantity of bitter compounds passing from solid parts into liquid (Crupi and Rispoli 2002).

4.8 Bitterness in Citrus

Taste is one of the sensory quality attributes that, together with color and flavor determines food selection. Bitter or astringent tastes tend to be rejected by the consumer, for that reason early season orange juice or orange juice from immature fruits must be subjected to an appropriate treatment to reduce bitterness. Limonoids and flavonoids are the main groups of bitter compounds in citrus. Limonin, nomilin, and nomolinic acids are triterpene derivates compounds that occur gradually in certain varieties of citrus after juice processing giving a "delayed bitterness" (Puri et al. 1996). Limonin is the most representative compound in this group (Kimball and Kimball 1991). Among flavonoids some are bitter while others are not, depending on the type of chain glycosides. The neohesperidose flavanones (rhamnosyl-α-1,2 glucose), such as naringin, neohesperidin, and neoeriocitrin are responsible for the bitter taste in grapefruit and bitter orange, while rutinoside flavanones (rhamnosyl-α-1,6 glucose) such as hesperidin, narirutin, and didymin are tasteless (Horowitz 1986). Naringin is not transported after being synthesized in the fruit or leaves. Limonin and naringin co-exist in most citrus cultivars, but in different amounts, and their threshold levels are different. The limonin in orange juice has been detected by taste panels at a minimum concentration of 6 ppm (Guadagni et al. 1973; Kimball and Norman 1990a, b). Naringin is found in the membranes and albedo of the fruit and is extracted into the juice, giving it an "immediate" bitterness when their levels exceed 20 ppm (Fisher and Wheaton 1976).

4.8.1 Debittering Process

Some citrus fruits contain a significant amount of bitterness components (such as naringin and limonin). To reduce/remove bitterness in citrus juices below the threshold level for consumer acceptability, a number of chemical treatments, physical separation processes, blending with nonbitter citrus juice and sugars, and enzyme treatments have been described. Although many different methods have been used since the early 1970s the current debittering technology uses adsorption onto cellulose acetate or macroporous resin beads or cross-linked styrene divinylbenzene resins (Shaw et al. 2000). The effect of different debittering processes on citrus juice quality was given in Table 4.

The fresh industrial squeezed orange juice (FOJ) and corresponding orange juice (OJ) after the debittered orange juice (DOJ) were taken. No significant differences were found in acidity, pH, and total soluble solids. Hue and lightness were lower

Table 4 Effect of different debittering processes on citrus juice quality

References	Methodology	Juice	Quality parameters/ Bioactive compounds	Results
Fernández-Vázquez et al. (2013)	Resin	Orange	General	No significant differences were found in acidity, pH and total soluble solids
			Phenolic	Decrease in total phenolic compounds due to the debittering (24 %, from 716 to 547 mg/L)
			Color	L* and h*values decreased, a* value increased, b* and C* values not changed after the debittering process
			Flavor compound	All the aroma compounds decreased significantly in debitterred samples. Percentages of reductions were from 16 to 61 %. The main compound in both groups was limonene decreased from 1,667 to 793 mg/L after the process
			Sensory	According to sensory analyses; there were significant differences between color, aroma and taste of fresh and debittered juices
Olsen and Hill (1964)	Naringinase	Grapefruit	Naringin	After increasing the enzyme concentration to 0.08 % in concentrated juice, the reduction of naringin was increased to 85.2 % at 27 °C. Naringin and the bitterness decreased as the naringinase concentration and the storage temperature and time were increased
Stéphane et al. (2007)	Polymer	Orange	Limonin	Low molecular weight (LMW) poly (vinyl chloride) (PVC) plasticized with dioctyl adipate (DOA) gave the best results for both limonin sorption and low DOA migration. Thick films did not significantly sorb more limonin than thin films
Stinco et al. (2013)	Resin	Orange	Biactive compounds	The results suggest that debittering decreases the nutritional quality concerning the vitamin C and phenolic compounds, but not the provitamin A

after debittering. Odor profile (limonene, a-pinene, ethyl butanoate, octanal, linalool, citral, and terpineol) decreased significantly (from 16 to 61 % on average). Significant differences not only in taste but also in color and aroma between FOJ and DOJ; however, preference was not clear. In conclusion, total phenolic compounds and the entire aroma compounds analyzed decreased significantly due to debittering process. Color parameters also changed significantly in DOJ, being these orange juices more reddish and darker. Sensory analyses showed that panellists were able to discriminate between FOJ and DOJ in color, taste, and odor (Fernández-Vázquez et al. 2013).

Concentrated grapefruit juice (55° Brix) or reconstituted juice (10.5° Brix) was treated with naringinase D-100 at levels ranging from 0.0025 to 0.08 %. After being stored at temperatures of −8°, 40°, 60°, 80°, and 122 °F. Naringin in reconstituted juice was reduced by naringinase much more rapidly than that in concentrated juice even at lower temperatures. When the enzyme concentration, the storage temperature, or the duration was increased, the rate of reduction of naringin, and consequently that of debittering, also increased. The greatest reduction of 88.9 % of naringin in grapefruit concentrate was obtained with 0.08 % naringinase when the concentrate was held for 32 h at 80 °F. Reconstituted grapefruit juice containing 0.018 % naringinase and stored at 40 °F was almost completely debittered in 15 days. At −8 °F there was no apparent naringinase activity in 55° Brix grapefruit concentrate over a 1 year period (Olsen and Hill 1964). Stéphane et al. (2007) monitored the limonin concentration decrease in orange juice over 72 h for the selected polymer films. For practical purposes, since at least 90 % of the equilibrium concentration was reached after 5 h, and since all plasticized LMW PVC films could decrease limonin concentration by 40–85 % in this time, they did not consider longer durations for any other films (LDPE, HDPE, PET, PVA, and SB) because of their lower debittering performance. Stinco et al. (2013) studied on removing bitter components in orange juice by physical adsorption in a resin. The levels of bioactive compounds (carotenoids, ascorbic acid, and phenolics), antioxidant activity and the color in the fresh orange juices (non-debittered) and in the debittered counterparts were measured. The results demonstrated that the carotenoid contents were not significantly affected by the treatment. However, the debittered orange juices showed a reduction of 26 % in ascorbic acid, 32 % in hydroxycinnamic acids, 28 % of flavones, and 41 % of flavanones in comparison with the nontreated juices. The antioxidant activity was significantly higher in untreated juice than in debittered juices. Some color parameters (L*, a*, and h*) were also affected.

5 Affecting Factors on Quality of Juice

Citrus fruit composition varies significantly due to fluctuating effects from rootstock, fruit size, variety, maturity, storage, horticultural conditions (scion, fertilization, frequency of irrigation, date of harvesting, age of tree, tree spacing, position of fruit on the tree and place of growing) (Sandhu and Minhas 2006) and climate, suggesting that nutrient and constituent analysis only provide within- and between-variety estimates, and general conclusions are difficult to form (Kefford and Chandler 1970). Degreening treatment (with ethylene) can be used to enhance early season citrus peel color, with minimal effect on nutritional quality. Ethylene treatment had no significant influence on levels of ascorbic acid, carotenoids, total phenolics, and radical scavenging activity, while it had differential effects on limonoids, coumarins, and flavonoids (Chaudhary et al. 2012).

6 Citrus Juice Processing

6.1 Harvesting

Because citrus fruits grow throughout the year and because the ripening period can last for a total of 5 months, the fruits obtained at one particular harvest are different from fruits obtained at other periods of the harvest, particularly with respect to acid, sugar, and pigment content. The different types of fruit, depending on type and ripening conditions, also differ in other respects (vitamin C, pectin, carotenoids, flavonoids, etc.). The characteristic properties of the raw product exert a great influence on the final product. Today, the assessment of juice quality is simplified by statistical indices obtained from genuine raw product samples. Fruits of suitable quality may be harvested manually using clippers or mechanically, depending upon the facilities available. Manual harvesting may be preferred in the countries where cheap labor is available and comparatively small acreage of orchards is managed. In developed countries, mechanical harvesting is practiced and a number of abscission chemicals are applied to facilitate detachment of fruits from the tree. Care should be taken to avoid any damage to fruit during handling. The fruits are packed in bags or bins and transported to the processing factory (Sandhu and Minhas 2006).

The citrus fruit must have reached an appropriate degree of development and ripeness, account being taken of criteria proper to the variety, the time of picking and the growing area. Maturity of citrus fruit is defined by the following parameters (Table 5) specified for each species.

Table 5 Affecting quality parameters for citrus fruit

1. Soluble solids (°Brix) are primarily sugars; sucrose, fructose and glucose. Citric and ascorbic acids and minerals in the juice also contribute to the soluble solids	*6. Acid (titratable acidity)* The predominant acid naturally occurring in citrus juice is citric acid. There are small amounts of malic acid and tartaric acid present
2. °Brix/acid ratio depending on the cultivar and maturity of a quality criterion for the best. This value does not change after the harvest, because citrus does not mature after harvest	*7. Juice color* The orange–yellow color of citrus juice is due to carotenoids found in the flavedo, or peel, and juice vesicles which get into the juice during processing. The range of colors seen in orange juice are due to cultivar, season of production, processing methods, and maturity of the fruit
3. Juice yield	
4. Good flavor	
5. Low level of bitter compounds	*8. Potential PME activity*

6.2 Receiving

Fruit destined for fresh fruit markets is generally harvested into bins to protect the fruit; however, open trailers, which may contain 20 tons of fruit, usually convey fruit destined for juice. Fruit can be unloaded using bin dumpers if it comes to the plant in bins. If the fruit comes in trucks, the truck usually mounts a tilted ramp and is unloaded from the side or rear of the trailer. Instead of a ramp, hydraulic lifts can elevate the trailer fruit can be directly unloaded into the plant for immediate processing, or it can be stored in large fruit storage bins (Kimball et al. 2004).

6.3 Washing

The source of the fruit affects the needed fruit preparation prior to juice extraction. Fruit coming directly from field, should have had extraneous leaves and twigs removed from it during unloading. This type of trash can also be eliminated when the fruit is drawn from the bins prior to processing. Often fruit from the field is just plain dirty and even may contain pesticide residue undesirable for the processed product (Kimball 1999).

The use of rotating brushes under a water spray with an added cleaner is a common practice for washing fruit just prior to processing. This is followed with a potable water rinse that may include a sanitizer. Fruit washers also often eliminate much of the trash and damaged fruit that had not been previously removed. Fruit washing is best done outside the plant due to the fact that the debris has a tendency to attract insects. Washed fruit enters the plant clean and free from extraneous trash, insects and excessive levels of microorganisms.

Fruit-washing in the production of unpasteurized citrus juice must be thorough and scrupulous. It is common for fruit to undergo several washing or sanitizing steps prior to juice extraction. Some companies provide a heat treatment of the clean fruit surface (hot water or steam) just prior to the juice extraction step. This heat treatment provides the required 5-log reduction needed to comply with the FDA Juice HACCP Rule (Kimball et al. 2004). The fruit goes through inspection lines for removal of bruised or damaged fruits by qualified inspector. The sorted fruits are conveyed to storage bins and sufficient quantity is accumulated for continuous operation of the processing plant (Sandhu and Minhas 2006).

6.4 Sorting

A system of conveyor belts feeds the section in which citrus fruit is squeezed. Before squeezing, the fruit passes through sizing machines. Generally roller machines that provide sizing. Thus, separating fruit in classes according to diameter. This operation is necessary for better rendering of extractors and for a good yield and quality of juice and essential oil.

Processing plant mechanical sizers allow conveyed fruit to pass through rollers or openings dependent on the fruit diameter and sizer clearance. Sizer adjustment is possible during operation to allow for variations in the incoming fruit streams. Smaller fruits have more juice/fruit mass, and the juice has a bigger higher Brix than large fruits. This allows some juice blending opportunities, provided that proper sizing is performed. It also should be mentioned that growing conditions (temperature, rainfall, etc. placement on a tree can affect the average seasonal fruit size and quality. Obviously, a close match between extractor cup diameter and fruit size advantageous and will improve juice yield and quality (Braddock 1999).

6.5 Extraction

There is a significant difference in the processing of citrus fruits as compared to the processing of pomace fruit, stone fruit, and small fruits. When extracting the juice from oranges, grapefruits, lemons, and limes, the objective is to obtain the best possible separation of juice, peel, and peel oil. It therefore does not make sense to completely crush the fruit and press the resulting mash. For this reason, special machines were developed for extracting the juice from citrus fruits. Each fruit can be processed individually with these machines.

The peel contains the valuable peel oils. These peel oils are the principal factors which affect the taste of the juice. The proportion of oil in the juice should be between 0.01 and 0.03 % if the juice is to have a balanced taste and if the juice is to be preserved for any length of time. The choice of machinery depends upon the capacity, yield, and quality of the product required by the processor. For small-scale work, halving and burring machines, plunger type press, continuous screw expeller press, superfine pulper, and cup-type extractor can be used for large-scale commercial production, automatic plants are being used (Sandhu and Minhas 2006). Squeezing separates the citrus fruit in to three basic parts: essential oil, juice, and peel. The systems can be subdivided in two groups: a system separating oil and juice in an only operation and systems in which oil extraction follows juice extraction.

6.5.1 FMC Extractor (JBT, Since 2008)

JBT extractor is part of the systems separating oil and juice in an only operation. These extractors are common in citrus processing industry of the world and built by Food Machinery Corporation of Lakeland, Florida, USA (www.jbtfoodtech.com). JBT extractors simultaneously separate raw juice, peel, core, and essential oil. The core is constituted by two peel discs and by the central part of the fruit, included rags, pulp, and seeds. Essential oil is removed under water sprays in form of an emulsion. JBT FoodTech Citrus Extractor models are capable of handling 2–7 tons of citrus per hour, depending upon fruit size. JBT commercializes for each model two varieties of extractors characterized by a different speed, at 75 and 100 strokes per minute. A five-headed JBT extractor can theoretically process 500 fruit/min at the design rate of 100 strokes/min. The JBT juice extractor consists of three to eight upper and lower cups are formed by a series of finger-shaped elements. From the basic principle characterizing the in-line JBT extractors other constructors have drawn their inspiration; for example, Bertuzzi Food Processing, in Milano, Italy (Citrostar MU1, MU3, MU6), (www.bertuzzi.com.it). JBT juice extractor principle illustrated four steps were given Fig. 5.

In order for the JBT extractor to function properly, the fruit must be sized prior to entering the extractor in order to match the cup size of the machine. If the fruit is too large for the cup, the upper cup will chop, rather than squeeze, the fruit. If the fruit is

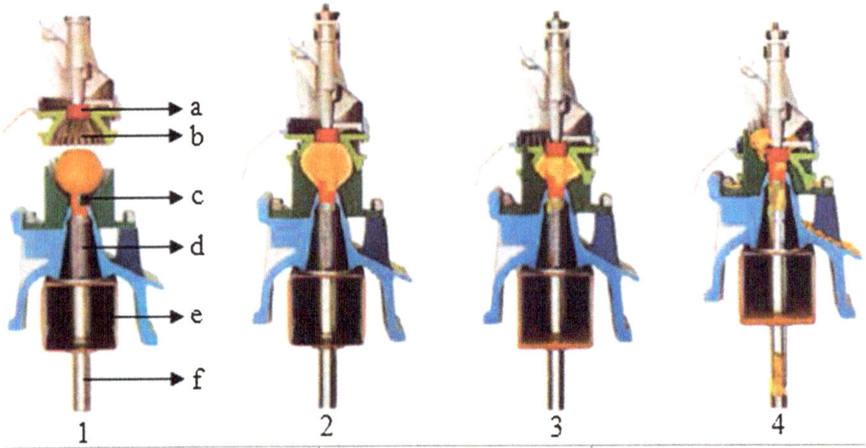

1. ***Components*** *a. Upper Cutter* cuts a plug In the top of the citrus to permit the separation of the peel from the internal portions of the fruit *b. Upper Cup and Lower Cup* supports the exterior of the citrus throughout the squeeze cycle to prevent bursting *c. Lower Cutter* cuts a plug in the bottom of the citrus to allow the internal portions of the fruit access to the prefinisher tube *d. Prefinisher Tube* separates, based on particle size, the internal elements of the citrus *e. Juice Manifold* collects juice and juice sacs. *f. Orifice Tube* generates pressure inside the prefinisher tube and collects and discharges membrane and seeds	***Implementation*** **2.** In this early phase of the extraction cycle, the upper cup moves downward to cause pressure on the citrus so that the top and bottom plugs begin to be cut. Due to the profile of the cups, the citrus is totally supported, and thus will not burst but will get an even squeeze **3.** As the extractor cycle continues, pressure increases on the citrus causing the internal portions of the fruit to be forced through the bottom plug and into the prefinisher tube. The between the upper cup and peel is now being discharged cutter	**4.** Upon completion of the extraction cycle, the internal portions of the citrus are located in the prefinisher tube. At this time, the orifice tube moves upward, placing pressure on the contents of the prefinisher tube. This causes the juice and juice sacs, due to their small particle size to flow through the holes of the prefinisher tube and into the juice manifold Those internal portions of the citrus, whose particle sizes are larger than the holes in the prefinisher tube, are forced through an opening in the orifice tube and are discharged out the bottom

Fig. 5 JBT juice extractor principle

too small for the cup, the upper cup will smash, rather than squeeze, the fruit. A typical array of JBT extractors will consist of machines containing various size cups fed by previously sized fruit. Generally, three- to six-cup sizes are adequate for most processing plants. Improper sizing will have a significant effect on juice yield and quality.

The Model 291B/391B Citrus Extractor has 5 cups and operates at 100 strokes per minute or 500 fruit per minute. The 291B/391B is designed to use a cup size of 76 mm to handle fruit from 44 mm to 83 mm diameter, or a cup size of 102 mm to handle fruit from 83 mm to 108 mm diameter, or it can also be changed over to 60 mm cup if fruit size is predominantly under 57 mm diameter. The Model 191B Citrus Extractor has 8 cups, operates at a speed of 100 strokes per minute or 800 fruit per minute. The cup size is 60 mm and will efficiently handle fruit ranging from 25 mm to 60 mm maximum. The Model 491B Citrus Extractor has 3 cups, operates at 75 strokes per minute or 225 fruit per minute. Designed primarily to use a 127 mm cup typically used for large grapefruit ranging in size from 108 mm to 140 mm in diameter. The 491B can also be equipped with smaller cups to extract oranges or smaller grapefruit.

It has been found that most of the oil in the juice comes from the flavedo of the peel plug or the portion of the peel cut out by the sharp edges of the 1-in. diameter holes in the fruit in the upper and lower cups. If using a smaller diameter cutter reduced the amount of flavedo cut out, the amount of oil found in the juice can be greatly reduced. This is of greater significance to not-from-concentrate (NFC) juice and fresh-squeezed juice processors who would normally produce single-strength juices with excessive oil levels. Excess oil in juice destined for evaporation is removed during evaporation and is not affected by excessive oil levels (Kimball et al. 2004).

6.5.2 Brown Extractors

Brown method is part of the systems in which oil extraction follows juice extraction. A lot of models of Brown extractors built by Automatic Machinery and Electronics, Inc., Winter Haven, FL, USA (www.brown-intl.com).

Before juice extraction with this system, fruit oil is removed from peel. The oil removal is achieved by lightly puncturing the entire surface of the fruit with over three million sharp stainless steel points configured in the form of rotating rolls. After the oil extraction from the whole fruit peel, the fruits are cut two halves. The juice is then extracted from each halve by means of a rotating squeezing head. This system is simply a mechanized and automatic version of the simple manual citrus press found in every home.

The Brown Models 720, 620, 520, and 570 Extractors are commercial citrus juice extractors. These extractors can be equipped with components to accommodate oranges, grapefruit, lemons, limes, and tangerines in several size ranges from 2 to 6 in. and process, depending on the Model, up to 750 fruit per minute. This family of Brown extractors is typically installed in lines of 10–14 machines of which several are configured to accommodate each fruit size. Each size extractor tolerates some overlap of fruit sizes making efficient use of all the extractors on each processing line. Each line generally processes from 45 to 80 tons of citrus fruit per hour.

Whole pre-sized citrus fruit is fed into the extractor for processing. The fruit enters a hopper and is picked up one at a time by a feeder mechanism. Two opposing endless chains are each composed of fruit processing cups, which receive whole fruit from the feeder. These cups first pass by a halving knife to cut the fruit in two. They carry the fruit halves to a reaming area where the juice is reamed from the fruit half in much the same manner as a kitchen hand juicer. The reamed peel halves are then ejected from the cups allowing for the reception of another whole fruit to begin another cycle.

Brown extractors are designed to extract juice, cell sacs, seeds, and some of the membranes from the peel by a combined pressing and wiping action. This gentle method leaves a sound extracted peel half, which is not crushed or mangled. Reamed juice from a Brown extractor has greater whole cell sac retention and very little bitter tasting oils and peel extracts. Reamed juice is free of all Albedo and flavedo. A significant advantage of the reaming method is that numerous undesirable components inherent in the peel are not liberated into the juice. The degree of juice extraction is determined by the amount of air pressure applied to the cups and by adjustable stops which limits the cups' clearance with the reamers. This feature provides positive control of reamer penetration and allows accommodation for peel thickness variations. Adjustments for extraction pressure can be made quickly, and while the extractor is in operation.

6.5.3 Fratelli-Indelicato Extractor

This system is part of the systems in which oil extraction fallows juice extraction. The use of revolving drums, which press the half fruit against a grid, is a system utilized for a long time in industrial extraction of citrus juice. The extractors of this class are preceded by systems for the extraction and recovering of essential oil; they are generally paired to machines rasping fruit peel. On this principle are based the juice extractors "Polycitrus" "Fratelli-Indelicato" (Giarre, Catania) (www.indelicato.it) built in Italy and diffused above all in the Mediterranean area.

POLYCITRUS Spellalbedo juice extractor automatically extracts the juice from citrus fruit (oranges, lemons, mandarins, limes, kinnows, grapefruits) with a diameter from 25 mm to 140 mm, already de-oiled by Polycitrus (Polycitrus M6—M10—M15—Citrus oil extractors), and cleans the internal part of peels for candies fruit production. The process is made, without any previous grading of sizes, with considerable savings in initial investments, space required, and power installed. The squeezing pressure can be easily adjusted according to their variety, ripeness, and peel thickness. The working capacity can match the capacity of the oil extractor; for instance, in case of oranges it can reach 20 t/h.

The fruit go into the juice extractor, where every fruit is cut into two halves by a fixed knife placed between two counter rolling cylinders covered by a rasping sheet; each half fruit is squeezed against a perforated stainless steel wall, adjustable in height, placed under the two rasping cylinders in order to go with their outer contour. The distance between the wall and the drums decreases as the fruit is squeezed so that the squeezing can be carried out smoothly and progressively.

Birillatrice-Sfumatrice AZS 204

These are machines in which juice is extracted by means of revolving reamers by citrus fruit previously cut in halves and laid down on plastics cups. The principle is ancient but still valid and it is used in Italy and in other Mediterranean countries above all in lemon processing, where essential oil has a pre-eminent value. By spent peel, essential oil is recovered by means of "sfumatrici" or screw presses. Even if continuously updated, they remain suitable for special processing, not for mass-production.

7 Essential Oil Extraction from the Citrus Peel

Citrus peel contains essential oils. These oils are formed in the cells of the so-called oil glands, and they are some of the most important byproducts obtained from the processing of citrus fruit. Essential oils are in demand as taste and aroma agents for the food industry and for the production of odorous substances. Several methods are available for recovering essential oil.

The oil-water emulsion obtained in modern fruit processing lines contains approximately 70–90 % of the essential oil naturally present in the peel. The peel wash water contains between 2 and 5 kg essential oil as an emulsion per 100 kg of raw material processed, depending on the type of fruit and the processing technology used.

Juice and oil are extracted in one process step where JBT extractors are used for processing citrus fruit. The peel of the fruit is cut and squeezed during the processing procedure. This releases the oil which is contained in the peel. The oil is obtained from the flavedo during this mechanical treatment. The valuable essential oils are then washed out and separately collected during this process step. The JBT FoodTech citrus extraction system is simultaneously recovers oil as well as juice during the same extraction cycle. This feature not only minimizes the space and energy required to recover oil but also delivers high yields. Water usage and waste disposal are minimized through the use of water recycle systems.

The Brown Oil Extractor (BOE) liberates the essential oil from whole citrus fruit. BOE method, either the peel of the citrus fruit is pricked before the extracting process and the oil which appears is washed away with water, or the outer section of peel (flavedo) is rasped before being mixed with water. For complete coverage of an elongated fruit such as a lemon, the rolls not only rotate, but also oscillate horizontally in opposite directions causing the entire peel surface to be punctured, releasing the oil. The puncturing operation takes place beneath the surface of a shallow pool of water contained in the BOE to insure that there is no loss of oil to the atmosphere. Obtained liquid is added to the oil/water stream leaving the BOE. After leaving the BOE, the oil/water mixture passes through a fine screening operation in preparation for centrifugation. A first centrifuge produces a stream of rich oil. The water phase from this centrifuge is recycled back to the BOE to eliminate losses. Without costly freezing or enzyme treatment, the rich oil stream passes through a second centrifuge which continuously produces pure polished oil. The BOE Citrus Oil Extractor uses approximately 49 l less of water per metric ton than competitor's oil recovery systems. This feature is very important for saving water consumption.

Another oil extraction method is developed by INDELICATO. This method is principally similar Brown oil extractor. The peel oil extractions occurs by water sprays in a closed, cylindrical drum environment as the fruit flavedo is scarified, or rasped, on rotating cylinders of sharp, punch-perforated stainless steel.

8 Finishing Operations

The finishing process removes the excess of pulp, bits of peel, rag, and seeds. The yield is important to the grower who wants the highest return of his fruit and to the processor who is responsible for the quality of the finished product (Sandhu and Minhas 2006). Finisher is used to balance juice quality and yield with hard squeeze or soft-squeeze extractor setting. Besides viscosity, finisher performance can affect juice cloud, color, flavor, flavonoid glycosides, pectin, and pulp level. There is evidence that finisher setting do not affect the level of oil in the juice, which is function of the juice extractor (Braddock 1999). And also finishing operations are important for residual PME activity. Coarse pulp of raw squeezed juice is removed as soon as possible because of high content and heat resistant isoenzymes of PME. Pectin methyl esterase (PME) is major impact in orange juice causing losses in the quality if not inactive. During orange juice processing, pectin ester molecules rapidly reduced causing increase in PME activity, deterioration of the natural turbidity of orange juice, and gelation of concentrated orange juice (Marshall et al. 1985; Kimball 1991a, b; Basak et al. 2001).

The orange juice is discharged from both extraction systems with approximately 11°Brix and a pulp content of between 20 and 25 % (by volume). The pulp is then reduced to approximately 10–12 % (by volume) in downstream finishers (horizontally-mounted screen drums).

The finishers usually consist of cylindrical screens through which the juice is forced by means of a helical worm screw. The juice enters through an end and while the liquid, together with the finest pulp, is filtrated through the perforated wall, the roughest parts are pushed forward and discharged by the other end. Another finisher system is made up by a horizontal perforated cylinder, or by a slightly leaned one, in which a shaft rotates, bringing a series of sheet metal paddles. These push the juice against the perforated wall and determine its screening. A third finishing system uses screens moved by a double movement, rotating and vibrating. They are constituted by three horizontal screens laid upon one another, with different meshes. The raw juice is separated in coarse particles, cells, and juice with fine pulp (Crupi and Rispoli 2002).

JBT and Brown extracting systems require primary and secondary finishing prior to entering evaporators. The reason for this is that the solid pulpy material has a tendency to hang up in restricted spaces and adhere to heated surfaces.

The JBT uses its orifice tube as a primary finisher followed by an external secondary finisher. The Brown extractor usually employs two external finishers. Both JBT and Brown finishers commonly use a screw to force the juice and pulp against the finisher screen, but certain applications use paddle finishers. The unit operations of the Brown process require the first finisher to have openings of generally 0.040 in. and the secondary finisher 0.020 in. diameter (Kimball et al. 2004).

Finisher performance may be measured by a quality control procedure, termed the quick Fiber Test. This test is used to determine to relative dryness of pulp discharged from finisher. The correlation to juice yield is such that higher quick fiber wet pulp contains more free juice, thus a lower juice yield than dry pulp with lower quick fiber values. Automated control of Brown finishers by keeping a constant temperature difference (ΔT) between the feed and exit juices. The juice quality and pomace dryness determine the desired ΔT set point between the incoming feed juice and the discharged juice or pulp (Braddock 1999).

Discarded finisher pulp can be heat-treated to deactivate pectinase enzymes, frozen, and added to citrus juice products.

9 Centrifugation

Centrifuges in service for citrus processing are primarily solid bowl/sedimentation type of clarifiers used for solid-liquid or liquid-liquid separation applications. Some horizontal decanter-type machines are also in use for dewatering and clarification applicants, where continuous suspended solids removal is necessary. Citrus processing centrifuge applications are mostly clarification/solid recovery where the product is not solids-free (e.g., separation of pulp from juice), clarification where the product is clear liquid (e.g., peel oil recovery or very low pulp juice), and solid recovery where the product is the solid material (e.g., pulp dewatering) (Braddock 1999).

The centrifugal pre-clarification of the raw juice permits the production of juices with a high degree of cloud stability since centrifuges classify trub particles according to particle size. All coarse trub particles, which have a strong tendency to sediment, can be separated off. Only the extremely cloud stable fine particles remain in the juice. The risk of deposits forming is therefore reduced to a minimum (Pecoroni et al. 2013).

10 Pulp Washing

The discarded pulp can also be washed with water, mixed in a screw conveyer, and refinished to produce pulp wash or water-extracted soluble solids. In the production of pulp wash, the cells which have been separated out by the finisher are treated in a multistage counter-current washing system. With the aid of scroll mixers and finishers, it is possible to extract up to 90 % of the valuable juice components such as fructose, fruit acids, minerals, carotenes, etc., from the cells. The pulp discarded in so doing can be washed and refinished again to produce additional pulp wash. This can continue several times until 85 % of the available soluble solids have been extracted from the pulp. Generally, three to five stages with a water/pulp ratio of 1:2 are the most effective and efficient (Kimball 1991a, b).

The pulp wash obtained in this process has a Brix content of between 4 and 5° and still contains a high proportion of trub particles. These can be separated out with the aid of clarifiers. The preclarified product is concentrated to 60° Brix in an evaporator. Pulp wash can be marketed as a product in its own right. It is also usual and allowed in some markets to mix orange juice with the pulp wash before the concentration process. The maximum amount of pulp wash which may be blended with the first-class juice is governed by the statutes and standards in each individual country. The capacities of pulp wash clarification are roughly equivalent to the capacity of normal citrus juice clarification. This is true if the pulp content does not exceed 10–12 % (by vol.) upstream of the centrifuge (Pecoroni et al. 2013).

11 Pulp Recovery

The new citrus extraction system offers various methods of pulp recovery. Juice cell sac size can be custom tailored to the needs of the processor and their customers. In one configuration the JBT FoodTech extraction system can produce pure juice sacs of normal size typically used in concentrated juices and other products. Juice sacs from this system are primarily used for the marketing of not-from-concentrate juices. The juice cells sacs produced by this system range in size from 7 mm to 10 mm and are virtually free of peel and seed particles or embryonic seeds. Up to 90 % of these juice sacs retain their floating characteristics (Anon 2013a).

12 De-Aeration and De-Oiling

Juice de-oiling is a special application for the production of NFC juices. Depending on the extraction system used, more or less high quantities of peel oil enter the juice. When producing concentrate, this oil is evaporated which presents no problem. In the production of NFC juices, however, this oil cannot be thermally separated. Due to the fact that the oil remains in the juice, an unpleasant, bitter taste may develop. This makes it necessary to separate the oil in order to obtain a high-quality juice. The oil content is reduced to maximum 0.035 % (by vol.). Some producers demand values smaller than 0.03 % (by vol.).

This limitation of the oil content can be achieved by appropriately modifying the extraction systems. However, in most cases a compromise has to be accepted in terms of yield and hence the economy of the process. For this reason, there is a growing interest in centrifugal polishing of the oil from the partially pulped juice. The oil content is centrifugally separated gently so that values between 0.02 and

0.03 % (by vol.) can be obtained. Product losses are minimal. The machines used are separators (liquid–liquid separation). These machines are adapted to the very small contents of light phase (citrus oil). The separated oil is generally not used because of the low quantity. GEA Westfalia Separator (GEA Mechanical Equipment US, Inc. GEA Westfalia Separator Division, Northvale, NJ 07647 (www.wsus.com)) responded by presenting the ESE 500 for the high capacity range and by successfully putting this separator into operation. The separator type ESE 500 offers feasible economic possibilities for producing large volumes of NFC. The single strength lemon juice taste, even if stored at low temperature, can show significant degradations depending upon its essential oil content. The oxygen absorbed in the juice during technological operations is another cause of decay: ascorbic acid is oxidized and *d*-limonene is attacked with formation of terebenthic taste. The recent trend of industry is to reduce to a minimum these inclusions with the use of adequate equipments, like extractors modified in order to avoid oil excess; for what concerns oxygen, it is recommended to reduce the close contact between juice and air (Pecoroni et al. 2013).

Operating with specific de-aerator at 27 °C, more than 90 % of oil and more than 95 % of oxygen present in the juice is removed (Schofield 1994; Schofield and Riley 1998). The oxygen present in citrus fruit juices can be strongly reduced by means of glucoseoxidase which catalyzes the oxidation of glucose to gluconic acid (Crupi and Rispoli 2002).

13 Pasteurization Processes

13.1 Thermal Pasteurization

PME is more thermally resistant than that of most of the vegetative spoilage microorganisms. Therefore, the target of pasteurization of orange juice is the inactivation of PME (Chen and Wu 1998). Nowadays, in fruit juice industry, thermal pasteurization is a conventionally used method to inactivate PME (Eagerman and Rouse 1976). However, thermal processing has negative impacts on the quality of orange juice such as loss of fresh color, flavor, and nutrition value (Espachs-Barroso et al. 2006). Natural cloudy substances play an important role in turbidity, flavor, aroma, and the characteristic color of orange juices (Baker and Cameron 1999; Van Den Broeck et al. 2000). Therefore, in order to control the loss of quality caused by PME, orange juice must be pasteurized at high temperature (Cameron et al. 1994). In that case, process is done minimum time and temperature for citrus processing.

Yeom et al. (2000a, b) achieved 89 % inactivation of PME in orange juice by the heat treatment applied at 94.6°C for 30 s. Ağçam (2011) reported that PME residual activities of orange juice for 90 °C 10 s and 20 s heat pasteurization treatments were detected as 6.82 and 4.15 %, respectively. After 6-month storage at 4 °C, residual activities were detected as 9.40 and 8.64 %, respectively. İçdemir (2012) observed that no PME activities were detected after heat pasteurization at 85 °C 15 s for Star Ruby grapefruit juice and 95 °C 45 s for Marsh Seedless grapefruit juice.

And also, PME activity was not detected during the storage period (6 months at –25 °C) on the samples processed by pasteurization. Uçan (2013) was applied 95 °C 15 s on lemon juice PME and determined 92.6 % inactivation.

13.2 Pulsed Electric Fields Technology

Pulsed electric field (PEF) technology is a nonthermal food preservation technology that is based on the use of electric fields to exterminate food-borne pathogens and to control enzymes and spoilage microorganisms in foods. This technology is highly appreciated for its ability to extend the shelf life of food products without the application of heat, thus also preserving quality attributes such as sensory quality and nutritional value, as well as controlling the microbiological safety of food products. PEF food preservation is based on the ability of high-intensity pulsed electric fields to disrupt cell membranes, resulting in a lethal effect on the microorganisms. In this method, the food product is placed inside a treatment chamber wherein two conductive electrodes are mounted on a nonconductive material in which there is no electric flow from one electrode to the other through the chamber casing. A high-voltage pulse is then applied to the conductive electrodes to induce a high-intensity electric field pulse on the food product, which is located between the electrodes. The treated product is then removed or subjected to subsequent pulses until the treatment dose is complete. A typical PEF processing system consists of a high-voltage pulse generator, treatment chamber, fluid-handling system, and control and monitoring devices. The construction and characteristics of each component vary from model to model and among research groups, but the basic principles remain the same (Barbosa-Cánovas and Sepúlveda 2005).

13.2.1 Microorganisms Inactivation with PEF in Orange Juice

The exact mechanism by which PEF inactivate microorganisms and enzymes is not yet completely understood; however, much of the research in the field points toward damage of the cell membrane as the principal factor responsible for microbial inactivation (Heinz et al. 2002). An electromechanical model explaining the formation of pores on the cell membrane by the application of PEF was developed by Zimmermann et al. (1974). The formation of pores on the cell membrane (by dielectric breakdown of cell membrane) then cause swelling and rupturing of the cells.

13.2.2 PME Inactivation with PEF in Orange Juice

The mechanisms involved in the inactivation of enzymes by PEF are not fully understood. It is believed that PEF may cause denaturation of enzymes, probably by changing their conformational state. Enzymes are globular proteins whose catalytic activity relies on the native configuration of their active site and the conformation of the surrounding proteins. If the duration of the electric pulse is long enough, the effects of pulsed electric fields on proteins could entail: (1) polarization of the protein molecule, (2) dissociation of noncovalently linked protein sub-units involved in quaternary structures, (3) changes in the protein conformation so that buried hydrophobic amino acids or sulfhydryl groups are exposed, (4) attraction of polarized structures by electrostatic forces, and (5) hydrophobic interactions or covalent bonds forming aggregates (Castro et al. 2001; Perez and Pilosof 2004).

Cloud stability has traditionally been considered as an important quality parameter for orange juice, influencing juice grading and market acceptability. It provides turbidity, flavor, aroma, and the characteristic color of the juice. The loss of cloud is generally attributed to the action of pectin methylesterase (PME), which demethoxylates soluble pectins, allowing calcium pectates to precipitate and clarifying the juice. Commercial heat pasteurization for the production of shelf-stable orange juice is designed to inactivate PME, which is more thermally resistant than vegetative micro-organisms. A 90–100 % reduction of PME activity is normal in commercial heat-pasteurized orange juice (Irwe and Olsson 1994).

Effect of PEF on orange juice PME was given in Table 6. According to studies, PME inactivation was achieved between 34 and 95 % with PEF treatments. Decreased PME activity of orange juice by PEF treatment did not restore and remained constant during storage. It indicates that PEF caused irreversible inactivation of PME (Yeom et al. 2000a; Ağçam 2011).

13.2.3 Effect of PEF on Orange Juice Quality Parameters

The effect of PEF on quality parameters of orange juice was studied with different research groups and results were given in Table 7.

Table 6 Effect of PEF on PME activity of orange juice

References	Juice	Treatment conditions	Inactivation (%)
Zhang et al. (1996)	Orange	35 kV/cm, 200 μs	95
Rodrigo et al. (2001)	Orange-carrot blended	25 kv/cm, 340 μs	79
Rodrigo et al. (2003)	Orange-carrot blended	35 kv/cm, 1,500 μs	81
Vervoort et al. (2011)	Orange	23 kV/cm, 2 μs pulses	34
Yeom et al. (2000a)	Orange	35 kV/cm, 59 μs	90
Yeom et al. (2000b)	Orange	35 kV/cm, 59 μs	88
Yeom et al. (2002)	Orange	25 kV/cm, 250 ms	90
Sampedro et al. (2009)	Orange-milk blended	15–30 kV/cm, 50 μs (65 °C)	46–91
Elez-Martínez et al. (2007)	Orange	35 kV/cm, 1,500 μs	80
Ağçam (2011)	Orange	13.82–25.26 kV/cm, 1,033.9 and 1,206.2 μs	35–94

Table 7 Effect of PEF on the orange juice quality parameters

	Treatment conditions	Quality parameters/ Bioactive compounds	Results
Corteś et al. (2006)	25–40 kV/cm, 30–340 μs	Carotenoids	Total carotenoids decrease between 6.3 and 9.6 % and this result better than heat pasteurization treatment. There is a clear decrease in the concentrations of β-cryptoxanthin, α-carotene, and β-carotene when higher PEF treatment intensities are applied
		Vitamin A	In generally, vitamin A decreases slightly when the electric fields strength and treatment time increase
		Color	Color parameters of L* not affect, a* increase and b* value decrease
Elez-Martínez and Martín-Belloso (2007)	15–35 kV/cm, 100–1,000 μs	Vitamin C	Vitamin C was decreased after treatments. Degradation depended significantly on the electric field strength, the treatment time, the pulse frequency, the pulse width and the pulse polarity. Vitamin C retention higher than that of the heat-pasteurized
		Antioxidant activity	Not affected with treatments
Plaza et al. (2006, 2011)	35 kV/cm, 750 μs	Vitamin C	Degradation ratio 7.7 %. 40 days were successfully stored at 4 °C.
		Antioxidant activity	Processes did not modify significantly the antioxidant activity. At the end of refrigerated storage showed a significant quantitative increase
		Carotenoid	Total carotenoid of treated (thermal pasteurization (70 °C, 20 s) and PEF) juices showed a small losses (<11 %) comparing the content immediately after treatment and at the end of storage (4 °C, 40 days). Lutein, zeaxanthin, and α-cryptoxanthin concentrations were slightly decreased but, β-cryptoxanthin, α-carotene, and β-carotene concentrations were not changed during storage period
		Flavanone	For all treated (PEF and thermal) orange juices, flavanone content (naringenin and hesperidin) decreased significantly (around 50 %) during the first 20 days of storage at 4 °C
Rivas et al. (2006)	25 kV/cm, 280 and 330 μs	HMF	HMF content was very small and did not vary with any of the treatments applied. During storage at 2 °C, no variation was found in the content of HMF
		Color	Nonsignificant changes were found in L* and C* of the mixed orange and carrot juice after each of the treatments, whereas an increase in h* (changing toward yellow) was found after each treatment. Color parameters of the PEF-treated samples had better stability than thermal pasteurized (98 °C, 21 s) samples during storage (2 and 12 °C, 10 weeks)
		Sensory	Sensory characteristics of the PEF-treated juice were more similar to the untreated juice than the thermal pasteurized juice

Torregrosa et al. (2005, 2006)	25–40 kV/cm, 30–340 μs	Carotenoid	In this study, 15 carotenoid compounds were detected. PEF processing generally caused a significant increase in the concentrations of the carotenoids identified as treatment time increased. Concentration of 9-*cis*-violaxanthin and neoxanthin (32.9 %), antheraxanthin (3.9 %), α-cryptoxanthin (15.5 %), 9-*cis*-α-carotene (25.9 %), and 9-*cis*-β-carotene (27.3 %) were decreased significantly for thermal pasteurized juice (98 °C, 21 s)
		Vitamin A	PEF treatment at 25 and 30 kV/cm provided a vitamin A concentration higher than that found in the pasteurized juice
		Vitamin C	Ascorbic acid degradation rate was increased when increasing PEF treatment intensities. Degradation rate was less in PEF treatment at 25 kV/cm. PEF treatment at 25 kV/cm for 280–330 μs allows the half-life of the juice to reach 62 days when stored at 2 °C, whereas when the preservation temperature is 10 °C the half-life is 19 days. Shelf life of thermal pasteurized samples at 2 and 10 °C, 11 and 5 days, respectively
Vervoort et al. (2011)	23 kV/cm, 90 Hz–2 μs monopolar pulses	Sugar	Processing and storage had limited to no effect on sugar composition
		Organic acid	No significant differences in initial citric and malic acid concentration nor in evolution during storage (4 °C, 58 days) were detected
		Vitamin C	Ascorbic acid degradation of thermal pasteurized (72 °C, 20 s) orange juice was higher than PEF treated during storage
		HMF	HMF formation was not detected after treatment and storage period
		Carotenoid	In this study, 16 carotenoid compounds were identified. The total carotenoid content slightly decreased and also, during storage at 4 °C, only four carotenoids exhibited significant changes: *cis*-violaxanthin, antheraxanthin and 9Z-antheraxanthin decreased in concentration, while mutatoxanthin concentration increased
Zulueta et al. (2010)	15–40 kV/cm, 40–700 μs	Vitamin C	PEF processing caused no significant reduction in ascorbic acid concentrations and the minimum retention obtained was around 90 %. After heat pasteurization (90 °C, 20 s) ascorbic acid concentration was calculated around 86 % retention. 25 kV/cm at 280 μs treatment was applied for the shelf life study. Ascorbic acid degradations were calculated for PEF and heat pasteurization at end of storage 4 °C (42 days) 47.4 and 45.2 %, at end of storage 10 °C (35 days), 41 and 48 %; respectively. The shelf life for beverages stored at 4 °C was 52 and 47 days for pasteurized and PEF-treated beverages, respectively
		HMF	The HMF values were obtained 0.19 and 0.29 mg/L for pasteurized and also, 0.13 and 0.21 mg/L for the PEF-treated beverages stored at 2 and 10 °C, respectively
Ağçam et al. (2014)	13.82–25.26 kV/cm, 1033.9 and 1206.2 μs	Phenolic compounds	Except for syringic acid and neoeriocitrin, the concentration of the phenolic compounds indentified in the orange juice samples enhanced after the PEF or treatments. The samples treated with PEF had more stable flavonoids and phenolic acids than those treated with the thermal pasteurization (90 °C, 10 and 20 s)
		Sensory	The PEF-treated samples had higher sensory scores than the heat-treated samples

When the PEF is applied, the carotenoid concentrations generally tend to decrease or remain practically constant while the treatment time increases. The total color differences (ΔE*), which indicate the magnitude of the color difference between fresh and pulsed orange juices subjected to the electric field and time in which the highest treatment temperature was reached (25 kV/cm, 340 μs), were 7.32 ± 2.86 (Corteś et al. 2006). Orange juice retained 87.5–98.2 % of vitamin C after PEF treatments. There were no differences in antioxidant capacity between PEF-treated and untreated products, whereas heat-treated foods showed lower values of antioxidant capacity (Elez-Martínez and Martín-Belloso 2007). PEF may be technologies as effective as low pasteurization to retain antioxidant characteristics of orange juice during refrigerated storage (Plaza et al. 2006). For all treated orange juices, flavanone content decreased significantly (around 50 %) during the first 20 days of storage at 4 °C while carotenoid content showed a moderate decrease (less than 11 %) that took place during the last 20 days (Plaza et al. 2011). The effect of different PEF intensities and conventional thermal treatment (98 °C, 21 s) on quality characteristics of blended orange and carrot juice were investigated (Rivas et al. 2006). To establish the effect of PEF treatment, different field intensities were applied for different times, and in all cases, the results were compared with the results for untreated and thermally pasteurized (98 °C, 21 s) samples (Torregrosa et al. 2005). These researchers also determined ascorbic acid concentration changing of orange-carrot juice at the same PEF and thermal pasteurization conditions (Torregrosa et al. 2006). Vervoort et al. (2011) were studied effect of PEF and mild pasteurization (72 °C, 20 s) on orange juice quality. Ascorbic acid concentration of PEF treated sample slightly decreased with storage but mild pasteurized sample concentration decreased faster than. The ascorbic acid degradation kinetic of orange juice–milk beverage was determined after increasing PEF intensity treatments and compared with thermally pasteurization (90 °C, 20 s) (Zulueta et al. 2010). Ağçam et al. (2014) reported that samples treated with PEF had more stable flavonoids and phenolic acids than those treated with the thermal pasteurization (90 °C, 10 and 20 s) in Kozan Yerli orange (Turkish Cv.). Samples processed by PEF contained higher phenolic compound concentrations than those processed by the heat.

13.3 High-Pressure Processing

High-pressure processing (HPP) is defined as a method of food preservation that involves subjecting food to intense pressure in the range of 300–700 MPa, with or without addition of heat, to achieve microbial inactivation while achieving consumer-desired qualities, e.g., retention of freshness and nutritive values of food products. HPP causes little changes in the "fresh" characteristics of foods; in fact, it is possible to keep many foods longer and in better condition. Small molecules such as flavor compounds, vitamins, and pigments are typically not affected by pressure. A typical HPP system consists of a high-pressure vessel, a means to close the vessel off, a system for pressure generation, a system for temperature and pressure control, and a material handling system. Hydrostatic pressure is applied to a food product through a bath containing pressure-transferring medium (usually water) that surrounds the product. The hydrostatic pressure is transmitted to the product equally from all sides; this equal distribution of pressure is the reason why food is not crushed during treatment. This type of pressure also has little effect on covalent bonds and, as a result, the food being processed does not undergo significant chemical transformations (Rastogi 2010).

13.3.1 PME Inactivation with HPP in Orange Juice

High pressure processing constitutes an effective technology to stabilize fresh orange juice through reduction of PME activity. Preferably pressures higher than 500 MPa should be applied for economic and commercial viability of the process. High pressures in the range 400–600 MPa can be combined with mild heat (<50 °C) to accelerate PME inactivation (Nienaber and Shellhammer 2001a, b).

Effect of HPP on orange juice PME was given in Table 8. According to studies, PME inactivation was achieved between 10 and 96 % with different HPP treatments.

13.3.2 Effect of HPP on Orange Juice Quality Parameters

The orange juice quality parameters (°Brix, viscosity, total acid content, browning index, and color), ascorbic acid and β-carotene concentrations of the juices were not significantly affected by HPP (Table 9). HPP or thermal treatment (85 °C, 25 s) of both orange juices types did not significantly reduce the β-carotene content compared with the untreated juices (Bull et al. 2004).

High pressure treatment and storage up to 21 days at 4 °C caused no significant differences in antioxidant capacity, vitamin C, sugar, or carotene content. Vitamin C content of the orange juice processed at 500 and 800 MPa was not, or only insignificantly, reduced compared to unprocessed juice. During subsequent storage for 21 days, the L-ascorbic acid content of orange juice decreased 6–12 % in nontreated samples and 5–23 % in the pressurized ones, respectively. Retention was over 84 % of original antioxidant capacity after 21 days chilled storage for all studied cases.

Table 8 Effect of HPP on PME activity of orange juice

References	Juice	Treatment conditions	Inactivation (%)
Goodner et al. (1998)	Orange Grapefruit	600–900 MPa, 1 s 600–900 MPa, 1 s	10–93 50–87
Nienaber and Shellhammer (2001a)	Orange	400 MPa, 12 min, 50 °C	50
Nienaber and Shellhammer (2001b)	Orange	600–800 MPa, 1–5 min, 25 °C	92–96
Vervoort et al. (2011)	Orange	600 MPa, 1 min	92
Katsaros et al. (2010)	Orange	100–500 MPa, 20–40 °C	85

Table 9 Effect of HPP on the orange juice quality parameters

References	Treatment conditions	Quality parameters/ Bioactive compounds	Results
Bull et al. (2004)	600 MPa, 1 min	Color	Total color differences were calculated for HPP and thermally pasteurized (85 °C, 25 s) after the 12 weeks at 4 °C 12.31 and 7.40; at 10 °C 9.02 and 7.58; respectively
		Vitamin C	Ascorbic acid concentration slightly decreased with storage period
		Carotenoid	HPP or thermal treatment of both orange juices types did not significantly reduce the β-carotene content compared with the untreated juices. During storage at 4 and 10 °C, the β-carotene content did not significantly decrease
García et al. (2001)	500 and 800 MPa, 5 min	Vitamin C	Vitamin C was not, or only insignificantly, reduced compared to unprocessed juice. During storage (4 °C, 21 days) ascorbic acid degradation in 500 MPa treated samples were smaller than in those treated at 800 MPa, and in controls
		Antioxidant activity	Antioxidant capacity was slightly decreased but retention was over 84 % of after 21 days chilled storage
		Sugar	No significant changes were caused by pressurization on the contents of fructose, glucose, and sucrose in orange juice
		Carotenoid	α-carotene and β-carotene concentration of orange juice was not effect importantly after the HPP treatments and storage
		Sensory	Sensory quality parameters of untreated juice products changed with storage time, while overall quality of pressurized juices was less altered
Nienaber and Shellhammer (2001b)	500–800MPa, 1 and 5 min	Vitamin C	More than 80 % ascorbic acid was retained after 3 months storage at 4 °C or after 2 months at 15 °C which processed at 800 MPa 1 min
		Color	Color parameters did not change during storage except in the samples stored at 37 °C (L and b values decrease, a and ΔE values increase)
Parish (1998)	500 and 700 MPa, 60 and 90 s	Sensory	Pressure treated juices to be significantly closer to that of fresh/frozen juice than flavors of the thermally treated (75 and 98 °C, 10 s) juices. Storage at 8 °C was less favorable to sensory quality than 4 °C
Plaza et al. (2006; 2011)	400 MPa, 1 min	Vitamin C	Ascorbic acid concentration of sample decreased 4.74 % after treatment. HP juices maintained better the ascorbic acid content during more days of refrigerated storage than thermal pasteurized (TP) (70 °C, 30 s) juices.
		Antioxidant activity	After treatment, process did not modify significantly the antioxidant activity. HP juices showed no significant differences in terms of EC_{50} parameter with regard untreated juices. HP juices seem to maintain it better during more days of storage than TP juices
		Color	ΔE parameter not changed significantly after treatment and no differences determined between HP and TP at the end of storage (4 °C, 40 days)
		Carotenoid	HP juice showed a significant increase on total carotenoid (45.19 %). During refrigerated storage at 4 °C, HP juice showed higher carotenoid content than TP juice
		Flavanone	A significant increase on the extractability of each individual flavanone (naringenin hesperidin) concentrations after HP treatment (15.46 %). During first 20 days of storage at 4 °C, total flavanone content remained higher in HP juice with regard TP and untreated juices

No significant changes were caused by pressurization on the contents of fructose, glucose, and sucrose in orange juice. α-carotene and β-carotene concentration of orange juice was not effect importantly after the HPP treatments. Sensory quality of juices processed at 500 MPa was closer to untreated juices, and defects in odor and aroma were greater in juices pressurized at 800 MPa (García et al. 2001).

Ascorbic acid loss (processed at 800 MPa, 1 min) was less than 20 % after storage for 3 months at 4 °C or 2 months at 15 °C. Color values were stable during storage at 4, 15, and 26 °C (Nienaber and Shellhammer 2001b). The flavor of pressure treated juice stored at 4 °C up to 16 weeks or 8 °C up to 4 weeks was significantly closer to that of fresh/frozen juice than any of the thermal treatments (75 and 98 °C, 10 s) (Parish 1998).

Ascorbic acid concentration of sample decreased 4.74 % after HP treatment. In this study, regarding the total vitamin C content, within each type of treated (thermal pasteurization and HP) orange juice, after 40 days of storage at 4 °C, there was a significant quantitative decrease (23.84 and 18.55 %, respectively). Just after treatment, processes did not modify significantly the antioxidant activity of orange juices regarding the untreated ones. After 40 days of storage at 4 °C, the differences among treated juices were no significant in terms of ΔE parameter (Plaza et al. 2006). HP treated juice showed a significant increase on total carotenoid and flavanone content extracted (45.19 and 15.46 %, respectively) and on vitamin A value (30.89 %) with regard untreated juice, whereas no significant changes were observed for thermal pasteurized (TP) (70 °C, 30 s) juices. In general, during refrigerated storage, carotenoids and flavanones remained higher in HP juice than in TP. The same pattern was observed for all individual carotenoids (lutein, zeaxantin, α-cryptoxanthin, β-cryptoxanthin, α-carotene, and β-carotene). Just after treatment, HP juice presented a significant increase on the extractability of each individual flavanone with regard untreated juice and hence on total flavanone content (15.46 %). During first 20 days of refrigerated storage at 4 °C, total flavanone content remained higher in HP juice (Plaza et al. 2011).

13.4 Ultrasonication Process

In the last decade power ultrasound has emerged as an alternative processing option to conventional thermal approaches for pasteurization and sterilization of food products. While sonication alone may not be adequate for inactivation of various spoilage and harmful enzymes present in food, ultrasound in combination with mild heat treatment (thermosonication), pressure (manosonication) or both heat and pressure (manothermosonication) has shown potential for both enzyme and pathogen inactivation (O'Donnella et al. 2010). Ultrasound consists of vibrations like sound waves, but at frequencies above the range of human hearing (18 kHz–500 MHz). In liquids and wet biological material these vibrations produce cycles of compression and expansion and the phenomenon of cavitation. When cavitation bubbles implode they generate spots of extremely high pressure and temperature which can disrupt cell structure and inactivate microorganisms (Ohlsson and Bengtsson 2002).

13.4.1 PME Inactivation with Ultrasonication Process

Tiwari et al. (2009d) studied the effect of ultrasound treatments on PME inactivation kinetics in orange juice and reported that PME inactivation was increased by 5–62 % when the ultrasound intensities was increased at same times. The reduction of PME activity in sonicated lemon juice resulted in enhanced cloud stability during storage for 18 days at 4 °C compared to thermally processed lemon juice (Knorr et al. 2004). The improved cloud stability observed during storage could be due to the mechanical damage of the PME protein structure during sonication. Thermosonication reduced the activity of lemon PME by 83 % (50 °C, 63 min) while ultrasound alone (25 °C) reduced the activity of the enzyme by 30 % (Kuldiloke et al. 2007). PME present in oranges was treated by manothermosonication (MTS) at pressures of 200 kPa in citrate buffer and also in orange juice. At 35.5 °C the inactivation of the enzyme using MTS was estimated to be approximately seven times greater than with thermal treatment alone (Vercet et al. 2001).

13.4.2 Effects of Ultrasonication Process on Quality of Orange Juice

Sonication of juice did not cause statistically significant differences on panel-tested color, aroma, and flavor even at extreme treatment conditions. During storage at 4 and 10 °C, the differences in color values between control and sonicated juice maintained, storage temperature had not effect (Table 10). The sensory quality of the juice was slightly deteriorated after treatment, but during storage, it degraded faster for controls than for treated samples. Controls were rejected by the sensory panel after 6 days storage at 4 °C due to off-flavor, and ultrasonicated juice after 10 days due to off-odor. Ascorbic acid concentration decreased from 43.43 mg/100 mL to 40.71, 34.89 and 23.67 mg/100 mL for 6, 8, and 10 min treatments (at 20 kHz), respectively. Ascorbic acid degradated dramatically with storage period (4 and 10 °C, 10 days) (Gómez-López et al. 2010).

Sonicated orange juice samples show that amplitude and treatment time were the most significant factors influencing L*, a*, and b* values. Ascorbic acid was found to be influenced by processing temperature, amplitude, and treatment time. However, the magnitude of the reduction observed in ascorbic acid content was small with the mean loss found to be <1.5 % across the treatments (Tiwari et al. 2009a). Tiwari et al. (2009b) reported that during storage of sonicated and control orange juice at 10 °C significant changes in Hunter color values (L*, a*, b*), ΔE*, and ascorbic acid were observed. In general, L*, a*, b*, ΔE*, and ascorbic acid content were influenced by the factors investigated, i.e., storage period, amplitude level, and treatment time and their effects were either individual or cumulative. Greater ascorbic acid retention (74.5 %) was observed in sonicated samples (amplitude level of 100 % for 10 min) after storage period of 30 days compared to control (49 %). Ascorbic acid content was found to be influenced by storage time, amplitude level, and processing time. The degradation kinetics of vitamin C in sonicated samples followed first-order kinetics during processing. Predicted shelf life for sonicated

Table 10 Effect of ultrasonication process on the orange juice quality parameters

References	Treatment conditions	Quality parameters	Results
Gómez-López et al. (2010)	20 kHz, 2–10 min	Color	UP treated juices became slightly lighter, greener, and more yellow. During storage at 4 and 10 °C, the differences in color values between control and treated juice maintained, became lighter (higher L values), greener (lower a values) while the color parameter "b" remained constant
		Sensory	The sensory quality of the juice was slightly deteriorated after treatment, but during storage, it degraded faster for controls than for treated samples. Controls were rejected by the sensory panel after 6 days storage at 4 °C due to off-flavor, and ultrasonicated juice after 10 days due to off-odor
		Vitamin C	Ascorbic acid concentration decreased from 43.43 mg/100 mL to 40.71, 34.89, and 23.67 mg/100 mL for 6, 8, and 10 min treatments (at 20 kHz), respectively. Ascorbic acid degradated dramatically with storage period (4 and 10 °C, 10 days)
Tiwari et al. (2009a)	20 kHz, 2–10 min, 10–30 °C	Color	L*, a* and b* color parameters of fresh juice were 58.9, 11.9, and 56.3, while UP treated juice parameters were changed between 58.1 and 61.0, 8.4–11.3, and 56.2–60.5; respectively. ΔE* values of treated juices were changed between 1.9 and 5.3
		Vitamin C	Ascorbic acid content was small with the mean loss found to be <1.5 % across the treatments
Tiwari et al. (2009b)	20 kHz, 2–10 min amplitude 40–100 %	Color	Treatment time was significant for a*, b* and insignificant for L* and ΔE*. Amplitude level was found to be insignificant for color values during storage periods of up to 30 days at 10 °C
		Vitamin C	Greater ascorbic acid retention (74.5 %) was observed in sonicated samples (amplitude level of 100 % for 10 min) after storage period of 30 days compared to control (49 %). Ascorbic acid content was found to be influenced by storage time, amplitude level, and processing time
Tiwari et al. (2009c)	20 kHz, 2–10 min, acoustic energy 0.30–0.81 W/mL	Vitamin C	A significant decrease in orange juice ascorbic acid content was observed as a function of acoustic energy and treatment time. However at the highest energy value (0.81 W/mL) and treatment time (10 min), the largest reduction found in ascorbic acid less than 5 %. Sonicated orange juice ranged from 27 to 33 days compared to 19 days for thermally pasteurized (98 °C, 21 s) juice during storage at 10 °C
Valdramidis et al. (2010)	20 kHz, 2–10 min, 5–30 °C, amplitude 24.4–61.0 μm	Vitamin C	The largest reduction was observed at the highest amplitude (61.0 μm) and processing temperature (30 °C). However this reduction was less than 15 % loss of the initial ascorbic acid content of the unprocessed juice. Increase of temperature and amplitude resulted in higher ascorbic acid loss
Valero et al. (2007)	23 and 500 kHz, 120–600 W, 15 min	Limonin color	No ultrasound-related detrimental effects on the quality attributes of juice (limonin content, brown pigments, and color) were found

orange juice ranged from 27 to 33 days compared to 19 days for thermally pasteurized juice during storage at 10 °C. This work demonstrates that sonication results in improved retention of ascorbic acid in orange juice compared to thermal processing (Tiwari et al. 2009c).

A significant reduction in orange juice ascorbic acid content was observed as a function of treatment time. The degradation kinetics of ascorbic acid followed first order kinetics. The largest ascorbic acid reduction was observed at the highest amplitude (61.0 μm) and processing temperature (30 °C). However this reduction was less than 15 % loss of the initial ascorbic acid content of the unprocessed juice. Increase of temperature and amplitude resulted in higher ascorbic acid loss (Valdramidis et al. 2010). Valero et al. (2007) observed that no ultrasound-related detrimental effects on the quality attributes of juice (limonin content, brown pigments, and color) were found.

Ultrasound treatment has almost no effect on the evolution of color scoring in the analyzed orange juices.

13.5 Ultraviolet Light (UV) Processing

Antimicrobial treatment with UV can be regarded as a special case of pulsed light treatment (which contains 25 % UV), but without pulsing, with a minimum of visible light and at a much lower power intensity. The UV spectrum is commonly divided into three intervals: UV-A (λ 320–400 nm); UV-B (λ 280–320 nm); UV-C (λ 200–280 nm).

The main germicidal effect lies in the UV-C interval. At UV-C irradiation at intensities in the order of 1,000 J/m^2 or more, bacteria, yeast, and viruses suffer as much as a 4 log reduction. The mechanism causing cell death is related to the absorption of UV by DNA/RNA. The necessary treatment time is much longer than for the pulsed light treatment (Ohlsson and Bengtsson 2002).

The effect of the UV-C light on quality characteristics (microbiological, physicochemical and sensorial) and microbial stability of orange juice during storage (4 and 10 °C) were investigated by (Pala and Toklucu 2013). UV-C treatment (36.09 kJ/L) of orange juice resulted in 2.8 log and 0.34 log reductions in aerobic plate count and yeast and mold count, respectively. Ascorbic acid content did not change significantly after the UV-C treatment (36.09 kJ/L). Differences between untreated and UV-C treated (48.12 kJ/L) were small in terms of organic acids, antioxidant capacity and phenolics. Based on sensory analysis results, no significant differences were detected between fresh and UV-C-treated juices. UV-C treatment partially extended the shelf life of fresh juice during storage at the refrigerated conditions.

Torkamani and Niakousari (2011) reported that the shelf life was extended for 7 days for orange juice after being exposed to UV-C light (254 nm). Vitamin C content was measured 18 % reduction in juice. UV-C treatment, only 8 % was inactivated to pectin methyl esterase (PME) in orange juice. Less energy was consumed by UV treatment in comparison to thermal methods. No significant alteration was observed in juice pH and color.

Ultraviolet (UV) with a wavelength of 254 nm tends to inactivate most types of microorganisms. The shelf life of fresh squeezed orange juice was extended to 5 days with a limited exposure of UV (73.8 mJ/cm^2). The degradation of Vitamin C was 17 % under high UV exposure of 100 mJ/cm^2, which was similar to that usually found in thermal sterilization. UV processing does not inactivate PME. The energy required for UV treatment of orange juice was much smaller than that required in thermal treatment. The color and pH of the juice were not significantly influenced by the treatment (Tran and Farid 2004).

14 Concentration of Citrus Juice

14.1 Evaporation

Evaporation is employed in the food industry to reduce weight and volume of fluids, with a subsequent reduction of packaging, transportation, and storage costs. It generates products with a reduced water activity that will enhance the storage stability. Concentrated juices are more resistant than their single-strength form to microbial spoilage. It is used extensively the production of fruit (e.g., mainly grape, apple, and orange) and vegetable (e.g., mainly tomato products) juice concentrates. Essentially, it consists of a noncontact or indirect heat exchanger, enclosed in a large chamber that transfers heat from the steam to the liquid food. The product inside the evaporator is kept under vacuum; therefore, the product boils at a lower temperature. Vapors produced are conveyed through a vapor–liquid separator. Then they are condensed. If the vapors produced are discarded without further utilizing their inherent heat, they are called “single-effect” evaporators. If the vapors are reused as the heating medium in another evaporator chamber, the evaporator is called a “multiple-effect” evaporator. The vacuum is produced either by a mechanical pump or a steam ejector (Ramaswamy and Marcotte 2006). One of the problems in the evaporative concentration of fruit juices is the loss of volatile aroma components. Evaporators may be equipped with essence recovery systems for entrapping and concentrating the aroma, which is then added back to the concentrate (Mannheim and Passy 1974).

The TASTE evaporator (thermally accelerated short-time evaporation) provides, during preheating cycle and after first evaporation stage, the pasteurization and stabilization of juice. In the tube nest the juice is introduced as a turbolent fog and the heat transfer rate is several times higher than would be expected under any other conditions. For this reason the acceleration permits to reach a speed of approximately 700 km per hour into the tubes. The TASTE evaporator eliminates recycling of the juice during concentration, thereby minimizing the length of time at which the heat treatment is applied to the juice, resulting in a better quality. Simplified controls and special transfer pumps provide adequate flexibility for the entire evaporating cycle. TASTE evaporators have the highest existing evaporation rates per kg of steam. TASTE evaporators are the latest and the most desired evaporator for citrus concentrating industries. The TASTE evaporator permits a concentration up to

65/75 brix in a total cycle time of 2.50 min approximately. The evaporator can also be equipped with an aroma recovery system where the aromas are extracted from the juice and concentrated 150 times before pasteurization, thereby avoiding any alteration to the freshness of the aromas due to heat. Before leaving the evaporator, the concentrated juice passes through a fresh cooler which drops the temperature to 10 °C. And also, cleaning in place system is incorporated into the evaporator in order to provide high sanitation and cleaning in 30/45 min. The advantages of the TASTE evaporators are best quality product, very low operating cost, aroma recovery mounted on the machine, low capital investment, mechanical trouble free machine, minimum floor space requirement and completely automatic operation and control (Anon 2013b).

TASTE has been designed to use high temperature at first stage, which would simultaneously evaporate, pasteurize and enzyme-stabilize the juice. TASTEs have relatively low initial cost and ease of cleaning. The principal disadvantage is their relative inflexibility. It is not convenient to blend add-back concentrate with the feed juice, as might be done with other types of evaporators. The high temperatures cause more rapid fouling of heat exchanger tubes and necessitate more frequent cleaning (Chen et al. 1979). Changes in main constituents of orange juice were evaluated by Pino et al. (1987) during concentration on a TASTE. Mean retention of ascorbic acid was 93.5 % and concentration caused slight browning of juice, but sugar content was not affected. Viscosity reduction by homogenization of orange juice concentrate in a pilot TASTE was studied by Crandall et al. (1988). Controlling viscosity is critical for the efficient evaporation and pumping of citrus concentrates. Viscosity of an orange juice concentrate was reduced by installation of a commercial homogenizer between the third and fourth stages of a pilot plant TASTE.

Evaporation at low temperatures (vacuum operation) can prevent heat damage to food quality, but it has the disadvantage of lower heat transfer coefficients and longer residence time in the evaporator. High temperature-short time systems, such as the TASTE, are used for the concentration of citrus juices (Chen and Hernandez 1997).

14.2 Freeze Concentration

Freeze concentration is a technology that can be used in the food processing industry to concentrate sensitive fruit juices with high quality (Rahman et al. 2006). In conventional processes, such as evaporation, higher levels of concentration can be reached compared with freeze concentration or membrane techniques. However, the advantage of the freeze concentration technique is based on the quality of the product obtained due to the low temperatures used in the process which makes it a very suitable technology for processing of fruit juices. This process allows removal of water from a solution by freezing it until high-purity ice crystals are formed and separated to leave a concentrated juice. The nutritional and sensory quality of freeze-concentrated fruit juices is higher than those concentrated conventionally by means of evaporation due to the low processing temperatures that avoid undesirable

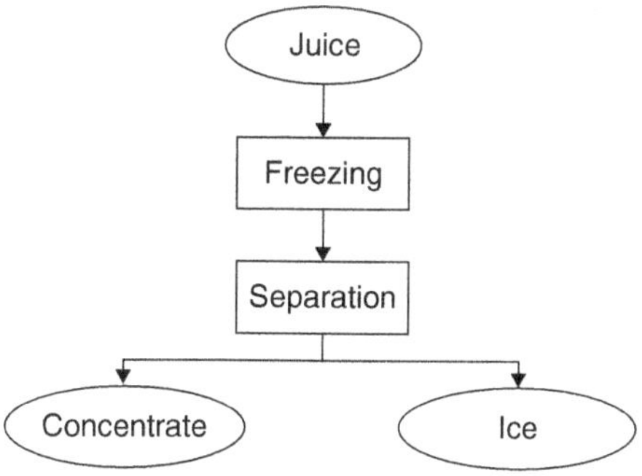

Fig. 6 Flow diagram of freeze concentration (Berk 2009)

chemical and biochemical changes and minimize the loss of organoleptic properties (Raventós et al. 2012). The freezing system consists of two stages: freezing (crystallization) and separation (Thijssen 1975). There are two basic methods for concentration with freezing: suspension and film freeze concentration.

The basic mechanism of freeze concentration is simple. When a solution or a liquid food is frozen, water is separated from the rest of the solution as crystals of pure water ice (Fig. 6). Subsequently, the mixture can be separated into ice and a concentrated solution. With respect to size and shape, large and near-spherical crystals are best. If the concentration ratio is substantial, the quantity of ice to be crystallized and removed and with it, the proportional loss of solids would be very large. For example, in order to concentrate citrus fruit juices at the concentration ratios practiced by evaporative concentration (e.g., six-to-one), 5 kg of ice have to be formed and removed for every 6 kg of juice. The quantity of the viscous concentrate adhering to the crystals would be considerable. For this reason, freeze concentration is preferably applied to processes requiring moderate concentration ratios. Another solution is multistage freeze concentration, where a small proportion of the ice is removed at each stage. In practice, the crystallization unit is usually a swept-surface freezer and the separation unit is a centrifuge (Van Pelt 1975). In one variation of the process, a "recrystallizer" is inserted after the freezer. The very small crystals disappear while the larger crystals grow further.

Solid–liquid centrifuges, such as the basket centrifuges are the standard equipment for the separation of the ice crystals. Obviously, the crystals should be washed in order to recover the concentrate entrapped in the ice cake. If the centrifuge is not insulated, the water released by the melting ice participates in the washing. Another device for carrying out the separation step is the wash column (Van Pelt 1975), consisting of a vertical cylinder equipped with a piston at the bottom. The column is filled with "magma" from the recrystallizer. The concentrate is recovered by a combined action of pressing and washing.

Freeze concentration is currently being used in the fruit and vegetable juice industry mainly for orange and apple juice, as well as juice mixtures (Deshpande et al. 1984). Effect of freeze concentration conditions on orange (Thijssen 1986; Braddock and Marcy 1987) and lime (Nonthanum and Tansakul 2008) juices quality parameters were already studied.

14.3 Membrane Concentration

Membrane technologies follow evaporation as the second most commonly used method for concentrating or separating constituents. Main advantages of membrane concentration over evaporation are that (1) the food is not heated, (2) there is a negligible loss of quality, (3) there is less loss of volatiles, (4) it uses energy more efficiently because there is no phase change. Other advantages are simpler installation, lower labor and operating costs, and that no steam boilers are required (Ramaswamy and Marcotte 2006). The membranes can be applied in processing of fruit juices: for example, concentration, using reverse osmosis (RO) to produce fruit juice concentrates of greater than 42° Brix. Fruit juice concentration by reverse osmosis is potentially one of the largest applications of membrane technology. The major benefit of RO is that it avoids thermal damage of delicate aroma components, although these effects are more noticeable with some products (Cheryan 2007).

Producer of membrane reported that fruit juice concentrates of 45 to 55° Brix were obtained commercially and up to 70° Brix was obtained in pilot trials (Walker 1990). Cellulosic RO membranes can concentrate centrifuged juice up to 25° Brix at 5–7 °C and 30–70 bar. Although a high retention of sugars, acids, phenolics, nitrogen compounds, and ash was obtained, the retention of volatile aroma compounds was lower (Peri 1974). Reverse osmosis with a polyamide composite membrane can concentrate juice without a significant loss of aroma, sugar, or acids (Medina and Garcia 1988). No losses of acids, vitamin C, limonene, or pectin were observed. Glucose and fructose amounts in permeate were 0.02–0.4 % and there was no loss of sucrose. There were more aroma losses by vaporization than by permeation through the membrane (Braddock et al. 1988). However, conventional RO is limited by osmotic pressure and viscosity considerations to less than 30° Brix. Therefore RO can be used as a preconcentration step, with thermal evaporation completing the required concentration to 42° Brix. Adding RO ahead of the evaporators can increase evaporator capacity and reducing thermal treatment.

15 Filling

Citrus fruit juice, concentrated or single strength, can be stored and transported in bag-in-bins. The filling is sterile. The juice, after de-aeration, pasteurization, and cooling, is sent to the collecting tank under nitrogen and then to the aseptic filling machine. Sterilized bags, made of food grade polyethylene and metallic film, are used as a barrier for oxygen (Crupi and Rispoli 2002).

16 Storage

The conditions for the low temperature storage depend on kind of juice, concentration degree and use purpose of use. The most common options are the storage of concentrate juices in drums, tanks, and bag-in-bins.

The drums are metallic containers with a high density polyethylene double bag. Juice is previously cooled at 5 °C and stored under −18 °C for at least a year. The storage in big tanks is the most common system. The tanks are built in stainless steel; presence of molybdenum, nickel, and chromium in alloy composition, makes the metal resistant to the attack of reducing acids as SO_2, while titanium increases the welding stability. The resistance to corrosion also depends on the conditions of the superficial metallic layer. Besides the surface passivation with nitric acid, the electro polishing permits the elimination of the small pores in which microorganisms can hide (Crupi and Rispoli 2002).

The tanks are refrigerated with refrigerating devices or placed in a refrigerating cell. The maintenance temperature is chosen according to the concentration degree of the juice and viscosity, in order to avoid that the content freezes inside the container or that it becomes uneasily extractible by means of a pump. For the orange juice at 60/65°Brix temperatures of about −10 °C are used. Nitrogen atmosphere is created in the tank to prevent oxidative variations. The very low storage temperature maintains the formation of off-flavor and color alteration. In concentrate orange juice, crystalline agglomerates of monopotassium citrate with small amounts of sugars, hesperidin. and pulp can be generated; their formation is favored by acidity and high concentration (Kimball 1991a, b; Filomena et al. 1998; Crupi and Rispoli 2002).

Color, cloud value, nonenzymatic browning and ascorbic acid content are important orange juice quality parameters (Meydav et al. 1977; Lee and Coates 1999; Zerdin et al. 2003) that are strongly influenced by various processing and storage conditions (Tiwari et al. 2009a, b, c, d). Tiwari et al. (2009a, b, c, d) reported that freshly squeezed orange juice was sonicated at various amplitude levels (40, 70, and 100 %), treatment times (2, 6 and 10 min) at a constant frequency of 20 kHz. Samples were stored for periods of up to 30 days at 10 °C. Storage of sonicated orange juice for periods of up to 30 days did not result in significant differences in °Brix and titratable acidity. However significant effects were observed in juice pH, color, nonenzymatic browning, cloud value, and ascorbic acid during storage. They reported that this work demonstrates that orange juice sonication results in improved ascorbic acid and cloud retention during storage.

Kanner et al. (1982) reported that orange juice concentrates were packaged aseptically by a "Dole" aseptic canning machine using 6 oz metal cans. The final juice products (11°, 34°, 44°, 58° Brix) were stored between −18° and 36 °C and tested periodically for nonenzymatic browning, ascorbic acid destruction, furfural and sensory changes. Nonenzymatic browning, the main deterioration phenomena in these products, was satisfactorily retarded at 12 °C or lower. Ascorbic acid destruction rate constant was dependent on temperatures between 5 and 25 °C and was affected by degree of juice concentration. Furfural accumulation in juice was higher than that in 58° Brix concentrate. Orange juice concentrate of 58° Brix did not show flavor changes after storage at 5 or 12 °C for 17 or 10 months, respectively, when evaluated after reconstitution to 11° Brix.

Sensory properties of orange juices are highly related to their levels of d-limonene. Decreases in sensory quality (overall scores for color, appearance, aroma, and flavor) found during storage in glass bottles, are greater at higher storage temperature and with exposure to light. Significant deterioration in sensory quality occurred after 3 months at ambient temperature, and after 1–2 months at 30 °C. Changes in bitterness are similar to those of oxidized flavor, but less pronounced, while no significant differences were found for sourness. Sorption of d-limonene by plastic packaging is affected by the external factors such as temperature, relative humidity, and other storage conditions. A gradual decrease in several flavor components, 1-penten-3-one, hexanal, ethyl butyrate, octanal, neral, and geranial, is observed during storage. Contents of some volatile components (α-terpineol and ethyl acetate) increase during storage (Moshonas and Shaw 1989; Sandhu and Minhas 2006).

Microbial growth and ascorbic acid concentration are sensitive to variations in d-limonene concentration, within the range of values typically observed in commercial orange juice. Consequently, packaging materials that absorb d-limonene, potentially influence microbial stability and ascorbic acid content in single-strength orange juices. However, scalping of flavor volatiles by LDPE, PET, polyamide, and ethylene (co)vinyl alcohol does not result in significant differences in flavor of high oil, typical oil, low oil, and thermally abused orange juice samples made from high-quality orange juice concentrate (Sadler et al. 1997; Sandhu and Minhas 2006).

Grapefruit juice is probably the most stable of the citrus group, though the changes in flavor are probably masked to some extend by high acidity and bitterness due to naringin that has not been found in significant amount in other citrus juices. Under normal storage temperature and conditions, properly processed juice should retain its normal flavor and appearance for about 9 months; it should still possess a good flavor for about 15 months, after which definite off-flavor and off-colors develop. The most significant changes in stored canned juice are a decrease in limonene and an increase in linalool monoxide, α-terpineol, and furfural. Canned or bottled grapefruit juice will retain its normal flavor almost indefinitely when stored at 0 °C (Pino 1986; Sandhu and Minhas 2006).

Effects of adding SO2, Sn^{2+}(tin) or cysteine to concentrated lemon juice on its color during storage have been investigated (Nunez et al. 1989). The results show that at 45 °C, browning was inhibited by greater than 125 ppm SO2, and the degradation of ascorbic acid, formation of furfural and hydroxymethylfurfural by greater than 250 ppm SO2. Browning rate was reduced by Sn^{2+}(tin) depending on the concentration used: at 1,000 mg Sn^{2+}(tin)/kg juice, browning was reduced to about one-third of the initial rate. Cysteine inhibited color formation only slightly at high concentration and affected the aroma of the juice at concentration greater than 500 mg/kg. Use of Sn^{2+} (tin) was promising because of its low toxicity and high legal tolerance levels, especially as its concentration would be reduced when the juice was diluted for use (Sandhu and Minhas 2006).

Browning is the most common quality problem of many concentrated fruit juices (Toribio and Lozano 1984) and causes loss of nutrients and the formation of intermediate undesirable compounds, like furfural and 5-hydroxymethylfurfural (Buedo et al. 2001).

Nonenzymatic browning, accompanied by undesirable off-taste and off-flavor, is considered as one of the major causes of quality loss during processing and storage of orange juice. It is accelerated by temperature and time of processing and storage (Dinsmore and Nagy 1972; Nagy and Randal 1973; Lee and Nagy 1988a,b; Roig et al. 1999; Shinoda et al. 2004, 2005). Furfural and 5-hydroxymethylfurfural

(HMF) are indicated as the principal degradation products from ascorbic acid and sugar breakdown, the main sources of this browning. Because of their correlation with browning reactions, furfural and HMF are recognized as useful indicators for temperature abuse during processing and storage and for quality deterioration in general (Dinsmore and Nagy 1972; Nagy and Randal 1973; Lee and Nagy 1988b; Shinoda et al. 2004, 2005; Vervoort et al. 2011).

In citrus juices, nonenzymatic browning is due to the reactions of sugars, amino acids and ascorbic acid (Johnson et al. 1995). However, the decomposition of ascorbic acid is reported to be the major deteriorative reaction occurring during the storage of orange juice (Solomon et al. 1995). Lee and Nagy (1988a) also reported a high correlation between the percentage loss of ascorbic acid and an increase in browning in grapefruit juices. On the other hand, sugar–amino acid reactions of the classical Maillard type are of minor importance in citrus juice browning because of the high acidity (pH 2.0–4.0) involved (Clegg 1964).

However, the presence of amino acids in ascorbate systems is also considered a major contributor to the development of browning (Robertson and Samaniego 1986). This is illustrated by the fact that the main degradation product of juices with pH values below 4.0 is furfural (Huelin et al. 1971). Furfural is known to undergo polymerization and, as an active aldehyde, may combine with amino acids and contribute to the browning of the juice (Solomon et al. 1995). Likewise, HMF concentration has a high correlation with the level of browning in lemon juice and therefore plays an important role in the formation of brown pigments (Robertson and Samaniego 1986; Koca et al. 2003).

17 Cloud Stabilization

Cloud loss is a major quality defect in orange juice. Citrus juices are generally preferred in a more natural opaque form. The cloud in juice originates from organelle and other cytoplasmic material found in the juice cell, and is mixed during extraction to provide a juice with an opaque appearance. Orange juice cloud is responsible for much of the color, flavor, turbidity, and aroma of orange juice (Baker and Cameron 1999). If the cloud is removed from orange juice, the result is a mostly clear serum that is unacceptable to consumers in taste and appearance.

Even though pectin constitutes a minor portion (5 %) of the cloud, it has a dominating effect on the behavior of the cloud (Bennett 1987). Baker and Bruemmer (1969) suggests that serum pectin does not stabilize cloud, but may in fact play a role in cloud instability. Baker and Bruemmer (1972) found that the floc contained pectates and hesperidin. The floc clarified cloud that had been suspended in serum, but soluble pectin added to the solution inhibited the clarification. In the presence of active PME enzyme, pectin forms calcium pectate complex, which causes precipitation of cloud particles (Croak and Corredig 2006). If natural pectinesterase enzymes are not deactivated by heat treatment, they will proceed to break down the methoxy ester groups to produce acidic galacturonic acid groups. This increase in the acidity of the pectin results in aggregation of cloud material that then will either fall out leaving a clarified juice or will produce interlinkage with divalent cations to form gelation in citrus concentrates. Either result is undesirable for citrus juices.

Protein is the most abundant material in citrus juice cloud; it contributes approximately 52.4 % of the cloud of commercial orange juice (Klavons et al. 1991). In the protein constituent, 53 % is insoluble protein, 30 % is complexed with low molecular weight cloud components, and 17 % is covalently bonded with components such as hemicellulose. Commercial orange juice cloud contains approximately 4.5 % pectin, and in the pectin constituent, 60 % is associated with insoluble protein, 25 % is calcium pectate, and 15 % is protopectin (Klavons et al. 1994).

Thermolabile isozymes is clarified very slowly, however, thermostable isozyme is rapidly clarified the juice (Versteeg et al. 1980). The cations can accelerate clarification in orange juice. Higher levels of cations caused clarification more quickly than the lower levels. If PME was left out of the system, the cations did not clarify the juice. If both PME and cations are absent from pasteurized juice, the juice remained stable at 4 ° C (Wicker et al. 2002, 2003). Croak and Corredig (2006) concluded that cloud particles are destabilized even in a calcium depleted environment with the addition of PME (15 and 24 units/mL). Stevens et al. (1950) observed little change in the rate of flocculation over the pH range of 2.8–3.8. However, the rate of flocculation drastically decreases at pH levels of lemon juice (pH 2.2–2.4). Rouse and Atkins (1952) found that as the pH of orange juice is decreased, the temperature required for complete inactivation of PME also decreases. Croak and Corredig (2006) showed that juice is more stable against the addition of PME at pH 6 than at pH 3.8.

The most stable clouds of citrus juice have particle sizes of 2 μm and smaller (Mizrahi and Berk 1970). Corredig et al. (2001) concluded that particle size is dependent on the amount of pectin present in the orange juice.

Cloud stability decreases with increasing amounts of pulp. Juices and concentrates containing 3 % pulp have been found to be more stable than ones with higher pulp content of about 9 %. Combination of homogenization and pasteurization gives better stability than either of these treatments alone. The combination of pasteurization and addition of stabilizers such as pectin and gum acacia give good stability to both juices and concentrates (Ahmad and Bhatti 1971).

Orange juice cloud is commercially stabilized by pasteurization at 90–95 ° C for 15–60 s (Chen et al. 1993). The high temperatures are not necessary for microbial destruction but because there is more than one form of PME in cloud that is inactivated only by these high temperatures (Baker and Cameron 1999). The problem with the high temperatures of pasteurization is browning and flavor deterioration in the orange juice (Chen et al. 1993). A thermal treatment in the range of 50–60 °C has been found to be satisfactory during minimal heat processing for the retention of cloud stability.

Mizrahi and Berk (1970) used a mild heat treatment on fresh juice and saw an increase in cloud stability. The authors concluded that the increase in cloud stability could not be due to the inactivation of PME. They concluded that the stabilizing effects of the heat treatment cannot solely be explained by the increase in serum viscosity.

Methods to determine the cloud stability of citrus juices are based on measuring juice pectinesterase activity, by adding juice samples to concentrated pectin solutions and then measuring the change in pH. If acid groups are formed too quickly, the pH will drop quickly, or a large amount of NaOH will be needed to maintain a pH, meaning pectinase enzymes are active and insufficient heating has occurred (Kimball 1991a, b). After heat treatment, cloud instability may occur even though pectinase activity is low. Citrus concentrates can be inspected for gelation, and single-strength juices can be inspected for cloud separation. Most modern evaporators and pasteurizers heat juice sufficiently to deactivate pectinase enzymes found in citrus juices (Kimball et al. 2004).

Ellerbee (2009) studied on clarification mechanisms of citrus juice cloud were evaluated in fresh juice and orange juice serum model systems. In a model orange juice serum, the addition of hesperidin and protein (heated and unheated citrus protein and soy protein) showed that hesperidin interacted with protein. Higher concentrations of hesperidin (1.0–0.1 mg/mL) increased in turbidity in the presence of protein.

18 Orange Juice: Types and Their Characteristics

Juice is the cell sap that is present in the cell vacuoles and expressed from sound fruits by squeezing. Orange juice is consumed in a natural cloudy state. Different types of orange juices are available in the market. The chilled single-strength orange juice has limited shelf life and requires installation of expensive refrigerated tanks. The conventional pasteurized single-strength orange juice in cans is widely used, but the frozen concentrated orange juice (FCOJ) is now a commodity, which is traded worldwide. Concentrated juices are distributed in large containers as a base for the manufacture of a variety of soft drinks. The same is reconstituted to single-strength juice for direct consumption. Comminuted orange products are prepared for use in beverages. Dehydrated juices in powder form are also available in the market (Sandhu and Minhas 2006).

18.1 Fresh Orange Juice

The manufacture of fresh squeezed, unpasteurized orange juice has become more common and fills this consumer preference. Increased consumer awareness has also contributed to the increasing demand for orange juice which is as natural as possible and without extensive processing (Lee and Coates 1999). Refrigerated, fresh squeezed orange juice is reported to have a relatively short shelf life up to 14 days (Fellers 1988) based on subjective flavor evaluation. The absence of pasteurization and lack of preservatives allow for the opportunity for the growth of bacteria and yeasts, spoilage, certain enzyme activity causing off-favors, and oxidation (Attaway et al. 1989).

18.2 Pasteurized Juice

Pasteurizing technologies have received renewed interest in the citrus industry due to continuous growth of market share for pasteurized orange juice as not from concentrate (NFC). Citrus juices are pasteurized to inactivate pectin methyl esterase for preventing cloud loss and to inactivate spoilage microorganisms for enhancing storage stability. NFC orange juice is preferred over chilled orange juice reconstituted from evaporated concentrate due to consumer perception of reduced heat degradation of juice flavors. The design for thermal pasteurization of orange juice is based on the thermal destruction characteristics of the thermostable form of pectin methyl esterase which is more thermal resistant than vegetative microorganisms (Chen and Wu 1998). Pasteurization is important to the stability of citrus juice during transportation and marketing for food safety and quality requires. Many important nutrients in citrus juices including carotenoids, ascorbic acid, phenolic compounds, sugar, acid, minerals, pectin, and other components are quite heat stable under the conditions of pasteurization. Thus, inactivation of PME by nonthermal treatments becomes a matter of interest to avoid quality degradation of orange juice by thermal processing (Elez-Martínez et al. 2007).

18.3 Aseptic Single-Strength Juice

Now the technology is available on a large scale to extract, process, and store single-strength juice in bulk aseptic refrigerated tanks, minimizing microbial spoilage, and product quality deterioration. This technology enables provision of blended juices to consumers on a year-round basis, when the fruit is not in season. Depending on the processing capacity of the plant, number of tanks of capacity 950–3,800 m^3 each, are installed in refrigerated rooms or insulated with refrigeration. With proper nitrogen blanketing and mixing, the juice quality may be maintained for a year or more (Wilke 2002; Sandhu and Minhas 2006). After juice is processed by juice extraction line, finished and chilled, the juice is fed to an aseptic processing system. The sterilized juice is sent to bulk storage tanks where it is stored. From these storage tanks, juice can be exported to filling lines with optional blending and aseptic flavor injection. Depending on the product consumers are looking to produce, citrus pulp can also be aseptically added to the juice by a pulp injection unit before finally being packed in a aseptic system.

18.4 Single-Strength Juice from Concentrate

A significant amount of orange juice is packaged from reconstituted concentrate as chilled juice. Because of the economics of storing large bulk quantities of concentrated citrus juice and the consumer preference for a ready-to-serve product, the volume of this product is large now. Pasteurized juice is packaged in cartons or glass containers and is microbiologically stable. The flavor of juice from reconstituted concentrate is not comparable with single-strength juice because of the two thermal treatments and the loss of volatiles during the concentration process. Addition of aromas and essences can improve the quality of the finished product (Ranganna et al. 1983; Sandhu and Minhas 2006).

18.5 Frozen Concentrated Juices

Concentrated orange juice with soluble solids content of 65° Brix is now largely produced in the world. The primary water removal technology is high temperature short-time evaporation, although freeze concentration and membrane processes are also used. The concentration process is accompanied by aroma recovery. The concentrate is blended with a small amount (less than 0.01 %, v/v) of cold-pressed oil to mask the off-flavors that develop during storage.

The small quantity of fresh juice can also be added back to concentrate to make up the losses of flavor during concentration process. The concentrate is chilled to −9 °C by passing through heat exchanger and pumped to large stainless steel tanks maintained at desirable temperature in cold rooms. This concentrate is blanketed with nitrogen and carefully monitored for quality characteristics, so that the juice with different characteristics may be accurately blended to produce a uniform-quality finished product. Under these conditions, the concentrate can be stored for over a year with little loss in quality (Ranganna et al. 1983; Sandhu and Minhas 2006).

19 Clear Citrus Juices

In the field of citrus processing the clear juices have a minor relevance. The bigger part is represented by lemon juices and lime juices. Clarified lemon juice is a suitable acidifying agent that can substitute citric acid and provide more "natural" products that will fit the new market for ecological products and new alternatives to the manufacturing of lemon (Saura et al. 2012). Clarified lemon juice can be consumed directly such as fruit juice and also production of nectar, industry of soft drink, production of jam, marmalade and sweet, as a lemon sauce in ready meals sector, production of candy, industry of alcoholic beverages, as acidity regulator in fruit juice industry and has recently started to be used as filling liquids in canned fruit. Saura et al. (2003, 2004, 2012) reported that use of lemon juice as a substitute for citric acid in canned fruit, e.g., peach halves, will yield a "natural" product that will fit the new market for ecological products.

The production of clear lemon juice used to be very difficult, time-consuming, and associated with a high level of risk in terms of quality. Using traditional methods, the clarification process involved allowing the juice to stand. In view of the lengthy amount of time necessary for the sedimentation process, the juice had to be protected against oxidation and also against undesirable microbiological activity. This has been achieved by making use of traditional preclarifiers combined with modern high-performance clarifying centrifuges (g force: 15,000) and also by making use of newly developed enzyme preparations (Pecoroni et al. 2013).

After the juice is discharged from the extractor, the solids are reduced by means of a finisher and preclarifier. A maximum pulp content of 2 % (by vol.) in the discharge of the preclarifier is desirable. The preclarified juice is then heated to between 25 and 40 ° C (optimum enzymation temperature, depending on the pH value of the juice). It is not easy to choose a suitable enzyme preparation and optimum reaction conditions for reducing pectin at pH values of around 2.8. The juice is treated with enzymes in the downstream reaction tank. The correct enzyme dosage depends on the desired reaction time. If the process does not exceed 3 h, then it is not necessary to add SO_2. If the process does exceed 3 h, then adequate protection against oxidation must be provided by adding SO_2 in quantities of up to 500 mg/kg. When the enzyme reaction has finished (this is checked by a quick alcohol test), then the product is refined with Kieselsol. This process causes the fine trub particles which are still in suspension to agglomerate (Pecoroni et al. 2013).

Uçan (2013) is produced clarified lemon juice by using Novozym 33095 (200 mL/ton (20 μL/100 mL)) at 35 °C. Enzymation was carried out in a 6 ton-tank, for 30 min. Clarification fining agents were added in amounts of bentonite (5 %), gelatin (1 %), and kieselsol (Levasil 200/30 %) which were determined as 35 L/ton, 2 L/ton, and 2 L/ton lemon juice, respectively.

The clear juices are usually then further processed into concentrates. Instead of the high performance centrifuges a membrane filtration plant can be used. A filtration step is required to produce clear juices. Colloidal trub material as well as particles have to be removed to prevent the juice or concentrate from subsequently becoming turbid. As more stringent requirements are placed on the type of membrane in these applications, the ceramic membrane has been tried and tested in the field (Pecoroni et al. 2013).

The use of membrane processes, such as microfiltration (MF) and ultrafiltration (UF) in the clarification of citric juices has lately gained importance over some conventional treatments including diatomaceous earth, paper filters, bentonite, etc. (Jiao et al. 2004; Cassano et al. 2007). This is due to the fact that membrane processes have the advantage that separation occurs at room temperature (without loss of aromatic volatile substances), high selectivity—implying the reduction of microbial load, no need of additives to boost separation, and very low electricity consumption. However, the major disadvantage of this process is membrane fouling during permeation caused by the retention of some components over the surface of the membrane, causing a rapid decrease of flux (Mondor et al. 2000; De Bruijn et al. 2002; Carneiro et al. 2002; Cassano et al. 2003; Espamer et al. 2006). Flux and product quality are two important aspects to consider when selecting the membrane clarification process. High flux is essential for a practical and economic filtration. The quality of the product must reach the level of at least other standard methods (Chornomaz et al. 2013).

The core element of these units are ceramic membranes with a pore size of 20–200 nm. They are extremely resistant to temperature, pressure and chemicals, easy to clean and, compared to polymer membranes, have a very long service life. Ceramic membrane makes it possible to concentrate retentate until there is no free juice left. This allows operators to extend filter cycles considerably and to minimize product loss. Combining filtration with centrifugal separation allows product losses to be minimized (Pecoroni et al. 2013).

20 Acid Reduction

Some orange juice consumers prefer a reduced acid version of the juice. Resins have been used to process orange juice for the reduction of acid content. In ion exchange process, the juice is sent through a column containing an ion exchange resin. In this way fruit acids are exchanged onto an absorbent medium. The resin has to be regenerated, when the capacity of the resin has been exhausted so that it can be reused. Stainless steel columns are used to contain the resins. First, the pulp is removed from the juice via centrifugation. The most efficient way of doing this, is to reduce the pulp level prior to the adsorption process to below 1 % by means of a centrifuge (Pecoroni et al. 2013).

The pulp free juice can then be passed through a weak base anion column to reduce the citric acid content of the juice. The resin also retains the ascorbic acid and folic acid of the juice. This can be monitored by following the pH of the juice effluent coming off the column. Below a pH of 4.6, the ascorbic and folic acid portions are washed through the resin. The processed deacidfied juice effluent can then be blended back to unprocessed juice to achieve the right balance of reduced acid and fresh juice tastes. The pulp can also be added back, or the juice can be sold as a reduced pulp juice. Juices from four cultivars of sour orange, i.e., Seville, Bigaradier, Sour, and Bittersweet were treated with neutral (XAD-16) and weak base anionic exchange (IRA-93) resins for increasing palatability of juice (Couture and Rouseff 1992). Average acidity was reduced 57–87 % using IRA-93 before depletion and sensory acceptability of the juice increased (Sandhu and Minhas 2006).

21 Dehydrated Citrus Juices

Many technologies have been proposed for the production of dehydrated citrus fruit juices; the main inconvenient which industry has always had to face is the hygroscopicity of the finished product. This characteristic involves on one hand great difficulties to bring the juice at a residual humidity lower than 3 %, and on the other the necessity of confectioning the dehydrated juice in limited dimension containers, in a dry atmosphere and in presence of dehydrating agents. To improve dehydrated juice stability, before proceeding to drying, more or less relevant quantities of corn syrup or maltodextrin can be added. The so-obtained product cannot be legally declared as "juice" and is used for particular necessities. In any case, the preparation of dehydrated citrus fruit juices is an activity which interests a limited part of the market, also because of the rather high processing cost. From the standpoint of quality, the best dehydrated juice is that obtained through freeze-drying, in which water removal occurs by sublimation (Crupi and Rispoli 2002).

22 Beverages Made from Orange Juice

Ready to serve drinks, cordial, squash, crush, and orange syrup are prepared from orange juice or concentrate, squash being the most commonly prepared product.

References

Abeysinghe DC, Li X, Sun CD, Zhang WS, Zhou CH, Chen KS (2007) Bioactive compounds and antioxidant capacities in different edible tissues of citrus fruit of four species. Food Chem 104:1338–1344

Ağçam (2011) Effects of pulsed electric fields and thermal process treatments on quality and shelf life of orange juices (in Turkish). Dissertation, Çukurova University

Ağçam E, Akyıldız A, Akdemir Evrendilek G (2014) Comparison of phenolic compounds of orange juice processed by pulsed electric fields (PEF) and conventional thermal pasteurisation. Food Chem 143:354–361

Ahmad N, Bhatti MB (1971) Studies on the stabilization of cloud in orange juices and concentrates. Agric Pak 22(1):41–47

Altan A (1995) Determination of some technological characteristics of five cultivars of oranges grown in the Çukurova region for the juice industry. Gida 20(4):215–225

Anagnostopoulou MA, Kefalas P, Kokkalou E, Assimopoulou AN, Papageorgiou VP (2005) Analysis of antioxidant compounds in sweet orange peel by HPLC-diode array detection-electrospray ionization mass spectrometry. Biomed Chrom 19(2):138–148

Anon (2013a) JBT Citrus juice extractor-pulp recovery. http://www.jbtfoodtech.com/en/Solutions/Equipment/Citrus-Juice-Extractor. Accessed Oct 2013

Anon (2013b) The thermally accelerated short time evaporator (TASTE) http://www.jbtfoodtech.com/en/Solutions/Equipment/TASTE-Evaporator. Accessed Oct 2013

Arena E, Fallico B, Maccarone E (2001) Evaluation of antioxidant capacity in blood orange juices as influenced by constituents, concentration process and storage. Food Chem 74:423–427

Attaway JA, Carter RD, Fellers PJ (1989) Trends in citrus industry-the production and handling of fresh-squeezed, unpasteurized orange juice. Fluss Obst 56:613–616

Baker RA, Bruemmer JH (1969) Cloud stability in the absence of various orange juice soluble components. Proc Fla State Hort Soc 82:215–220

Baker RA, Bruemmer JH (1972) Influence of pectate hesperidin floc on orange juice clarification. Proc Fla State Hort Soc 85:225–229

Baker RA, Cameron RG (1999) Clouds of citrus juices and juice drinks. Food Technol 53(1):64–69

Balasundram N, Sundram K, Samman S (2006) Phenolic compounds in plants and agri-industrial by-products: antioxidant activity, occurrence, and potential uses. Food Chem 99(1):191–203

Barbosa-Cánovas GV, Sepúlveda D (2005) Present status and the future of PEF technology. In: Barbosa-Cánovas GV, Tapia MS, Cano MP (eds) Novel food processing, 1st edn. CRC Press, Boca Raton, pp 1–2

Barros HR, Ferreira TA, Genovese MI (2012) Antioxidant capacity and mineral content of pulp and peel from commercial cultivars of citrus from Brazil. Food Chem 134:1892–1898
Basak S, Ramaswamy HS, Simpson BK (2001) High pressure inactivation of pectin methyl esterase in orange juice using combination treatments. J Food Biochem 25(6):509–526
Benavente-Garcia O, Castillo J (2008) Update on uses and properties of citrus flavonoids: New findings in anticancer, cardiovascular, and anti-inflammatory activity. J Agric Food Chem 56:6185–6205
Benavente-García O, Castillo J, Marín FR, Ortuño A, Del Río JA (1997) Uses and properties of citrus flavonoids. J Agric Food Chem 45:4505–4515
Bennett RD (1987) Paper presented at the meeting of the citrus products technical committee. U.S. Department of Agriculture Fruit and Vegetable Laboratory, Pasadena, CA
Bennett RD, Albach RF (1981) The nature of freeze-induced white spots on orange segment walls. J Agric Food Chem 29:511–514
Berhow MA, Hagegawa S, Manners GD (2000) Citrus limonoids: functional chemicals in agricultural and foods. American Chemical Society, Washington, DC
Berk Z (2009) Freeze drying (lyophilization) and freeze concentration. In: Food process engineering and technology. Elsevier, New York, p 522
Bermejo A, Llosá MJ, Cano A (2011) Analysis of bioactive compounds in seven citrus cultivars. Food Sci Technol Int 17(1):55–62
Bocco A, Cuvelier ME, Richard H, Berset C (1998) Antioxidant activity and phenolic composition of citrus peel and seed extracts. J Agric Food Chem 46:2123–2129
Braddock RJ (1995) Byproducts of citrus fruits. Food Tech 49:74–77
Braddock RJ (1999) Handbook of citrus by-products and processing technology. Wiley, New York
Braddock RJ, Marcy JE (1987) Quality of freeze concentrated orange juice. J Food Sci 52:159–162
Braddock RJ, Nikdel S, Nagy S (1988) Composition of some organic and inorganic compounds in reverse osmosis-concentrated citrus juices. J Food Sci 53:508–512
Buedo AP, Elustondo MP, Urbicain MJ (2001) Non-enzymatic browning of peach juice concentrate during storage. Innovat Food Sci Emerg Technol 1:255–260
Bull MK, Zerdin K, Howe E, Goicoechea D, Paramanandhan P, Stockman R, Sellahewa J, Szabo EA, Johnson RL, Stewart CM (2004) The effect of high pressure processing on the microbial, physical and chemical properties of Valencia and Navel orange juice. Innovat Food Sci Emerg Technol 5:135–149
Calabrò ML, Galtieri V, Cutroneo P, Tommasini S, Ficarra P, Ficarra R (2004) Study of the extraction procedure by experimental design and validation of a LC method for determination of flavonoids in *Citrus bergamia* juice. J Pharmaceut Biomed Anal 35:349–363
Cameron RG, Niedz RP, Grohmann K (1994) Variable heat stability for multiple forms of pectin methylesterase from citrus tissue culture cells. J Agric Food Chem 42:903–908
Carneiro L, Gomez FS, Matta VM, Cabral LMC (2002) Cold sterilization and clarification of pineapple juice by tangential microfiltration. Desalination 148:93–98
Cassano A, Drioli E, Galaverna G, Marchelli R, DiSilvestro G, Cagnasso P (2003) Clarification and concentration of food and carrot juices by integrated membrane processes. J Food Eng 57:153–163
Cassano A, Donato L, Drioli E (2007) Ultrafiltration of kiwifruit juice: operating parameters, juice quality and membrane fouling. J Food Eng 79:613–621
Castro AJ, Swanson BG, Barbosa-Cánovas GV, Dunker AK (2001) Pulsed electric field denaturation of bovine alkaline phosphatase. In: Barbosa-Cánovas GV, Zhang QH (eds) Pulsed electric fields in food processing. Fundamental aspects and applications. Technomic, Lancaster, pp 83–103
Chandler B, Nicol K (1975) Some relationships of naringin: their importance in orange juice bitterness. CSIRO Food Res Q 35:79–88
Chaudhary P, Jayaprakasha GK, Porat R, Patil BS (2012) Degreening and postharvest storage influences 'Star Ruby' grapefruit (*Citrus paradisi* Macf.) bioactive compounds. Food Chem 135(3):1667–1675

Chen SC, Hernandez E (1997) Design and performance evaluation of evaporators. In: Valentas KJ, Rotstein E, Singh RP (eds) Handbook of food engineering practice. CRC Press, New York
Chen CS, Wu MC (1998) Kinetic models for thermal inactivation of multiple pectinesterases in citrus juices. J Food Sci 63(5):747–750
Chen CS, Carter RD, Buslig BS (1979) Energy requirements for the TASTE citrus juice evaporator. In: Changing energy use futures. Lecture, pp 1841–1848
Chen CS, Shaw PE, Parish ME (1993) Orange and tangerine juices. In: Chen CS, Shaw PE, Nagy S (eds) Fruit juice processing technology. AG SCIENCE, Auburndale, FL, pp 110–165
Cheryan M (2007) Membrane concentration of liquid foods. In: Heldman DR, Lund DB (eds) Handbook of food engineering, 2nd edn. CRC Press, Boca Raton, pp 590–595
Chornomaz PM, Paglierob C, Marchesea J, Ochoa NA (2013) Impact of structural and textural membrane properties on lemon juice clarification. Food Bioprod Process 91:67–73
Clegg KM (1964) Nonenzymic browning of lemon juice. J Sci Food Agric 15:878–885
Corredig M, Kerr W, Wicker L (2001) Particle size distribution of orange juice cloud after addition of sensitized pectin. J Agric Food Chem 49(5):2523–2526
Corteś C, Torregrosa F, Esteve MJ, Frígola A (2006) Changes of colour and carotenoids contents during high intensity pulsed electric field treatment in orange juices. Food Chem Toxicol 44:1932–1939
Cortés C, Esteve MJ, Frígola A (2008) Effect of refrigerated storage on ascorbic acid content of orange juice treated by pulsed electric fields and thermal pasteurization. Eur Food Res Technol 227:629–635
Couture R, Rouseff R (1992) Debittering and deacidifying sour orange (*Citrus aurantium*) juice using neutral and anion exchange resins. J Food Sci 57(2):380–384
Crandall PG, Davis KC, Carter RD, Sadler GD (1988) Viscosity reduction by homogenization of orange juice concentrate in a pilot TASTE evaporator. J Food Sci 53:1477–1481
Croak S, Corredig M (2006) The role of pectin in orange juice stabilization: Effect of pectin methylesterase and pectinase activity on the size of cloud particles. Food Hydrocoll 20(7):961–965
Crupi F, Rispoli G (2002) Citrus juices technology. In: Dugo G, Di Giacomo A (eds) The genus citrus. Taylor & Francis, London, pp 77–113
De Bruijn J, Venegas A, Borquez R (2002) Influence of crossflow ultrafiltration on membrane fouling and apple juice quality. Desalination 148:131–136
Deshpande SS, Sathe SK, Salunkhe DK (1984) Freeze concentration of fruit juices. Food Sci Nutr 20:173–248
Dhuique-Mayer C, Caris-Veyrat C, Ollitrault P, Curk F, Amiot MJ (2005) Varietal and interspecific influence on micronutrient contents in citrus from the Mediterranean area. J Agric Food Chem 53:2140–2145
Di Giacomo A, Calvarano M, Tribulato E (1977) Sul contenuto di limonina del succo diarancia. Nota IV. Ruolo del portinnesto sulle cultivar 'Valencia Late' e Moro. Essenz Deriv Agrum 47:156–166
Dias MG, Filomena M, Camões GFC, Oliveira L (2009) Carotenoids in traditional Portuguese fruits and vegetables. Food Chem 113:808–815
Dinsmore HL, Nagy S (1972) Colorimetric furfural measurement as an index of deterioration in stored citrus juices. J Food Sci 37:768–770
Eagerman BA, Rouse AH (1976) Heat inactivation tempurature-time relationships for pectinesterase inactivation in citrus juices. J Food Sci 41:1396–1397
Elez-Martínez P, Martín-Belloso O (2007) Effects of high intensity pulsed electric field processing conditions on vitamin C and antioxidant capacity of orange juice and gazpacho, a cold vegetable soup. Food Chem 102:201–209
Elez-Martínez P, Suárez-Recio M, Martín-Belloso O (2007) Modeling the reduction of pectin methyl esterase activity in orange juice by high intensity pulsed electric fields. J Food Eng 78:184–193
Ellerbee LM (2009) Orange juice cloud stability and the influence of calcium and hesperidin. Dissertation, University of Georgia

Erazo GS, Beuchemin CLF, Abbot CFJ (1984) Preliminary study on processing of oranges of the cv. Washington, Thompson, Chilena and Valencia. Alimentos 9(2):9–16
Espachs-Barroso A, Van Loey A, Hendrickx M, Martin-Belloso O (2006) Inactivation of plant pectin methylesterase by thermal or high intensity pulsed electric field treatments. Innovat Food Sci Emerg Technol 7:40–48
Espamer L, Pagliero C, Ochoa NA, Marchese J (2006) Clarification of lemon juice using membrane process. Desalination 200:565–567
FAO (2011) Food and Agricultural Organization of the United Nations (FAOSTAT). Availablefrom: http://faostat.fao.org/site/567/DesktopDefault.aspx?PageID=567#ancor. Accessed Oct 2013
Fayoux SC, Hernandez RJ, Holland RV (2007) The debittering of navel orange juice using polymeric films. J Food Sci 72(4):144–154
Fellers PJ (1988) Shelf life and quality of freshly squeezed unpasteurized polyethylene-bottled citrus juices. J Food Sci 53:1699–1702
Fernández-Vázquez R, Stinco CM, Hernanz D, Heredia FJ, Vicario IM (2013) Industrial orange juice debittering: effect on volatile compounds and overall quality attributes. Int J Food Sci Technol 48:1861–1867
Filomena M, Valim CFA, Menezes HC (1998) Formation of crystalline agglomeratesduring storage of concentrated orange juice. Fruit Process 8:232–236
Fisher JF, Wheaton TA (1976) A high-pressure liquid chromatographic method for the resolution and quantitation of naringin and naringenin rutinoside in grapefruit juice. J Agric Food Chem 24:898–899
Fong CH, Hasegawa S, Herman Z, Ou P (1990) Limonoid glucosides in citrus juices. J Food Sci 54:1505–1506
Fraser PD, Bramley PM (2004) The biosynthesis and nutritional uses of carotenoids. Prog Lipid Res 43:228–265
Galati EM, Monforte MT, Kirjavainen S, Forestieri AM, Tripodo MM (1994) Biological effects of hesperidin, a citrus flavonoid: antiinflammatory and analgesic activity. Farmaco 49(11):709–712
Gama JJT, Sylos CM (2005) Major carotenoid composition of Brazilian Valencia orange juice: identification and quantifiation by HPLC. Food Res Int 38:899–903
García AF, Butz P, Bognàr A, Tauscher B (2001) Antioxidative capacity, nutrient content and sensory quality of orange juice and an orange-lemon-carrot juice product after high pressure treatment and storage in different packaging. Eur Food Res Technol 213:290–296
Gardner PT, White TAC, McPhail DB, Duthie GG (2000) The relative contributions of vitamin C, carotenoids and phenolics to the antioxidant potential of fruit juices. Food Chem 68(4):471–474
Garg A, Garg S, Zaneveld LJD, Singla AK (2001) Chemistry and pharmacology of the citrus bioflavonoid hesperidin. Phytothery Res 15:655–669
Gattuso G, Barreca D, Gargiulli C, Leuzzi U, Caristi C (2007) Flavonoid composition of citrus juices. Molecules 12:1641–1673
Ghasemi K, Ghasemi Y, Ebrahimzadeh MA (2009) Antioxidant activity, phenol and flavonoid contents of 13 Citrus species peels and tissues. Pak J Pharm Sci 22(3):277–281
Gómez-López VM, Orsolani L, Martínez-Yépez A, Tapia MS (2010) Microbiological and sensory quality of sonicated calcium-added orange juice. Food Sci Technol 43:808–813
González-Molina E, Domínguez-Perles R, Moreno DA, García-Viguera C (2010) Natural bioactive compounds of *Citrus limon* for food and health. J Pharmaceut Biomed Anal 51:327–345
Goodner K, Braddock RJ, Parish ME (1998) Inactivation of pectinesterase in orange and grapefruit juices by high pressure. J Agric Food Chem 46:1997–2000
Gorinstein S, Belloso OM, Park Y, Haruenkit R, Lojek A, Ciz M et al (2001) Comparison of some biochemical characteristics of different citrus fruits. Food Chem 74:309–315
Gorinstein S, Cvikrova M, Machackova I, Haruenkit R, Park Y, Jung S et al (2004) Characterization of antioxidant compounds in Jaffa sweeties and white grapefruits. Food Chem 84:503–510

Gross J, Gabai M, Lifshitz A (1972) A comparative study of the carotenoid pigments in juice of Shamouti, Valencia, and Washington oranges, three varieties of *Citrus sinensis*. Phytochemistry 11:303–308

Guadagni DG, Maier VP, Turnbaugh JG (1973) Effects of some citrus constituents on taste thresholds for limonin and naringin bitterness. J Sci Food Agric 24:1199–1205

Hashinaga F, Itoo S (1981) Seasonal changes of limonoids in the component parts of several citrus fruits. Proc Int Soc Citricult 2:901–905

Heinz V, Alvarez I, Angersbach A, Knorr D (2002) Preservation of liquid foods by high intensity pulsed electric fields-basic concepts for process design. Trends Food Sci Technol 12:103–111

Hertog MGL, Feskens EJM, Hollman PCH, Katan MB, Kromhout D (1993) Dietary antioxidant flavonoids and risk of coronary heart-disease- the zutphen elderly study. Lancet 342(8878): 1007–1011

Hofsommer HJ, Fisher-Hayloff-Cook KP, Radke HJ (1991) Neue technologische aspekte–zur enterbitterung von citrussaeften. Fluessiges Obst 58:62–64

Horowitz RM (1986) Taste effects of flavonoids. Prog Clin Biol Res 213:163–175

Huelin FE, Coggiola IM, Sidhu GS, Kenneth BH (1971) The anaerobic decomposition of ascorbic acid in the pH range of foods and in more acid solutions. J Sci Food Agric 22:540–542

İçdemir G (2012) Grapefruit juice production and determination of quality parameters (in Turkish) Dissertation, University of Çukurova

Irwe S, Olsson I (1994) Reduction of pectinesterase activity in orange juice by high pressure treatment. In: Singh RP, Oliveria FA (eds) Minimal processing of foods and process optimization. CRC Press, Ann Arbor, pp 35–42

Jiao B, Cassano A, Drioli E (2004) Recent advances on membrane processes for the concentration of fruit juices: a review. J Food Eng 63:303–324

Johnson JR, Braddock RJ, Chen CS (1995) Kinetics of ascorbic acid loss and nonenzymatic browning in orange juice serum: experimental rate constants. J Food Sci 60:502–505

Jordan MJ, Goodner KL, Castillo M, Laencina J (2005) Comparison of two headspace solid phase microextraction fibres for the detection of volatile chemical concentration changes due to industrial processing of orange juice. J Sci Food Agric 85:391–396

Kandaswami C, Perkins E, Soloniuk DS, Drzewiecki G, Middleton E (1991) Antiproliferative effects of citrus flavonoids on a human squamous cell carcinoma in vitro. Cancer Lett 56:147–152

Kanner J, Fishbein J, Shalom P, Harel S, Ben-Gera I (1982) Storage stability of orange juice concentrate packaged aseptically. J Food Sci 47:429–431

Karim MR, Hashinaga F (2002) Isolation and characterization of limonoid glucosyltransferase from pummelo albedo tissue. Food Chem 76:431–436

Katsaros GI, Tsevdou M, Panagiotou T, Taoukis PS (2010) Kinetic study of high pressure microbial and enzyme inactivation and selection of pasteurisation conditions for Valencia orange juice. Int J Food Sci Technol 45:1119–1129

Kaul TM, Middleton E, Ogra PL (1985) Antiviral effect of flavonoids on human viruses. J Med Virol 15:71–79

Kefford JF, Chandler BV (1970) The chemical constituents of citrus fruits. Academic, New York, p 94

Kelebek H, Selli S (2011) Determination of volatile, phenolic, organic acid and sugar components in a Turkish cv. Dortyol (*Citrus sinensis* L. Osbeck) orange juice. J Sci Food Agric 91(10):1855–1862

Kimball DA (1991a) Citrus processing quality control and technology. Van Nostrand Reinhold, New York

Kimball DA (1991b) Citrus processing, quality control and technology. AVI Book Van Nostrand Reinhold, New York, pp 117–125, 272–278

Kimball DA (1999) Citrus processing. A complete guide, 2nd edn. Aspen, Maryland

Kimball D, Kimball D (1991) Bitterness in citrus juices. In: Citrus processing. Springer, Netherlands, pp 136–161

Kimball DA, Norman SI (1990a) Changes in California navel orange juice during commercial debittering. J Food Sci 55(1):273–274

Kimball DA, Norman SI (1990b) Processing effects during commercial debittering of California navel orange juice. J Agric Food Chem 38(6):1396–1400
Kimball D, Parish M, Braddock R (2004) Oranges and tangerines. In: Barrett DM, Somogyi LP, Ramaswamy HS (eds) Processing fruits: science and technology, 2nd edn, Major processed products. CRC Press, Boca Raton, pp 617–638, 841
Kırca A, Cemeroğlu B (2003) Degradation kinetics of anthocyanins in blood orange juice and concentrate. Food Chem 81(4):583–587
Klavons JA, Bennett RD, Vannier SH (1991) Nature of the protein constituent of commercial orange juice cloud. J Agric Food Chem 39(9):1545–1548
Klavons JA, Bennett RD, Vannier SH (1994) Physical/chemical nature of pectin associated with commercial orange juice cloud. J Food Sci 59(2):399–401
Knorr D, Zenker M, Heinz V, Lee D (2004) Applications and potential of ultrasonics in food processing. Trends Food Sci Technol 15(5):261–266
Koca N, Burdurlu HS, Karadeniz F (2003) Kinetics of nonenzymatic browning reaction in citrus juice concentrates during storage. Turk J Agric For 27:353–360
Krinsky NI (1989) Carotenoids and cancer in animal models. J Nutr 119:123–126
Kuldiloke J, Eshtiaghi M, Zenker M, Knorr D (2007) Inactivation of lemon pectinesterase by thermosonication. Int J Food Eng 3:art. 3
Ladaniya MS (2008) Citrus fruit: biology, technology and evaluation. Academic Pr Elsevier, San Diego, p 558
Lee HS, Coates GA (1999) Vitamin C in frozen, fresh squeezed, unpasteurized, polyethylene-bottled orange juice: a storage study. Food Chem 65:165–168
Lee HS, Nagy S (1988a) Quality changes and nonenzymic browning intermediates in grapefruit juice during storage. J Food Sci 53:168–172
Lee HS, Nagy S (1988b) Relationship of sugar degradation to detrimental changes in citrus juice quality. Food Technol 42:91–94
Lee CY, Downing DL, Iredale HD, Chapman JA (1976) The variations of ascorbic acid content in vegetable processing. Food Chem 1:15–22
Lee HS, Castle WS, Coates GA (2001) High-performance liquid chromatography for the characterization of carotenoids in the new sweet orange (Earlygold) grown in Florida, USA. J Chromatogr A 371–377
Maccarone E, Campisi S, Fallico B, Rapisarda P, Sgarlata R (1998) Flavor components of Italian orange juice. J Agric Food Chem 46:2293–2298
Maier VP, Beverly GD (1968) Limonin monolactone, the nonbitter precursor responsible for delayed bitterness in certain citrus juices. J Food Sci 33:488–496
Maier VP, Hasegawa S, Hera E (1969) Limonin D-ring-lactone hydrolase. A new enzyme from citrus seeds. Phytochemistry 8:405–407
Maier VP, Bennett RD, Hasegwa S (1977) Limonin and other limonoids. In: Nagy S, Shaw PE, Veldius MK (eds) Citrus science and technology. The AVI Publishing Company Inc, Westport, pp 355–396
Mannheim CH, Passy N (1974) In: Spicer A (ed) Advances in preconcentration and dehydration. Applied Science, London
Manthey JA, Grohmann K (2001) Phenols in citrus peel byproducts. Concentrations of hydroxycinnamates and polymethoxylated flavones in citrus peel molasses. J Agric Food Chem 49:3268–3273
Marshall MR, Marcy JE, Braddock RJ (1985) Effects of total solids level on heat inactivation of pectinesterase in orange juice. J Food Sci 50:220–222
Matsumoto H, Ikoma Y (2012) Effect of different postharvest temperatures on the accumulation of sugars, organic acids, and amino acids in the juice sacs of satsuma mandarin (*Citrus unshiu* Marc.) fruit. J Agric Food Chem 60:9900–9909
Medina BG, Garcia A (1988) Concentration of orange juice by reverse osmosis. J Food Process Eng 10:217–230
Melendez-Martinez AJ, Vicario IM, Heredia FJ (2004) Importancia nutricional de los pigmentos carotenoides. Arch Latinoam Nutr 54:149–154

Melendez-Martinez AJ, Vicario IM, Heredia FJ (2007) Analysis of carotenoids in orange juice (review). J Food Compos Anal 20:638–649

Meléndez-Martínez AJ, Britton G, Vicario IM, Heredia FJ (2008) The complex carotenoid pattern of orange juices from concentrate. Food Chem 109:546–553

Meydav S, Saguy I, Kopelman IJ (1977) Browning determination in citrus products. J Agric Food Chem 25:602–604

Miller EG, Gonzales-Sanders AP, Couvillon AM, Binnie WH, Hasegawa S, Lam LKT (1994) Citrus limonoids as inhibitors of oral carcinogenesis. Food Technol 110–114

Mizrahi S, Berk Z (1970) Physico-chemical characteristics of orange juice cloud. J Sci Food Agric 21(5):250–253

Mondor M, Girard B, Moresoli C (2000) Modeling flux behavior for membrane filtration of apple juice. Food Res Int 33:539–548

Moshonas MG, Shaw PE (1989) Changes in composition of volatile components in aseptically packaged orange juice during storage. J Agric Food Chem 37(1):157–161

Moshonas MG, Shaw PE (1994) Quantitative determination of 46 volatile constituents in fresh, unpasteurized orange juices using dynamic headspace gas chromatography. J Agric Food Chem 42:1525–1528

Moulehi I, Bourgou S, Ourghemmi I, Tounsi MS (2012) Variety and ripening impact on phenolic composition and antioxidant activity of mandarin (*Citrus reticulate* Blanco) and bitter orange (*Citrus aurantium* L.) seeds extracts. Ind Crop Prod 39:74–80

Mouly PP, Arzouyan CR, Gaydou EM, Estienne JM (1994) Differentiation of *Citrus* juices by factorial discriminant analysis using liquid chromatography of flavanone glycosides. J Agric Food Chem 42:70–79

Mouly PP, Gaydou EM, Corsetti J (1999) Determination of the geographical origin of Valencia orange juice using carotenoid liquid chromatographic profiles. J Chromatogr A 844:149–159

Nagy S, Randal V (1973) Use of furfural content as an index of storage temperature abuse in commercially processed orange juice. J Agric Food Chem 21:272–275

Nienaber U, Shellhammer TH (2001a) High-pressure processing of orange juice: combination treatments and a shelf life study. J Food Sci 66:332–336

Nienaber U, Shellhammer TH (2001b) High-pressure processing of orange juice: kinetics of pectinmethylesterase inactivation. J Food Sci 66:328–331

Nisperos-Carriedo MO, Shaw PE (1990a) Comparison of volatile flavor components in fresh and processed orange juices. J Agric Food Chem 38:1048–1052

Nisperos-Carriedo MO, Shaw PE (1990b) Volatile flavor components of fresh and processed orange juices. Food Technol 44:134–138

Nonthanum P, Tansakul A (2008) Freeze concentration of lime juice. Int J Sci Technol 1:27–37

Nunez JM, Laencina J, Saura D (1989) Effect of adding chemical reducing agents for storing concentrated lemon juice. Essenze Deriv Agrum 59(4):386–387

O'Donnella CP, Tiwarib BK, Bourkec P, Cullen PJ (2010) Effect of ultrasonic processing on food enzymes of industrial importance. Trends Food Sci Technol 21:358–367

Ohlsson T, Bengtsson N (2002) Minimal processing of foods with nonthermal methods. In: Minimal processing technologies in the food industry. CRC Press, Boca Raton, pp 54–55

Olsen RW, Hill EC (1964) Debittering of concentrated grapefruit juice with naringinase. Florida Agric Exp Station J Ser 321–325

Olson JA (1999) Carotenoids and human health. Arch Latinoam Nutr 49:7–11

Ozaki Y, Miyake M, Inaba N, Ayano S, Ifuku Y, Hasegawa S (2000) Limonoid glucosides of Satsuma mandarin (*Citrus unshiu Marcov.*) and its processing products. ACS symposium series, pp 107–119

Pala ÇU, Toklucu AK (2013) Microbial, physicochemical and sensory properties of UV-C processed orange juice and its microbial stability during refrigerated storage. Food Sci Technol 50:426–431

Parish ME (1998) Orange juice quality after treatment by thermal pasteurization or isostatic high pressure. Food Sci Technol 31:439–442

Pecoroni S, Flocke R, Gunnewig W (2013) Separators, decanters and process lines from GEA Westfalia separator for citrus processing. www.wsus.com
Perez OE, Pilosof AMR (2004) Pulsed electric fields effects on the molecular structure and gelation of β-lactoglobulin concentrate and egg white. Food Res Int 37:102–110
Perez-Cacho PR, Rouseff RL (2008) Fresh squeezed orange juice odor: a review. Crit Rev Food Sci Nutr 48:681–695
Peri C (1974) Concentration of orange juice by reverse osmosis. Sci Technol Degli Alim 4:43–47
Peterson JJ, Beecher GR, Bhagwat SA, Dwyer JT, Gebhardt SE, Haytowitz DB, Holden JM (2006) Flavanones in grapefruit, lemons, and limes: a compilation and review of the data from the analytical literature. J Food Compos Anal 19:74–80
Pichaiyongvongdee S, Haruenkit R (2009) Comparative studies of limonin and naringin distribution in different parts of pummelo [*Citrus grandis* (L.) Osbeck] cultivars grown in Thailand. Kasetsart J (Nat Sci) 43:28–36
Pino J (1986) Changes caused by storage temperature on volatile constituents of concentrated grapefruit juice. Technologia Quimica 7(2):67–72
Pino J, Ramos M, Sanchez S, Torricella R (1987) Change in orange juice during production frozen concentrate and ways to increase its quality. Technologia Quimica 8(6–13):81
Plaza L, Sánchez-Moreno C, Elez-Martínez P, de Ancos B, Martín-Belloso O, Cano MP (2006) Effect of refrigerated storage on vitamin C and antioxidant activity of orange juice processed by high-pressure or pulsed electric fields with regard to low pasteurization. Eur Food Res Technol 223:487–493
Plaza L, Sánchez-Moreno C, de Ancos B, Elez-Martínez P, Martín-Belloso O, Cano MP (2011) Carotenoid and flavanone content during refrigerated storage of orange juice processed by high-pressure, pulsed electric fields and low pasteurization. Food Sci Technol 44:834–839
Polydera AC, Stoforos NG, Taoukis PS (2003) Comparative shelf life study and vitamin C loss kinetics in pasteurised and high pressure processed reconstituted orange juice. J Food Eng 60:21–29
Pupin AM, Dennis MJ, Toledo MCF (1998) Flavanone glycosides in Brazilian orange juice. Food Chem 61(3):275–280
Puri M, Marwaha SS, Kothari RM, Kennedy JF (1996) Biochemical basis of bitterness in citrus fruit juices and biotech approaches for debittering. Crit Rev Biotechnol 16:145–155
Rahman MS, Ahmed A, Chen XD (2006) Freezing-melting process and desalination: I. Review of the state-of-the-art. Sep Purif Rev 35:59–96
Ramaswamy H, Marcotte M (2006) Seperation and concentreation In: Food processing: principles and applications. CRC Press, Boca Raton, pp 320–325
Ranganna S, Gobindarajan VS, Ramanna KV (1983) Citrus fruits. II. Chemistry, technology and quality evaluation. B. Technology. CRC Crit Rev Food Sci Nutr 19:1–98
Rapisarda P, Tomaino A, Lo Cascio R, Bonina F, De Pasquale A, Saija A (1999) Antioxidant effectiveness as influenced by phenolic content of fresh orange juices. J Agric Food Chem 47:4718–4723
Rapisarda P, Lo Bianco M, Pannuzzo P, Timpanaro N (2008) Effect of cold storage on vitamin C, phenolics and antioxidant activity of five orange genotypes *Citrus sinensis* (L.) Osbeck. Postharv Biol Technol 49:348–354
Rastogi NK (2010) Food processing on physicochemical changes of foods: a review. In: Devahastin S (ed) Physicochemical aspects of food engineering and processing. CRC Press, Boca Raton, pp 107–108
Raventós M, Hernández E, Auleda JM (2012) Freeze concentration applications in fruit processing. In: Rodrigues S, Fernandes FAN (eds) Advances in fruit processing technologies. CRC Press, Boca Raton, pp 263–270
Rivas A, Rodrigo D, Martínez A, Barbosa-Cánovas GV, Rodrigo M (2006) Effect of PEF and heat pasteurization on the physical–chemical characteristics of blended orange and carrot juice. Food Sci Technol 39:1163–1170
Robbins RJ (2003) Phenolic acids in foods: an overview of analyticalmethodology. J Agric Food Chem 51:2866–2887

Robertson GL, Samaniego CML (1986) Effect of initial dissolved oxygen levels on the degradation of ascorbic acid and the browning of lemon juice during storage. J Food Sci 51:184–187
Rodrigo D, Ruíz P, Torregrosa F, Martínez A, Barbosa-Cánovas GV, Rodrigo M (2001) Pulsed electric field inactivation of pectin methyl esterase in blended orange juice and carrot juice. EUROCAFT, Berlin
Rodrigo D, Barbosa-Cánovas GV, Martínez A, Rodrigo M (2003) Weibull distribution function based on an empirical mathematical model for inactivation of *Escherichia coli* by pulsed electric fields. J Food Prot 66:1007–1012
Roig MG, Bello JF, Rivera ZS, Kennedy JF (1999) Studies on the occurrence of non-enzymatic browning during storage of citrus juice. Food Res Int 32:609–619
Rothschild G, Karsenty A (1974) Influence of holding time before pasteurization, pasteurization and concentration on the turbidity of citrus juices. J Food Sci 39:1042–1044
Rouse AH, Atkins CD (1952) Heat inactivation of pectinesterase in citrus juices. Food Tech 6:291–294
Rouseff RL, Martin S, Youtsey F, Charles O (1987) Quantitative survey of narirutin, naringin, hesperidin, and neohesperidin in citrus. J Agric Food Chem 35(6):1027–1030
Rouseff R, Gmitter F, Grosser J (1994) Citrus breeding and flavour. In: Piggott JR, Peterson A (eds) Understanding natural flavors. Blackie Academic, London, pp 113–127
Sadler G, Parish M, Clief D, Davis J (1997) The effect of volatile absorption by packaging polymers on flavor, microorganisms and ascorbic acid in reconstituted orange juice. Food Sci Technol 30(7):686–690
Sampedro F, Geveke DJ, Fan X, Zhang HQ (2009) Effect of PEF, HHP and thermal treatment on PME inactivation and volatile compounds concentration of an orange juice–milk based beverage. Innovat Food Sci Emerg Technol 10:463–469
Sandhu KS, Minhas KS (2006) Oranges and citrus juices. In: Hui YH, Cano MP, Gusek T, Sidhu JS, Sinha NK (eds) Handbook of fruits and fruit processing: science and technology. Blackwell, Ames, pp 309–358
Saura D, Laencina J, Pérez-López AJ, Lizama V, Carbonell-Barrachina AA (2003) Aroma of canned peach halves acidified with clarified lemon juice. J Food Sci 68(3):1080–1085
Saura D, Marti N, Laencina J, Lizama V, Carbonell-Barrachina AA (2004) Sensory evaluation of canned peach halves acidified with clarified lemon juice. J Food Sci 69(2):74–78
Saura D, Marti N, Valero M, Gonzalez E, Carbonell A, Laencina J (2012) Separation of aromatics compounds during the clarification of lemon juice by cross-flow filtration. Ind Crop Prod 36:543–548
Schofield TF (1994) Aroma improvement by means of the spinning cone column. Fruit Process 4:144–147
Schofield TF, Riley P (1998) Developments with the spinning cone column to extract natural concentrated aromas. Fruit Process 8:52–55
Scott WC, Hearn CJ (1966) Processing qualities of new citrus hybrids. Proc Florida State Horticult Soc 79:304–306
Selli S, Kelebek H (2011) Aromatic profile and odour-activity value of blood orange juices obtained from Moro and Sanguinello (*Citrus sinensis* L. Osbeck). Ind Crop Prod 33:727–733
Selli S, Canbas A, Cabaroglu T (2004) Volatile flavour components of orange juice obtained from the Kozan variety of Turkey. J Food Compos Anal 17:789–796
Shaw PE (1991) Fruit. In: Maarse H (ed) Volatile compounds in food and beverages. Marcel Dekker, New York, p 305
Shaw PE, Moshonas M (1997) Quantification of volatile constituent in orange juice drinks and its use for comparison with pure juices by multivariate analysis. LWT Food Sci Technol 30:497–501
Shaw PE, Baines L, Mitnes BA, Agmon G (2000) Commercial debittering processes to upgrade quality of citrus juice products. ACS Symp Ser Citrus Limonoid 758:120–131

Shaw PE, Moshonas M, Buslig BS, Barros SM, Widmer WW (2005) Discriminant and principal component analyses to classify commercial orange juices based on relative amounts of volatile juice constituents. J Sci Food Agric 79:1949–1953

Shinoda Y, Murata M, Homma S, Komura H (2004) Browning and decomposed products of model orange juice. Biosci Biotechnol Biochem 68:529–536

Shinoda Y, Komura H, Homma S, Murata M (2005) Browning of model orange juice solution: Factors affecting the formation of decomposition products. Biosci Biotechnol Biochem 69:2129–2137

Solomon O, Svanberg H, Sahlstrom A (1995) Effect of oxygen and fluorescent light on the quality of orange juice during storage at 8 °C. Food Chem 53:363–368

Stéphane CF, Ruben JH, Robert VH (2007) The debittering of navel orange juice using polymeric films. J Food Sci 72:143–154

Stevens JW, Pritchett DE, Baier WE (1950) Control of enzymatic flocculation of cloud in citrus juices. Food Technol 4:469–473

Stinco CM, Fernández-Vázquez R, Hernanz D, Heredia FJ, Meléndez-Martínez AJ, Vicario IM (2013) Industrial orange juice debittering: impact on bioactive compounds and nutritional value. J Food Eng 116:155–161

Sun J, Chu Y-F, Wu X, Liu RH (2002) Antioxidant and antiproliferative activities of common fruits. J Agric Food Chem 50:7449–7454

Sun Y, Wang J, Gu S, Liu Z, Zhang Y, Zhang X (2010) Simultaneous determination of flavonoids in different parts of Citrus reticulata 'Chachi' fruit by high performance liquid chromatography—photodiode array detection. Molecules 15:5378–5388

Tariq AM, Chaudry MS, Qureshi MJ (1974) Effect of processing and storage on the development of bitterness in the orange juice. Pak J Sci Ind Res 17(1):27–28

Thijssen HA (1975) Current developments in the freeze concentration of liquid foods. In: Goldblith SA, Rey L, Rothmayr WW (eds) Freeze drying and advanced food technology. Academic, London

Thijssen HAC (1986) The economics and potentials of freeze concentration for fruit juices. In: International federation of fruit juice producers, XIX scientific technical commission, symposium, Den Haag, May 12–15, The Hague, pp 97–103

Tiwari BK, Muthukumarappan K, O'Donnell CP, Cullen PJ (2009a) Inactivation kinetics of pectin methylesterase and cloud retention in sonicated orange juice. Innovat Food Sci Emerg Technol 10:166–171

Tiwari BK, O'Donnell CP, Muthukumarappan K, Cullen PJ (2009b) Effect of sonication on orange juice quality parameters during storage. Int J Food Sci Technol 44:586–595

Tiwari BK, O'Donnell CP, Muthukumarappan K, Cullen PJ (2009c) Effect of low temperature sonication on orange juice quality parameters using response surface methodology. Food Bioprocess Technol 2:109–114

Tiwari BK, O'Donnell CP, Muthukumarappan K, Cullen PJ (2009d) Ascorbic acid degradation kinetics of sonicated orange juice during storage and comparison with thermally pasteurised juice. Food Sci Technol 42:700–704

Tomás-Barberán FA, Clifford MN (2000) Review: flavanones, chalcones and dihydrochalcones-nature, occurrence and dietary burden. J Sci Food Agric 80:1073–1080

Topuz A, Topakci M, Canakci M, Akinci I, Ozdemir F (2005) Physical and nutritional properties of four orange varieties. J Food Eng 66:519–523

Toribio JL, Lozano JE (1984) Nonenzymatic browning in apple juice concentrate during storage. J Food Sci 49:889–892

Torkamani AE, Niakousari M (2011) Impact of UV-C light on orange juice quality and shelf life. Int Food Res J 18:1265–1268

Torregrosa F, Cortés C, Esteve MJ, Frígola A (2005) Effect of high-intensity pulsed electric fields processing and conventional heat treatment on orange-carrot juice carotenoids. J Agric Food Chem 53:9519–9525

Torregrosa F, Esteve MJ, Frígola A, Cortés C (2006) Ascorbic acid stability during refrigerated storage of orange–carrot juice treated by high pulsed electric field and comparison with pasteurized juice. J Food Eng 73:339–345
Tran MTT, Farid M (2004) Ultraviolet treatment of orange juice. Innovat Food Sci Emerg Technol 5:495–502
Trifiro A, Gherardi S, Bigliardi D, Bazzarini R, Castaldo D (1984) Limonin, l-malic acid and d-isocitric acid contents of Italian Tarocco and Sanguinello oranges. Industria Conserve 59(1):12–17
Uçan F (2013) Production of clear and naturally cloudy lemon juice concantre and determination of quality parameters. (in Turkish) Dissertation, University of Çukurova
Valdramidis VP, Cullen PJ, Tiwari BK, O'Donnell CP (2010) Quantitative modelling approaches for ascorbic acid degradation and non-enzymatic browning of orange juice during ultrasound processing. J Food Eng 96:449–454
Valero M, Recrosio N, Saura D, Muñoz N, Martí N, Lizama V (2007) Effects of ultrasonic treatments in orange juice processing. J Food Eng 80:509–516
Van Den Broeck I, Ludikhuyze LR, Van Loey AM, Hendrickx ME (2000) Inactivation of orange pectinesterase by combined high pressure and temperature treatments: a kinetic study. J Agric Food Chem 48(5):1960–1970
Van Pelt WHJM (1975) Freeze concentration of vegetable juices. In: Goldblith SA, Rey L, Rothmayr WW (eds) Freeze drying and advanced food technology. Academic, London
Vercet A, Sanchez C, Burgos J, Montanes L, Lopez Buesca P (2001) The effects of manothermosonication on tomato pectic enzymes and tomato paste rheological properties. J Food Eng 53:273–278
Versteeg C, Rombouts FM, Spaansen CH, Pilnik W (1980) Thermostability and orange juice cloud destabilizing properties of multiple pectinesterases from orange. J Food Sci 45:969–971
Vervoort L, Van der Plancken I, Grauwet T, Timmermans RAH, Mastwijk HC, Matser AM, Hendrickx ME, Van Loey A (2011) Comparing equivalent thermal, high pressure and pulsed electric field processes for mild pasteurization of orange juice Part II: Impact on specific chemical and biochemical quality parameters. Innovat Food Sci Emerg Technol 12:466–477
Voragen AGJ, Pilnik W, Thibault JF, Axelos MAV, Renard MGC (1995) Pectins. In: Stephen AM (ed) Food polysaccharides and their applications. Marcel Dekker, New York, pp 287–339
Walker JB (1990) Membrane process for the production of superior quality fruit juice concentrates. In: Proceedings of the 1990 international congress on membranes and membrane processes, vol 1, pp 283–285
Wang H, Cao GH, Prior RL (1996) Total antioxidant capacity of fruits. J Agric Food Chem 44:701–705
Wang YC, Chuang YC, Ku YH (2007) Quantitation of bioactive compounds in citrus fruits cultivated in Taiwan. Food Chem 102(4):1163–1171
Wang YC, Chuang YC, Hsu HW (2008) The flavonoid, carotenoid and pectin content in peels of citrus cultivated in Taiwan. Food Chem 106(1):277–284
Wicker L, Ackerley JL, Corredig M (2002) Clarification of juice by thermolabile Valencia pectinmethylesterase is accelerated by cations. J Agric Food Chem 50:4091–4095
Wicker L, Ackerley JL, Hunter JL (2003) Modification of pectin by pectinmethylesterase and the role in stability of juice beverages. Food Hydrocoll 17:809–814
Widmer W (2000) Determination of naringin and neohesperidin in orange juice by liquid chromatography with UV detection to detect the presence of grapefruit juice: collaborative study. J AOAC Int 83(5):1155–1165
Wilke B (2002) Aspetic packaging of fruit juices in thermoformed containers. Fruit Process 13(1):13–16
Xu G, Liu D, Chen J, Ye X, Maa Y, Shi J (2008) Juice components and antioxidant capacity of citrus varieties cultivated in China. Food Chem 106:545–551
Yeom H, Streaker C, Zhang Q, Min D (2000a) Effects of pulsed electric fields on the quality of orange juice and comparison with heat pasteurization. J Agric Food Chem 48:4597–4605

Yeom HW, Streaker CB, Zhang QH, Min DB (2000b) Effects of pulsed electric fields on the activities of microorganismis and pectin methyl esterase in orange juice. J Food Sci 65(8): 1359–1363
Yeom HW, Zhang QH, Chism GW (2002) Inactivation of pectin methyl esterase in orange juice by pulsed electric field. J Food Sci 67:2154–2159
Yoo KM, Lee KW, Park JB, Lee HJ, Hwang IK (2004) Variation in major antioxidants and total antioxidant activity of yuzu (Citrus junos Sieb ex Tanaka) during maturation and between cultivars. J Agric Food Chem 52:5907–5913
Yusof S, Ghazali H, King GS (1990) Naringin content in local citrus fruits. Food Chem 37(2):113–121
Zaare-Nahandi F, Hosseinkhani S, Zamani Z, Asadi-Abkenar A, Omidbaigi R (2008) Delay expression of limonoid UDP-glucosyltransferase makes delayed bitterness in citrus. Biochem Biophys Res Commun 371:59–62
Zerdin K, Rooney ML, Vermue J (2003) The vitamin C content of orange juice packed in an oxygen scavenger material. Food Chem 82:387–395
Zhang QH, Sastry SK, Yousef AE (1996) Integrated processing and aseptic packaging system using high voltage pulsed electric field technology. IFT Annual Meeting, New Orleans
Ziegler RG (1989) A review of epidemiologic evidence that carotenoids reduce the risk of cancer. J Nutr 119:116–122
Zimmermann U, Pilwat G, Riemann F (1974) Dielectric breakdown of cell membranes. Biophys J 14:881–899
Zulueta A, Esteve MJ, Frigola A (2010) Ascorbic acid in orange juice–milk beverage treated by high intensity pulsed electric fields and its stability during storage. Innovat Food Sci Emerg Technol 11:84–90

Nutritional Quality Assessment in Dairy Products: A Perspective

A. Adnan Hayaloglu and Mehmet Güven

1 Introduction

Milk is a biological liquid secreted by the female of all mammalian species to feed their babies. Although more than 4,000 species are present in the world, only four species including cow, sheep, goat, and buffalo produce milk for human consumption and their milk have technological importance (Fox et al. 2000). The consumption of milk as liquid form is the most beneficial form for human; however, it is very difficult to maintain milk as liquid form for a long time. So, milk has been converted to other products by concentration, fermentation, drying techniques, etc. During these processes, many changes take place with respect to milk nutrients and also these products gain new characteristics which may be preferable for consumers (Kosikowski and Mistry 1997; Hayaloglu and Ozer 2011). Changes in milk constituents during processing of dairy foods are summarized in Table 1. In general, enzymes and proteins are sensitive to heat treatment, homogenization, etc. However, the manufacture of some dairy products has been achieved by these dairy operations. For examples, fermented milks including yogurt, kefir, ayran (a drinkable yogurt, is very common in Turkey) have been produced by conversion of lactose to lactic acid and the lactic acid formed changes the biochemical and textural properties of milk (Kocak and Avsar 2010; Hayaloglu and Karagul-Yuceer 2011).

Cheese is produced by enzymatic coagulation of milk protein (casein) by milk-clotting enzymes and the milk protein and milk fat may be concentrated about tenfold by drainage of whey in the curd (Hayaloglu and Ozer 2011). Cream is manufac-

A.A. Hayaloglu (✉)
Department of Food Engineering, Engineering Faculty, Inonu University, Malatya 44280, Turkey
e-mail: adnan.hayaloglu@inonu.edu.tr

M. Güven
Faculty of Agriculture, Department of Food Engineering, Cukurova University, Adana 01330, Turkey

A. Malik et al. (eds.), *Food Processing: Strategies for Quality Assessment*, Food Engineering Series, DOI 10.1007/978-1-4939-1378-7_4

Table 1 Dairy operations and changes in milk constituents

Processes	Milk constituents	Benefits	Loses
1. Cooling to 4 °C	Microorganisms and enzymes	– Inhibit microorganisms and increases shelf-life of milk	–
2. Heat treatment	Microorganisms, enzymes, proteins, salts, vitamins	– Kill harmful bacteria at pasteurization temperature – Inactivate some enzymes	– Serum proteins become insoluble after 90 °C and is problem in cheese making. This temperature is preferred in yogurt production. – All vegetative microorganisms and enzymes may be destroyed – All microorganisms and enzymes are inactivated at 120 C for 15–20 min (sterilization) and browning reaction can be occurred. – Changes in flavor
3. Homogenization	Milk fat	– Uniform milk fat distribution in milk and dairy foods – Prevention of cream formation on the surface	– Lipolysis may be enhanced – Damages in proteins
4. Separation	Milk fat	– Production of low-fat or reduced fat milk or milk products and standardization of liquid milk – Production of cream	–
5. Evaporation	Water	– Increasing of total solid in milk and standardized production of milk product	– Lose of water – Changes in pH – Changes in flavor
6. Membrane process	All constituents based on membrane technique	– Concentration of milk constituents – Reduction of microorganisms – Production of some dairy by-products	–
7. Fermentation	Lactose, proteins, pH	– Production of dairy foods including cheese, yogurt, kefir, kimiz, ayran, etc. – Increasing shelf-life – Flavor development	–
8. Enzyme application	Proteins	– Development of flavor and texture in cheese – Acceleration of ripening – Improvement of texture – Changes in microstructure – Crosslinking between protein molecules – Structural changes and hydrolysis in caseins	– Probably off-flavor – Probably some textural defects

tured by separation of milk fat from whole milk and butter is produced by churning of cream. Cream and butter contain about 35 and 80 % of milk fat, respectively. Ice cream is a frozen dairy-based dessert and is produced by using milk and cream and by combining some fruits and other ingredients. All of the above products show different nutritional value, physical, and biochemical characteristics. In this article, the products from milk have been shortly reviewed and changes during processing are discussed on nutritional aspects (Walstra et al. 2006).

2 Milk Constituents

Milk is a polydispersed food and the milk constituents are distinguished in this colloidal solution due to their physicochemical nature (polarity, surface hydrophobicity, molecular size, etc.) (Metin 2001). Representative demonstration of the milk constituents are shown in Fig. 1.

The milk constituents are present in three phases including emulsion, colloidal dispersion, and true solution. Lactose, organic acids, organic and inorganic salts, vitamins, and other small molecules result in a true solution. In this aqueous solution, proteins at various molecular levels (the whey proteins and caseins) are dispersed, while milk fat exists in an emulsified form as globules with varying diameter ranging from 0.1 to 20 μm. The milk pH is about 6.7 at room temperature. The viscosity of milk is low (2 Cp), about twice that of water. The dissolved substances give milk an osmotic pressure of about 7 bar and a freezing point depression close to −0.52 °C. The water activity is high, about 0.995, which means that the milk is good substrate for bacteria and it may be deteriorated in a short time. Milk density is about 1.029 g/mL at 20 °C; it varies especially with fat content (Ucuncu 2005; Walstra et al. 2006).

Some physical, microbiological, and chemical changes occur during milk processing or dairy operations including cooling, heat treatment, pumping, homogenization, separation, evaporation, membrane processes, microbial fermentation, enzyme applications, high pressure, etc. Proteins, lactose, and fat are macroconstituents in milk and are well characterized due to their technological importance as well as nutritional value. Minor constituents like vitamins, salts (calcium, phosphate, citrate, etc.), and organic acids are also important in terms of nutritional requirements for human. Figure 2 shows the chemical composition of cow's milk (as averaged values). Milk is a perfect food for the young calf, and it can also provide good nutrition to humans. It contains virtually all nutrients, most of these in significant quantities. However, it is poor in iron and vitamin C contents. It contains no antinutritional factors, but it lacks dietary fiber.

2.1 Proteins

The protein in milk is the most important constituent for the manufacture of cheese and the level of protein in milk shows large interspecies differences (Fox and McSweeney 1998). The level of protein in milk affects the growth of baby of the

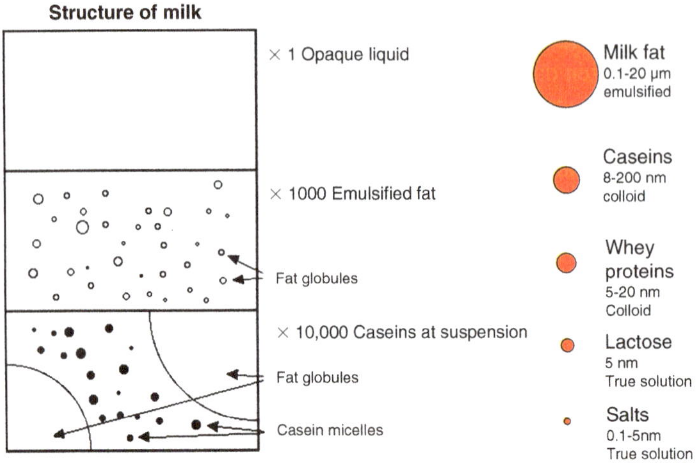

Fig. 1 Representative demonstration of the milk constituents

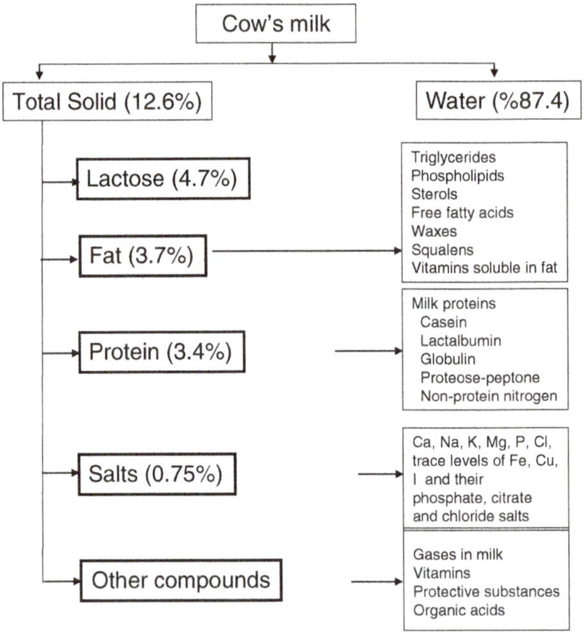

Fig. 2 Chemical composition of cow's milk

species; for example, babies of man (1.0 % protein), cow (3.5 % protein), and dog (7.3 % protein) access double birth-weight after 180, 50, and 7 days, respectively. These differences in the growth are due to the levels of protein in each species (Fox and McSweeney 2003). The milk proteins can be categorized as shown Fig. 3

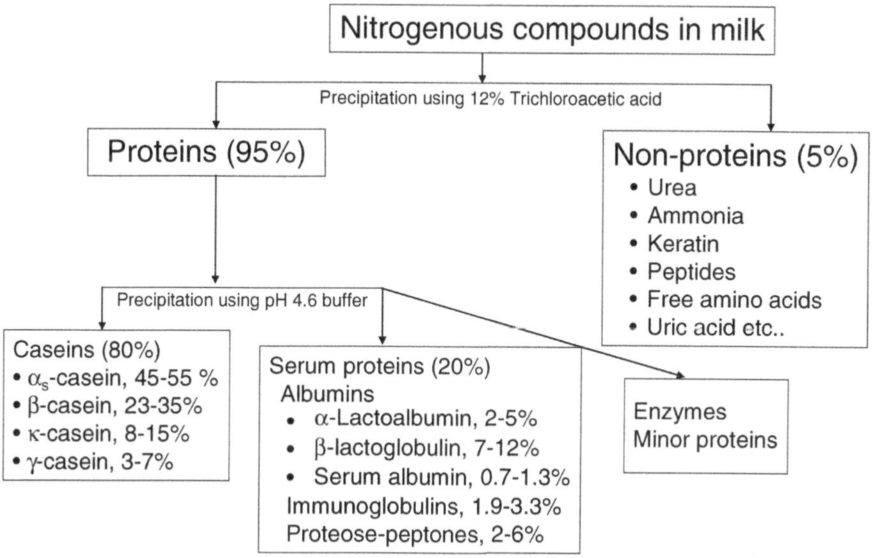

Fig. 3 Basic classification of nitrogenous compounds in milk

and the main categorization can be done by pH 4.6 solubility. pH 4.6-insoluble fraction of the milk are caseins and it represents a 80 % of the proteins (is valid for cow, sheep, goat, and buffalo milks). Bovine milk casein consists of four types of fractions including α_{s1}-, α_{s2}-, β-, and κ-caseins with levels of 38, 10, 34, and 15 % in whole casein, respectively. The whey protein fractions also contain four principal proteins, e.g., β-lactoglobulin, α-lactalbumin, blood serum albumin, and immunoglobulins (Fox et al. 1996).

Nutritional value of the proteins is mainly related to essential amino acids (Ile, Leu, Lys, Met, Phe, Thr, Trp, and Val). The digestion of proteins in human body is well and about 100 %; however, heat-induced changes may alter the digestibility of protein. Protein efficiency ratio of the milk proteins (the caseins or serum proteins) is better than that of other foods. Also, protein efficiency ratio is high when the presence of a mixture of casein and serum protein instead of each one separately due to some essential amino acids (Walstra et al. 2006). Casein is relatively high in Tyr and Phe, serum protein in the sulfur-containing Cys and Met. Presence of calcium and phosphate is another nutritional superiority for milk protein and these minerals are important for human in bone growth. Milk proteins contain bioactive peptides which play functional role in human body and these peptides yield during digestion in the intestine (Fox et al. 2004). Dairy foods, especially cheese, are source of bioactive peptides that are released by digestive enzymes or that are produced by fermentation. So, these exhibit their bioactive properties and beneficial effects in human body. Bioactive peptides are divided into various groups including casomorphins, angiotension-converting enzyme (ACE)-inhibiting peptides, phosphopeptides, immunopeptides, casoplateins, antimicrobial, opioid, mineral binding, etc.

Due to their hypertensive effect, ACE-inhibitory peptides are of special interests and many workers focused on them within the last decade (Sieber et al. 2010). Milk proteins, one or more, may cause some allergic reactions including some dermatological problems, diarrhea, asthma, etc., for some individuals; however, it is not common. Heat denaturation or hydrolysis of some proteins may have effects on reducing the allergic symptoms, as suggested some literatures (Korhonen and Pihlanto-Leppälä 2001; Hayaloglu and Ozer 2011).

2.2 *Lipids*

Due to its high price in technological standpoint, high energy, and high nutritional value, the chemical and physical properties of milk fat have been characterized in detail. The lipid content in milk varies widely interspecies, as in protein contents, the fat contents of cow, sheep, goat, and buffalo milk is *ca.* 3.5, 6.5, 3.5, and 7.0 %, respectively. The fat contents are changeable within the same species and many factors including age, lactation stage, animal health, milking intervals, nutrition, etc., affect the level of fat in milk. Lipids are fatty acids esters and majority (about 98 % of the lipid fraction) of the lipids are tricylglycerols (triglycerides). The remaining lipidic fractions (about 2 %) are fatty acids, mono- and diglycerides, phospholipids, sterols, and fat-soluble vitamins including A, D, E, and K (Walstra et al. 2006).

Nutritional value of milk fat is originated from its high energy, fat-soluble nutrients, and bioactive lipids for mammals. Bioactive lipids in milk include triacylglycerides, diacylglycerides, saturated and polyunsaturated fatty acids (PUFA), and phospholipids. Milk lipids play a role as anticancer, antimicrobial, anti-inflammatory, and immunosuppression properties. The major mammalian milk that is consumed by humans as a food commodity is that from bovine whose milk fat composition is distinct due to their diet and the presence of a rumen. Bovine milk fat is the most complex fat and it contains about 200 different fatty acids, even if majority of them are trace. In this lipid complex in bovine milk, a total of 12 fatty acids are predominantly present and nutritionally considered for human (German and Dillard 2006). Bovine milk fat is lower in PUFA and higher in saturated fatty acids than human milk. The physical properties of bovine milk fat that result from its composition including its plasticity make it a highly desirable property in butter or as food ingredient. Among the 12 major milk fatty acids, only three (lauric, myristic, and palmitic) have been associated with raising total cholesterol levels in plasma, but their individual effects are variable—both toward raising low-density lipoproteins (LDL) and raising the level of beneficial high-density lipoproteins (HDL) (German and Dillard 2006). Milk fat is digested well, with exception of long-chain fatty acids, in human body and homogenization, butter-making or other processes are not negative effect on its digestibility (Fox and McSweeney 2006). However, some scientists claim that any changes in the structure of fat by homogenization can affect the biological properties when consumed. Milk fat is carriers for vitamins (A, D, E, K) and also contains essential fatty acids (linoleic and linolenic acids) which are important for human nutrition. To able to product some hormones, the essential fatty acid plays a precursor role. It was claimed that presence of conjugated linoleic acids (CLA) in milk fat may exhibit anticancer, antimutagenic, and immune modulating effects. However, some scientist point out that a high intake of milk fat can be cause certain types of cancer (Donmez et al. 2005; German and Dillard 2006).

2.3 Lactose

Lactose is a reducing disaccharide consisting of glucose and galactose linked by a β 1-4 glycosidic bond (Fox et al. 2000). Milk is only source of lactose and other sugars including glucose, fructose, glucosamine, galactosamine, and neuraminic acid are scarcely found in milk and majority of them are covalently bound to milk proteins. The levels of lactose in milk vary widely among species and the breed, udder infection, and lactation stages are important factor on lactose concentration within the same species' milk. Lactose plays an important role in the manufacture of fermented milks and cheese. Ripe cheese is almost free of lactose or very low level; however, lactic fermentation or starter activity (use of certain level of lactose and convert lactic acid) is a crucial role during cheese making (McSweeney and Fox 2009).

In industry, lactose is produced from whey (which consists about 70 % lactose of the total solids) by crystallization. It is used as a filler material in the production of many foods and also uses in pharmaceuticals. Hydrolysis, oxidation, reduction, and mutarotation are the main biochemical reactions that occur in lactose. Other reactions are caramelization and Maillard reactions which occur during heat treatment. Maillard reaction occurs in the presence of amino groups, especially the ε-amino group of Lys residues in proteins. These reactions can lead to formation of flavor compounds and brown pigments and to a decrease in the nutritionally available Lys (Walstra et al. 2006).

Lactose is hydrolyzed by lactase (β-galactosidase) enzyme secreted in small intestine. The enzyme is present in suckling babies and is sufficient for hydrolysis of lactose. However, the amount of lactase or its activity decreases with age and some adults suffer from some pains in their intestine due to poor metabolization of lactose. About 60–70 % of the adults in the world have difficulties in digestion of lactose and the syndrome is called lactose intolerance and the people are called lactose malabsorbers. Milk is consumed in their diet in most of Europe, India, and some parts of Asia and Africa; so, a few people are lactose malabsorbers in these countries. The lactose-intolerant people are very common in rest of the countries in the world. Use of lactase-treated milk or consuming cheese or some fermented milks are alternative diet for people who are lactose malabsorbers.

2.4 Salts

Salts occur in solution in milk serum or casein and the most important ones are Ca, Na, K, and Mg (Metin 2001). However, these salts are present as phosphates, citrates, chlorides, and caseinates. The most abundant salts are Ca and K in milk and their levels decreased at the end of lactation. In udder infection, the concentration of chloride increased with decreasing of lactose and the milk gives a salty taste. The most important salts or ions are calcium and phosphate for cheese-making technology (Fox et al. 2000). Milk contains about 1.0–1.2 g Ca in per liter of milk and some of them are insoluble and occurs mainly associated with the casein micelles as micellar or colloidal calcium phosphate. A part of Ca is present as ionic form in milk and is of major importance for rennet coagulation milk.

Milk and dairy foods are good sources of Ca (125 mg/100 mL of milk) for human nutrition and the utilization of the Ca in the body is well. Calcium is essential for the healthy growth and maintenance of teeth and bones and plays a vital function in blood clotting and muscle contraction. Phosphorus is also essential for healthy bones and teeth as well as cell membrane structure, tissue growth, and regulation of pH levels in the body. The molar ratio Ca/P (including organic phosphate) is about 0.9 in milk; this is quite a high ratio when compared to other foods. Some people suffer from osteoporosis and this ratio is considered to be beneficial for these people (Fox et al. 1996). Milk is a good source for many other minerals, including trace elements especially zinc. Zinc is a constituent of many enzymes in the body; its role is to fight infections, growth development, and sexual development. Iron prevents the growth of pathogenic bacteria in babies. At birth, a human infant or a calf has a sufficient level of iron. Magnesium is essential for skeletal development, protein synthesis, muscle contraction, and nerve function (Walstra et al. 2006).

2.5 *Vitamins, Enzymes, and Miscellaneous Compounds*

Vitamins are organic substances and essential for normal life. Milk is a good source of some vitamins including A, B_1, B_2, and B_{12}; however, milk is not good source of vitamins C, D, and E. The latter vitamins should be taken from other foods for a balanced diet. Carotenes are pro-vitamin A and give a yellow color in milk fat, as well as riboflavin give a fluorescent appearance for the yellowish color of the whey. Tocopherols play an antioxidative role and prevent off-flavor in milk fat (Fox et al. 1996).

Enzymes in milk are secreted in mammary gland and also released by bacteria. The enzymes produced in mammary gland are called as indigenous, while the other are called bacterial enzymes. Bacterial enzymes (e.g., proteinases, lipases) are released after lysis of the bacteria. Indigenous milk enzymes are present in solution or associated with the fat globule membranes or the casein micelles. Several enzymes including peroxidase, catalase, phosphatase, xanthine oxidase, etc., in milk are utilized for the quality control and as marker in the testing of milk (Ucuncu 2005).

Proteinases and lipases in milk are very important enzymes in the technology of liquid milk and processed dairy foods. Plasmin is an indigenous milk proteinase and is largely associated with casein micelles. Plasmin hydrolyzes casein at alkaline pH optima (7.5) and produces γ-caseins and proteose-peptones from β-casein. It is heat stable and may cause gelation in ultrahigh temperature (UHT)-treated milk.

Catepsin D is also an indigenous proteinase in milk and originated from somatic cells. The enzyme is not stable at higher temperature like plasmin and its pH optima is low (4.0). These two enzymes contribute to proteolysis in cheese more or less. Lipase is also present in milk and is associated with casein micelles. The action of lipase is weak; however, some processes including homogenization, agitation, etc., may enhance lipolysis in milk (Kelly et al. 2006).

Trace levels of organic acids are present in milk; however, their functions are not negligible. Citric, lactic, pyruvic, orotic, oxalic, hippuric, oxalic, propionic acids, etc., are present and contribute the pH of milk more or less. Milk contains several hormones also and the most known hormones are estrogen, prolactin, progesterone, somatotropin, etc. (Metin 2001). Somatic cells are normally present in milk and their counts are increased when udder is infected. The majority of somatic cells are leukocytes, white blood cells, which become present in increasing numbers in milk usually as an immune response to a mastitis-causing pathogen. The somatic cells also include a small number of epithelial cells, which are milk-producing cells shed from inside of the udder when an infection occurs. Essentially, a lower somatic cell count indicates better animal health, as somatic cells originate only from inside the animal's udder (Walstra et al. 2006). When the number of somatic cells increases, milk yield is likely to drop, primarily due to the damage to milk-producing tissue in the udder caused by mastitis pathogens. In EU countries, somatic cell count limit is 400,000 per mL of milk for human consumption.

3 Dairy Foods

3.1 Liquid Milk

Milk is the primary source of nutrients for young mammals before they are able to digest other types of food. At the early stage of lactation, the chemical and nutritional composition of the milk is completely different and it is called colostrum, which carries the mother's antibodies to the baby and can reduce the risk of many diseases. It also contains many other nutrients and the composition of the colostrum is converted to normal milk after a week of birth (Walstra et al. 2006). Normal liquid milk is produced and moved to market places as raw, pasteurized, and sterilized forms. Marketing of raw milk is impermissible in many countries due to potential risks for public health and food safety. It is possible to preserve the milk by pasteurization technique and the pasteurized milk is safe for people and the milk gains a shelf-life 1 week or longer at 4 °C in a suitable package (Fig. 4).

Most of pathogenic microorganisms especially (*Mycobacterium tuberculosis, Salmonella* spp., enteropathogenic *E. coli, Campylobacter jejuni, Listeria monocytogenes*, etc.) and contaminants in milk during milking and storage can be eliminated by pasteurization. Also, some enzymes which catalyze lipolytic reactions may be inactivated at heat treatment at certain levels. Homogenization of drinking milk is usually mandatory because many consumers dislike such a fat layer in milk

Fig. 4 Packages used for UHT-treated milk

Table 2 Loses (%) in some nutrient in some types of beverage milk (from Walstra et al. 2006)

Types of beverage milk	Lysine	Vitamin B_1	Vitamin B_6	Vitamin B_9	Vitamin B_{12}	Vitamin C
Pasteurization	0	5–10	0–5	3–5	3–10	5–20
UHT sterilization, just after	0	5–15	5–10	10–20	10–20	10–20
UHT sterilization, after 3 months	2	10–20[a, b]	20–50[a]	30–100[b]	20–50[b]	30–100[b]
In-bottle sterilization	5–10	20–40	10–20	30–50	30–60	30–60
Boiling	10–20	5–8	20	15	15–20	10–20

[a]Dependent on exposure to light
[b]Dependent on O_2 concentration

and the homogenized milk is highly susceptible to lipolysis because of its readily accessible substrate (Walstra et al. 2006). Extension of shelf-life of drinking milk (until 1 month) may be achieved by UHT treatment, bactofugation, and microfiltration techniques. Some bacterial and indigenous milk enzymes are still active and these enzymes restrict the shelf-life of the milk produced. To produce longer shelf-life of beverage milk (4–6 months at room temperature, not necessary for refrigerated storage condition), it should be sterilized by UHT or in bottle sterilization techniques. In bottle sterilization, browning in color, cooked taste, off-flavors, loses in vitamins may be observed, but these defects may be diminished, in some extent, by using UHT techniques. The loss of some nutrients in milk based on heat treatment are shown in Table 2. The level of lactulose (not more than 600 mg per liter of UHT milk) and residual plasmin activity (not more than 1 %) are markers to characterize the UHT-treated milk by means of chemical changes during heating. Sterilized milk must be aseptically filled with no head space in oxygen and light-impermeable packs to prevent enzymatic or oxidative deterioration during storage. In this condition, milk sterilized by UHT technique may be stored about 6 months at room temperature (Metin 2001).

3.2 Fermented Milks

Fermented milks are produced by lactic or alcoholic fermentation (in a lesser extent) of milk under certain conditions. In fermentation, mesophilic or thermophilic lactic acid bacteria and yeasts and mould play a major role during fermentation. The microorganisms result in some metabolites, predominantly lactic acid and acetic acid, acetaldehyde, diacetyl, carbon dioxide, and other compounds produced by fermentation. The main fermented milk is yogurt and the others are kefir, kumis, ayran, cultured buttermilk, acidophilus milk, probiotic fermented milk, ymer, filmjölk, langfil, etc.

Yogurt is a fermented dairy product obtained by lactic acid fermentation of milk by the action of yogurt starter bacteria (*Streptococcus thermophilus* and *Lactobacillus delbrueckii* subsp. *bulgaricus*) and is the most popular fermented milk throughout the world. The highest production or consumption of yogurt is in Mediterranean, Asian countries, and in central Europe (Tamime and Deeth 1980; Tamime and Robinson 1999). The origin of the yogurt is not known definitely; however, the historical records say the origin of yogurt was Middle East and it was firstly made by Turks when they were in Middle Asia and it was named as "yogurt." Now, the product has gained an international recognition with this word, and many other countries use "yoghurt" (Hayaloglu et al. 2007). Different fat contents reflect the source of milk the yogurt is made from. To increase its consumption and variety, fruit flavors and cereals can also be added. After 1990s probiotic bacteria and prebiotics are added to yogurt and it is often used as a carrier for probiotic or prebiotic.

Recently, a drinkable fermented milk (called *ayran*) began to be produced and its popularity has grown increasingly. Originally, it is produced in Turkey and exported to neighboring countries including EU members. It is the most popular drinkable fermented product in Turkey and it is claimed that ayran was discovered by Gokturks during having gone on a war, where yogurt was diluted with water to reduce its sour taste (Kocak and Avsar 2010). Presently, ayran is produced by three different methods as follows: (1) dilution of yogurt with drinkable water (1:1, v/v), (2) fermentation of milk which is standardized to 7 or 8 % dry matter by using water, (3) dairy byproduct obtained during traditional manufacture of butter from yogurt, after dilution of yogurt with water and churning (Hayaloglu and Karagul-Yuceer 2011). The most popular method is fermentation of diluted milk and the method is used in well-mechanized dairy factories. After completion of fermentation, salt is added to ayran at the level of 0.5–1.0 % with regardless of the production method mentioned above. Yogurt bacteria, *Streptococcus thermophilus* and *Lactobacillus delbrueckii* subsp. *bulgaricus,* are used in the fermentation of ayran milk. To increase its viscosity, exopolysaccharide-producing cultures have been used for its production. The flavor formation and textural properties are strongly dependent on the activities of these bacteria. Ayran contains the same nutrients as in yogurt and it is also carrier for probiotics. It is served cold and often as an accompaniment to grilled meat, chicken, kebab, bulgur products, etc., especially in the summer months.

Kefir is a fermented and self-carbonated (contains carbon dioxide) fermented dairy beverage originated from northern Caucasus, which began to interest more other countries including Turkey, Sweden, Norway, Finland, Germany, Russia, and some European countries in east. The name "kefir" comes from the Turkish language, i.e., the word "keyif" meaning good feeling, stimulating, encouraging by self (Hayaloglu and Karagul-Yuceer 2011). Kefir has its distinctive flavor to a mixture of lactic acid, ethanol, carbon dioxide, and other flavor products such as acetaldehyde and acetoin. Its lactic acid and alcohol contents are 0.7–1.0 % and 0.05–1 %, respectively; however, the alcohol content is not over 0.5 %. The unique flavor is the result of the symbiotic metabolic activity of a number of bacterial and yeast species. Because of the ability of kefir to establish healthy intestinal flora, it is beneficial in preventing many gastrointestinal disorders. Kefir contains certain healthy bacteria that is not available in yogurt, including *Lactobacillus caucasus*, *Leuconostoc*, *Acetobacter*, *Streptococcus*, *Saccharomyces kefir*, and *Torula kefir*. These beneficial microorganisms may help to support digestive health and prevent the growth of harmful bacteria in the intestines. Vitamins, such as vitamin K and B_{12}, are produced in the gut and the probiotics in kefir may potentially help facilitate the production.

Kumis is a fermented dairy product similar to kefir and is traditionally produced using mare's milk in Russia and western part of Asia. Mare's milk contains higher sugar (about 7 %) than those of cow's milk (4.5 %). Kumis has higher levels of alcohol (*ca.* 2 %) than kefir (*ca.* 0.5 %). In addition of the fermentation of lactose by yogurt starter bacteria and *Torula kumis*, proteolysis is observed during manufacture of kumis due to complex microflora and the proteolysis contributes to flavor of kumis. Both kefir and kumis are good alternative for lactose malabsorbers due to conversion of lactose to lactic acid and alcohol. Kumis cure has been applied in Russia for patients for treatment of intestinal and chronic deceases (Tamime and Marshall 1997).

By consuming of fermented milks, some benefits may be gained as follows:

- A slight lower energized, lower lactose (good for lactose malabsorbers)
- Lower pH of products (good for diminishing pathogens)
- Probably higher levels of probiotic bacteria and prebiotics (these fermented milks play a carrier role)

3.3 Cheese

Cheese evolved in the "Fertile Crescent" between the Tigris and Euphrates rivers, about 8,000–9,000 years ago. This area now forms part of Turkey, Iraq, and Iran. More than 1,000 varieties of cheese are produced around the world. Cheese is made using coagulating of milk by milk-clotting enzymes or organic acids and then by removal of the whey (Hayaloglu and Ozer 2011). The resultant curd is ripened for a period (2 weeks to 2 years) with or without selected microbial cultures (Fox et al. 2004). The manufacture of cheese has a number of steps which are specific for cheese variety; however, the main production methodology of cheese has been illustrated in Fig. 5.

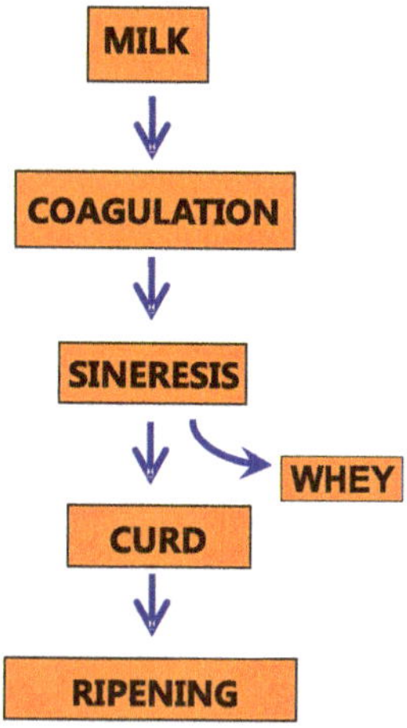

Fig. 5 Main steps in cheese manufacture with regardless of variety

Table 3 Chemical composition (in 100 g) of some varieties of cheese (from Fox et al. 1996; Hayaloglu and Ozer, 2011)

Cheeses	Water (g)	Fat (g)	Protein (g)	Cholesterol (mg)	Energy (Kcal)
Beyaz peynir	58.7	19.8	17.9	53.4	250
Cheddar	36.0	34.4	25.5	100	412
Cottage	79.1	3.9	13.8	13	98
Edam	43.8	25.4	26.0	80	333
Emmental	35.7	29.7	28.7	90	382
Feta	56.5	20.2	15.6	70	250
Gouda	40.1	31.0	24.0	100	375
Gruyere	35.0	27.2	33.3	100	409
Kashar	44.3	28.3	24.6	73.3	353
Mozzarella	49.8	21.0	25.1	65	289
Parmesan	18.4	32.7	39.4	100	452
Roquefort	41.3	32.9	19.7	90	375
Stilton	38.6	35.5	22.7	105	411
Tulum	50.5	27.4	19.4	92.6	324

Cheese contains a high level of fat and protein (about tenfold when compared to milk). The fat and protein levels in cheeses are dependence of variety and values ranges from 3.9 to 35.5 % and 13.8 to 39.4 %, respectively (Table 3). Cow's, sheep's, goat's, and buffalo's milk have been used for the manufacture of cheese; however, the most common source is cow's milk (about 80–90 %) due to scarce availability (during all season of year) of other species' milk in the world (Fox et al. 2004).

Cheese contains essential nutrients (Table 3) which are influenced by type of milk (milk species, lactation stage, milk constituents), manufacturing, and ripening methods. Although caseins, colloidal minerals, fat, fat-soluble vitamins are retained in curd, whey proteins, lactose, water-soluble vitamins, and minerals are lost in the whey (Fox et al. 1996; Hayaloglu and Ozer 2011).

So, sulfur-containing proteins are also lost in whey; consequently, the biological value of cheese protein (essential amino acid index is 91–97) is slightly less than that of whole milk protein (essential amino acid index is 100). Cheese contains only small amounts of carbohydrate (mainly lactose) and the residual lactose in the curd is almost completely converted to lactic acid by starter or non-starter microorganisms. This makes the cheese a suitable dairy food for lactose malabsorbers. Fat in cheese serves a vitamin and dietary fat carrier and it affects textural and microstructural properties of cheese. Cheese fat contains about 66 % saturated, 30 % monounsaturated, and 4 % PUFA. So, it represents a significant dietary source of both total fat and saturated fatty acids. Cholesterol content in cheese is highly dependent on variety and it ranges from 13 to 105 mg/100 g (Table 3). Cholesterol content of cheese has much less importance on blood cholesterol levels. Although fat-soluble vitamins are retained in cheese fat, water-soluble vitamins (B vitamins) are lost in the whey. However, B vitamins are synthesized by some microorganisms during cheese ripening. High levels of calcium, phosphorus, and magnesium are present in cheese and the levels of calcium in a hard-type cheese (800 mg/100 g) meet the recommended daily requirement for adults. A 100 mL of whole milk contains about 125 mg of calcium and the bioavailability of calcium is equivalent to that from milk (about 25 % is absorbed in the body). Dietary iron content of cheese is insufficient and fortifying with iron may enhance their nutritional value. Consumption of salty cheeses (especially brine-ripened varieties) may cause a high intake level of sodium which give rise to hypertension, in turn, coronary heart disease. However, cheese contributes only about 5–8 % of total sodium intake daily and the other foods in the diet should be taken into consideration for their sodium contents (Walstra et al. 2006; Hayaloglu and Ozer 2011).

Some studies showed that cheese prevents tooth decay and eliminate the effects of acids formed by bacteria. Cheese stimulates the secretion of saliva and the saliva rinse the mouth from acids and sugar. Also, the casein, calcium, and phosphor protect tooth enamel.

Mycotoxins may be present in cheese and these are originated from the animals' feeds or contamination of cheese by mycotoxin-producing fungi. Aflatoxin B1 may be present in animals' feed and the toxin is excreted in milk as aflatoxin M1 which is stable during cheese making. However, the availability of the toxin has been decreased due to strict control in animals' feedstuff due to legislations in many countries. Some strains *Penicillium* spp. including *P. roqueforti* and *P. camemberti* which are used as culture in blue-type or surface-ripened cheese may produce some toxins; however, these strains do not produce aflatoxins. The toxins, including patulin, pencillic acid, and PR toxin, have not been determined in commercial blue-type cheeses, probably these were not at detectable levels (Hayaloglu and Ozer 2011).

Nonvolatile low molecular weight aliphatic and heterocyclic amines, called biogenic amine, such as histamine, tyramine, tryptamine, putrescine, cadaverine, and phenylethylamine may be present in cheeses. Biogenic amines can be formed in cheese during ripening by the action of decarboxylase-positive microorganisms, especially lactobacilli. If these microorganisms were inactivated by pasteurization, no amine is formed in cheese. High concentrations of biogenic amines can be toxic for human and these influence neurotransmitters, changes of perception, smooth muscle contractions, headache, etc. Even the ingestion of large concentrations of biogenic amines does not signal any toxicity symptoms in most individuals due to conversion of amines to aldehydes by mono- and diamine oxidases or carboxylic acids by oxidative deamination. If the mono- and diamine oxidases are impaired either due to genetic defects or administration of inhibitory drugs, the amines cause adverse reactions in the body (Hayaloglu and Ozer 2011).

3.4 Butter and Cream

Milk is separated to make a cream. Single cream contains about 18 % fat, whereas double cream is about 48 % fat and whipping cream about 40 %. These creams are mainly used for the production of butter and ice cream. Butter is a water-in-oil emulsion made from "cream by phase inversion" by churning process. Butter is about 80 % fat and also contains fat-soluble vitamins, e.g., A (butter is good source), D, E, K. Butter can be manufactured using sweet or acidified (cultured) creams (Anonymous, no date). Lipolytic and oxidative changes occur and accordingly off-flavors may be observed in butter. In Turkey, butter is made from yogurt, but this method is not common in recent years due to unmechanized production technology. Butter contains lactose at trace levels, but high level of cholesterol. Recently, reduced fat spreads based on dairy or non-dairy components are often more popular. Spreads are often used to carry "functional" ingredients such as cholesterol-lowering plant-derived sterols. Diacetyl is a principal flavor component of butter and other carbonyl compounds and organic acids also contribute to the flavor of butter (Ucuncu 2005).

3.5 Milk Powder

Milk powder, powdered milk, or dried milk is manufactured by evaporating of milk to dryness. The resultant product can be stored for long periods (may be up to 3 years when produced from skimmed milk) without refrigerated storage due to low moisture content. There are no significant deterioration/changes in flavor and nutritive values during storage. Powdering process gives an advantage in storage and transportation conditions due to reduction of volume/bulk. In the production of powdered milk technology, dry whole milk, non-fat dry milk, dry buttermilk, dry whey products, and dry dairy blends may be produced with suitable source(s) (Caric 1994). Chemical composition of the milk powder from different sources is varied as shown in Table 4. The chemical composition and physical properties (appearance, solubility, particle size, etc.) are varied due to differences in the composition of raw

Table 4 Chemical composition (mean values, %) of some types of powdered milk (from Walstra et al. 2006)

Constituents	Whole milk	Skimmed milk	Whey	Buttermilk
Fat	26	1	1	5
Lactose	38	51	73	46
Casein	19.5	27	0.6	26
Other proteins	5.3	6.6	8.5	8
Ash	6.3	8.5	8	8
Water	2.5	3	3	3

material used in manufacture. Regardless of the source, the produced powders can be used for different food industries including (Walstra et al. 2006):

- Bakery industry to increase the volume of bread and increase water holding capacity
- Bakery and confectionery industry (including caramel and candy) to replace eggs
- Chocolate industry to produce milk chocolate
- Meat industry as a filling material in sausages and similar products
- Infant formulas to substitute mother's milk
- Dairy industry to produce ice cream or increase milk non-fat dry solid in yogurt production to prevent serum separation
- Dairy industry to produce drinking milk by reconstituting about 10 % (w/v) of the water. It is delivered by United Nation (UN) aid in developing countries due to lower transport and storage costs.

Nutritional value of milk powder is noticeable. If the products are stored at suitable conditions, nutritional losses are not observed. Whole milk powder contains all amino acids including majority of essential ones, calcium and potassium. Due to the oxidation of milk fat in the powder and its gradual deterioration, whole milk fat is not preferred for human consumption. The shelf-life of resultant milk powder can be extended by addition of antioxidant, packaging under a system which is impermeable to oxygen and light or inert gas. Under these conditions, chemical or oxidative reactions in milk powder take place slowly and the nutritive value is protected during storage (Ucuncu 2005; Walstra et al. 2006).

3.6 *Ice Cream*

Ice cream is a mixture of water, sugar, flavor substances, and other components, which are partly frozen and beaten to form a rigid foam (Marshall and Arbuckle 2000). There are four types of ice cream according to the ingredients used: (a) ice cream made exclusively from milk or its products (dairy-based ice cream), (b) ice cream containing vegetable fat, (c) sherbet made using fruit juice and milk, (d) water ice made using water, sugar, and fruit concentrate. Ingredients used for ice cream are fat (12 %), milk solids-non-fat (11–11.5 %), sugar (10–18 %), water, emulsifiers (0.3–0.5 %), stabilizers (0.2–0.5 %), flavoring and coloring agents. Fat is originated milk or vegetable fat and it is the flavoring and structure-forming agent in ice cream (Anonymous, no date; Ucuncu 2005). It has a significant role on the consistency, appearance, and freezing and melting behavior of ice cream. Milk solid-non-fat consist of proteins, lactose, and salts in milk; these are responsible for depression of freezing point, increasing of viscosity, stabilizing of the foam formed, and the formation of fat globule membrane after homogenization. Sugar is the main ingredient in ice cream making and it gives a sweetness which consumers prefer. Different types of sugar can be used in ice cream manufacturing, e.g., cane or beet sugar, glucose, lactose, invert sugar, glucose syrup, etc. Lack of sugar may cause the formation of ice too much. Sorbitol is used in the manufacture of ice cream for diabetics. Stabilizers, including pectin, gelatin, xhanthan gum, alginate, carrageenan, guar gum, locust bean gum, and carboxymethylcellulose, have high importance in ice cream making. These affect the consistency and consequently heat transfer during freezing, structure, water-binding, forming a network, and crystallization (Ucuncu 2005; Walstra et al. 2006). Emulsifiers help to reduce the surface tension and the proteins are sufficient to overcome the task. That is, no necessary to use an emulsifier in ice cream making in the practice. Egg yolk, monoglycerides, and sorbitan esters can be used as emulsifiers if necessary. Flavoring and coloring agents are optional and the most commonly used flavors are vanilla, nougat, chocolate, strawberry, nut, and peanut (Marshall and Arbuckle 2000).

In Turkey, a high quality ice cream has been manufactured and it has a geographical indication by Turkish Patent Institute with the name "Maras ice-cream." The distinctive characteristics of this type of ice cream are the hard texture, the longer melting time at room temperature, and the use of goats' milk. The strong and compact texture of this ice cream are due to a special stabilizer called salep, in Turkish, which comes from a wild orchid genus such as *Orchis* (*Orchis anatolica*, *O. macula*, *O. spitzelii*, *O. morio*, and so on) *Serapias*, *Plantanthera*, and *Dactylorhiza* (Hayaloglu and Karagul-Yuceer 2011).

Ice cream is a frozen dessert and it has high levels of nutrients including milk fat, fat-soluble vitamins, low lactose (good for lactose malabsorbers), omega-3 fatty acids and antioxidants (when included nut), and some water-soluble vitamins and minerals (when included fruits).

4 Conclusions

Milk and dairy foods contain many nutrients which are essential for children and adults. In this chapter, chemical composition and nutritional aspects of milk and dairy foods are shortly reviewed and nutritional changes during their manufacture are discussed. Milk is a good source for protein, fat-soluble vitamins, and salts especially calcium and phosphorus. By fermentation of milk, a number of dairy products are produced and sometimes these are highly preferred by consumer due to their different taste and high nutritive values, such as high levels of vitamins, low pH, and low level of lactose. Lactose in milk cannot be digested by about 60–70 % of world population and the alternative dairy products (such as some fermented milk and cheese) may be suggested for lactose malabsorbers. Some ingredients (e.g., some fruits, nuts, probiotics, prebiotics, etc.) contribute a functional or nutritive superiority in ice cream, yogurt, ayran, etc. Toxins or amines may be present in dairy products; however, their toxicity levels are quite low.

References

Anonymous (no date) Dairy handbook. Alfa-Laval AB, Lund

Caric M (1994) Concentrated and dried dairy product. VCH, New York

Donmez M, Seckin AK, Sagdic O, Simsek B (2005) Chemical characteristics, fatty acid compositions, conjugated linoleic acid contents and cholesterol levels of some traditional Turkish cheeses. Int J Food Sci Nutr 56:157–163

Fox PF, McSweeney PLH (1998) Dairy chemistry and biochemistry. Chapman & Hall, London

Fox PF, McSweeney PLH (2003) Advanced dairy chemistry: volume 1 proteins, 2nd edn. Kluwer Academic, New York

Fox PF, McSweeney PLH (2006) Advanced dairy chemistry: volume 2, lipids, 2nd edn. Springer, New York

Fox PF, O'Connor TP, McSweeney PLH, Guinee TP, O'Brien NM (1996) Cheese: physical, chemical, biochemical and nutritional aspects. Adv Food Nutr Res 39:163–328

Fox PF, Guinee TP, Cogan TM, McSweeney PLH (2000) Fundamentals of cheese sciences. Aspen, Gaithersburg

Fox PF, McSweeney PLH, Cogan TM, Guinee TP (2004) Cheese: chemistry, physics and microbiology, vol 1, 3rd edn, General aspects. Elsevier, London

German JB, Dillard CJ (2006) Composition, structure and absorption of milk lipids: a source of energy, fat-soluble nutrients and bioactive molecules. Crit Rev Food Sci Nutr 46:57–92

Hayaloglu AA, Karagul-Yuceer Y (2011) Utilization and characterization of small ruminants' milk and milk products in Turkey: current status and new perspectives. Small Rumin Res 101:73–83

Hayaloglu AA, Ozer BH (2011) Peynir Biliminin Temelleri [Principles of cheese sciences]. Sidas Medya, Izmir, Turkey

Hayaloglu AA, Karabulut I, Alpaslan M, Kelbaliyev G (2007) Mathematical modeling of drying characteristics of strained yoghurt in a convective type tray–dryer. J Food Eng 78:109–117

Kelly AL, O'Flaherty F, Fox PF (2006) Indigenous proteolytic enzymes in milk: a brief overview of the present state of knowledge. Int Dairy J 16:563–572

Kocak C, Avsar YK (2010) Ayran: microbiology and technology. In: Yildiz F (ed) Development and manufacture of yogurt and other functional dairy products. CRC Press, Boca Raton

Korhonen H, Pihlanto-Leppälä A (2001) Milk protein-derived bioactive peptides-novel opportunities for health promotion. IDF Bull 363:17–26

Kosikowski FV, Mistry VV (1997) Cheese and fermented milk foods, vol 1. F.V. Kosikowski LLC, Westport, CT

Marshall RT, Arbuckle WS (2000) Ice cream, 5th edn. Aspen, Gaithersburg

McSweeney PLH, Fox PF (2009) Advanced dairy chemistry, vol 3, 3rd edn, Lactose, water, salts and minor constituent. Springer, New York

Metin M (2001) Sut Teknolojisi [Milk technology]. EU Muhendislik Fakultesi Yayinlari, Izmir, Turkey

Sieber R, Butikofer U, Egger C, Portmann R, Walther B, Wechsler D (2010) ACE-inhibitory activity and ACE-inhibiting peptides in different cheese varieties. Dairy Sci Technol 90:47–73

Tamime AY, Deeth HC (1980) Yoghurt: technology and biochemistry. J Food Prot 43:939–977
Tamime AY, Marshall VME (1997) Microbiology and technology of fermented milks. In: Law BA (ed) Microbiology and biochemistry of cheese and fermented milk. Blackie Academic and Professional, London
Tamime AY, Robinson RK (1999) Yoghurt: science and technology, 2nd edn. Woodhead, Cambridge, UK
Ucuncu M (2005) Sut ve Mamulleri Teknolojisi [Technology of milk and dairy products]. Meta Basim, Izmir, Turkey
Walstra P, Wouters JTM, Geurts TJ (2006) Dairy science and technology, 2nd edn. Taylor & Francis, Boca Raton

Meat Products and Byproducts for Value Addition

Saghir Ahmad and Abdol Ghafour Badpa

1 Introduction

Meat refers to skeletal muscle and associated fat and other tissues, but it may also describe other edible tissues such as offal. Conversely, meat is sometimes also used in a more restrictive sense—the flesh of mammalian species (pigs, cattle, lambs, buffalo, etc.) raised and prepared for human consumption, to the exclusion of fish and other seafood, poultry, and other animals.

The word meat comes from the Old English word mete, which referred to food in general. The term is related to mad in Danish, mat in Swedish and Norwegian, and mature in Icelandic and Faroese, which also mean food. The word "mete" also exists in Old Frisian (and to a lesser extent, modern West Frisian) to denote important food.

Meat is defined as the edible portion of the muscles of terrestrial animals, particularly cattle, sheep, pigs, etc. In practice this definition is restricted to a few dozen of the 3,000 mammal species, but it is often widened to include edible organs like kidney, liver, brain, and heart as well.

Adult mammalian muscle flesh consists of roughly 75 % water, 19 % protein, 2.5 % intramuscular fat, 1.2 % carbohydrates, and 2.3 % other soluble non-protein substances. These include nitrogenous compounds such as amino acids and inorganic substances such as minerals (Lawrie and Ledward 2006; Hodgson et al. 2006). Muscle proteins are either soluble in water (sarcoplasmic proteins, about 11.5 % of total muscle mass) or in concentrated salt solutions (myofibrillar proteins, about 5.5 % of the mass) (Lawrie and Ledward 2006).

S. Ahmad (✉) • A.G. Badpa
Faculty of Agricultural Sciences, Department of Post Harvest Engineering and Technology, Aligarh Muslim University, Aligarh, UP 202002, India
e-mail: alsaghir@rediffmail.com

A. Malik et al. (eds.), *Food Processing: Strategies for Quality Assessment*, Food Engineering Series, DOI 10.1007/978-1-4939-1378-7_5

Meat can be broadly classified as red or white depending on the concentration of myoglobin in muscle fibers. When myoglobin is exposed to oxygen, reddish oxymyoglobin develops making myoglobin-rich meat appear red. The meat of adult mammals such as cows, sheep, goats, and horses is generally considered red, while domestic chicken and turkey breast meat is generally considered white.

An animal is not all meat, the animal in view of industry point consists of two parts: carcass and offal; the carcass in turn consists of meat and trim, the latter being mainly bones and excess fat. The definition of carcass is variable, but the definition of offal has been always the same: it is everything other than the carcass. In general beef and sheep carcasses do not include the head, feet, tail, skin, blood, and internal organs; poultry is similar but include the (de-feathered) skin, while pig carcasses may include everything other than the outer layers of skin, the hair, toenails, blood, and internal organs.

Some offal is edible (e.g., Liver, tripe, and kidneys), some is inedible (e.g., Hides and hooves) and some can be used as pharmaceuticals (e.g., Heparin and insulin). All parts of the offal can be sold, and the profitability of the meat packaging business has traditionally depended heavily on its ability to market byproducts.

1.1 Status of Meat

The world buffalo population is estimated at 185.29 million, spread in some 42 countries, of which 179.75 million (97 %) are in Asia (FAO 2008). India has 105.1 million and they comprise approximately 58 % of the total world buffalo population. During the last 10 years, the world buffalo population increased by approximately 1.49 % annually, by 1.53 % in India, 1.45 % in Asia, and 2.67 % in the rest of the world (CIRB 2009); refer to Table 1.

Indian buffalo meat production is growing significantly. Although no official production statistics are available, industry sources and export data indicate that continued strong export demand is triggering an expansion in buffalo meat supplies in India. As a result, new slaughter houses are emerging, providing farmers with a new market for nonproductive buffalo heifers, bulls, and bull calves. In the current year (CY-2013) Indian buffalo meat production is thus forecast to rise to a record of

Table 1 World wide status of buffalo population

Country	1997	2008	Percent change
India	89.91 (56.48)	105.1 (56.7)	1.53
Pakistan	20.83 (13.08)	29.0 (15.65)	3.56
China	21.73 (13.65)	23.27 (12.55)	0.64
Asia	154.91 (97.31)	179.75 (97.01)	1.45
Rest of the world	4.28 (2.69)	5.54 (2.99)	2.67
World	159.19	185.29	1.4

Source Fao.org/stat (2008)

4.16 million tons (on a carcass weight equivalent basis), up 14 % from CY 2012. CY 2012 buffalo meat production is estimated at 3.64 million tons (up 12 % from CY 2011), and CY 2011 production has been slightly revised up to 3.24 million tons. Industry and government sources have indicated that cattle supplies will remain robust over the next decade, but will level off as more productive dairying technology is adopted and inefficient dairy producers exit the market.

1.2 *Trends in Export of Buffalo Meat*

India is a net exporter of buffalo meat (deboned frozen buffalo meat). In the last 2 years, exports have grown to record levels, making India the fourth country in the world to export more than one million tons of bovine meat annually. India's growing exports are the result of its low cost of production (relative to international competitors). Production costs are low due to herd growth of strong dairy demand and new incentives from slaughter facilities to salvage previously underutilized animals. As a result, calendar year (CY) 2013 buffalo meat exports was forecast at 2.15 million tons (on a carcass weight equivalent basis), around 30 % over 1.66 million tons in CY 2012 (Singh and Wright 2012; Kumar et al. 2012). CY 2011 buffalo meat exports are also revised up to a record of 1.29 million tons, based on trade data. Year-on-year export growth for 2011 is 41 %, while 2012 growth is 28 %. Given its tremendous export growth, India is likely to become the world's largest beef (buffalo meat) exporter by 2013, if not sooner. Import of beef from all sources is restricted and as such imports are set at nil.

As a price-based competitor, India has seen export increases in the previous 2 years to Middle Eastern, African, and Southeast Asian countries (Singh and Wright 2012; Kumar et al. 2012). Table 2 presents Indian buffalo meat exports and Table 3 average annual quantity and value of export of buffalo meat from India.

Table 2 Indian buffalo meat exports (Metric tonnes)

	Partner country	2007	2008	2009	2010	2011
	World total	484,499	480,339	434,704	654,985	924,177
1	Vietnam	39,151	92,258	111,166	90,773	254,959
2	Malaysia	52,325	50,553	48,214	79,865	96,646
3	Egypt	26,560	43,567	28,590	69,779	67,804
4	Saudi Arabia	35,310	25,599	24,274	47,634	62,553
5	Jordan	20,835	15,465	12,503	41,742	61,883
6	Philippines	51,365	53,036	41,479	43,828	47,196
7	United Arab Emirates	26,887	18,416	16,594	31,757	38,184
8	Algeria	30	0	0	18,158	33,381
9	Iran	12,238	12,112	5,391	18,023	27,131
10	Iraq	5,370	2,982	7,436	17,652	23,902
11	Thailand	99	77	0	3,855	22,996

(continued)

Table 2 (continued)

	Partner country	2007	2008	2009	2010	2011
	World total	484,499	480,339	434,704	654,985	924,177
12	Angola	47,863	29,870	17,082	19,916	21,287
13	Kuwait	40,315	31,203	28,363	36,234	20,619
14	Syria	0	1,702	6,496	13,929	18,838
15	Georgia	11,214	8,189	6,563	9,427	16,704
16	Oman	11,962	11,501	8,027	11,351	12,564
17	Congo	10,576	13,997	12,148	12,377	12,472
18	Myanmar	12	0	286	19,800	7,684
19	Gabon	7,653	6,234	5,837	7,857	7,459
20	Lebanon	5,068	6,448	6,830	8,823	7,308
21	Qatar	2,904	4,076	4,116	4,624	7,011
22	Azerbaijan	3,271	4,241	2,823	3,819	4,974
23	Armenia	4,346	5,570	3,717	2,793	4,780
24	Senegal	7,825	5,119	4,814	5,054	4,262
25	Pakistan	13,049	3,701	2,939	2,967	3,822
26	Ghana	9,489	6,425	3,674	4,492	3,416
27	Mauritius	3,361	4,001	2,974	3,189	3,348
28	Equatorial Guinea	1,366	1,266	1,098	2,056	3,016
29	Bahrain	2,692	2,322	1,932	2,849	2,551
30	Cote d Ivoire	5,025	4,176	3,783	2,430	2,477
31	Brunei Darussalam	1,641	1,972	954	1,231	1,804
32	Yemen	3,163	1,368	1,314	2,450	1,769
33	Comoros	1,702	2,007	1,786	1,945	1,677
34	Tajikistan	2,790	1,762	2,196	1,681	1,644
35	Uzbekistan	897	624	196	577	1,608
36	Liberia	449	452	254	655	1,594
37	South Africa	1	0	1	598	1,539
38	Canada	0	0	0	280	1,445
39	Namibia	393	1,181	281	813	1,291
40	Turkey	943	306	158	836	1,153
41	Maldives	548	311	186	530	938
42	Afghanistan	3,390	1,745	464	405	838
43	Hong Kong	572	83	4,063	360	802
44	China	948	630	363	27	709
45	Sierra Leone	838	503	255	263	581
46	Benin	0	58	29	316	309
47	Seychelles	53	333	219	336	266
48	Singapore	0	0	0	525	210
49	Albania	1,110	1,425	726	141	84
50	United States	0	121	67	193	2
51	Greece	1,560	512	0	0	0
52	Mozambique	1,237	57	0	252	0
53	Germany	37	12	7	450	0

(continued)

Table 2 (continued)

	Partner country	2007	2008	2009	2010	2011
	World total	484,499	480,339	434,704	654,985	924,177
54	United Kingdom	1	0	0	433	0
55	Indonesia	0	0	0	683	0
	Others	4,065	771	2,036	1,952	2,687

Source Global Agricultural Information (2012)

Table 3 Average annual quantity and value of export of buffalo meat from India

Period (TE)	Quantity (laces tonnes)	Value (million US $)	Unit value of buffalo meat (US $/kg)
1992–93	7.6	68.9	0.91
1995–96	12.6	119.6	0.94
1998–99	16.3	178.3	1.10
2001–02	23.3	234.61	1.00
2004–05	32.6	333.00	1.02
2007–08	47.9	745.8	1.55
2010–11	55.6	1351.3	2.40

Source Basic data from DGCI&S, Government of India (2011)

1.3 Nutritional Value of Meat

The quality of fresh meat can be judged by color, marbling, firmness, and texture. High-quality fresh meat has bright colors, slight marbling, firm structure, and fine texture. The nutritional value of meat comes from its proteins, vitamins, minerals, and fats. Although nutritionists no longer advise meat at every meal, meat is a good source of calories, proteins, and fats.

Red meat such as beef, veal, lamb, and goat is packed with important nutrients required for good health. Beef, veal, lamb, and goat meat has the highest iron and zinc content of commonly consumed protein foods. Iron and zinc in red meat are well absorbed by the body. The red meat has the highest iron content (Casey 1992). Table 4 shows the iron and zinc in 150 g uncooked meat.

A 150 g serve of uncooked red meat (average of beef, veal, and lamb) is a source of omega-3, providing 50 mg of omega-3. It contributes 22 % to the amount of omega-3 (500 mg) which is recommended for the prevention of chronic diseases (NHMRC 2006).

1.3.1 Protein Content

Meat products are an excellent source of complete proteins. Complete proteins are proteins that contain all of the amino acids that our body needs to function properly. Adequate protein intake is important for producing and maintaining healthy muscles, bones, skin, hair, blood, organs, and glands. Human body also uses proteins to repair damaged cells and to manufacture new cells (Moffett 2013).

A 150 g serve of uncooked red meat (average of beef, veal, and lamb) provides more than 50 % of the recommended dietary requirements for key nutrients. Component of Protein and vitamin in uncooked meat is presented in Table 5.

1.3.2 Vitamin Content

Meat products contain traces of several different vitamins. This includes vitamin E and vitamins B1, B2, B3, and B6. Vitamin E serves as an antioxidant, helping to stop free radicals from damaging cells. All B vitamins work together to convert the food into energy. Vitamin B1 helps promote healthy muscles, nerves, and a healthy heart; Vitamin B2 helps to manufacture red blood cells; Vitamin B3 helps promote good digestion; and Vitamin B6 helps to manufacture proteins.

Table 4 Iron and zinc in 150 g uncooked meat

150 g uncooked	Iron (mg)	% RDI	Zinc (mg)	% RDI
Beef	3	25	6	50
Veal	1.7	14	6.2	52
Lamb	3	25	6	50
Goat	3.8	32	6.5	54
Chicken	0.8	6.6	2	17
Pork	1.5	12.5	3.5	29
White Fish	0.9	7.5	2.3	19
Oily fish	1.7	14	0.9	7.5

SOURCE OF OMEGA-3
RDI Recommended daily intake

Table 5 Protein and vitamin in 150 g uncooked meat

Per 150 g uncooked	Protein (g)	%RDI	Vitamin B12 (ug)	% RDI
Average: Red meat	35	59	1.8	90
Beef	34	58	1.5	75
Lamb	33	56	1.5	75
Veal	37	63	2.4	120
Goat	33	56	1.5	75

1.3.3 Mineral Content

Meat products contain several minerals including magnesium, iron, and zinc. Magnesium is needed to build and maintain strong bones. Iron helps in transport of oxygen through blood. Zinc is important for maintaining a healthy immune system.

1.3.4 Unsaturated Fat Content

Some meat products contain high amounts of healthy unsaturated fats like omega-3 fatty acids. Omega-3 fatty acids are polyunsaturated fats that can have a positive impact on cardiovascular health. Adding omega-3 fatty acids to diet can help lower LDL cholesterol levels, blood pressure levels, and triglyceride levels. This may decrease risk of developing cardiovascular disease or arrhythmia. According to the American Heart Association, fish species like salmon, herring, and trout have the highest levels of omega-3 fatty acids.

1.3.5 Saturated Fat Content

Some meat products such as pork, lamb, beef, sausage, lunch meats, and organ meats like liver contain high levels of unhealthy saturated fats. According to the US Department of Agriculture, at 3.5-Oz. serving of lean beef contains less than 4.5 g of saturated fat. Eating too much saturated fat raises risk of developing high cholesterol and cardiovascular disease. The fatty parts of meat cuts contain the most saturated fat. To reduce fat intake, choose leaner cuts of meat like roasts, round steaks, pork loin, skinless chicken breasts, turkey cutlets, lean ground beef, and low-fat lunch meats.

1.3.6 Nutrient Composition of Organs

Table 6 provides a comparison of the nutrient content of liver, kidney, heart, brains, and tripe from beef and lamb. From this table the following general statements can be made:

- All organ meats (except tripe) are extremely rich in vitamin B12, with more than 100 % RDI in 100 g.
- Liver is a rich source of protein, iron, zinc, riboflavin, niacin, vitamin A, and folate.
- Kidney is rich in protein, thiamin, riboflavin, iron, and a source of folate.

Table 6 Selected nutrients (per 100 g) in raw liver, kidney, heart, brain and tripe[a]

	Liver		Kidney		Heart		Brain	Tripe
	Beef	Lamb	Beef	Lamb	Beef	Lamb	Beef	Beef
Protein (g)	20.0	21.4	18.2	17.1	18.2	17.8	12.3	13.2
Fat (g)	8.6	7.5	1.6	2.5	3.0	5.6	8.0	2.1
Saturated fat (g)	2.8	2.2	0.6	0.9	1.2	2.3	2.2	0.9
Long-chain omega-3 fat (mg)	561	361	47	103	54	102	574	20
Cholesterol (mg)	271	433	313	338	103	129	1,352	82
Thiamin (mg)	0.23	0.24	0.40	0.56	0.50	0.61	0.14	0
Riboflavin (mg)	4.80	2.80	3.60	2.10	1.50	1.10	0.40	0.10
Niacin (mg)	9.4	10.9	6.5	7.6	6.9	5.9	5.1	0.2
Folate (µg)	290	230	98	28	3	2	3	5
Vitamin B12 (µg)	59	90	28	52	9	10	11	1
Retinol equivalents (µg)	13,877	31,400	155	93	10	0	0	0
Zinc (mg)	3.6	4.3	1.8	2.6	1.6	1.6	1.1	1.1
Iron (mg)	5.8	9.5	5.4	9.8	5.0	3.9	1.7	0.4
Magnesium (mg)	15	19	15	16	17	17	12	6
Sodium (mg)	78	67	160	190	91	82	120	100
Potassium (mg)	320	300	250	260	280	260	340	23

[a]Folate values of US data; all other values from NUTTAB (2006)

- Heart is a good source of iron and zinc, but not as well as liver and kidney.
- Brains and tripe are not particularly good sources of vitamins or minerals.
- All organs meats are high in cholesterol, especially brains, and mostly low in sodium.
- Liver is such a rich source of retinol that consumption of large amounts is not recommended in pregnancy.

1.3.7 Red Meat and the Heart Foundation Tick

The biggest determinant of fat content of red meat is the amount of fat trimmed. Today, there is a wide choice of trimmed red meat cuts available in supermarkets and butcher shops. Research shows that most Australians either buy red meat trimmed or remove the fat, if any, before consumption (Drouz et al. 2006; Williams and Droulez 2010; Williams et al. 2003) (Table 7).

When trimmed of visible fat, red meat has less than 4 % saturated fat, earning it the Heart Foundation's Tick Approval. To trim red meat, use a sharp knife and cut away any visible fat. Improvements in blood cholesterol, blood pressure, and weight loss have been reported in studies where trimmed red meat is consumed as part of a healthy diet and lifestyle (Delbridge et al. 2009). Table 7 shows the fat and fatty acids in 100 g uncooked red meat.

Table 7 Fat and fatty acids content in 100 g uncooked red meat

100 g uncooked red meat (Average of beef, veal, and lamb)	
Total fat (g)	4
Saturated (g)	1.3
Monounsaturated (g)	2.1
Polyunsaturated (g)	0.6
Omega 3 (EPA + DHA) (mg)	35

1.3.8 Health and Nutritional Benefits of Meat

A study conducted in Perth, Australia showed that the per capita consumption of red meat had fallen steadily, while white meat consumption rose. One of the factors that contributed to this pattern was raising concerns about fat, cholesterol, nutritional value, calories, fiber, waste, and artificial additives. Lean meat is rapidly becoming the preferred choice, but it is often more expensive (Droulez et al. 2006).

Meat definitely has great nutritional value, but consumers must be reminded that too much of a good thing can be harmful. Human population studies show that red and processed meats are associated with an increased risk of colorectal cancer, but white meat is not. The evidence is not yet conclusive, but consuming a lot of preserved meat probably increases the risk of colorectal cancer.

Scientific studies have demonstrated that the key factor in disease outcomes is the dietary patterns rather than the individual food components. Dietary patterns that mainly contain processed meats are more associated with some diseases, including CVD, colon cancer, and type II diabetes than dietary patterns that mainly contain poultry.

1.4 Why Meat: As a Food

Meat and their product contain protein of high biological value due to their essential amino acid composition, notably tryptophan, methionine, and lysine, which the body cannot produce. The essential amino acids are well balanced, i.e., in the ratio that the body needs.

Meat and their products are natural source of vitamins and minerals which are essential to healthy growth and function, notably, zinc, iron, selenium, phosphorus, vitamins of the B group, and vitamin D.

Meat products account for significant amounts of micronutrients (>15 % RDA) in the diet of broad groups of the population and have been regularly consumed as a part of the so-called Mediterranean diet over the centuries. Table 8 shows world population, production, and consumption of meat.

- 780 million people out of 4 billion living on insufficient diet
- Protein energy malnutrition affects 34 % of all preschool children in third world
- 56 % of child death in 53 developing countries
- 205 million Children are underweight in Asia and Africa

Table 8 World population, production and consumption of meat

Region	Population (mill.)	Production (mill.mt.)	Consumption (per capita) (kg)
N. America	264.2	28.2	106.8
W. America	377.4	31.0	82.0
E. Europe and Russia	392.4	27.1	69
Africa	449.0	5.0	11.2
Latin America	406.0	15.3	37.6
Far East	1,379.4	6.3	4.6
Developing countries	3,632.8	52.1	14.3
Developed countries	1,201.5	94.8	78.3

- Only 25 g of meat will supply 45 % of a child's daily need for protein and half of the vit.B-12
- 100 g meat to the average diet will increase
  - Protein—50 %
  - Iron—12 %
  - Niacin—40 %
  - Energy—25 % (Table 8)

1.5 Meat Product and Byproduct

The new Code provides several definitions for meat and meat products. However, it does not provide definitions for minced meat, comminuted meat, mechanically separated meat, frozen meat, or poultry. There is a general legal requirement that food sold by manufacturers and retailers must be safe and suitable for human consumption. Processors of raw meat products must also comply with relevant hygiene and processing standards to produce a safe and suitable product.

1.5.1 Meat

In the new Code meat means the whole or part of a carcass of any buffalo, camel, cattle, deer, goat, hare, pig, poultry, rabbit, or sheep that is slaughtered other than in the wild state or any other animal that is permitted for human consumption.

The definition of meat does not include eggs or fish because these foods are regulated elsewhere in the new Code.

1.5.2 Red Meat

Red meat in traditional culinary terminology is meat which is red when raw, and not white when cooked. Red meat includes the meat of most adult mammals and some fowl (e.g., ducks). In gastronomy red meat is dark-colored meat, as contrasted with white meat. The exact definition varies by time, place, and culture, but the meat of adult mammals such as cows, sheep, and horses is invariably considered red, while chicken and rabbit are invariably considered white. The meat of young mammals such as milk-fed veal calves, sheep, and pigs is traditionally considered white, while the meat of duck and goose is considered red (Aberle et al. 2001).

Red meat contains large amounts of iron, creatine, minerals such as zinc and phosphorus, and B-vitamins: (niacin, vitaminB_{12}, thiamin, and riboflavin). Red meat is the richest source of alphalipoic acid, a powerful antioxidant.

Red meat contains small amounts of vitamin D. The liver contains much higher quantities than other parts of the animal.

1.5.3 White Meat

White meat or light meat refers to the light-colored meat of poultry as contrasted with dark meat. In a more general sense, white meat may also refer to any light-colored meat, as contrasted with red meats like beef.

1.5.4 Meaty Flesh

Meaty flesh is defined as skeletal muscle to distinguish it from other parts of a carcass of meat such as offal, bone, and bone marrow. Meaty flesh includes any attached fat, connective tissue, rind, nerves, blood vessels and blood, and skin (if poultry).

1.5.5 Chicken

Chicken is the most common type of poultry in the world and is prepared as food in a wide variety of ways, varying by region and culture. Modern varieties of chicken such as the Cornish Cross are bred specifically for meat production, with an emphasis placed on the ratio of feed to meat produced by the animal.

Chickens raised specifically for food are called broilers. In the USA, broilers are typically butchered at a young age. Modern Cornish Cross hybrids, for example, are butchered as early as 8 weeks for fryers and 12 weeks for roasting birds.

1.5.6 Offal

Offal is defined to distinguish it from the meat, flesh, and to assist with provisions relating to offal such as labeling requirements. Offal means parts of a carcass such as blood, brain, heart, kidney, liver, pancreas, spleen, thymus, tongue, and tripe, but excludes meat flesh bone and bone marrow. The use of offal in meat products is restricted subject to specific labeling requirements.

1.5.7 Processed Meat

Processed meat as a category is a continuum of products ranging from meat products with a minimum of 30 % meat to products that are all meat flesh. The meat must have undergone a method of processing other than boning, slicing, dicing, mincing, or freezing. It includes manufactured meat and cured and/or dried meat, flesh in whole cuts or pieces. Examples of processed meat containing between 30 % and 66 % meat would include some sausages and some Frankfurt's, whereas processed meats that contain more than 66 % meat would include products like ham or prosciutto. The definition for processing meat encompasses the processes of smoking, drying, salting, curing, fermenting, pickling, cooking, and forming. Processed meat may contain other ingredients, but must contain no less than 300 g/kg meat, i.e., they must consist of at least 30 % meat.

1.5.8 Cured and/or Dried Meat Flesh in Whole Cuts or Pieces

Cured and/or dried meat flesh in whole cuts or pieces is meat flesh including attached bone and must contain at least 160 g/kg meat protein on a fat-free basis, i.e., it must have at least 16 % protein on a fat-free basis. Note that fat-free meat flesh is measured analytically by determining the amount of meat protein present; it does not mean meat flesh without visible fat.

Manufactured Meat

Manufactured meats are a category of processed meats. They are products made from meat and are usually processed with other foods. They must contain at least 660 g/kg meat. Manufactured meat also includes any cured and/or dried meat flesh products in whole cuts or pieces that have had other foods added to them. Manufactured meats are not exempt from percentage labeling requirements.

Sausages

Sausages are a category of processed meat. They are minced processed meat and/or comminute meat, which may be combined with other foods and are encased or formed into discrete units. They do not include meat formed or joined into the assemblance of cuts of meat (Fig. 1). A separate definition for sausages assists with referring to the other provisions in the Code that specifically regulate sausages including compositional requirements and food additive permissions for sausages.

Meat Pies

A meat pie must contain at least 250 g/kg of meat based on the total weight of the pie including pastry, which specify that the meat content be based on the filling only, i.e., minus the pastry.

Fig. 1 Emulsion sausage

1.6 Processing of Sausage

An important concept to recognize is that processing of sausage is a continuous sequence of events in which each step is an integral part of the whole; thus, it is not practical to consider any one step separately or to assign more importance to one step than to another. Nevertheless, in studying sausage processing, it is convenient to separate the process into definite steps or categories. The operational processing of sausage begins with grinding of the meat ingredient and proceeds through packaging.

1.6.1 Grinding

The grinding of meat and fat ingredients has largely been practiced for many years and is still in use, mainly by small processors, particularly in the manufacture of specialty products. The fist-size chunks of lean meats are initially ground by running them through a 3–6 mm grinder plate, while fat trimmings or fatty tissues are reduced through a 6–9 mm grinder plate. The tripe and the filler meats are preferably ground twice: first, through a 3–4 mm and then through a finer grinder plate. Grinding through a coarser plate increases the capacity of the machine and heats the meat less. Particularly in bull meat, grinding through a finer plate is considered to give a product with better binding and emulsifying properties. The curing salts are then added and the batch is mixed in a mechanical mixer to ensure that the ingredients are well dispersed. The curing process may take place either overnight in a chiller at 1–4 °C (this practice is being increasingly abandoned) or after final chopping in the cutter with other ingredients and stuffing, i.e., prior to or during the smoking. During many emulsion-type sausage procedures, a pre-comminution in the grinder is followed by chopping because it contributes to a better and more uniform size reduction in the cutter. In many cases comminution is not too finely done.

1.6.2 Mixing

Cylinders of fat and lean obtained by grinding are tumbled in a mixer to give a uniform distribution of fat and lean particles. Mixing also aids in extracting and coating of the fat particles with salt-soluble protein. This can be used for coarse-ground sausage or for emulsion-type sausage by utilizing a chopper or emulsifier and with suitable additions of the required ingredients to obtain the desired texture and uniformity of composition. The mixer should never be overloaded, since it prevents good mixing. Filling the mixer only to the top of the paddles or blades assures proper mixing.

1.6.3 Chopping

The grinding of meat has by large been replaced by cutter chopping which renders a fine meat–fat mixture, usually known as an "emulsion" (Fig. 2). During the chopping process the meat is cut to a very fine particle size which encourages protein extraction. Proteins have the function of binding the water surrounding fat droplets and keeping them dispersed. Preparation of sausage emulsion is basically in two phases. First, the lean meat, either previously ground or not, is placed on the cutter and chopped. This is done by the simultaneous addition of all the curing salts (kitchen salt, nitrite), phosphate, and/or citrate for the total batch and one-third of the total amount of finely crushed ice or water. Increased salt concentration in the water phase of the mixture will result in a greater extraction of the meat protein and is of paramount importance in forming a stable emulsion. The extraction of protein is more effective when the meat is near freezing point, but the emulsification process is adversely affected by low temperature. Since protein extraction is increasing with the time of chopping, the lean meat should be chopped for a sufficient period, normally not less than 6–8 min. After this time, fat trimmings and other fat meats, then spices, and the remaining two-thirds of the total water are added. Chopping is then continued until the batch is thoroughly chopped or the temperature of the meat mass reaches

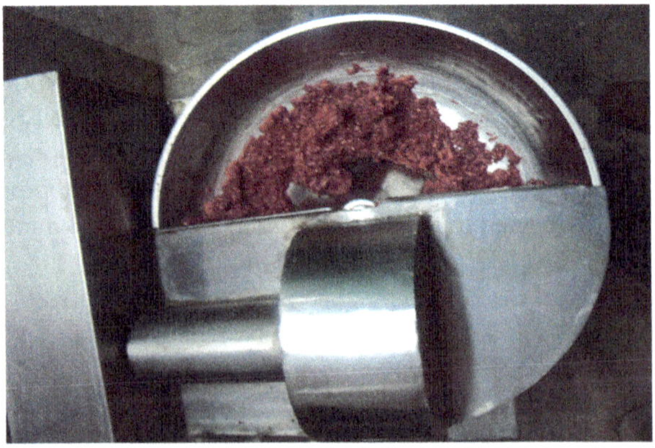

Fig. 2 Bowl chopper mincing the meat

not more than 18 °C (second phase). In the course of this time, all water is taken up by the disintegrated and homogenized meats. If sodium ascorbate is used, it is also added to the very end of the chopping operation. Preparation procedures which provide for one-phase simultaneous meat and fat cutter treatment are increasingly popular. The polyphosphate and curing ingredients should be dissolved in a small quantity of warm water before being added early enough to enable the effect of polyphosphate on actomyosin and the action of salt and nitrite on water binding properties and color of the meat. With the high salt content and the longest cutter process, more salt-soluble proteins are extracted and the binding quality of the finished product is improved. If hot boned meat is used, the addition of phosphates is not necessary.

1.6.4 Emulsifying

The emulsifier machine combines the principle of grinding and chopping. Emulsifiers handle large volumes of meat rapidly to produce a desired texture. The emulsifier is uniformly mixed because it rapidly passes increments of meat mixes through an orifice that may have 0.907 kg or less of product. In the course of a few seconds 100 lb of meat will pass through an emulsifier. Due to the high velocity of the rotating blades, a heat rise of 8–15 °F in the product is common. If the rotating blades are not sufficiently tight or the meat particles sufficiently fine, excessive heat can be transferred to the product. No more than 10 % of heat rise should occur through each plate. Where dual plates are used the meat should always pass through the coarse plate first. The heat rise can be helpful if the meat has been frozen and hydroflaked (cut into frozen flakes) just prior to chopping. It is particularly useful in making emulsion-type products from frozen mechanically deboned poultry meat. The advantage in the use of emulsifiers has been the speed of handling materials, the high degree of disintegration of meat tissues, and the ease of obtaining desired textures.

1.6.5 Stuffing

The sausage emulsion also known in the trade as mix, sausage dough, or batter is transferred to stuffers for extruding into casings. At this point the size and shape of the product are determined. Three types of stuffing pumps are used: (1) Piston, (2) Auger or screw, (3) Rotary. The piston type stuffer is essentially a large barrel or cylinder that has a moving plate. The plate is raised by air pressure and pushes the meat mixture through a stuffing lock and finally through a tubular structure called a stuffing horn. The horn size is selected in relation to the size and type of casing to be used. Usually a horn of as large a diameter as possible is used to reduce smearing of the emulsion. The piston-type stuffer is recommended for coarse-ground sausage and those having fat chunks, olives, pimento, and pickles because these items may be damaged by the impeller-type pumps which usually have feedback and pop-off connectors and are satisfactory for stuffing other emulsion type products such as frankfurters or Bologna.

The stuffer that combines features of the piston and pump usually have a volumetric delivery and are used for small sausages and for stuffing uniform weight rather than catch weight products. Air pressure of 125 psi is used with many stuffers. The pumps, however frequently work on a continuous basis, the bypass valve handling the cycling of the emulsion when stuffing is not being carried out. A special system for stuffing and linking skinless frankfurters has been developed which allows automatic handling of casings. Approximately 3,000 lb or more of product per hour can be stuffed and linked with this machine.

Linking and Threading

After the emulsion is stuffed into casings, the encased mass is tied with thread or fastened with metal clips. In the case of small sausage such as frankfurters, stuffed casing are twisted or drawn together to produce links either by hand or with mechanical devices.

Large sausage items are tied or clipped at one end with a hanging tie and suspended from a smoke stick or hook so the entire surface is free from contact with the equipment. This permits a good flow of air around the sausage in the smokehouse and prevents touch marks and spotting due to contact with adjacent hanging products with long Bologna 48–60 in. At length the tendencies are to process in uniform cellulosic casings and place the encased Bologna on a screen in horizontal position. This horizontal processing aids in retaining a uniform cylinder of meat. If the heat capacity of the smokehouse is adequate 25–30 % more products can be placed in the smokehouse compared to the amount held in the vertical hanging positions. This has considerable interest where slicing bologna or luncheon meats are prepared.

For frankfurters and other small sausages, hand-linking is rarely done today. Machines that stuff and link are now the accepted practice. For 10 to the lb frankfurters, hand-linking of 100 lb per hour was considered excellent. Present machines will stuff and link from 600 to 3,600 lb/h. These high speed linkers set the production economics of the sausage industry today.

Sausage links of the 10-to-the lb size are draped on smokes ticks, eight or nine links forming a loop. A frankfurter emulsion stuffed into 25 mm casings 84 ft long gives 186 links and 18.6 lb of finished sausage. This corresponds to approximately 23 loops of 8 links or 21 loops of 9 links (Pearson and Gillett 1997).

1.6.6 Filling

Before filling into casings, oxygen should be excluded from the mixture (by vacuum-filling devices) and the temperature of the mix should not exceed 2 °C. Natural casing (made from the intestines of slaughter animals) as well as casings made from modified collagen or cellulose are most frequently used. Four undried products, synthetic casings are also used. Evacuation of air from the product enhances color stability and the visual effect of the sausages. It also reduces fat oxidation and bacterial action and prevents proteolysis. A longer shelf life of sausages is therefore achieved by vacuum filling (Essien 2003).

1.6.7 Smoking and Cooking

The draped smoke sticks are placed on smoke trees or trolleys with 12–18 sticks per tree. The filled trees are transferred to the smokehouse and whole houses of two or four trees may be used, the trend is to larger houses holding at least 10 trees. In the continuous type of smokehouse the sausage is draped on hooks, or the smoke stick is conveyed on a double belt or a moving screen conveyor carries the product without use of sticks (Pearson and Gillett 1997).

The smokehouse operation is essentially a specialized drying and cooking operation in which sausage emulsion is coagulated. The important factors relating to smokehouse performance are as follows: (1) dimension, (2) time cycle, (3) temperature range, (4) thermal requirements, (5) relative humidity, (6) air flow, (7) air follows a pattern, and (8) smoke density. These factors control the environment to which the sausage will be exposed during smoking and cooking.

Encased sausage at the time of introduction into the smokehouse usually has an internal temperature of 60–70 °F. During cooking this rises to 155–160 °F. A rise of approximately 100 °F is usually needed requiring at least 10,000 Btu for each 100 lb sausage to be cooked. The rate at which sausage cooks is influenced to a large extent by the air velocity in the smokehouse. The greater the velocity, the faster the internal temperature of the sausage will be rising. The cooking rate is much less affected by the level of humidity in the smokehouse. At high air velocities (4,000 ft/min), there is practically no difference in the heating rate of frankfurters cooked in both high and low humidity atmospheres. However at the lower air velocities (2,000 ft/min) frankfurters do cook slightly faster in a high humidity atmosphere. A four cage smokehouse 9 ft long 9 ft wide and 8fit high with a smoke producing unit attached has a 600,000 Btu/h rating 40,000 Btu being supplied by the smoke generator. This smokehouse holds from 1,000 to 2,000 lbs of product, and which 8–11 changes of air per minute cooks and smokes frankfurters in 1 h or less and large diameter Bologna in 6–8 h. The four cages totaling 10,000 lb of sausage require 100,000 Btu.

However the high heat exchange potential is important when rapid heating of the environment within the smokehouse is required. After the temperature level of the smokehouse is reached, only 15–20 % of the heat is required to maintain this temperature level.

Airflow patterns are important to production performance when the variables in size, shape, and method of holding within a smokehouse are considered. Ham, bacon, bologna, and frankfurters, because of the size and shape difference, may require different air velocities to achieve optimum heat exchange.

Smoke-density control is necessary to obtain uniformly smoked products. Smoke density can be measured with an electric eye instrument; 30 or 40 % transmission of light as recorded on the instrument has been demonstrated to provide an acceptable level of smoke on frankfurters made in the continuous processing event. Some humidity (35–40 %) is required to transfer smoke through cellulose casings to the product. Both humidity and smoke should be applied early in the cooked cycle to obtain transfer of the smoke components to the product, which imports color and flavor. Excessively high humidity (above 45 %) can cause fatting out especially in beef products, which appear to be more prone to greasing out than pork or poultry product.

1.6.8 Chilling

After smoking and cooking the product is showered with the cold water and chilled by refrigeration. On large volume continuous operation chilling is frequently done with a brine solution by dipping or spraying the product. Six percent of salt brine is reasonably close to osmotic balance with the sausage. This brine permits lower chill temperature and rapid cooling of the product. A 10-to the lb size frankfurter can be chilled to an internal temperature of 40 °F in 7–8 min. The balanced brine inhibits leaching of salt from the sausage and imbibitions of water by the sausage.

1.6.9 Peeling and Packaging

After properly chilling the product usually to an internal temperature of 35–40 °F, the cellulosic casings on frankfurters and slicing bologna are removed. This is known as the peeling operation. In the past cellulosic casings were removed from frankfurters by hand, with 100–125 lb of sausage per hour considered a good rate. Modern machines remove the casings from 5,000 lb of frankfurters per hour. Peeled frankfurters are collated and unit packed usually in 1-lb units by special packaging machinery; 6–10 lb bulk packages are assembled by hand. Large slicing bolognas are peeled, sliced, and packaged, 6 Oz to 1 lb packages being the most popular sizes.

1.7 Composition of Sausages

Adults and children consume significant amounts of processed meat products (including sausages), often using them as substitutes for meat flesh. Therefore, it is important to maintain the nutritional profile of these products. To assist in this, the standard requires that sausages must contain at least 50 % fat-free meat flesh. The standard also requires that they have a maximum of fat allowed. This maximum is 50 % of the fat-free meat flesh. For example, if you make a sausage with 600 g fat-free meats, flesh per kg of final sausage, you are allowed to have up to 300 g per kg of fat in the final sausage (i.e., 50 % of the 600 g fat-free meats flesh per kg).

Note that fat-free meat flesh is measured analytically by determining the amount of meat protein present; it does not mean meat flesh without visible fat.

1.8 Fresh Sausage

Fresh sausage as the name suggests are meat products sold fresh without prior heat treatment. The entire heat treatment of the product occurs at the very last point in the supply chain: the consumer's home. Generally, those sausages are grilled in the oven or barbecued. Fresh sausage mostly stored and sold in chilled form, although some are sold frozen. The frozen product is bought in the shop or supermarket. In household without a refrigerator, it is then defrosted at the room temperature, which is quite commonly between 25 and 40 °C in many tropical countries, and then heat treated. Defrosting also takes place in refrigerator or microwave.

A wide variety of fresh sausages are available containing many different tastes and cuts of meat as well as countless different flavors. The most common materials utilized are beef, fork, mutton, and poultry such as chicken and turkey.

Fresh sausage is commonly available in two basic types: either as fine emulsion-type products or as coarsely minced and mixed products.

1.8.1 Breakfast Sausage

Breakfast sausage is a type of fresh pork sausage usually served at breakfast in the USA. It is a common breakfast item in traditional American "farmer" or "country" breakfasts, as it originated as a way for farmers to make use of as much of their livestock (usually pigs) as possible. Often, scraps and trimmings that would ordinarily have been fed to a dog or discarded were instead ground, seasoned, refrigerated, and later consumed by the farmer as an inexpensive, high-protein morning meal.

It can often be found in rural regions, especially in the southern states, where it is either fresh patties or in links with either natural or synthetic casings as well as smoked. This sausage is most similar to English style sausages and has been made in the USA since colonial days. It is essentially highly seasoned ground meat, so it does not keep and should be stored and handled appropriately. Newer variations made from pork and beef mixtures as well as poultry can now be found. There are also vegetarian varieties that use textured vegetable protein in place of meat. In America, the predominant spices used for seasoning are pepper and sage, although there are varieties also seasoned with cayenne pepper or even maple syrup. Some breakfast sausage is flavored with cured ham.

1.8.2 Kielbasa

Kielbasa, kołbasa, and kubasa are common North American Anglicizations for a type of Eastern European sausage. Synonyms include Polish sausage, Ukrainian sausage, etc. In English, these words refer to a particular genre of sausage, common to all Eastern European countries but with substantial regional variations. In the Slavic languages, these are the generic words for all types of sausage, local or foreign (Barber 2006).

1.8.3 Bockwurst

Bockwurst is a German sausage invented in 1889 by restaurant owner R. Scholtz of Berlin. It is one of the most popular varieties within Germany and can be found abroad. The sausage is traditionally made from ground veal and pork (tending more toward veal, unlike bratwurst). In modern Germany, however, it is made from different types of ground meat, such as pork, lamb, turkey, chicken and in rare cases even from horse meat. In Northern Germany there is also a version of bockwurst which is made from fish. Bockwurst is flavored with salt, white pepper, and paprika. Other spices, such as chives and parsley, are often also added and in Germany itself bockwurst is often smoked as well. Bockwurst was originally eaten with bock beer and it is usually served with mustered. A natural casing sausage, it is usually cooked by simmering although it may also be grilled. When thoroughly cooked, its casing, usually splits open. Ideally, one stop cooking just before that occurs because the split casing may look unappetizing and the sausage may then lose flavor to the cooking water.

1.8.4 Bratwurst

A bratwurst is a sausage usually composed of veal, pork, or beef. The name is derived from Old High German *Bratwurst*, from *brat-*, which is finely chopped meat and Wurst or sausage. Though the brat in bratwurst described the way the sausages are made, nowadays Germans associate it with the German verb "braten," which means to pan fry or roast. Bratwurst is usually grilled or pan fried and sometimes cooked in broth or beer. Bockwurst made in America, also from veal and pork, bears more resemblance to the Bavarian Weisswurst in color and taste, albeit parsley is rarely used in this version.

1.8.5 Goetta

Goetta is a breakfast sausage of likely German-American origin that is popular in the greater Cincinnati area. It is primarily composed of ground meat (pork, or pork and beef) and steel-cut oats. Goetta was originally a peasant dish, meant to stretch out servings of meat over several meals to conserve money.

The modern popularity of goetta in Cincinnati has led to it being called "Cincinnati Caviar." Glier's Goetta, the largest commercial producer of goetta, produces more than 1,000,000 lb (450 metric tons) annually, around 99 % of which is consumed locally in greater Cincinnati.

Composition

While goetta comes in a variety of forms, all goetta is based around ground meat combined with pin head or steel-cut oats. Usually goetta is made from pork shoulder or "Cali," but occasionally contains equal parts pork and beef.

Goetta is typically flavored with bay leaves, rosemary, salt, pepper, and thyme. It contains onions and sometimes other vegetables. While similar to scrapple in that it contains a grain product and meat for the purpose of stretching out the meat over several days, goetta looks very different. Scrapple is made with the meal, while goetta uses steel-cut or chopped oats. The oats in goetta are much coarser than the fine powder used in scrapple and add texture to the dish.

Preparation

Goetta is typically formed into small loaves and then cut into squares and fried, often in the oil left over from browning the meats or in bacon drippings. Traditionally a breakfast food, goetta is often served with apple butter, ketchup, mustard, syrup, grape jelly, honey, or eaten by itself.

More recently, goetta has become an all-purpose food eaten with any meal. New goetta products in the Cincinnati area include goetta burgers, goetta dogs, and goetta pizza. As the meat in goetta is precooked during the process of making the loaves, goetta can be kept in the freezer.

1.8.6 Falukorv

Falukorv is a large Swedish sausage made of a grated mixture of pork and beef or veal with potato starch flour and mild spices.

Typical falukorv meal

- Sliced and fried with boiled, fried, or mashed potato
- Sliced and fried with elbow macaroni
- Sliced and fried, served with baked Swedish brown beans and fried egg
- Gratinated whole, partially sliced, in the oven with cheese and mustard, often with slices of onion tucked in between the slices. Accompanied by roasted or mashed potatoes
- Substitute for beef in beef stroganoff (is then called korvstroganoff which means "sausage stroganoff")

1.8.7 Kranjska Klobasa

Kranjska klobasa (Australian English: Kransky, German: Krainerwurst) is a Slovenian sausage most similar to what is known as kielbasa or Polish sausage in North America. Kranjska klobasa contains at least 75–80 % pork (aside from bacon) and at most 20 % bacon It may contain as much as 5 % water, sea salt from Sečovlje salt pans, little garlic, saltpeter, and black pepper. No other ingredients are permitted. The meat must be cut in small pieces 10–13 mm and bacon 8–10 mm. The filling is stuffed into pork intestine with a diameter of 32–36 mm. They are formed in pairs of 12–16 cm lengths and a weight of 180–220 g. Pairs are linked together with a wooden skewer. The sausages are hot smoked and heat-cured at about 70 °C (158 °F).

1.9 Dry Sausage

Fermented sausages can be either dry or semi-dry. The most well-known dry sausages, such as Genoa salami, dry salami, and pepperoni, originated from Italy (Ricke and Keeton 1997). In general, dry sausages have a final pH of 5.0–5.3, lactic acid percent of 0.5–1.0 %, and an MPR of <2.3:1. The moisture loss is between 25 and 50 % and the final moisture percent on average is <35 % with water activity (aw) ranging between <0.85 and 0.91. However, these values may be different due to government and company specifications. German and hard salami are required by FSIS to have an MPR of 1.9:1, with the exception of Genoa salami, which is required to have an MPR of 2.3:1 (FSIS 1986). The pH range of these products is 4.7–4.9, whereas pepperoni has a slightly lower pH range of 4.5–4.8 and a lower MPR of 1.6:1 as required by FSIS (1986). The moisture content for these products ranges from 25 to 39 %. All of these products, because of their moisture-to-protein ratios, are considered shelf stable (FSIS 1986; Ricke and Keeton 1997). Semi-dry sausages such as summer sausage, servlet, and Mettwurst typically have a final pH between 4.7 and 5.1, a lactic acid percent of 0.5–1.3 %, and an MPR of >2.3:1 but <3.7:1. The moisture loss ranges from 8 to 15 % and the moisture percent ranges from 45 to 50 %. The water activity range is 0.90–0.94. Again, values will vary depending upon the government and company specifications. For instance, a summer sausage will have a final pH <5.0, a lactic acid percent to 1 %, and an MPR of 3.1:1 with a moisture percent of 41–51 %. Lebanon Bologna is unique in that it contains a higher moisture content of 56–62 % (Ricke and Keeton 1997). Due to the higher moisture-to-protein ratios, semi-dry sausages are required to be refrigerated.

1.9.1 Salumi

Salumi are Italian cured meat products and predominantly made from pork. It comes from the Italian word paul salumi "salted meat." The term salumi also encompasses bresaola, which is made from beef, and also cooked products such as mortadella and prosciutto cotto. Salami is a specific type of salumi.

1.9.2 Salami

Salami is cured sausage, fermented and air-dried meat, originating from one of the variety of animals. Historically, salami were popular among Southern European peasants because it can be stored at room temperature for periods of up to 30–40 days once cut, supplementing a possibly meager or inconsistent supply of fresh meat. Varieties of salami are traditionally made across Europe.

Ingredient of Salami

Salami is made from one or more of the following meats: Pork or, especially in Kosher and Halal meat, (beef particularly veal) venison, and poultry (mostly turkey because of dietary limitations, but also goose salami is traditional in some areas of Northern Italy).

Additional ingredients may include salt, spices usually white pepper, garlic, minced fat, wine, various herbs, and vinegar.

The raw meat mixture is usually allowed to ferment for a day, then the mixture is either stuffed into an edible natural or inedible cellulose casing and hung to cure. Heat treatment to about 40 °C may be used to accelerate further fermentation and start the drying process. Higher temperatures about 60° Care used to stop the fermentation when the desired pH is reached, but the product is not fully cooked (75 °C or higher). The casings are often treated with an edible mold (*Penicillium*) culture as well. The mold is desired as it imparts flavor, helps the drying process, and prevents spoilage during the curing process.

1.9.3 Pepperoni

Pepperoni is an Italian-American variety of salami (a dry sausage) usually made from cured pork and beef although poultry may also be used. Pepperoni is characteristically soft, slightly smoky, and bright red in color. Thinly sliced pepperoni is a popular pizza topping in American-style pizzerias, a filling in the West Virginia pepperoni roll, and is used to make some varieties of sub sandwiches.

Ingredients

Ingredient like peppers, garlic, fennel, or mustard seeds can be included in the production of pepperoni to provide different flavors and levels of spiciness. Likewise, the type of meat used to produce pepperoni can vary. Pepperoni may be substituted by similar cured meats like Genoa salami, soppressata, or chorizo.

1.9.4 Soppressata

Soppressata is an Italian dry salami. Two principal types are Alyde, a cured dry sausage typical of Basilicata, Apulia, and Calabria, and very different uncured salami, native to Tuscany and Liguria. Each of these varieties qualifies for prodotto agroalimentare tradizionale (PAT) status.

Preparation

Soppressata can be made of fresh hams, as well as other cuts. Pork is the traditional meat used, though it is sometimes made using beef. The meat is either coarsely pressed or ground as with other salamis. Pressing gives it an uneven, rustic appearance when sliced. Soppressata is a specialty of southern Italy and often includes hot pepper (though, as with all salami, seasonings vary). The sausage is hung up to dry for anywhere between 3 and 12 weeks, depending on the diameter, and loses about 30 % of its original weight. Cured soppressata is often stored in jars of olive oil. It is commonly sliced thin and placed on crackers or sandwiches or eaten by itself. Soppressata is becoming a popular alternative topping to papperoni for pizza in some pizzerias in the USA.

1.10 Smoked Sausage

1.10.1 Andouilla

Andouille in US English; French pronunciation: [anduj] is a smoked sausage made using pork, originating in France and which was taken to the USA through Louisiana by French immigrants. It is distinguished in some varieties by its use of the entire gastrointestinal system of the pig.

In the USA the sausage is most often associated with Cajun cooking, where it is a coarse-grained smoked sausage made using pork, garlic, pepper, onions, wine, and seasonings.

1.10.2 Linguiça

Linguicais a form of smoke-cured pork sausage seasoned with garlic and paprika in Portuguese-speaking countries.

Linguiça, like many other sausages, generally serves as part of a heavy meal, typically accompanied by rice, beans, and other pork products. Feijoada, for example, is a traditional Brazilian dish, also common in Angola that incorporates linguiça with beans, ham hocks, and other foods. Linguiça is also used in Francesinha, a traditional Portuguese dish, from Porto. The linguica is incorporated in its sauce, giving it a distinct flavor (Histor 2008).

1.10.3 Hungarian Sausages

The cuisine of Hungary produces a vast number of types of sausages. The most common smoked Hungarian sausages are Gyulai kolbasz, Csabai kolbász, Csemege kolbász, Házi kolbász, Cserkész kolbász, lightly smoked, like Debreceni kolbász or (Debrecener) and Lecsókolbász, a spicy sausage made specifically for serving as part of the dish a vegetable stew with peppers and tomatoes. Hungarian boiled sausages are called "hurka," liver sausage, "májas," and blood sausage, "véres." The main ingredient is liver and rice or blood and rice. Spices, pepper, and salt are added (Gyulai 2008).

Different regions in Hungary may have their own sausage recipes and tastes. The Hungarian sausages may be boiled, fresh, or dried and smoked, with different spices and flavors, "hot" or "mild." These sausages may be eaten like a cold cut or used in the main courses. The Hungarian cuisine uses the different types of sausages in many ways, in vegetable stews, soups, potato stews like "paprikas krumpli" (paprika-based stew with spicy sausage and potatoes), bean soups like Jokai bableves, in meat stews, in some goulash soup variations, pastry dishes, and even salads. The smoked sausages may contain bacon, ground pork, beef, boar or lamb, paprika, salt, garlic, black pepper, allspice, white pepper, caraway, nutmeg, zest, marjoram, cayenne pepper, sugar, white wine, or cognac. The meat is coarsely ground and salted. If garlic is added, it is mashed in water to produce slurry and added to the meat along with spices. The sausage is then stuffed into natural casings usually using the small intestine of the pig. The sausage is then hung overnight to allow the flavors to meld and some of the grease to drip out. It is now ready to be used fresh and unsmoked. Fresh sausages may have additional ingredients like liver, mushroom, bread, rice, lemon juice, eggs, cream, or milk. The unsmoked sausages are typically roasted with sauerkraut or red or green cabbage and served with mashed potatoes.

1.11 Cooked Sausage

A cooked sausage is an important meat product and exists in two forms. One consists of a homogeneous fine mass with no visible particles of meat or fat, while the other has a fine homogeneous mass at the base but also has visible particle fat and meat. Well-known products such as frankfurters, Vienna sausage, hot dog, beer ham, meat loaf, and many others are eaten daily all over the world. Any type of meat can be used to make cooked sausage including exotic meats such as crocodile and kangaroo. Cooked sausage is commonly sold as portioned products, such as hot dogs, and are consumed either hot or cold or sliced.

A cooked sausage is a complex mix of different system, including solutions of dissolved materials such as protein and salts, suspensions of larger particles in added water, gels made from muscular protein, and emulsion that contain stabilized fat in a gel made from protein and fat which is partially present in liquid form.

1.11.1 Hot Dog

A hot dog is a sausage, typically served in a sliced bun. It is often garnished with mustard, ketchup, onions, mayonnaise, relish, cheese, chili, and/or sauerkraut. Common hot dog ingredients include meat trimmings, fat, flavoring, and preservatives. Pork and beef are the traditional meats used in hot dogs. Less expensive hot dogs are often made from chicken or turkey, using low cost mechanically separated poultry. Hot dogs often have high sodium, fat and nitrite content, ingredients linked to health problems. Changes in meat technology and dietary preferences have led manufacturers to use turkey, chicken, vegetarian meat substitutes, and to lower the salt content (Lavin 1980).

1.11.2 Cumberland Sausage

Cumberland sausage is a form of sausage that originated in the ancient county of Cumberland, England now part of Cumbria. They are traditionally very long (up to 50 cm), and sold rolled in a flat, circular coil but within western Cumbria they are more often served in long curved lengths. Sometimes they are made shorter, like ordinary British sausages, and sometimes they are coated in breadcrumbs.

The meat is pork, and seasonings are prepared from a variety of spices and herbs, though the flavor palate is commonly dominated by pepper, both black and white, in contrast to the more herb-dominated flavors of sausage varieties such as those from Lincolnshire. There are traditionally no colorings or preservatives added. The distinctive feature is that the meat is chopped, not minced, giving the sausage a chunky, meaty texture (Traditional Comberland Sausage 2011).

Another type of cooked sausage are battered sausage, blach pudding, bologna, botifarra, cervelat, kielbasa mortadella, and savely.

1.12 Some Traditional Meat Products

1.12.1 Patty

A patty is a flattened, usually disc-shaped, serving of ground meat or meat alternatives. The meat is compacted and shaped, cooked if applicable, and served. Patties can be eaten with a knife and a fork, in dishes like Salisbury steak but typically serve in a sandwich called a hamburger if made from ground beef. The patty itself is also called a burger, whether or not it's served in a sandwich.

1.12.2 Kebab

Kebab is a wide variety of skewered meals originating in the Middle East and later on adopted in the Balkans, the Caucasus, other parts of Europe, as well as Central and South Asia that are now found worldwide. In English, kebab with no qualification generally refers more specifically to shish kebab (Turkish: "sis Kebap") cooked on a skewer. In the Middle East, however, kebab refers to meat that is cooked over or next to the flames, large or small cuts of meat, or even ground meat; it may be served on plates, in sandwiches, or in bowls. The traditional meat for kebab is lamb, but depending on local tastes and religious prohibitions, it may now be beef, goat, chicken, or fish. Like other ethnic foods brought by travelers, the kebab has become a part of everyday cuisine in many countries.

1.12.3 Shami Kabab

Shami kebab or shami tikka is a popular Iranian, Indian variety of kebab also found in Pakistani cuisine. Shami Kabab is a delicious lamb/meat preparation, very popular among all types of Kababs in Indian subcontinent. Easy to prepare and can be stored in refrigerators for a real long time. A good accompaniment for any kind of meals and even served as snacks with evening tea.

1.13 Meat Byproducts

Livestock is often described using such terms as single, dual, or triple-purpose meaning that they produce one (e.g., Beef from beef cattle), two (e.g., Wool and mutton from sheep), or three (e.g., Milk, draft power, and meat from camels) commercial products; but in truth they are all multipurpose. The number of byproducts harvested from animals in addition to the primary product (meat) is almost limitless, and they are used, generally unknowingly, by almost everyone in all aspects of their daily lives. Meat industry byproducts can be divided for convenience into three major groups: edible, inedible, and pharmaceutical products. Among the edible byproducts are organs such as the heart, liver, kidneys, and poultry giblets, as well as natural sausage casings, tripe, and the enzyme rennet used in making junket. The list of inedible products is almost endless: traditionally the inedible offal was rendered (ground and heated to remove the fat) and the fat portion used for tallow, candles, and soap making, while the remainder, a high protein dried meat meal, was used as a protein supplement for animal feed or as anitrogen fertilizer. Current commercial uses of inedible byproducts in addition to those already mentioned include ornaments and buttons from horns and hooves, soaps, toothpaste, shaving cream, lotions, lipsticks, deodorants, glues, paints varnishes, polishes, antifreeze, surfactants, and even explosives from glycerin; artists' brushes from hair and bristles; surgical sutures, violin, and racquet strings from intestinal walls; bone china, bone black, and hardened steel from bone-ash; rubber and plastic polymerization, fabric softeners, lubricants, plasticizers, and biodiesel from fat. Even the undigested food from the alimentary tract can be composted and sold. The list of pharmaceuticals collected during harvesting has included progesterone, estrogen, cortisone, ACTH, insulin, thyroxin, heparin, glucagon, thrombin, trypsin, and many others but these days many of these can be synthesized in the laboratory or derived biologically using recombinant DNA technology. For a comprehensive list of meat industry byproducts the reader is referred to the World Wide Web and to such publications as

"Encyclopaedia of Meat Sciences" and "The Meat We Eat." There is no doubt that human kind continues to have a very close association with a wide variety of animal products on a daily basis (Gregory 2007).

2 Conclusion

Meat and meat products make important nutrition contribution to the diet of the people significant percentage of recommended dietary allowances for proteins, vitamins B, magnesium, phosphorous, iron, and zinc are contributed by red meat and poultry.

Meat products are considered to be more concentrated in nutrition than vegetable food; even the mankind in early days survived on animal foods. It is a well-known food with high nutrition and with high quality proteins, a good balance of essential amino acids. Meat products are consumed by majority of the people in the world.

References

Aberle ED, Forrest JC, Gerrard DE Mills E W (2001) An introduction to muscle physiology and meat science. In: Principles of meat science. Kendall Hunt, Lowa, p 354

Barber K (2006) Fall, Kielbasa. Retrieved 18 Mar 2012, from wikipedia: http://en.wikipedia.org/wiki/Kielbasa

Casey NH (1992) Goat meat in human nutrition. In: Proceedings international conference on goats. Indian council of Agricultural Research, New Delhi

CIRB (2009) Population of buffalo. Buffalo pedia. Retrieved 18 Feb 2013, http://www.buffalopedia.cirb.res.in/index.php?option=com_content&view=article&id=51:population&catid=46:buffalo&Itemid=59&lang=en

Delbridge EA, Prendergast LA, Pritchard JE, Proietto J (2009) One-year weight maintenance after significant weight loss in healthy overweight and obese subjects: does diet composition matter. Am J Clin Nutr 90:1203–1214

DGCI&S, Government of India (2011). Directory of Indian exporters. Retrieved 10-03-2013. http://www.dgciskol.nic.in/directory_of_indian_exporters.asp.

Droulez V, Williams P, Levy G, Stobaus T, Sinclair AJ (2006) Composition of Australian red meat 2002, 2, fatty acid profile. Food Aust 58(7):335–341

Essien E (2003) Sausage manufacture principles and practice. Woodhead, Cambridge, pp 45–50

FAO (2008) Food Outlook. June 2008. Rome

FSIS. (1986). Standards and Labeling Policy Book. Food Safety and Inspection Service. Standards and Labeling Division. Washington, DC

Global Agricultural Information (2012). Livestock and products annual. Foreign agricultural service, New Delhi

Gregory NG (2007) Describes aspects of good and poor welfare in meat animals and their effects on meat quality as well as descriptions of methods used in harvesting livestock. In: Animal welfare and meat production. CABI, UK, p 299

Gyulai K (2008) Hungarian sausages. Retrieved 10 Apr 2013, Wikipedia http://en.wikipedia.org/wiki/Hungarian_sausages

Histor BF (2008) Linguica. Retrieved 10 May 2013, wikipedia http://en.wikipedia.org/wiki/Lingui

Hodgson JM, Burke V, Beilin LJ, Puddey I (2006) Partial substitution of carbohydrate intake with protein intake from lean red meat lowers blood pressure in hypertensive persons. Am J Clin Nutr 83(4):780–787

Kumar A, Jee S, Yadav C (2012) Export of buffalo meat from India: performance and prospects. Indian J Anim Sci 82(12):1578–1583
Lavin C (1980) Hot dog. Retrieved 13 Apr 2013, wikipedia http://en.Wikipedia.Org/wiki/Hot_dog
Lawrie RA, Ledward DA (2006) Lawrie's meat science, 7th edn. Woodhead, Cambridge, pp 75–155
Moffett T (2013). Nutritional Value of Meats. Retrieved 10 Nov 2013, http://www.livestrong.com/article/335280-nutritional-value-of-meat-products
NHMRC (2006) Nutrient reference values for Australia and New Zealand including recommended dietary intakes, commonwealth of Australia
NUTTAB (2006) Food Standards Australia New Zealand. Online database of the nutritional composition of Australian foods. FSANZ, Canberra, 2007. Available at: http://www.foodstandards.gov.au/monitoringandsurveillance/nuttab2006/%20onlineversionintroduction/index.cfm, http://www.foodstandards.gov.au/monitoringandsurveillance/nuttab2006/onlineversionintroduction/index.cfm. Cited 10 April 2013
Pearson AM, Gillett TA (1997) Sausage formulation. In: Processed meats. CBS Publisher, New Delhi, pp 242–290
Ricke SC, Keeton JT (1997) Fermented meat poultry and fish products. In: Doyle MP, Beuchat LR, Montville TJ (eds) Food microbiology: fundamentals and frontiers, 1st edn. ASM Press, Washington, DC, pp 610–628
Singh R, Wright T (2012). Livestock and products annual. Foreign agricultural service, New Delhi, IN2116
Traditional Cumberland sausages win protected status in Europe (2011) Department for environmental food and rural affairs. Comberland sausage, Retrieved 18 Mar 2013. http://en.wikipedia.org/wiki/Cumberland_sausage
US Department of Agriculture, USDA National Nutrient Database for Standard Reference (2007) Release 19. USDA, Washington DC. Available at: http://www.nal.usda.gov/fnic/foodcomp/search/. Cited 10 April 2013
Williams P, Droulez V (2010) Australian red meat consumption – implications of changes over 20 years on nutrient composition. Food Aust 62(3):87–94, University of Wollongong Research Online
Williams PG, Yeatman H, Zakrzewski S, Aboozaid B, Henshaw S, Ingram K, Rankine A, Walcott S, Ghani F (2003) Nutrition and related claims used on packaged Australian foods – implications for regulation. Asia Pac J Clin Nutr 12(2):138–150, University of Wollongong Research Online

Surface Decontamination Treatments for Improving the Safety of Meat and Poultry

Hakan Benli

1 Introduction

Contamination of beef carcass surfaces with *Escherichia coli* O157:H7 and *Salmonella* occur during the slaughter process due to contact with feces and the hide which are the most likely sources of contamination (Kochevar et al. 1997; Phebus et al. 1997; Huffman 2002; Keeton and Eddy 2004; Edwards and Fung 2006). Similarly, various processing steps have been reported to contribute to contamination or cross-contamination of the poultry carcasses including live receiving, immobilization, bleeding, scalding, feather removal, evisceration, and chilling (U.S. Department of Agriculture 2008a, b). To reduce the risk of foodborne diseases related to meat and poultry, the United States Department of Agriculture's Food Safety Inspection Service (USDA-FSIS) issued the Pathogen Reduction: Hazard Analysis and Critical Control Point (PR/HACCP) regulation on July 25, 1996. PR/HACCP established pathogen reduction requirements applicable to meat and poultry establishments to reduce the occurrence and numbers of pathogens in meat and poultry products (U.S. Department of Agriculture 1996; Mead et al. 1999). However, an estimated 9.4 million episodes of foodborne illnesses, 55,961 hospitalizations, and 1,351 deaths have been attributed to 31 major pathogens in the USA each year. Norovirus caused the most of illnesses (58 %) followed by nontyphoidal *Salmonella* spp. (11 %), *Clostridium perfringens* (10 %), and *Campylobacter* spp. (9 %). Nontyphoidal *Salmonella* spp. (35 %), norovirus (26 %), *Campylobacter* spp. (15 %), and *Toxoplasma gondii* (8 %) were the leading causes of hospitalization. Nontyphoidal *Salmonella* spp. (28 %), *T. gondii* (24 %), *Listeria monocytogenes* (19 %), and norovirus (11 %) were the leading causes of death (Scallan et al. 2011).

H. Benli (✉)
Department of Food and Engineering, Cukurova University, 01330 Adana, Balcali, Turkey
e-mail: hbenli@cu.edu.tr

A. Malik et al. (eds.), *Food Processing: Strategies for Quality Assessment*,
Food Engineering Series, DOI 10.1007/978-1-4939-1378-7_6

In response to demands from consumers for safer meat and poultry products and implementation of government regulations, numerous studies testing possible interventions have been conducted by the industry and researchers including physical and chemical decontamination treatments.

2 Physical Decontamination Treatments

A variety of thermal and nonthermal decontamination systems are currently available for physical decontamination of carcasses. Physical decontamination treatments are designed to reduce or eliminate the numbers of microorganisms on the carcasses by destroying (thermal or nonthermal methods) or removing pathogens (e.g., by washing or spraying). Some of these physical systems include water-based treatments for instance washing, spray-washing, hot water pasteurization, stem pasteurization, and steam-vacuuming. Others may include carcass trimming, irradiation, and HPP systems.

2.1 Water-Based Treatments

Washing is a fundamental unit operation in the processing of meat and poultry and mainly used to remove visible contaminants such as soil, feathers, and other debris and fecal contamination from the surface of the carcasses. However, water wash as a single carcass intervention has been reported to slightly reduce the bacterial load on the carcasses (Hardin et al. 1995; Castillo et al. 1998a; Northcutt et al. 2003b, 2005; Smith et al. 2005, 2007). The removal of the contaminants by washing with water could be increased using a rinse, spray, and immersion bath or steam treatment.

In commercial poultry processing, spraying or some form of rinsing is used for carcass washing at pressures sufficient enough to remove visible contamination, usually in a whole carcass inside–outside washer. Furthermore, washing might include multiple sprays of water from bleeding through chilling contributing to the reduction of *Salmonella* prevalence on carcasses by 50–90 % (Bolder 2007; Buncic and Sofos 2012). Loretz et al. (2010) indicated that cold and warm water treatments including immersion chilling reduced *Salmonella* by 0.6–1.3 log units. Similarly, inside–outside cabinet washer has been reported to slightly reduce aerobic plate counts, as well as *Salmonella*, *Campylobacter*, *Escherichia coli*, and coliforms (Fletcher and Craig 1997; Byrd et al. 2002; Jimenez et al. 2002, 2003; Li et al. 2002; Northcutt et al. 2003b, 2005; Smith et al. 2005, 2007). Benli et al. (2011) also reported only a 0.3 log CFU/ml reduction in *Salmonella* following water spray of inoculated poultry carcasses. Likewise, spray washing poultry carcasses with water alone has been reported ineffective for reducing either *Salmonella* or the total bacterial load on carcasses in several studies (Hwang and Beuchat 1995; Li et al. 1997; Sakhare et al. 1999;

Northcutt et al. 2003b, 2005; Mehyar et al. 2005). Furthermore, Lillard (1988) proposed that water immersion of poultry carcasses during processing forms crevices on the skin in which bacteria lodge and are protected from effects of saline and other solutions of varying ionic strength or surfactants. This hypothesis was suggested to explain the persistence of salmonellae on poultry carcasses and the ineffectiveness of some antimicrobial applications for reducing salmonellae. However, Morrison and Fleet (1985) reported that immersion treatment of inoculated chicken carcasses with hot water (60 °C) for 10 min reduced *Salmonella* Typhimurium by 2 logs. Conversely, Berrang et al. (2000) found that a second scald applied after defeathering either as an immersion treatment at 60 °C (28 s immediately or 30 min after defeathering) or as a spray treatment at 71–73 °C (20 s immediately or 30 min after defeathering) was not effective for reducing *Campylobacter, E. coli*, and coliforms on chicken carcasses. Sanchez et al. (2002) compared immersion chilling and air chilling for reducing microbiological load, the incidence of *Salmonella* spp. and *Campylobacter* spp. on broiler carcasses. They found no significant differences between immersion chilling and air chilling for total aerobic counts (3.38 log and 3.31 log CFU/ml, respectively), generic *E. coli* (1.17 log and 1.43 log CFU/ml, respectively), or coliforms (1.72 log and 1.97 log CFU/ml, respectively). Counts of psychrotrophs were significantly higher for immersion chilled carcasses than air-chill carcasses (3.20 log and 1.91 log CFU/ml, respectively). The incidence of *Salmonella* spp. and *Campylobacter* spp. were reported lower in air-chilled broilers due to a higher prevalence of crosscontamination among immersion-chilled broilers.

Huffman (2002) noted that hot water applications also have potential of reducing bacterial counts on beef carcasses by 1–3 log cycles. Barkate et al. (1993) reported that when the surface temperature of beef carcasses was raised to 82 °C for about 10 s using hot water sprays (95 °C), a significant reduction in bacterial numbers was observed between control and hot water-treated carcass surfaces. Castillo et al. (1998b) likewise reported that a water wash followed by hot water spray (95 °C) reduced levels of for *Escherichia coli* O157:H7, *S.* Typhimurium, APC, and coliforms by 3.7, 3.8, 2.9, and 3.3 log, respectively, on carcass surfaces. Spray-washing (26 °C, 276 kPa followed by 1,000 kPa) followed by hot-water rinsing (>77 °C, 138–152 kPa, 2.5–8 s) and knife-trimming followed by a second spray-wash also have been shown to be an effective beef carcass decontamination method (Delmore et al. 1997). Gorman et al. (1995) concluded that hot water (74 °C at the surface of the sample) applied as a spray washing process onto the beef surfaces caused reductions in bacterial counts exceeding 3.0 log CFU/cm^2 when compared to the combination of hand-trimming and spray-washing with colder (<35 °C) water. In summary, hot water treatment (>74 °C) of beef carcasses is a common practice in the industry and the data indicate that hot water applications to carcasses have been effective to reduce bacterial counts by 1–3 log cycles. However, water temperature, water pressure, carcass coverage, and dwell time are needed to take into consideration to effectively implement and validate hot water as a decontamination step (Huffman 2002).

Application of steam to accomplish thermal destruction of bacteria on the surface of meat carcasses has been considered as an alternative to hot water spraying.

A commercial antimicrobial carcass intervention process called steam pasteurization which was approved by the FDA in 1995 for whole carcasses as well as parts of carcasses that are to be further processed has been adopted by the industry (Chen et al. 2012). Nutsch et al. (1997) evaluated effectiveness of a steam pasteurization process for reducing naturally occurring bacterial populations on freshly slaughtered beef sides in a large commercial facility. The results indicated that steam pasteurization is very effective in a commercial setting for reducing overall bacterial populations on freshly slaughtered beef carcasses. Steam pasteurization process includes exposing meat carcasses and meat products to water steam at 82–97 °C inside a chamber or a tunnel at atmospheric pressure for 6–12 s. The treatment consists of three steps; water removal, steam pasteurization, and rapid chilling (Aymerich et al. 2008; Chen et al. 2012). Phebus et al. (1997) reported that the steam pasteurization consistently produced numerically greater pathogen reductions than knife-trimming or hot water/steam vacuum spot cleaning on beef carcasses and the reductions for all three treatments ranged from 2.5 to 3.7 log CFU/cm^2. All three treatments were also more effective than water washing (35 °C) which gave only a reduction of 1.0 log CFU/cm^2. Likewise, Nutsch et al. (1998) evaluated a steam pasteurization system in a commercial beef processing facility and found significant reductions in total aerobic plate counts and *E. coli* counts at five separate anatomical locations on the carcasses. Avens et al. (2002) reported application of flowing steam at 98 °C for 3 min virtually destroyed aerobic bacteria on the skin of naturally contaminated poultry carcasses. In another study, Whyte et al. (2003) exposed the broiler carcasses to atmospheric steam at 90 °C for 24 s which was provided reductions of 0.75, 0.69, and 1.3 log CFU/g in total viable counts, *Enterobacteriaceae*, and *Campylobacter* counts, respectively. However, they also reported visible damages to the outer epidermal skin tissue of carcasses following the steam pasteurization.

A variation of the steam pasteurization called steam vacuuming has been developed and adopted by the industry to remove fecal and visible contamination which is less than 2.54 cm at its greatest dimension on carcasses. Steam vacuum systems consist of two sequential steps including steam or hot water spray (82–88 °C) on a small, designated carcass area and then vacuuming using a handheld device developed for this purpose. Thus, combined effect of removing and inactivating surface contamination can be achieved on the carcasses (Huffman 2002). Dorsa et al. (1997) examined application of steam vacuum and hot water washes on beef carcass surfaces. They concluded that the use of steam vacuum and hot water effectively reduces bacterial populations from beef carcass tissues immediately after treatment with reductions of up to 2.7 log CFU/cm^2 for APC, lactic acid bacteria, and *L. innocua* and as much as 3.4 log CFU/cm^2 for *E. coli* O157:H7. Castillo et al. (1999a) indicated that steam vacuuming reduced the number of different indicator organisms around 3.0 log cycles but also spread the bacterial contamination to areas of the carcass surface adjacent to the contaminated sites. However, they suggested that a combined treatment including steam vacuuming followed by spraying with hot water and then lactic acid effectively reduced the relocated contamination.

2.2 Carcass Trimming

Although trimming has been considered for completely removing the physical contamination including fecal and other visible contaminants and pathogens, under the commercial slaughtering conditions, trimming has been reported to be a highly variable process since the efficacy of the trimming primarily related to the skill and carefulness of the individual who is applying the trimming. In addition, spreading the contamination due to improperly sanitized equipment used during trimming and holding the carcasses for trimming at the warm slaughter room temperature raise questions about the actual efficacy of trimming as a method of reducing pathogen contamination of carcasses (Reagan et al. 1996; Castillo et al. 1998a; Edwards and Fung 2006). In a comparative study, Reagan et al. (1996) examined treatment procedures included trimming, washing, and trimming followed by washing under the industrial conditions. Their results indicated trimming followed by washing produced approximately 2 log CFU/cm^2 reduction in aerobic bacteria, while trimming alone which was done by industry personnel at normal slaughtering speeds and operating practices, reduced contamination by approximately 1.3 log CFU/cm^2. The mechanism of removing visible contamination by use of the trimming followed by washing was explained as a combination of physical removal by trimming with additional removal of debris and foreign material by washing. They also indicated that some visible contaminates were left on the carcasses following the trimming alone due to the possible accidental recontamination. Similarly, Delmore et al. (1997) reported that decontamination of beef carcasses could be achieved by knife trimming followed by spray washing or by spray washing followed by hot water rinsing. Likewise, water wash or trimming as a single beef carcass intervention was reported not sufficient for significantly reducing pathogens on beef carcasses (Castillo et al. 1998a). Laster et al. (2012) reported that trimming of external fat surfaces during the normal fabrication process may reduce contamination of *E. coli* O157:H7. However, fat and lean surfaces that were not inoculated became contaminated during the fabrication process. They concluded that trimming external surfaces reduced levels of pathogens, but under normal fabrication processes, pathogens were still spread to newly exposed surfaces.

2.3 Irradiation

Ionizing radiation has been described as radiation that has enough energy for removing electrons from atoms, thus leading to the formation of ions. While there are different types of ionizing radiation, gamma-rays produced from the radioisotopes Cobalt 60 (1.17 and 1.33 MeV) and Cesium 137 (0.662 MeV), X-rays generated from a machine operated at or below 5 MeV, and machine-generated electron beams (maximum energy 10 MeV) are permitted for food irradiation to inactivate microorganisms including pathogens (Dincer and Baysal 2004; Pillai 2004; Chen et al. 2012).

Cobalt 60 is used by the majority of facilities in the industry due to stronger gamma ray producing ability and lack of water solubility. Alternatively, electron beams produced by commercial electron accelerators have the advantage of switching the system on and off like any other electrical apparatus and they can be used for removing surface contamination of meat and poultry products. Lastly, producing X-rays requires slamming fast moving electrons into a metal objects. Strong X-ray that has an energy superior to 1 MeV can be produced using tantalum or platinum targets with the possibility of processing packaged meat products in large quantities (Aymerich et al. 2008).

Regardless of the source and the facility generating irradiation, the main target of the irradiation is the molecular bonds in the microbial DNA. In addition to damages to DNA, denaturation of enzymes and cell membrane alteration may also occur with the irradiation. RNA is likewise a target for ionizing radiations since lethal effects of irradiation on RNA containing viruses have been observed. Nucleic acids can also be damaged by an ionized adjacent molecule such as water that produces a lethal product for the genetic material. Water molecules lose an electron due to ionizing radiation and produce H_2O^+ and e^-. A number of compounds including hydrogen, hydroxyl radicals, molecular hydrogen, oxygen, and hydrogen peroxide are then produced with the reactions of water molecules. The most reactive of them are the hydroxyl radicals (OH•) and hydrogen peroxide (H_2O_2). All of these byproducts react with other water molecules, nucleic acids, and other biologically sensitive molecules. Although biological systems have a repair capacity of both single- and double-stranded breaks of the DNA backbone, ionizing radiation at doses used in food irradiation is probably causing damages so extensive that bacterial repair of the damages become nearly impossible (Pillai 2004; Aymerich et al. 2008).

Application of ionizing radiation to frozen and chilled poultry with doses of 3–5 and 1.5–2.5 kGy, respectively, reduced *Salmonella* by about 3 log units (Corry et al. 1995; Farkas 1998; Buncic and Sofos 2012). Similarly, aerobic bacteria were reduced around 3.0 log CFU/g on chicken legs by irradiation of 1 kGy (Loretz et al. 2010). Sarjeant et al. (2005) stated that electronic beam irradiation reduced inoculated *S.* Typhimurium on fresh chicken breasts about 4 log units with doses of 1, 2, or 3 kGy. Likewise, gamma radiation was reported effective for inactivating psychrophiles and *Enterococcus* bacteria on turkey breast samples which were irradiated at doses from 1 to 3 kGy (Henry et al. 2010). Exposure of poultry viscera to a higher dose of gamma radiation (20 kGy) rendered the viscera sterile, while 5 and 10 kGy decreased the total bacterial count by 4 and 6 log cycles, respectively, and eliminated the coliforms to <1 CFU/g of tissue (Jamdar and Harikumar 2008). Arthur et al. (2005) reported that low-dose, low-penetration electron beam irradiation (dose of approximately 1 kGy with a depth penetration of 15 mm) reduced *E. coli* O157:H7 on beef carcass surfaces by at least 4 log CFU/cm^2 indicating potential use of the treatment as an antimicrobial intervention on beef carcasses during processing.

2.4 *High Pressure Processing Systems*

High pressure processing (HPP) also called hydrostatic pressure (HHP) or ultrahigh pressure processing (UHP) is primarily applied as a batch process to prepackaged food products using a chamber surrounded by water or another pressure-transmitting fluid. The food products usually are vacuum-packaged in a flexible package and placed in the pressure vessel and then submitted to pressures ranging from 100 to 900 MPa. However, the pressure levels of 400–600 MPa are mostly used in commercial applications depending on the product for 3–5 min. Following the pressure treatment, the processed product is removed from the vessel and stored or distributed in a conventional manner. Several factors have been reported to contribute to the inactivation of microorganisms by HPP including changes in the cell membranes, cell wall, proteins, and enzyme-mediated cellular functions. The primary sites damaged by the pressure are cell membranes with subsequent alterations of cell permeability, transport systems, loss of osmotic responsiveness, organelle disruption, and inability to maintain intracellular pH (Simpson and Gilmour 1997a, b; Campus 2010).

HPP application of 700 MPa at 15 °C for 1 min was reported causing up to 5 log reduction of *E.coli* O157:H7 in raw minced meat and increasing the shelf-life of the raw minced meat under refrigerated conditions (Gola et al. 2000). Similarly, Morales et al. (2008) found that multiple-cycle treatments of HPP resulted with a higher *E. coli* O157:H7 lethality than single-cycle treatments since the single-cycle treatments at 400 MPa and 12°C ranged from 0.82 log CFU/g for a 1 min cycle to 4.39 log CFU/g for a 20 min cycle while multiple-cycle treatments produced reduction of 4.38 log CFU/g with four 1 min cycles at 400 MPa and 12°C and 4.96 log CFU/g with three 5-min cycles. Garriga et al. (2004) reported that the safety risks associated with *Salmonella* and *L. monocytogenes* in sliced marinated beef loin stored up to 120 days at 4 °C was reduced with HPP treatment at 600 MPa and 31 °C for 6 min.

3 Chemical Decontamination Treatments

Consumers, especially in developed countries, demand high quality and safe meats nowadays. To meet the consumer demand and as a response to the USDA-FSIS mandate to increase the safety of meat and poultry products, numerous chemical compounds have been evaluated as decontamination agents. The chemical interventions include various food-grade chemicals that are usually applied to the meat surface, to inhibit or kill microorganisms. The mode of action of the chemicals is mainly due to their ability to disrupt cellular membranes or other cellular constituents and interrupt physiological processes (Loretz et al. 2010). Chemical compounds must be proven effective and approved for use by the U.S. Food and Drug Administration (FDA) and the USDA-FSIS before used as a decontamination agent. Once proven to be safe, antimicrobials may be applied to carcass or product surfaces. These compounds that are naturally derived or manufactured should not conceal spoilage, but should extend shelf life and prevent pathogen growth as a consequence of their bactericidal or bacteriostatic activity (Keeton and Eddy 2004). This section will focus on chemical decontamination treatments currently available to the meat industry, such as organic acids, chlorine-based treatments, trisodium phosphate, electrolyzed water, acidic calcium sulfate, epsilon polylysine, and lauric arginate.

3.1 Organic Acids

Solutions of organic acids (1–3 %), such as lactic and acetic acids, are commonly used for beef and lamb (Chen et al. 2012). An organic acid spray is the most commonly used means of chemical decontamination of beef carcasses in combination with steam vacuuming, hot water washing, or steam cabinets (Huffman 2002). Bolton et al. (2001) reported that organic acids such as lactic or acetic acid are usually applied using a spray cabinet and recommended critical limits including at least 500 mL of a 2.5–10 % (v/v) acid (to allow for dilution when applied to the carcass) maintained at a pH of $\leq$2.8 and temperature of 25°–55 °C and sprayed for 35 s at 13.8–27.6 Pa for an organic acid spray application. Even though the mechanism for the antimicrobial activity of organic acids is not completely known, it is generally believed that the undissociated form of the acid, or its ester, is responsible for the activity. Weak acids penetrate the bacterial cell membrane and accumulate in the cytoplasm, and the protonated acid acidifies the cytoplasm if the intracellular pH is higher than the pKa of the acid, resulting in cell injury or death (Keeton and Eddy 2004).

Hardin et al. (1995) found that beef carcass washing followed by warm acid sprays (55 °C) of lactic acid or acetic acid performed better than trimming or washing alone for reducing *Salmonella* and *E. coli* O157:H7 and that lactic acid was more effective than acetic acid for *E. coli* O157:H7 reduction. In another study, both a water wash and trimming combined with sanitizing treatments of hot water (95 °C) or warm (55 °C) 2 % lactic acid spray or a combination of these two sanitizing methods resulted in reductions of more than 4.0 log CFU/cm^2 for *S.* Typhimurium and *E. coli* O157:H7 on beef carcasses (Castillo et al. 1998a). Further, Castillo et al. (2001b) found that prechill treatment of 2 % lactic acid spray (250 ml, 55 °C) reduced the counts of *S.* Typhimurium and *E. coli* O157:H7 on beef carcass surfaces that had been inoculated. In addition, a 4 % L-lactic acid spray at 55 °C prior to fabrication has also been suggested for chilled beef carcasses which were previously subjected to a hot water spray followed by a lactic acid spray prior to chilling (Castillo et al. 2001a). In a comparative study, King et al. (2005) reported that a peroxyacetic acid spray was not an effective intervention for *S.* Typhimurium and *E. coli* O157:H7 reduction on chilled beef carcasses when compared to carcasses treated with 2 % L-lactic acid spray before chilling or 4 % L-lactic acid spray after chilling. Similarly, Dorsa et al. (1998a) suggested that a 2 % lactic acid or 2 % acetic acid wash during beef carcass processing could lower the bacterial counts in ground beef.

Sakhare et al. (1999) found that acetic acid (0.5 %) or lactic acid (0.25 %) treatments applied by either dipping or spraying after scalding, defeathering, and evisceration of chicken carcasses were more effective than spray washing with water alone to decrease crosscontamination and improve microbial quality. In a comparative study Sinhamahapatra et al. (2004) reported that lactic acid dip and hot water dip were the most effective for reducing aerobic plate counts by 1.36 log and 1.28 log/cm^2 on broiler carcasses. Treatments with acetic acid or lactic acid by either dipping or spraying after scalding, evisceration, and defeathering have been claimed to decrease crosscontamination and improve the microbial quality of chicken carcasses (Sakhare et al. 1999). Yoder et al. (2012) studied eight antimicrobial compounds (acetic acid, citric acid, lactic acid, peroxyacetic acid, acidified sodium chlorite (ASC), chlorine dioxide, sodium hypochlorite, and aqueous ozone) applied at various concentrations with small, handheld spraying equipment for suitable to use in very small meat plants. Relative antimicrobial effectiveness of the compounds was determined as organic acids > peroxyacetic acid > chlorinated compounds > aqueous ozone. A comparative study of acetic, citric, lactic, malic, mandelic, propionic, and tartaric acids against *S.* Typhimurium attached to broiler skin found that concentrations of greater than or equal to 4 % of the acids were required to kill greater than or equal to 2 log number of the pathogen (Tamblyn and Conner 1997).

3.2 *Chlorine-Based Treatment*

Chlorinated water is used to control microbial contamination and growth in the meat industry. The chlorine levels do not normally exceed 50 ppm, which results in a reduction in microbial load of 1 log cycle (Bolder 1997). However, Northcutt et al. (2005) conducted a study to investigate the microbiological impact of spray washing broiler carcasses with chlorinated water (0 or 50 ppm) at different temperatures (21.1, 43.3, or 54.4 °C). They concluded neither adding chlorine nor elevating the water temperature during spray washing in an inside–outside bird washer did enhance the removal of bacteria from broiler carcasses. Similarly, Keeton and Eddy (2004) indicated that use of chlorinated water (20–50 ppm) to reduce the pathogen load on poultry carcasses at the prechill washer or in the chill tank have had mixed results and chlorinated water is less effective than other compounds such as hypochlorite, chlorine dioxide, ASC, and cetylpyridinium chloride (CPC).

Efficacy of 200 ppm hypochlorite on the bacterial counts has been reported on beef carcasses with some residual effect during refrigerated storage. However, its use as a carcass decontamination agent is limited due to effectiveness of other chemicals (e.g., organic acids) against pathogens which is generally more than hypochlorite (Keeton and Eddy 2004). In contrast, chilling poultry carcasses with 20 ppm sodium hypochlorite solution reduced coliforms, *Campylocater, E. coli,* and *Salmonella* counts by 1.2, 1.3, 1.4, and 0.5 logs, respectively (Northcutt et al. 2003a).

Disinfecting public water supplies using chlorine dioxide is a common practice in the industry. Chlorine dioxide is also promising as a decontamination agent for carcass surfaces. The mode of action of chlorine dioxide is due to the irreversible damage to fatty acids and proteins in the bacterial cell membrane, resulting in the loss of permeability and the destruction of the transmembrane ionic gradient (Keeton and Eddy 2004). Beef trimmings were treated with 200 ppm chlorine dioxide to reduce inoculated *E. coli* and *S.* Typhimurium but only 0.71 and 0.61 log CFU/g declines were observed, respectively, in ground beef manufactured from trimmings (Stivarius et al. 2002). Likewise, Cutter and Dorsa (1995) indicated that spraying beef carcass tissues with chlorine dioxide at a concentration of 20 ppm was not effective for reducing fecal contamination on beef regardless of duration of the spraying. Berrang et al. (2011) reported that application of 50 ppm of chlorine dioxide during defeathering of poultry carcasses produced significantly lower numbers of *Campylobacter* and *E. coli* and a lower prevalence of *Salmonella* than carcasses treated with the water spray as control defeathering.

ASC is an acid-activated, broad spectrum antimicrobial approved by the USDA-FSIS as an antimicrobial agent for use on poultry and beef (Keeton and Eddy 2004; U.S. Department of Agriculture Food Safety and Inspection Service 2013). As a processing aid ASC does not require labeling with insignificant residue levels. In poultry operations, ASC is mostly applied at the end of the evisceration line before or after carcass chilling. Similarly ASC is generally applied as a carcass rinse after evisceration or to trimmings immediately before grinding in red meat operations. Effectiveness of ASC has been shown against pathogens (*E. coli* O157:H7, *Listeria, Campylobacter, Salmonella*), viruses, fungi, yeast, molds, and some protozoa. Acidification of $NaClO_2$ forms $HClO_2$(ASC) and then ASC and organic matters react to form several oxychlorous intermediates which are broad-spectrum germicides. Then these compounds break oxidative bonds (sulfide and disulfide linkages) on the bacterial cell membrane surface and kill the cell. The reaction residues are primarily chloride and chlorate salts (Keeton and Eddy 2004). ASC must be used in combination with any GRAS acid at a level sufficient to achieve a pH of 2.3 or 2.9 depending on the meat or poultry product (U.S. Department of Agriculture Food Safety and Inspection Service 2013). In a comparative study, Sinhamahapatra et al. (2004) tested the effects of hot water (70 °C for 1 min), 2 % lactic acid (30 s), 1,200 ppm ASC (5 s), and 50 ppm chlorine solution (5 min) applied to broiler carcasses as an immersion or spray treatment. The lactic acid dip and hot water dip were the most effective for reducing aerobic plate counts by 1.36 log and 1.28 log/cm^2, respectively, whereas ASC and a hot water dip reduced presumptive coliforms counts by 1.37 log and 1.34 log/cm^2. Similarly, Del Rio et al. (2007) reported that 1,200 ppm ASC immersion solutions were effective on chicken legs for reducing microbial population including mesophilic aerobic counts, psychrotrophs, *Enterobacteriaceae*, coliforms, *Micrococcaceae*, enterococci, *Brochothrix thermosphacta*, pseudomonads, lactic acid bacteria, molds, and yeasts during 5 days of storage at 3 °C. Kemp et al. (2000) indicated that ASC treatment was an effective method for significantly reducing naturally occurring microbial contamination on carcasses and the highest antimicrobial activity was achieved with prewashing and then exposing to a 5 s dip in a solution containing phosphoric acid- or citric acid-activated ASC. Their results showed that a 5 s dip in 500–1,200 ppm ASC reduced total aerobes by 82.9–90.7 %,

E. coli by 99.4–99.6 %, and total coliforms by 86.1–98.5 % on poultry carcasses before chilling. In another study, both *E. coli* O157:H7 and *S.* Typhimurium counts on beef carcasses were reduced by 3.8–3.9 log and 4.5–4.6 log with a water wash followed by a phosphoric acid-activated acidified sodium chloride spray or a citric acid-activated ASC spray, respectively (Castillo et al. 1999b).

CPC is a quaternary ammonium compound which is a water-soluble, colorless, and neutral pH. Levels of 0.05–0.5 % CPC are used to reduce or inhibit gingivitis and biofilm and plaque formation in mouthwashes (Cutter et al. 2000; Keeton and Eddy 2004). CPC penetrates and destroys bacterial cell walls and cell membrane and kills bacteria by the interaction of basic cetylpyridinium ions reacting with the acid groups of bacteria to form weakly ionized compounds that inhibit bacterial metabolism (Li et al. 1996; Keeton and Eddy 2004). In a model system, 4.87 logs CFU/cm^2 of *S.* Typhimurium reduction was observed following 0.4 % of CPC application to chicken skin for 3 min (Breen et al. 1997). Li et al. (1996) found that spraying contaminated poultry skin with 0.1 % CPC reduced *Salmonella* by 0.9–1.7 logs CFU/cm^2, similar reductions were also obtained (1.0–1.6 logs CFU/cm2) when the poultry skin was immersed in CPC. Cutter et al. (2000) determined the effectiveness of 1 % CPC spray (862 kPa, 15-s, 35 °C) against pathogens associated with lean and adipose beef surface. *S.* Typhimurium and *E. coli* O157:H7 were immediately reduced by a 1 % CPC solution on lean beef from 5 to 6 logs CFU/cm^2 to undetectable levels while pathogen counts were reduced to <2.5 logs CFU/cm^2 on fat tissues. The pathogen counts on lean tissue were undetectable following 35 days of storage at 4 °C but counts on fat tissue remained at <1.3 log CFU/cm^2. Even though CPC was proven to effectively reduce pathogenic bacteria on beef tissues residual CPC levels following any of the treatments exceeded those for human consumption.

3.3 Trisodium Phosphate

As a very alkaline (pH 12–13) antimicrobial ingredient, trisodium phosphate (TSP) has been approved for use as a spray or dip for on raw poultry carcasses and giblets (U.S. Department of Agriculture Food Safety and Inspection Service 2013). TSP is applied to poultry carcasses or parts up to 15 s using an 8–12 % solution within a temperature range of 18.3°–29.4 °C. Similarly, the giblets are sprayed or dipped for a minimum of 30 s with an 8–12 % solution. TSP's antimicrobial effect is apparently the result of disruption of cytoplasmic cell membrane followed by the leakage of internal contents and phase separation of the cytoplasm into dark and light zones due to water solubility of the bacterial DNA at high pH (Mendonca et al. 1994; Keeton and Eddy 2004).

Kim et al. (1994) reported that trisodium phosphate reduced *Salmonella* by 1.6–1.8 logs when post-chill poultry carcasses were dipped into a 10 % solution at 50 °C for 15 s. In a different study, experiments were conducted to determine the effect of TSP treatment on reducing salmonellae recovery from broiler carcasses immediately after chilling or following 7 days of storage. Carcasses were subjected to a 5-s dip in 10 % TSP solution. The results indicated that a prechill trisodium phosphate treatment reduced salmonellae-positive samples immediately after chilling or following 7 days of storage on broiler carcasses (Bourassa et al. 2004). In a comparative study, Li et al. (1997) tested 0.85 % sodium chloride, 5 or 10 % trisodium phosphate, 5 or 10 % sodium bisulfate, 0.1 % CPC, or 1 % lactic acid sprays on prechilled chicken carcasses. They reported that spraying 10 % trisodium phosphate for 90 s reduced *S.* Typhimurium by 3.7 logs while 2.4, 1.6, and 1.6 logs of reductions were obtained for 10 % sodium bisulfate, 0.1 % cetylpyridinium, and 1 % lactic acid, respectively.

Potential use of TSP solutions as a decontamination treatment has investigated in a few studies for beef carcasses. Kim and Slavik (1994) evaluated TSP for removing attached *E. coli* O157:H7 and *S.* Typhimurium from beef surfaces. Fat and fascia surfaces inoculated with *E. coli* O157:H7 and *S.* Typhimurium rinsed with a 10 % TSP (10 °C) solution for 15 s. Compared to controls, the levels of *E. coli* O157:H7 were 1.35 and 0.92 logs lower on TSP treated fat and fascia surfaces, respectively while *S.* Typhimurium were 0.91 and 0.51 logs lower, respectively. Ramirez et al. (2001) tested a water rinse followed by either a 2 % lactic acid (9 s, at 55 °C) or a 12 % trisodium phosphate (60 s, at 55 °C) dip or a combination of these treatments. Both treatments alone or in combination were effective for reducing *E. coli* O157:H7 by more than 1.6 log/cm^2 on lamb breast tissue. In contrast, Dorsa et al. (1998b) tested the effect of 2 % lactic acid, 2 % acetic acid, 12 % TSP, and water washes at 72 and 32 °C for reducing pathogens and other bacterial populations on beef carcass surfaces and cuts held for up to 21 days (4 °C) under vacuum. They reported that TSP was not as effective as organic acid treatments for growth suspension on beef surfaces and in some cases the effect was similar to untreated samples. Conversely, 10 % trisodium phosphate or 0.5 % CPC treatment applied by tumbling significantly reduced *E. coli* O157:H7 and *S.* Typhimurium and improved the redness of ground beef (Pohlman et al. 2002).

3.4 Other Chemical Treatments

Antimicrobial activity of some other chemicals including electrolyzed water, acidic calcium sulfate, ε-polylysine, and lauric arginate were also evaluated for the decontamination of meat and poultry in the literature. Electrolyzed water (EW) is getting popular as a sanitizer in the food industries for reducing bacterial populations on foods and processing surfaces. A dilute sodium chloride solution is dissociated by electrolysis into acidic electrolyzed water (AEW) and basic electrolyzed water (BEW). AEW has a pH of 2–3, an oxidation–reduction potential of >1.100 mV and an active chlorine content of 10–90 ppm, whereas BEW has a pH of 10–13 and an oxidation–reduction potential of −800 to −900 mV. It was reported that AEW reduced vegetative cells of various bacteria in suspension more than 6.0 log CFU/ml. However, reductions were limited for chicken carcasses ranging from about 0.8 to 3.0 orders of magnitude (Hricova et al. 2008; Loretz et al. 2010). Park et al. (2002) evaluated effectiveness of EW for killing *Campylobacter jejuni* on chicken wings.

They found that EW was as effective as chlorinated water in reducing *Campylobacter jejuni* on poultry meat by about 3 log CFU/g. Similarly, Northcutt et al. (2007) reported that washing poultry carcasses with EW is slightly better (total aerobic bacteria and *E. coli*) or equivalent to (*Campylobacter* and *Salmonella*) washing with sodium hypochlorite in an inside–outside bird washer. In a comparative study, it was demonstrated that EW could reduce *S.* Typhimurium on poultry surfaces following extended refrigerated storage and could provide poultry establishments with an inexpensive and easy alternative to chlorine treatments to control pathogens during processing (Fabrizio et al. 2002).

A blend of organic acid–calcium sulfate, known as acidic calcium sulfate (ACS), is a very acidic (pH 1.0–1.5) decontamination agent for meat and poultry products that is approved by USDA-FSIS (Keeton and Eddy 2004; U.S. Department of Agriculture Food Safety and Inspection Service 2013). A combination of ACS plus organic acids has been reported to disable the proton pumps in bacterial membranes and thus serve as a metabolic inhibitor (Keeton et al. 2002). The effectiveness of ACS as a surface decontamination agent for reducing pathogens on beef or poultry carcasses or RTE meat products has been reported in several studies (Huffman 2002; Keeton et al. 2002, 2006; Dickens et al. 2004; Nunez de Gonzalez et al. 2004; Zhao et al. 2004; Keeton and Eddy 2004; Luchansky et al. 2005). Another antimicrobial, ε-polylysine (EPL), is a cationic homopolymer of 25–35 L-lysine residues connected at the ε-amino and α-carboxyl group juncture (Geornaras et al. 2007). EPL is an edible, water-soluble agent with a wide range of antimicrobial activity that includes both Gram-positive and Gram-negative bacteria (Yoshida et al. 2002; Yoshida and Nagasawa 2003; Geornaras and Sofos 2005; Geornaras et al. 2007). EPL has been reported to be nontoxic in an acute oral toxicity study in rats with no mortality at concentrations up to 5 g/kg body weight. It was not observed to be mutagenic in bacterial reversion assays and is confirmed safe as a food preservative (Hiraki et al. 2003). Lauramide arginine ethyl ester (LAE), also known as lauric arginate, is an antimicrobial compound derived from lauric acid and arginine with a broad spectrum of antimicrobial activity (Rodriguez et al. 2004; Bakal and Diaz 2005). LAE has been verified to be nontoxic and is metabolized rapidly to naturally occurring amino acids, largely arginine and ornithine after consumption (Ruckman et al. 2004). LAE affects the cyptoplasmic membranes of microorganisms by causing a disruption or instability of the plasma membrane lipid bilayer thus further altering the metabolic process and detaining the cellular cycle (Bakal and Diaz 2005). LAE was confirmed as GRAS by the USDA-FSIS and is considered a safe and suitable ingredient when used in the production of meat and poultry products (U.S. Department of Agriculture Food Safety and Inspection Service 2013).

Dickens et al. (2004) found that spraying with ACS solution (1:1 solution of deionized water and ACS; 4 ml/wing) increased the shelf-life of chicken wings from 7 days to 10 days. Geornaras and Sofos (2005) compared antimicrobial activity of EPL with sodium diacetate, sodium lactate, lactic acid, and acetic acid, against different foodborne pathogens including reduced *E. coli* O157:H7, *S.* Typhimurium, and *L. monocytogenes* in a culture broth medium. They concluded that EPL has minimum inhibitory concentrations of 0.02 % for *E. coli* O157:H7 and *L. monocytogenes*, and 0.04 % for *S.* Typhimurium that EPL inhibited growth of these foodborne pathogens at 24 °C. EPL also has been reported to have enhanced antimicrobial activity when combined with glycine, vinegar, ethanol, and thiamine laurylsulfonate (Yoshida and Nagasawa 2003). Rodriguez et al. (2004) exposed *S.* Typhimurium and *Staphylococcus aureus* to their minimal inhibitory concentrations of 32 and 8 μg/ml of LAE, respectively. They observed alterations mainly in the outer membrane of *S.* Typhimurium and in the cytoplasm of *S. aureus* after exposure to LAE. Further, the proportions of damaged cells after 24 h contact time were reported as 97 and 56.3 % for *S.* Typhimurium and *S. aureus*, respectively.

Benli et al. (2011) evaluated the concept of applying more than one antimicrobial to poultry carcasses to obtain greater reductions than one treatment alone due to different modes of action of individual antimicrobials. They reported that sequential spray applications of 300 mg of EPL per liter followed by 30 % ACS and of 200 mg of LAE per liter followed by 30 % ACS produced the highest *Salmonella* reductions on inoculated chicken carcasses, by 2.1 and 2.2 log CFU/ml, respectively. Similarly, Njongmeta et al. (2011) also reported that sequential application of warm ACS, followed by EPL significantly reduced inoculated levels of *S.* Typhimurium, *E. coli* O157:H7, and *L. monocytogenes* with an extended effect over 7 storage days. These studies indicated that using sequential, multi-hurdle interventions might be a better strategy than applying single decontamination treatment to obtain significant reductions in pathogen numbers on poultry and beef carcasses.

4 Conclusion

Safety of meat and poultry products is an important public health concern in most of the countries. Although maintenance of good hygiene practices is an important part of the meat and poultry production, preventing carcasses from pathogen contamination cannot be guaranteed. Application of sequential or multiple-intervention decontamination systems including physical or chemical decontamination treatments or emerging technologies such as hot water washing, steam application, steam vacuuming, carcass trimming, irradiation, HPP, organic acids, chlorinated water, hypochlorite, chlorine dioxide, ASC, CPC, trisodium phosphate, electrolyzed water, acidic calcium sulfate, ε-polylysine, and lauric arginate can greatly reduce or eliminate the pathogens in meat and poultry products.

References

Arthur TM, Wheeler TL, Shackelford SD, Bosilevac JM, Nou XW, Koohmaraie M (2005) Effects of low-dose, low-penetration electron beam irradiation of chilled beef carcass surface cuts on *Escherichia coli* O157: H7 and meat quality. J Food Prot 68(4):666–672

Avens JS, Albright SN, Morton AS, Prewitt BE, Kendall PA, Sofos JN (2002) Destruction of microorganisms on chicken carcasses by steam and boiling water immersion. Food Control 13(6–7):445–450. doi:Pii S0956-7135(01)00073-1

Aymerich T, Picouet PA, Monfort JM (2008) Decontamination technologies for meat products. Meat Sci 78(1–2):114–129. doi:10.1016/j.meatsci.2007.07.007

Bakal G, Diaz A (2005) The lowdown on lauric arginate: food antimicrobial hammers away at plasma membrane, disrupting a pathogen's metabolic process. http://www.foodquality.com/mag/02032005/02032005_ST1-2.html. Accessed Feb 2008

Barkate ML, Acuff GR, Lucia LM, Hale DS (1993) Hot-water decontamination of beef carcasses for reduction of initial bacterial numbers. Meat Sci 35(3):397–401. doi:10.1016/0309-1740(93)90044-I

Benli H, Sanchez-Plata MX, Keeton JT (2011) Efficacy of epsilon-polylysine, lauric arginate, or acidic calcium sulfate applied sequentially for *Salmonella* reduction on membrane filters and chicken carcasses. J Food Prot 74(5):743–750. doi:10.4315/0362-028x.Jfp-10-463

Berrang ME, Dickens JA, Musgrove MT (2000) Effects of hot water application after defeathering on the levels of *Campylobacter*, coliform bacteria, and *Escherichia coli* on broiler carcasses. Poult Sci 79(11):1689–1693

Berrang ME, Meinersmann RJ, Cox NA, Fedorka-Cray PJ (2011) Application of chlorine dioxide to lessen bacterial contamination during broiler defeathering. J Appl Poult Res 20(1):33–39. doi:10.3382/japr.2010-00178

Bolder NM (1997) Decontamination of meat and poultry carcasses. Trends Food Sci Technol 8(7):221–227

Bolder NM (2007) Microbial challenges of poultry meat production. World Poult Sci J 63(3):401–411. doi:10.1017/S0043933907001535

Bolton DJ, Doherty AM, Sheridan JJ (2001) Beef HACCP: intervention and non-intervention systems. Int J Food Microbiol 66(1–2):119–129

Bourassa DV, Fletcher DL, Buhr RJ, Berrang ME, Cason JA (2004) Recovery of *Salmonellae* from trisodium phosphate-treated commercially processed broiler carcasses after chilling and after seven-day storage. Poult Sci 83(12):2079–2082

Breen PJ, Salari H, Compadre CM (1997) Elimination of *Salmonella* contamination from poultry tissues by cetylpyridinium chloride solutions. J Food Prot 60(9):1019–1021

Buncic S, Sofos J (2012) Interventions to control *Salmonella* contamination during poultry, cattle and pig slaughter. Food Res Int 45(2):641–655. doi:10.1016/j.foodres.2011.10.018

Byrd JA, Hargis BM, Corrier DE, Brewer RL, Caldwell DJ, Bailey RH, McReynolds JL, Herron KL, Stanker LH (2002) Fluorescent marker for the detection of crop and upper gastrointestinal leakage in poultry processing plants. Poult Sci 81(1):70–74

Campus M (2010) High pressure processing of meat, meat products and seafood. Food Eng Rev 2(4):256–273. doi:10.1007/s12393-010-9028-y

Castillo A, Lucia LM, Goodson KJ, Savell JW, Acuff GR (1998a) Comparison of water wash, trimming, and combined hot water and lactic acid treatments for reducing bacteria of fecal origin on beef carcasses. J Food Prot 61(7):823–828

Castillo A, Lucia LM, Goodson KJ, Savell JW, Acuff GR (1998b) Use of hot water for beef carcass decontamination. J Food Prot 61(1):19–25

Castillo A, Lucia LM, Goodson KJ, Savell JW, Acuff GR (1999a) Decontamination of beef carcass surface tissue by steam vacuuming alone and combined with hot water and lactic acid sprays. J Food Prot 62(2):146–151

Castillo A, Lucia LM, Kemp GK, Acuff GR (1999b) Reduction of *Escherichia coli* O157:H7 and *Salmonella* Typhimurium on beef carcass surfaces using acidified sodium chlorite. J Food Prot 62(6):580–584

Castillo A, Lucia LM, Mercado I, Acuff GR (2001a) In-plant evaluation of a lactic acid treatment for reduction of bacteria on chilled beef carcasses. J Food Prot 64(5):738–740

Castillo A, Lucia LM, Roberson DB, Stevenson TH, Mercado I, Acuff GR (2001b) Lactic acid sprays reduce bacterial pathogens on cold beef carcass surfaces and in subsequently produced ground beef. J Food Prot 64(1):58–62

Chen JH, Ren Y, Seow J, Liu T, Bang WS, Yuk HG (2012) Intervention technologies for ensuring microbiological safety of meat: current and future trends. Comprehen Rev Food Sci Food Safety 11(2):119–132. doi:10.1111/j.1541-4337.2011.00177.x

Corry JE, James C, James SJ, Hinton M (1995) *Salmonella*, *Campylobacter* and *Escherichia coli* O157:H7 decontamination techniques for the future. Int J Food Microbiol 28(2):187–196
Cutter CN, Dorsa WJ (1995) Chlorine dioxide spray washes for reducing fecal contamination on beef. J Food Prot 58(12):1294–1296
Cutter CN, Dorsa WJ, Handie A, Rodriguez-Morales S, Zhou X, Breen PJ, Compadre CM (2000) Antimicrobial activity of cetylpyridinium chloride washes against pathogenic bacteria on beef surfaces. J Food Prot 63(5):593–600
Del Rio E, Panizo-Moran M, Prieto M, Alonso-Calleja C, Capita R (2007) Effect of various chemical decontamination treatments on natural microflora and sensory characteristics of poultry. Int J Food Microbiol 115(3):268–280. doi:10.1016/j.ijfoodmicro.2006.10.048
Delmore LRG, Sofos JN, Reagan JO, Smith GC (1997) Hot-water rinsing and trimming washing of beef carcasses to reduce physical and microbiological contamination. J Food Sci 62(2):373–376
Dickens JA, Ingram KD, Hinton A (2004) Effects of applying Safe$_2$O poultry wash to broiler wings on shelf life, *Listeria monocytogenes*, *Pseudomonads*, *Staphylococcus* species, and psychrotrophic bacteria levels after three, seven, and ten days of storage. Poult Sci 83(6):1047–1050
Dincer AH, Baysal T (2004) Decontamination techniques of pathogen bacteria in meat and poultry. Crit Rev Microbiol 30(3):197–204. doi:10.1080/10408410490468803
Dorsa WJ, Cutter CN, Siragusa GR (1997) Effects of steam-vacuuming and hot water spray wash on the microflora of refrigerated beef carcass surface tissue inoculated with *Escherichia coli* O157:H7, *Listeria innocua*, and *Clostridium sporogenes*. J Food Prot 60(2):114–119
Dorsa WJ, Cutter CN, Siragusa GR (1998a) Bacterial profile of ground beef made from carcass tissue experimentally contaminated with pathogenic and spoilage bacteria before being washed with hot water, alkaline solution, or organic acid and then stored at 4 or 12 °C. J Food Prot 61(9):1109–1118
Dorsa WJ, Cutter CN, Siragusa GR (1998b) Long-term effect of alkaline, organic acid, or hot water washes on the microbial profile of refrigerated beef contaminated with bacterial pathogens after washing. J Food Prot 61(3):300–306
Edwards JR, Fung DYC (2006) Prevention and decontamination of *Escherichia coli* O157:H7 on raw beef carcasses in commercial beef abattoirs. J Rapid Methods Automat Microbial 14(1):1–95
Fabrizio KA, Sharma RR, Demirci A, Cutter CN (2002) Comparison of electrolyzed oxidizing water with various antimicrobial interventions to reduce *Salmonella* species on poultry. Poult Sci 81(10):1598–1605
Farkas J (1998) Irradiation as a method for decontaminating food. A review. Int J Food Microbiol 44(3):189–204
Fletcher DL, Craig EW (1997) An evaluation of on-line processing on visual contamination and microbiological quality of broilers. J Appl Poult Res 6:436–442
Garriga M, Grèbol N, Aymerich MT, Monfort JM, Hugas M (2004) Microbial inactivation after high-pressure processing at 600 MPa in commercial meat products over its shelf life. Innov Food Sci Emerg Technol 5(4):451–457
Geornaras I, Sofos JN (2005) Activity of ε-polylysine against *Escherichia coli* O157:H7, *Salmonella* Typhimurium, and *Listeria monocytogenes*. J Food Sci 70(9):M404–408
Geornaras I, Yoon Y, Belk KE, Smith GC, Sofos JN (2007) Antimicrobial activity of ε-polylysine against *Escherichia coli* O157:H7, *Salmonella* Typhimurium, and *Listeria monocytogenes* in various food extracts. J Food Sci 72(8):M330–334
Gola S, Mutti P, Manganelli E, Squarcina N, Rovere P (2000) Behaviour of *E. coli* O157: H7 strains in model system and in raw meat by HPP: microbial and technological aspects. High Pres Res 19(1–6):481–487
Gorman BM, Sofos JN, Morgan JB, Schmidt GR, Smith GC (1995) Evaluation of hand-trimming, various sanitizing agents, and hot water spray-washing as decontamination interventions for beef brisket adipose tissue. J Food Prot 58(8):899–907
Hardin MD, Acuff GR, Lucia LM, Oman JS, Savell JW (1995) Comparison of methods for decontamination from beef carcass surfaces. J Food Prot 58(4):368–374

Henry FC, Silva TJP, Franco RM, Freitas MQ, De Jesus EFO (2010) Effect of gamma radiation on frozen turkey breast meat quality. J Food Saf 30(3):615–634. doi:10.1111/j.1745-4565.2010.00229.x
Hiraki J, Ichikawa T, Ninomiya S, Seki H, Uohama K, Seki H, Kimura S, Yanagimoto Y, Barnett JW (2003) Use of ADME studies to confirm the safety of ε-polylysine as a preservative in food. Regul Toxicol Pharmacol 37(2):328–340
Hricova D, Stephan R, Zweifel C (2008) Electrolyzed water and its application in the food industry. J Food Prot 71(9):1934–1947
Huffman RD (2002) Current and future technologies for the decontamination of carcasses and fresh meat. Meat Sci 62(3):285–294
Hwang CA, Beuchat LR (1995) Efficacy of selected chemicals for killing pathogenic and spoilage microorganisms on chicken skin. J Food Prot 58(1):19–23
Jamdar SN, Harikumar P (2008) Radiation decontamination of poultry viscera. Radiat Phys Chem 77(4):467–472. doi:10.1016/j.radphyschem.2007.07.004
Jimenez SM, Salsi MS, Tiburzi MC, Pirovani ME (2002) A comparison between broiler chicken carcasses with and without visible faecal contamination during the slaughtering process on hazard identification of *Salmonella* spp. J Appl Microbiol 93(4):593–598
Jimenez SM, Tiburzi MC, Salsi MS, Pirovani ME, Moguilevsky MA (2003) The role of visible faecal material as a vehicle for generic *Escherichia coli*, coliform, and other enterobacteria contaminating poultry carcasses during slaughtering. J Appl Microbiol 95(3):451–456
Keeton JT, Eddy SM (2004) Chemical methods for decontamination of meat and poultry. In: Beier RC, Pillai SD, Phillips TD, Ziprin RL (eds) Preharvest and postharvest food safety - contemporary issues and future directions. Blackwell, Ames, IA, pp 319–336
Keeton JT, Acuff GR, Nunez de Gonzalez MT, Ringer LJ, Lucia LM (2002) Antimicrobial effects of surface treatments and ingredients on cured RTE meat products. http://www.mionix.com/safe20_studies/food_studies.htm. Accessed Jul 2013
Keeton JT, Ricke S, Anderson R, Miller D, Azefor NNL (2006) Application of novel hurdle technologies to meat carcasses and trimmings for reduction of pathogens. http://www.fsis.usda.gov/PDF/New_Technology_Final_Report_C-14.pdf. Accessed Jul 2013
Kemp GK, Aldrich ML, Waldroup AL (2000) Acidified sodium chlorite antimicrobial treatment of broiler carcasses. J Food Prot 63(8):1087–1092
Kim JW, Slavik MF (1994) Trisodium phosphate (Tsp) treatment of beef surfaces to reduce *Escherichia coli* O157:H7 and *Salmonella* Typhimurium. J Food Sci 59(1):20–22. doi:10.1111/j.1365-2621.1994.tb06887.x
Kim JW, Slavik MF, Pharr MD, Raben DP, Lobsinger CM, Tsai S (1994) Reduction of *Salmonella* on post-chill chicken carcasses by trisodium phosphate (Na_3PO_4) treatment. J Food Saf 14(1):9–17. doi:10.1111/j.1745-4565.1994.tb00580.x
King DA, Lucia LM, Castillo A, Acuff GR, Harris KB, Savell JW (2005) Evaluation of peroxyacetic acid as a post-chilling intervention for control of *Escherichia coli* O157:H7 and *Salmonella* Typhimurium on beef carcass surfaces. Meat Sci 69(3):401–407
Kochevar SL, Sofos JN, LeValley SB, Smith GC (1997) Effect of water temperature, pressure and chemical solution on removal of fecal material and bacteria from lamb adipose tissue by spray-washing. Meat Sci 45(3):377–388
Laster BA, Harris KB, Lucia LM, Castillo A, Savell JW (2012) Efficacy of trimming chilled beef during fabrication to control *Escherichia coli* O157:H7 surrogates on subsequent subprimals. Meat Sci 90(2):420–425. doi:10.1016/j.meatsci.2011.08.011
Li YB, Kim JW, Slavik MF (1996) Cetylpyridinium chloride (CPC) treatment on poultry skin to reduce attached *Salmonella*. J Food Prot 59(5):533–533
Li YB, Slavik MF, Walker JT, Xiong H (1997) Pre-chill spray of chicken carcasses to reduce *Salmonella* Typhimurium. J Food Sci 62(3):605–607
Li Y, Yang H, Swem BL (2002) Effect of high-temperature inside-outside spray on survival of *Campylobacter jejuni* attached to prechill chicken carcasses. Poult Sci 81(9):1371–1377
Lillard HS (1988) Effect of surfactant or changes in ionic strength on the attachment of *Salmonella* Typhimurium to poultry skin and muscle. J Food Sci 53(3):727–730

Loretz M, Stephan R, Zweifel C (2010) Antimicrobial activity of decontamination treatments for poultry carcasses: a literature survey. Food Control 21(6):791–804. doi:10.1016/j.foodcont.2009.11.007

Luchansky JB, Call JE, Hristova B, Rumery L, Yoder L, Oser A (2005) Viability of *Listeria monocytogenes* on commerically-prepared hams surface treated with acidified calcium sulfate and lauric arginate and stored at 4°C. Meat Sci 71(1):92–99

Mead PS, Slutsker L, Dietz V, McCaig LF, Bresee JS, Shapiro C, Griffin PM, Tauxe RV (1999) Food-related illness and death in the United States. Emerg Infect Dis 5(5):607–625

Mehyar G, Blank G, Han JH, Hydamaka A, Holley RA (2005) Effectiveness of trisodium phosphate, lactic acid and commercial antimicrobials against pathogenic bacteria on chicken skin. Food Prot Trends 25(5):351–362

Mendonca AF, Amoroso TL, Knabel SJ (1994) Destruction of gram-negative food-borne pathogens by high Ph involves disruption of the cytoplasmic membrane. Appl Environ Microbiol 60(11):4009–4014

Morales P, Calzada J, Avila M, Nunez M (2008) Inactivation of *Escherichia coli* O157: H7 in ground beef by single-cycle and multiple-cycle high-pressure treatments. J Food Prot 71(4):811–815

Morrison GJ, Fleet GH (1985) Reduction of *Salmonella* on chicken carcasses by immersion treatments. J Food Prot 48(11):939–943

Njongmeta NLA, Benli H, Dunkley KD, Dunkley CS, Miller DR, Anderson RC, O'Bryan CA, Keeton JT, Nisbet DJ, Crandall PG, Ricke SC (2011) Application of acidic calcium sulfate and ε-polylysine to pre-rigor beef rounds for reduction of pathogens. J Food Saf 31(3):395–400. doi:10.1111/j.1745-4565.2011.00312.x

Northcutt JK, Berrang ME, Dickens JA, Fletcher DL, Cox NA (2003a) Effect of broiler age, feed withdrawal, and transportation on levels of coliforms, *Campylobacter*, *Escherichia coli* and *Salmonella* on carcasses before and after immersion chilling. Poult Sci 82(1):169–173

Northcutt JK, Berrang ME, Smith DP, Jones DR (2003b) Effect of commercial bird washers on broiler carcass microbiological characteristics. J Appl Poult Res 12:435–438

Northcutt JK, Smith DP, Musgrove MT, Ingram KD, Hinton A (2005) Microbiological impact of spray washing broiler carcasses using different chlorine concentrations and water temperatures. Poult Sci 84(10):1648–1652

Northcutt J, Smith D, Ingram KD, Hinton A, Musgrove M (2007) Recovery of bacteria from broiler carcasses after spray washing with acidified electrolyzed water or sodium hypochlorite solutions. Poult Sci 86(10):2239–2244

Nunez de Gonzalez MTN, Keeton JT, Acuff GR, Ringer LJ, Lucia LM (2004) Effectiveness of acidic calcium sulfate with propionic and lactic acid and lactates as postprocessing dipping solutions to control *Listeria monocytogenes* on frankfurters with or without potassium lactate and stored vacuum packaged at 4.5 °C. J Food Prot 67(5):915–921

Nutsch AL, Phebus RK, Riemann MJ, Schafer DE, Boyer JE, Wilson RC, Leising JD, Kastner CL (1997) Evaluation of a steam pasteurization process in a commercial beef processing facility. J Food Prot 60(5):485–492

Nutsch AL, Phebus RK, Riemann MJ, Kotrola JS, Wilson RC, Boyer JE, Brown TL (1998) Steam pasteurization of commercially slaughtered beef carcasses: evaluation of bacterial populations at five anatomical locations. J Food Prot 61(5):571–577

Park H, Hung YC, Brackett RE (2002) Antimicrobial effect of electrolyzed water for inactivating *Campylobacter jejuni* during poultry washing. Int J Food Microbiol 72(1–2):77–83. doi:10.1016/S0168-1605(01)00622-5

Phebus RK, Nutsch AL, Schafer DE, Wilson RC, Riemann MJ, Leising JD, Kastner CL, Wolf JR, Prasai RK (1997) Comparison of steam pasteurization and other methods for reduction of pathogens on surfaces of freshly slaughtered beef. J Food Prot 60(5):476–484

Pillai SD (2004) Food irradiation. In: Beier RC, Pillai SD, Phillips TD, Ziprin RL (eds) Preharvest and postharvest food safety - contemporary issues and future directions. Blackwell, Ames, IA, pp 375–387

Pohlman FW, Stivarius MR, McElyea KS, Waldroup AL (2002) Reduction of *E.coli*, *Salmonella* Typhimurium, coliforms, aerobic bacteria, and improvement of ground beef color using trisodium phosphate or cetylpyridinium chloride before grinding. Meat Sci 60(4):349–356

Ramirez AJ, Acuff GR, Lucia LM, Savell JW (2001) Lactic acid and trisodium phosphate treatment of lamb breast to reduce bacterial contamination. J Food Prot 64(9): 1439–1441

Reagan JO, Acuff GR, Buege DR, Buyck MJ, Dickson JS, Kastner CL, Marsden JL, Morgan JB, Nickelson R, Smith GC, Sofos JN (1996) Trimming and washing of beef carcasses as a method of improving the microbiological quality of meat. J Food Prot 59(7):751–756

Rodriguez E, Seguer J, Rocabayera X, Manresa A (2004) Cellular effects of monohydrochloride of L-arginine, Na-lauroyl ethylester (LAE) on exposure to *Salmonella* Typhimurium and *Staphylococcus aureus*. J Appl Microbiol 96(5):903–912

Ruckman SA, Rocabayera X, Borzelleca JF, Sandusky CB (2004) Toxicological and metabolic investigations of the safety of N-alpha-lauroyl-L-arginine ethyl ester monohydrochloride (LAE). Food Chem Toxicol 42(2):245–259. doi:10.1016/j.fct.2003.08.022

Sakhare PZ, Sachindra NM, Yashoda KP, Rao DN (1999) Efficacy of intermittent decontamination treatments during processing in reducing the microbial load on broiler chicken carcass. Food Control 10(3):189–194

Sanchez MX, Fluckey WM, Brashears MM, McKee SR (2002) Microbial profile and antibiotic susceptibility of *Campylobacter* spp. and *Salmonella* spp. in broilers processed in air-chilled and immersion-chilled environments. J Food Prot 65(6):948–956

Sarjeant KC, Williams SK, Hinton A Jr (2005) The effect of electron beam irradiation on the survival of *Salmonella enterica* serovar Typhimurium and psychrotrophic bacteria on raw chicken breasts stored at four degrees celsius for fourteen days. Poult Sci 84(6):955–958

Scallan E, Hoekstra RM, Angulo FJ, Tauxe RV, Widdowson MA, Roy SL, Jones JL, Griffin PM (2011) Foodborne illness acquired in the United States-major pathogens. Emerg Infect Dis 17(1):7–15. doi:10.3201/eid1701.P11101

Simpson RK, Gilmour A (1997a) The effect of high hydrostatic pressure on *Listeria monocytogenes* in phosphate-buffered saline and model food systems. J Appl Microbiol 83(2):181–188. doi:10.1046/j.1365-2672.1997.00215.x

Simpson RK, Gilmour A (1997b) The resistance of *Listeria monocytogenes* to high hydrostatic pressure in foods. Food Microbiol 14(6):567–573. doi:10.1006/fmic.1997.0117

Sinhamahapatra M, Biswas S, Das AK, Bhattacharyya D (2004) Comparative study of different surface decontaminants on chicken quality. Br Poult Sci 45(5):624–630

Smith DP, Northcutt JK, Musgrove MT (2005) Microbiology of contaminated or visibly clean broiler carcasses processed with an inside-outside bird washer. Int J Poult Sci 4(12): 955–958

Smith DP, Northcutt JK, Cason JA, Hinton A, Buhr RJ, Ingram KD (2007) Effect of external or internal fecal contamination on numbers of bacteria on prechilled broiler carcasses. Poult Sci 86(6):1241–1244

Stivarius MR, Pohlman FW, McElyea KS, Apple JK (2002) Microbial, instrumental color and sensory color and odor characteristics of ground beef produced from beef trimmings treated with ozone or chlorine dioxide. Meat Sci 60(3):299–305

Tamblyn KC, Conner DE (1997) Bactericidal activity of organic acids against *Salmonella* Typhimurium attached to broiler chicken skin. J Food Prot 60(6):629–633

U.S. Department of Agriculture Food Safety and Inspection Service (2013) Safe and suitable ingredients used in the production of meat and poultry, and egg products. http://www.fsis.usda.gov/wps/wcm/connect/bab10e09-aefa-483b-8be8-809a1f051d4c/7120.1.pdf?MOD=AJPERES. Accessed Sept 2013

U.S. Department of Agriculture, Food Safety and Inspection Service (1996) Pathogen reduction: hazard analysis and critical control point (HACCP) systems; final rule. Fed Regist 61(144):38806–38989

U.S. Department of Agriculture, Food Safety and Inspection Service (2008a) Public health risk-based inspection system for processing and slaughter - Appendix C - literature reviews. http://www.fsis.usda.gov/OPPDE/NACMPI/Feb2008/Processing_Appendix_C_041808.pdf. Accessed 19 May 2008

U.S. Department of Agriculture, Food Safety and Inspection Service (2008b) Public health risk-based inspection system for processing and slaughter - technical report. http://origin-www.fsis.usda.gov/OPPDE/NACMPI/Feb2008/Processing_Slaughter_Tech_Rpt_041808.pdf. Accessed Oct 2009

Whyte P, McGill K, Collins JD (2003) An assessment of steam pasteurization and hot water immersion treatments for the microbiological decontamination of broiler carcasses. Food Microbiol 20(1):111–117. doi:10.1016/S0740-0020(02)00084-9

Yoder SF, Henning WR, Mills EW, Doores S, Ostiguy N, Cutter CN (2012) Investigation of chemical rinses suitable for very small meat plants to reduce pathogens on beef surfaces. J Food Prot 75(1):14–21. doi:10.4315/0362-028X.JFP-11-084

Yoshida T, Nagasawa T (2003) ε-poly-L-lysine: microbial production, biodegradation and application potential. Appl Microbiol Biotechnol 62(1):21–26

Yoshida T, Hiraki J, Nagasawa T (2002) ε-poly-L-lysine. In: Fahnestock S, Steinbüchel A (eds) Biopolymers, vol 7. Wiley, Weinheim, pp 107–121

Zhao T, Doyle MP, Kemp MC, Howell RS, Zhao P (2004) Influence of freezing and freezing plus acidic calcium sulfate and lactic acid addition on thermal inactivation of *Escherichia coli* O157:H7 in ground beef. J Food Prot 67(8):1760–1764

Food Quality and Safety

Saghir Ahmad

1 Introduction

Quality makes a product what it is. It is the combination of attributes or characteristics of a product that have significance for determining the degree of acceptability of product to consumer. The quality is defined as degree of excellence and this excellence is contributed by nutritional, sensory and hygienic factors which are the real requirement of consumer in the present era. The sensory quality can be assured by subjective or objective evaluation methods. The quality and prices may not go together. We see a lot of processed food in the market but their quality is not related to price. For example the fast food which is costly, their quality may not be up to mark as per the definition. A cheap product like biscuit has desirable qualities in all respect. The quality factor in case of fresh fruit or vegetable may include appearances including size, shape, wholesomeness, colour and consistency, etc. Prediction of quality of the products and raw materials are quite different. The apple purchased from the market will be seen for colour, maturity and wholesomeness. However when the juice of the apple is extracted, we don't know what was the quality of fruit. The juices can be brought to acceptable form by processing method and it would be accepted by the consumer. The sensory properties evaluated subjectively give correct information of the product and differentiate between their sensory attributes. For example the flavour of a product is not only the aroma, but it is a composite characteristics giving information of taste, smell and feel. The feel factor is

S. Ahmad (✉)
Faculty of Agricultural Sciences, Department of Post Harvest Engineering and Technology, Aligarh Muslim University, Aligarh, UP 202002, India
e-mail: alsaghir@rediffmail.com

A. Malik et al. (eds.), *Food Processing: Strategies for Quality Assessment*,
Food Engineering Series, DOI 10.1007/978-1-4939-1378-7_7

important where the aroma or taste is same. Many food products are sweet like jaggery, dates, honey, etc. but the feel factor or its scores will vary according to choice of customers. Colour and texture can also be evaluated objectively. The colour measurement is done by Hunter Lab or Lovibond Tintometer. These instruments measure colour in three dimensional systems. Colour measurement is further quantified. Light reflected from a coloured object can be divided into three components as mentioned earlier and this can be described as value, hue and chroma. Value is explained as lightness/darkness of the colour that white vs. black. Numerical values can be accordingly assigned from 0 to 100. '0' for black and '100' for clear white. Hue of the colour determines what percentage of a particular colour is red, green, yellow, blue, etc. and chroma indicates the saturation or intensity of these colours (Potter and Hotchkiss 1993).

Texture is an associated property which is known by touch or feel. Texture is correctly determined by sensing that is perceived by soft and hard part of the mouth. Texture profile of a product is sensed by three stages: the initial, masticatory and residual (Brandt et al. 1963; Szczesniak et al. 1963; General Food Corporation 1970). The initial stage includes from 1 to 5, chews and it includes several properties of products viz. hardness, brittleness, springiness, crispness, etc. The second masticatory stage includes chewing of the food which is then swallowed. The third stage is theoretical and past experiences matching to the known product. The textural properties can be evaluated both subjectively and objectively (Sherman 1969; Szczesniak and Smith 1969). Subjective evaluation is done by group of panel while an objective evaluation is done by instrument.

Quality assessment of processed food has become on emerging issue in the present era. The quality factor has broadened and covers all aspects which satisfy consumer expectations. Subject of quality covers different sections like quality analysis, sensory quality, quality safety, quality assurance and quality standards and regulations. Quality analysis is an approach assisting the functioning of industries, establishing the standards and bringing the manufacturing process to establishment and successful. Every industry utilizes grades of raw materials permitted by Prevention of Food Adulteration (PFA). The analysis of food and food products requires proper sampling as the plant and animal tissues not only having variations in composition of different varieties/brands but also variations are observed in individuals of same variety. Sometimes there are variations in various parts of same fruits/vegetables/animal tissues. Food analysis is divided in two groups; namely proximate and ultimate analysis. The former gives the facts of nutritional/biochemical aspect, while the latter covers the information of particular element or organic compound (Ranganna 1993). Proximate analysis covers the determination of percentage of moisture, ash, crude fibre, acidity, proteins, lipid, sugars and carbohydrate. The ultimate analysis provides information of a particular element like calcium, sodium, iron, magnesium, etc., or vitamins, pigments, antioxidants, etc. Before analysis, sampling is very important to get the accurate information of analysis. The samples should be taken in sufficient quantity to compensate the variability. The amount of sample to be statistically determined when the extent of variability of the individual sample is known, the number of individual samples to be selected may be found from the following expressions:

$$n = c\sqrt{N},$$

where n is the number of individuals to be selected, c is a factor which represents the degree of accuracy desired in the sample and N is the lot size (Ranganna 2002).

The food analysis helps to standardize the formulation and gives information about the nutritional facts of the processed food. Sometime, it is essential to analyze particular constituents like protein which becomes important factor in manufacturing process. It not only needs a required standard but also provides functional property to a product, e.g. in a bakery industry the gluten content of wheat flour is important. It is the wheat flour protein consisting of two components glutenin and gliadin. If the gluten content lacks in wheat flour it begins to create defect while moulding in a rotary moulder. Due to this fact, it becomes essential to determine the glutenin content of wheat flour. Here manufacturing capability is more important than to keep the standards of nutrition.

Processing methods have considerable effects on the nutritional value of the food product. For example, thermal processing of any food material, rich in vitamin-C, leads to a loss of 70 % vitamin-C. The processed food must be fortified with the relevant vitamin. The food products at the end of packaging are labelled. Labelling is an important part of the manufacturing process. It gives information of the nutritional value and expiry date, etc. The following information are important for a packaged and processed food, viz. brand name, net quantity, company name, product date of manufacture (date of product was displayed), the date by which the product should be sold (sell by date), the last date of maximum quality (best use by date) and expiry date. Quality analysis department plays an important role of instructing the incoming raw material, operational efficiency of machine, temperature of stabilization and cooling. The department evaluates the organoleptic/sensory characteristics, chemical and physical evaluations. The sanitation and hygiene are also a part of the quality control work. The plant and machines are inspected for sanitary conditions before being used for production. The personal hygiene of a person working in the industry is also a part of quality control work. Raw material and final products both are microbiologically inspected. Equipments and machines are also evaluated for microbiological count. Waters used in the manufacturing process and also discharged as waste are determined for chlorination, biological oxygen demand (BOD), chemical oxygen demand (COD), hardness, desired oxygen and total solids.

Microorganisms if not taken care of, pose great threat to human health in terms of food spoilage and food intoxications. A number of outbreaks have been caused by botulism and other diseases. Food-borne disease is divided in two groups (a) food intoxication and food infection. In food intoxication, the intoxicating microorganisms produce toxins which are harmful and fatal to humans/animals. In food infection, the microorganism makes the food as vehicle and enters into human body.

These microorganisms are the cause of spreading diseases, e.g. *Salmonella typhi* which causes typhoid and *Vibrio cholerae* results in cholera in humans.

Food spoilage is caused in canned foods by different bacteria, aerobic spore forming, thermophilic anaerobes, mesophilic anaerobes, non-spore forming butyric anaerobe, spore forming anaerobes and name of spoilage caused by them are detected.

Possibilities of vacuum flat sour bacteria which are usually non-swelling burst due to production of carbon dioxide and hydrogen. In describing the abnormal situation of the shape of cans the following techniques are used. A hydrogen swell is the initial condition of swelling cans and it becomes normal after a little pressure. A can is flat, when both ends are concave. The vacuum is high enough to maintain the ends of its concave conditions. A can is in 'flipper' in which vacuum is so low that mechanical shock will produce distortion of one or are both ends. It is noted by the fact that if sharp blow is given against a massive object close to the top of the can, one or both ends may spring out. The can should not be dented while striking. A can is in 'springer' condition if one end is distorted and the other end is flat and the pressure on the convex end will cause the flat end to spring out when pressured. A can is in 'swell' condition in which both ends are convex. If sufficient pressure is applied it causes permanent distortion of both ends.

The group of microorganisms most likely to proliferate in a particular food item is determined by the following parameters

1. Moisture content of the food
2. pH of the food
3. Nutritional content of the food

Bacteria require high moisture and medium for the growth, while molds can grow even in low and medium acidic food and intermediate moisture level. Yeast requires mostly sugar.

According to perishability, entire food groups are divided into three categories: (a) perishable food, (b) semi-perishable food (c) non-perishable food. The perishable foods are those which have high moisture content (above 60 %); their shelf life is few hours to 1 day at room temperature. Semi-perishable foods like potato, onion can be kept for 1–3 weeks in safe condition; while non-perishable foods are grains with low moisture content (as low as 4–15 %).

Sensory quality of food is tested by sense organs. Signal resulting from these messages are transmitted to the brain, which consequently modified, sorts, classifies and interprets the information to provide a judgement. Part expresses and memories enable the brain to decipher the information. If a particular food drink product has not been previously consumed, then an individual is more likely to alert using the five senses together to bring the result with the help of relevant past experiences in conjunction with current experience.

Although there are margin features of good food sources, there are some general rules and some preferences may be in part genetically determined. Sensory response includes both physical and chemical characteristics of the food. When foods are evaluated by human assessors, there are relationship between the concentration of some chemicals in the foods and the responses of the evaluators/assessors thus with highest concentration of sodium chloride, there will be higher ratings of saltiness and similar relationship exist for response to more complex flavour stimuli or variations in physical properties. There is often a temptation to ascribe this sensory phenomenon to the food and chemical components themselves. However, the sensory attributes exist only in the interaction between the food and the person assessing. The behavioural responses in sensory evaluation are most important part of the evaluation.

In psychology, this type of relationship has been described as the stimulus responses or more accurately the stimulus organism–response relationship. The foods containing components at certain concentration form the stimuli and the responses are the action of assessors within that is a rating on a variety anchored scale a choice between alternative sayings whether a taste is present or describes a nature of taste or flavour. Although it might appear that this is simple transduction from stimulus to response. This is far from the case.

Although some variations in chemical composition are perceived, there remains the question of whether or not if so by how much it influences the choice of the food.

In the case of sugar intake, much of the early work centred on attempts to relate sugar preference to body weight and obesity, on the assumption that a higher preferences for sugar would lead to a greater intake of sugar and hence a higher body weight. Similar thinking with salt led to attempt to relate sensory responses to blood pressure and hypertension on the grounds that a higher preference for salt would lead to higher intake and hence higher blood pressure. There are some strong assumptions like that sugar intake will lead to a higher body weight as carbohydrate excess in the body to lead to fat accumulation. Salt intake will raise blood pressure.

2 Measuring Consumer Response

Data which measures the liking or disliking of samples can be controlled from consumer in a variety of different situation at home. Many food industries conduct acceptability studies on their own products by their technical staff to give an indication of likely consumer acceptability or to select a few samples from a larger range of samples for consumer test. Such people are likely to be representative of the consuming public as they form only small often limited section of the population. For example, the majority may be between 20 and 50 years. Thus people are also likely to have considerable knowledge about the company's product and may be able to pick it out in a blind test. Employees may also know about the development stages of a product and would therefore have information which could bias their evaluation to test the validity of a staff/panel in determining the acceptability of a product, one or more of a product allocated a low acceptability score should be included in the mass consumer trial. Several methods of sensory evaluation have been suggested, e.g. difference test, rating test, etc. Acceptability test measures how much a product is actually liked or disliked and the result from different samples can be compared. In preference testing, the samples are directly compared and judged on the basis of preference.

The instructions given and the format of the test are important in consumer testing. If the respondent has only a small amount of the product to taste, this will allow an immediate response but no evaluation of the product on continued eating will be determined. The situation of the list can also influence the results for example in a listing centre in a town hall respondents are unlikely to harm the time to spend evaluating a product in detail and are not in familiar surroundings or perhaps not particularly comfortable ones. Concentration, therefore likely to be minimal with many respondent wanting to finish the test quickly and resume their original task. The consumer expects that food should be wholesome, nutritious, safe and hygienic. The industries planning for food manufacture are covering enteric aspect of precautions to provide the safe and good quality food to the consumer.

3 Quality Control

The aims of quality control have been known well and discussed worldwide: (1) to ensure the safety with food supply, (2) to prevent economic fraud/deception to the consumer about nutritional content of food and (3) to inform consumer about the nutrient content of food.

Quality control department has the responsibility to discharge that it takes care of raw materials, washing, hygiene, nutrition and even future products plan. The raw material is first identified for its grade and if any defect is present, it is elaborated. Otherwise these defects will pass on to the final products and on the various operating machines working for their orderly in perfectness, for example grading, peeling, washing machines, etc. The other ingredients ready for blending are in right quality and right weight. Once the product is ready, it is to be packed or filled properly. The parameters like filling temperature, fill head space, close temperature and packages integrity are important at these stages. The Gross and net weight and cost are also mentioned on the package. Both sterilization and cooling are equally important. To get the response from the consumer panel for the quality, the rigorous laboratory tests are conducted related to sensory nutrition and microbiological characteristics. The flavour, which is a composite characteristic, comprises taste, smell and feel. Aroma, colour, texture, etc. are determined subjectively.

Quality analysis department has laboratory equipped with all the facilities of chemical, microbiological, sensory and instrumental analysis. Large food industries have a big quality control laboratory, with separate units/sections and experts technical staffs, who have the ability to analyse various constituents of food and also to run very sophisticated instruments like HPLC, Atomic Spectra, Texture Analyser, Viscometer, PCR Technique, Photo Flame Meter, Water Activity Meter,

Permeability Measuring Device, Water Vapour Permeability Cup, etc. Analysis performed is matched with standards. Standards have been laid down for convenience of the processors and also for the consumer. The food analysis helps to standardize the formulation and give information about the nutritional fact of the processed food. Sometime, it is essential to analyse a particular constituent like protein which becomes important factor in manufacturing process, because not only it needs a required standard but also provide functional property to a product. The standards are maintained with definite meanings. If raw material does not conform to the standard, it can affect both processor and consumer. For example in a bakery industry the wheat flour is tested for gluten content, the combined proteins consisting of glutenin and gliadin. These proteins impart elasticity and plasticity to dough. It helps in forming, shaping, fermenting and baking. If the gluten content is not up to the mark in the wheat flour, production in continuous processing would be interrupted.

The other example of the standard is the alcoholic acidity of wheat flour which indicates the storage time of wheat flour. The long stored wheat flour will have high alcoholic acidity developed by activity of enzymes (both protease and lipase). The maximum limits of the factors which represent the suitability and standards have been accordingly laid down.

There is a division for research and development in food industries. This department focuses attentions over the improvement in process and product quality. Any complaint received in the industry is seriously taken by the department and the customer relations are maintained by bringing improvement in the product quality. The noncompliance of different quality and safety parameters are seriously perfected by this department. Nutritional, microbiological and sensory characteristics of the products are determined periodically during storage. Quality deterioration is studied so that the shelf life of the product must be ascertained.

For chemical analysis total solids, soluble solids, insoluble, carbohydrates, protein, fat, minerals, vitamins, pigments, enzyme, pH and salts are analysed. Physical evaluation determines firmness, wholesomeness, viscosity, specific gravity, colour, mechanical damage and extraneous matter.

Sanitization and hygiene of the plant are very important part of the food industry. Healthful conditions during processing, preparing and handling of food are necessarily maintained. Sanitization is an application of science to provide wholesome food, handled in hygienic environment by healthy food handlers to prevent contamination with food poisoning microorganism and minimizing the chances of spoilage by microorganism. It also includes personal hygiene. Employee working in the industry should be taught about hygiene practices. They should be well groomed and free from contagious disease. Outside inspection include weed control, insect control, rodent control, building exteriors, etc. The microbiological safety includes total plate count, yeast mold count, drosophilae, insect fragments, rot fragments and the microbiological count of equipment and machines.

4 Quality Safety

The general subject of safety and food supply has been debated for three decades and the topic of issues also changes with time. At the end of twentieth century several legislation and regulations have been framed for given perfect solutions to food safety issues. In the UK, the food safety act 1990 had made significant changes to the food legislation provision in which the main conversation requirements of prohibiting the rate of food, which is either injurious to health or is of the nature substance or quality, demanded are retained, the act also includes provision requiring the registration of all premises and enabling clause that will permit the introduction of range of codes of practices and orders to embody European community legislation (Ranken and Kill 1993).

In the present age of increased awareness of the consumer, there is need for nutritional, energetic and safe food. Here safety means food must be free from toxin components and microbiologically safe. These toxins producing agents are classified as biological microorganism, i.e. bacteria (*Clostridium*, *Enterobacter*), viruses, parasites and molds. The toxins produced by bacteria are either neurotoxins or enterotoxins and the toxins produced by molds are called mycotoxins (in particular aflatoxin produced by *Aspergillus parasiticus* and *Aspergillus flavus*). Bacteria bring illness like typhoid, cholera, diarrhoea, liver cancer, respiratory failure, hypertension and dysentery.

Food industry has to adopt a well-documented food safety plan ensuring the freedom from the above discussed causative agents for the consumer. HACCP (Hazard Analysis Critical Control point) has become the most important tool to bring the safety plan successfully. Hazard brings risk if it is crossing the limit of critical controlled point. It is demarcation line for safety. The Hazard is always associated with the system but the capacity of the hazard to create a risk changes with the conditions. Hazard has been differentiated in relation to food safety and food quality. It is associated with product or its production and its intended use.

HACCP is a system which identifies, evaluates and control hazards which are significant for food safety. It is food safety plan developed for food industry. HACCP examines every step in food operation, identifying specific hazard as well as implementing effective control measures and verification procedures. HACCP is not a zero risk system. It is designed to minimize the risk and as such it is a risk management tool. The HACCP has unique features: it is systematic, directed by scientific evidence of risk, identifies specific hazards and measures for their control and focuses on prevention (rather than relying mainly on end product testing). It is capable of accommodating changes (e.g. advance equipment design, technological development). HACCP system can be applied throughout the food chain, 'from farm to fork', 'from paddock to plate'. It increases confidence in food safety. It requires full commitment and involvement of management and work forces. It requires multidisciplinary approach. It is compatible with implementation in quality management system. It is a system of choice for food safety management.

HACCP was originally developed to address fixed safety hazards in order to protect food against microbiological, chemical and other physical hazards.

5 Logic Sequences for Application of HACCP (Codex Guidelines)

The 12 steps are:

1. Assemble the HACCP Team
2. Describe the products
3. Identify intended use
4. Construct a flow diagram
5. On- site verification of flow diagram.

	The seven (7) HACCP Principles
6. List all potential hazards Conduct hazard analysis Determine control measures	**Principle 1**
7. Determine CCPs	**Principle 2**
8. Establish critical limits for each CCP	**Principle 3**
9. Establish a monitoring system for each CCP	**Principle 4**
10. Establish corrective action for deviations that may occur	**Principl e 5**
11. Establish verification procedures	**Principle 6**
12. Establish record keeping and documentation	**Principle 7**

Hazard (Food Safety): A biological, chemical or physical agent or the conditions of food with the potential to cause an adverse health effect, viz. food intoxication and food infection, emanating to several diseases as discussed earlier.

Hazard (Quality): A quality hazard is a factor that has potential to cause an adverse effect on product process quality and hence profitability.

Quality hazard includes product quality, environmental hazard welfare, production, occupational health and safety/personal hygiene and regulations. There are certain factors that affect the growth of the biological hazards (hazards caused by bacteria, viruses, fungi, parasites and algae).

The following factors affect the growth of biological hazard:

1. Intrinsic (Moisture content, nutrients, antimicrobial constituent, biological structure)
2. Extrinsic (temperature, humidity and gases in the environment)

The control of these biological factors can be planned by depriving the organisms to unsuitable conditions for growth, for example high or low temperature, low pH, low and moisture content (less than 20 %). Bacteria will not grow well at a pH below 4. Acidity has been commonly used means of preserving food for long time as in the case of pickles. Fermentation brings down the pH in milk, in pickles, in sauerkraut and soy-fermented producers.

Reduction of moisture content is the oldest method of preservation. The water requirement of microorganisms is defined in term of water activity (a_w) of a substance that can range from 0 to 1. Water activity of pure water is 1.0 and that of a food material is defined as the partial pressure of water in food divided by the vapour pressure of pure water. Mathematically it is expressed as

$$a_w = ERH / 100$$

where *ERH* is equilibrium in relative humidity and a_w may be reduced by removing water or adding solution, both salt and sugar to food. Bacteria require more water than yeast and mold for growth. The a_w range for the growth of bacteria is 1.0–0.75. a_w range for highly perishable (fresh) food and canned food, vegetables, meat, fish and milk, cooked sausage and breads is 1.0–0.95.

The following groups of bacteria are able to grow: *Bacillus, Clostridium perfringens, Clostridium botulinum, Escherichia, Lactobacillus pediococcus, Klebsiella, Pseudomonas, Proteus, Salmonella, Serratia, Shigella, Vibrio paralyticus* and also some yeast (Robertson 1993). Mostly bacteria grow in the range of a_w of 1.0–0.91, however, halophilic bacteria still can grow on the lower range of water activity (a_w 0.80–0.75). The mold and yeast require good amount of water for growth. However, they are able to grow in lower range of a_w, i.e. 0.91–0.60. The products, in which they proliferate, are fermented sausages, sponge cake, dry cheese, margarine, sweetened condensed milk, maple syrup, pulses (containing moisture content of 15–17 %), jam, jelly and marmalades. The food and food products having water activity of −0.5 or less do not allow microbial proliferation. From the above discussion, it is clear that water activity (a_w) plays an important role in determining the spoilage by microorganisms and also about the optimum storage conditions for a particular food in term of relative humidity and temperatures. As defined, water activity in *ERH*/100, indicates that the food material having a_w of 0.70 and if it is kept in an atmosphere with $RH = 60$ %, it will lose water. This is sorption isotherm of moisture vs. a_w (Fig. 1).

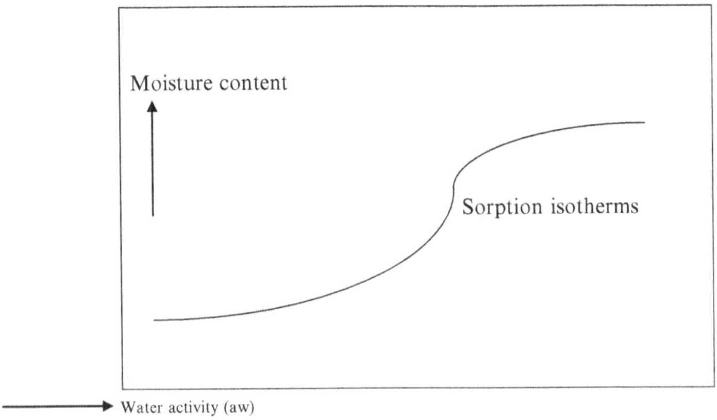

Fig. 1 Moisture sorption isotherm

Extrinsic factors are temperature, humidity and gases in the environment. *Salmonella* grows in chicken under a temperature range of 5–47 °C. Most of the bacteria fall in mesophilic group. In order to multiply rapidly, bacteria need both humidity and temperature.

In the storage environment, low concentration of oxygen in combination of carbon dioxide slows down the rate of respiration. Ozone is an oxidizer and it has the preservative effect on sweating food when added to storage environment. It is effective against a wide range of spoilage microorganisms and destroys naturally produce ethylene that advances fruit maturing. Both CO_2 and ozone are effective in inhibiting the spoilage microorganisms.

SO_2 is applied to food and beverage as a liquefied gas, more commonly in the form of sulphide, bisulphide or meta-bisulphide and salts to control the growth of unwanted microorganisms.

In spite of the biological factors discussed above, food-borne diseases are caused by viruses, fungi, parasites and algae. There are over 150 viruses which have potential to cause food spoilage. In food-borne outbreaks, viruses come either from food handlers or from sewage contaminated water. Hence it is important that only potable water should be used in food industry.

Fungi are classified as mold and yeast. Fungi can be beneficial even in the production of food products, like in production of cheeses. Some fungi are also used in manufacturing of some antibiotics like penicillins. A number of fungal species are capable of producing metabolites that are toxic to humans and domestic animals. These metabolites are called as mycotoxins. Mycotoxins have been found to be responsible for some cases of food-borne diseases throughout the history. Mycotoxins may be consumed directly by taking contaminated grains or indirectly by eating products, produced by animals that have consumed mycotoxins.

Parasites, like protozoa, also cause food-borne diseases. About 20 species of pathogenic protozoa have been transmitted to human through contaminated water or foods. Infection by protozoa is more common in tropical countries than are temperate zones.

Also, three types of algae Dinoflagellates, blue-green algae and golden brown algae can poison out food supply. In the right environmental conditions, a few algal cells can multiply into dense bloom in the river or sea water. Some species of algae produce toxin. When human eats shellfish that have fed on these algae, the human may suffer illness or even death. Unfortunately, poisonous shellfish do not look or taste differently from normal ones and cooking does not destroy toxins. There are six different types of shellfish which are poisonous in nature. These are (1) paralytic shellfish poisoning (PSP), (2) diarrheric shellfish poisoning (DSP), (3) amresic shellfish poisoning (ASP), (4) neurotomic shellfish poisoning (NSP), (5) cyrobacterial toxin and (6) cigratera.

Chemical Hazards: Chemical contamination of food stuffs can happen at any stage of their production, from growing of the raw materials to consumption of the finished products. The effect of food-borne chemical contamination of the consumer can be long term or immediate and acute, such as the effect of allergens in foods. The main chemical hazard issues in food products are described in the following paragraphs.

Chemical compounds are used frequently in the food supply chain and can present food safety risk if their use is not managed properly. The chemical hazards may be caused by any of the chemicals used in food supply chain, i.e. harming chemicals, pesticides, allergens, toxic metals, nitrites, nitrates and N-nitroso compounds and PCBs.

Further chemical hazards may also be caused by plasticizers and packaging migration, veterinary residues, chemical additives, seafood toxins, zootoxins and phyllotoxins.

Physical Hazards: Physical hazards in food are found to cause no illness or injury to the consumers. The following items may be included as a source of physical hazards, i.e. glass, metal, stone, twigs, leaves, wood, pests, jewellery, plastic, etc. Like biological and chemical hazards, physical hazards can enter in a product at any stage during production. Not all foreign objects found in the food will cause harm or illness. Although, it is very undesirable for a consumer to find hair, leaves, twigs, insects, etc. in their food.

Quality Hazards: Quality hazards cause the product to be considered as of poor quality by the customer, because it does not fulfil their requirements. Quality hazards differ from food safety hazards in that they do not cause illness in consumers. It can be classified as:

(a) Product quality hazard
(b) Environmental hazards
(c) Animal welfare hazards
(d) Production hazards
(e) Occupational health and safety hazards
(f) Regulatory hazards

Product quality hazard: Quality hazards may be caused by product quality that is affected due to poor agronomic/post-harvest practices. It may be due to not meeting the product specifications, i.e. incorrect size, shape, appearance, texture, flavour, smell, colour, nutritional value, etc. The quality may also be affected by wrong selection of raw material.

Environmental Hazards: Offensive smell emanates from intensive farming operations. Uncontrolled waste management practices are conducted for both liquid and solid waste.

Animal and Welfare Hazards: The stocking density of lying hens in cages is kept to uniform. During animal transport, overcrowding is avoided. Hygienic slaughter practices are adopted.

Production Hazards: Decrease in yield product size due to inaccurate application of chemical. Yield efficiency also decreases. Time is lost due to break down of equipment.

Occupation help and Safety Hazards: Injury, burns, cut falls to the stocks.

Regulatory Hazards: Sometime product weight remains less than the amount stated and ingredients labouring remain incomplete.

6 The HACCP Team Its Requirements

The team develops and derives the company of HACCP for food safety policy. Team ensures that the HACCP project will continue to move forward and remain valid. It elects the HACCP team leaders, reports on progress regulatory, ensures correct balance of technical/industrial experience, assesses the need for specialist knowledge and engages this resource as required.

Team must have specialised knowledge of raw materials, ingredients, finished products, and process requirements, processing procedure, pre-requisite programmes, the production environment, premises and surroundings. Specialist teams will have full knowledge of hazards associated with the raw material, the product and the process. The team should be trained and have a thorough technical knowledge of HACCP.

The team members need to be of equal status and be able to communicate not only with each other, but with people at all levels of the business. Their representative will also foster a sense of ownership among those who have implemented the HACCP system. HACCP team member should have attributes which will help to evaluate the data in a logical manner, can systematically solve the problems, able to think outside the box, able to delegate, cooperate with team players, have organizational time management, group meeting skills and also should have analytical and good communication skills. The size of team required is dependent on the size of business and variety of products they produce.

7 Joint FAO/WHO Standards Programme

The FPO specification for canned fruits and vegetables make suggestions about the food additive/preservative added to fresh fruits and the processed product. FPO licensing also covers sanitation and hygiene of industry processing fruits and vegetable products. The FAO/WHO, the joint venture, is known as CODEX ALIMENTARUS COMMISION (CAC). CAC food is a standard food programme and was established in February 1979 with 117 countries members, including India (Ranganna 2002). The purpose of this programme is to safeguard the health of consumers, to ensure honest practices in the food trade, to make co-ordination of the work on food standard decided by various agencies and to prepare the draft standard. These standards are to bring acceptance of the government, so that they may be published in codex alimentarius as either national or international standard. Another food safety programme was created with world trade organization in January 1995.

8 National and Regional Standards

In India, national standards like FPO fruits product order (1954) was designed for keeping the quality standard, nutritional standard, level of food preservatives and total soluble solids. Several other standards like Bureau of Indian Standards

(BIS) act (1954) are meant for preventing fraudulent practices and safeguard the interest of consumers.

Similarly in the USA, many national and regional standards are followed like federal regulations governing food additive. The main federal law governing food additive is the US federal food, drug and cosmetic act (FFDCA) (1938). The law had been amended many times. The latest amendments were the food quality protection act (FQPA) and food and drug administration and modernisation Act (FDAMA), had been passed respectively in federal register. It includes all federal regulations in 50 volumes. CFR is revised once each calendar year and updated after every 3 months. One more document is the US FDA policy guide which is accessible on internet (FDA 1998, http://www.rda.gov). The federal meat inspection Act (FMJA) and other poultry product inspection authorize are to regulate meat and poultry products (Federal Meat inspection Act, 1967 and poultry products inspection Act, 1957). USDA also regulates egg and egg products under egg product inspection Act 1970. USDA published new regulations in 1996 related to food safety inspection system.

In India, Food safety and standard Act, 2006 has been designed to consolidate the laws relating to food and establish the food safety and standard authority of India for laying down science-based standard for article of food and regulate their manufacture, storage, distribution, sale and import, to ensure availability safe and wholesome food. It is declared that it is expedient in the public interest that union government should take the 'food industry' under its control. The law is related to the following features in food system, viz. food adulteration, food additive, food business, food safety, etc. The central government has establish the body which is known as the Food Safety and Standards Authority of India to exercise the power conferred on and to perform function assign to it under this act. The food authority shall consist of a chairperson and following 22 members out of which one-third shall be women.

A pioneer approach has been adopted by government of India to safeguard the health of consumer by establishing the food safety and standard act 2006. The provision safeguards the interest of consumer to protect them against adulteration. The food sources mean whether processed, partially or semi-processed or unprocessed which is intended for human consumption, generally modified or engineered foods or food containing such ingredients, infant food, packaged drinking water, alcoholic non-alcoholic drinks, any processed food placed on the market for human consumption.

The 1990 Act was initially developed to consolidate and modernize the long series of Food Acts and Food and Drugs Acts. Two fundamental requirements of the earlier legislation were carried forward the new section-7 prohibits the sale of any food which is injurious to health and other section prohibits the sale of food which is not of the nature substance or quality demanded by the purchaser.

Section-8 prohibits the sale of any food which fails to comply with food safety requirement. Other important provision was included. Provision for regulation to be made which would require the obligatory registration of all premises upon which food is handled. Making the food legally responsible for the quality of goods brought into the country by them and no longer able to claim warranty from supplier located outside. The role of enforcement authorities and the manner in which they work varies throughout world and is suggested to deal only with situation as exist in the UK. Expert suggestions and advices are taken for other countries and from appropriate authorities in the country in question.

UK Food Safety act (1990) has increased the power of the local authorities to act, whereas in the part the authorities could act only if the food or the circumstances, in which it was being produced or sold, constituted a danger to the consumer. They have now increased the power to prohibit production at all stages and thus it is like the preventive measures to enhance public safety.

9 Food Safety

Nowadays the food industry has become global and there is a lot of competition in food business. There are industries to bring the cost down and to improve margin in order to survive. As a result, food safety experts are struggling to maintain adequate control of their firm activities to safeguard the interest of consumer and to prevent fraudulent practices. The food and beverage companies follow the food legislation and regulations for their brand names reputation.

Food processor has to go through all steps of productions; from grower to the consumer. Each step involves substantially different type of process and unit operations, which require monitoring for safety, compliance of FDA/HACCP and also profit optimization.

Mandate by FDA such as HACCP procedure and ISO-22000-based food safety management system are the basis for many compliance and quality programs in the food and beverage industry. Substandard products and their information may lead to severe consequences. Implementing statistical process control and manufacturing analysis, enable food companies to move from inspector based models to a process control model.

10 Conclusion

Food quality has become an important aspect of food business. The consumers now a days are well aware of the quality attributes including nutritional, sensory, textural and hygienic qualities. The composite quality is represented by these individual characteristics to fulfil the demand of consumer in the present era. Sensory phenomenon is related to food as an interaction between physical or chemical stimulus and a person. These are not simply mapping of physical and chemical properties.

Food safety ensures the prevention of any hazard and risk into the food. Food manufacturing process proceeds from farm to fork. HACCP has a great deal of promise for preventing the high risk. The technique is consistent with ISO 9000 quality system which requires the use of multidisciplinary team to develop quality plans and process control requirements to select and use suitable instrument to monitor and control and to record and analyze results and audit the control system.

Quality control, supported by statutory provision and food regulation PFA (Act 1954, Role 1955), was promulgated to check the adulteration of food stuffs. FPO 1955 was established for fruit and vegetable industries by ministry of food processing industry, government of India. Other food legislations are AGMARK 1937, Sugar control order 1956, Vegetable oil products control 1947 and Meat Food Products control order 1975.

The Codex Alimentarius commission is a joint venture of FAO/WHO. Commission made a valuable contribution in discussing the role of national and international exposure assignment by promoting a valuable and consistent frame work and recognising the legislation that prevent the risk and health analysis and analysing the principle and putting in practice across the globe.

References

Brandt MA, Skinner EZ, Coleman JA (1963) Texture profile method. J Food Sci 28:404–409. doi:10.1111/j.1365-2621.1963.tb00218.x

FDA (1998) http://www.rda.gov

General Food Corporation (1970) Food corporation sensory texture analysis manual. White Plants, New York

Potter NN, Hotchkiss JH (1993) Food safety, risk hazard. In: Food science. Chapman and Hall, Washington, DC, p 532

Ranganna S (1993) Handbook of analysis and quality control for fruit and vegetable products, 2nd edn. Tata McGraw-Hill, New Delhi

Ranganna S (2002) Examination of canned food, joint FAO/WHO standards programme. In: Handbook of analysis and quality control for fruit and vegetable products, 2nd edn. Tata McGraw-Hill, New Delhi, p 815

Ranken MD, Kill RC (1993) Food issues. In: Food industry manual. Leather Head Food Research Institution, New York, p 578

Robertson GL (1993) Water activity and food stability. In: Food packaging: principles and practice. Marcel Dekker, New York, p 295

Sherman PJ (1969) Texture is a sensory property. Food Sci 34:458

Szczesniak AS, Smith BJ (1969) Observations on strawberry texture. A three-pronged approach. J Texture Stud 1:65–89

Szczesniak AS, Brandt MA, Friendman HH (1963) Development of standard rating scales for mechanical parameters of texture and correlation between the objective and the sensory methods of texture evaluation. J Food Sci 28:397–403. doi:10.1111/j.1365-2621.1963.tb00217.x General

Food-Borne Microbial Diseases and Control: Food-Borne Infections and Intoxications

Sait Aykut Aytac and Birce Mercanoglu Taban

1 Introduction

World Health Organization (WHO) defined the term food-borne disease as any disease of an infectious or toxic nature caused by, or thought to be caused by, the consumption of food or water. Regardless of the latest advances in food hygiene and food manufacturing and processing, in addition with increase in consumer awareness, food-borne microbial diseases are the most widespread problem and still represent a significant threat to public health in the contemporary world, especially to the very young, the old, the very sick, and the immunocompromised people who are all more at risk of them.

Each year in the United States of America (the USA), for example, food-borne diseases result in an estimated 76 million illnesses, 325,000 hospitalizations and 5,000 deaths (WHO World Health Organization 2007). In 2011, Centers for Disease Control and Prevention (CDC) estimates the numbers of food-borne diseases caused by the main known pathogens and unspecified agents transmitted by food as 47.8 million/year, resulting in 127,839 hospitalizations and 3,037 deaths in the USA (Scallan et al. 2013). It has been also estimated that the risk of becoming ill as a result of microbial contamination of food is 100,000 times greater than the risk from pesticide contamination (Adams and Moss 2008).

Changes in farming practices, increase in international trading of foods, changes in food manufacturing and processing, increase in international movement of people, changing lifestyle of the population, and microbial evolution are the factors that contribute to food-borne microbial diseases (Adams and Moss 2008).

S.A. Aytac
Department of Food Engineering, Hacettepe University, 06532, Beytepe, Ankara, Turkey

B.M. Taban (✉)
Department of Dairy Technology, Ankara University, 06110, Diskapi, Ankara, Turkey
e-mail: btaban@ankara.edu.tr

A. Malik et al. (eds.), *Food Processing: Strategies for Quality Assessment*,
Food Engineering Series, DOI 10.1007/978-1-4939-1378-7_8

Food-borne microbial diseases are caused by the consumption of either food or water contaminated with alive pathogenic bacterial cells (or bacterial spores as in infant botulism) or food that include toxins produced by the toxigenic bacteria or moulds (Ray and Bhunia 2008). In other words, there are three types of food-borne microbial diseases: intoxication or poisoning, infection, and toxico-infection (Dhama et al. 2013). Food intoxication arises as a result of the consumption of pre-formed bacterial or mould toxin (mycotoxins) which has to be in active form in food. In this type of disease, there is no need of alive cells during the consumption of the food after the toxigenic bacteria or moulds have grown and produced the regarding toxin under favourable conditions in the food. In other words, toxins, not bacteria, cause disease. On the other hand, food infection arises due to the consumption of food or water contaminated with alive enteropathogenic bacteria or viruses. Lastly, toxico-infection arises as a result of the ingestion of many alive pathogenic bacterial cells via contaminated food and water and then they sporulate, colonize, and produce toxin(s) in the body (Ray and Bhunia 2008).

2 Food-borne Bacterial Pathogens

2.1 Bacterial Infections

2.1.1 *Campylobacter* spp.

The genus *Campylobacter* belongs to the family of Campylobacteriaceae that comprises 18 species of *Campylobacter* and 4 species of *Arcobacter* (Nachamkin 2001). *Campylobacter jejuni* subsp. *jejuni* (referred to subsequently as *C. jejuni*) is the most important member of this genus since it is known as one of the causative agents of food-borne microbial diseases. However, other *Campylobacter* species such as *C. coli* and *C. fetus* also cause infectional diseases called campylobacteriosis (FDA 2012).

Campylobacter spp. are Gram-negative, oxygen-sensitive microaerophilic (85 % nitrogen, 5–10 % carbon dioxide, and 3–5 % oxygen: oxygen concentrations lower than the atmospheric oxygen level), oxidase-positive, non-spore forming curved rods, or spiral-shaped bacteria and are motile with polar flagella on one or both ends that produce the corkscrew-like motility. They cannot ferment or oxidize sugars. All *Campylobacter* species grow at 37 °C, however, the optimum growth temperature for *C. jejuni* and *C. coli* is around 42–45 °C (thermoduric). They stay alive weakly at room temperature and do not grow below 30 °C. Under unfavourable conditions, they cannot be isolated by cultural detection methods but still remain infective [viable but non-culturable (VBNC) state] (Adams and Moss 2008; FDA 2012).

Campylobacter spp. are responsible for an estimated 2.5 million food-borne illness per year in the USA which means the 17 % of hospitalizations resulting from food-borne infections and an estimated 5 % of food-borne-related deaths in the USA (Mead et al. 1999). The infective dose of *Campylobacter* spp. given in the literature varies from 5×10^2 to 10^5 organisms due to the type of contaminated food consumed and the general health of the exposed person (Robinson 1981; Black et al. 1988; FDA 2012). Campylobacteriosis which is a generic term for a variety of syndromes caused by infection with *Campylobacter* spp., usually occurs in 2–5 days after ingestion of the contaminated foodstuff. Their symptoms include abdominal pain (which may mimic appendicitis), watery and/or bloody diarrhoea, malaise, fever, myalgia, and sometimes vomiting that last less than 1 week (FDA 2012). They also cause the serious neurological disease, Guillain–Barre syndrome (Adams and Moss 2008).

Since the optimum growth temperature of *C. jejuni* and *C. coli* is high as the body temperature of birds, the primary reservoir of *Campylobacter* spp. is the gastro-intestinal tract of wild and domesticated birds (Adams and Moss 2008). The main food sources are improperly handled or raw poultry products, raw or inadequately pasteurized milk, contaminated water, and cheeses made from unpasteurized milk (Horrocks et al. 2009; Jore et al. 2010; Schnider et al. 2010; Pitkänen 2013). In addition to these, they can form biofilm on the stainless steel surfaces over the time (Nguyen et al. 2010).

The ideal way to control the number of human infections by *Campylobacter* spp. would be to maintain hygienic conditions on farm and to avoid cross-contamination during slaughtering which lead to decrease in the number of *Campylobacter* spp. on poultry products. Raw or under-cooked foods, untreated drinking water, and unpasteurized milk should not be consumed. Hence, foods should be cooked completely. Post-pasteurization contamination should be avoided since it reintroduces these bacteria into milk. The personnel of the catering industry should not handle food if he/she suffers from campylobacteriosis or handle cooked food or ready-to-eat food with bare hands. Cutting boards and utensils used in handling uncooked poultry or other foods should be washed before being used for preparation of other foods that are eaten raw (Allos 2001).

2.1.2 *Cronobacter sakazakii*

The genus Cronobacter, which belongs to the family of Enterobacteriaceae, includes seven recognized species: *C. condimenti*, *C. dublinensis* (subsp. *dublinensis*, *lausannensis*, and *lactaridi*), *C. malonaticus*, *C. muytjensii*, *C. sakazakii* (formerly named *Enterobacter sakazakii* in honour of the eminent Japanese bacterial taxonomist Riichi Sakazaki), *C. turicensis*, and *C. universalis* (Adams and Moss 2008; Tall et al. 2013). Among them; *C. sakazakii* has recently considerable concern, although it is responsible for only sporadic individual cases or relatively small outbreaks of infections.

C. sakazakii is a Gram-negative, facultatively anaerobic, catalase-positive, oxidase-negative, non-spore forming rods, and is motile with peritrichous flagella that can grow at temperatures ranging from 6 to 47 °C with an optimum at 37 °C (mesophilic). Pasteurization at 72 °C for 15 s ensures more than a 10 log reduction in the number of survivor cells. It is moderately acid resistant compared to other Enterobacteriaceae. The minimum pH for its growth is 3.89 but the optimum pH ranges from 5.0 to 9.0. It produces yellow pigmented colonies on non-selective agar media (Lambert and Bidlas 2007; Adams and Moss 2008; Tall et al. 2013).

C. sakazakii is an emerging opportunistic pathogen that represents a significant risk to the health of neonates in hospital nurseries and neonatal intensive care units since it is related with occasional but life-threatening cases of meningitis, cerebritis, necrotizing enterocolitis (NEC), and septicaemia in neonates and infants with weakened immune systems, particularly premature and full-term infants with mortality rates of 40–80 % being reported. Even in the case of recovery, long-term neurologic sequela have been reported in affected infants. Wound infections, urinary tract infections, septicaemia, vaginitis, and aspiration pneumonia are the symptoms of this infection in adults and elderly people (Lai 2001; Giovannini et al. 2008; Healy et al. 2010; Tall et al. 2013).

C. sakazakii is regarded as ubiquitous organism that has been isolated from various foods. A high tolerance to osmotic and dry stress and high temperatures than other Enterobacteriaceae ensure the survival of it in powdered infant formulae (PIF) more than 2 years in which the contamination can occur during the manufacturing process or during post-manufacture reconstitution. Follow-on formula, weaning foods, milk and sodium caseinate powders, rice seed, dried herbs and spices, spiced meats, dried corn, soy, potato, wheat, and rice flours, dried infant and adult cereals, dried vegetables, grains, tofu, powdered herbal tea, mixed salad vegetables, tomato harvesting bins, chocolate and candied cough drops, and pastas are the other foods from which it has been isolated (Adams and Moss 2008; Healy et al. 2010; Tall et al. 2013).

The most important prevention and control way is to avoid poor hygiene during preparation and feeding of infants by applying regularly hand washing and proper washing of teats, bottles, cups, blender, and spoons with boiled water. In addition to these, it is better to avoid storage of reconstituted formulae over a prolonged period of time at refrigerated and at room temperatures and to discard unused formulae (Adams and Moss 2008; Tall et al. 2013).

2.1.3 Pathogenic *Escherichia coli*

Escherichia coli, which belongs to the family of Enterobacteriaceae, was firstly described by the German bacteriologist Thedor Escheric in 1885 during its isolation from childrens' faeces (Adams and Moss 2008).

E. coli is a Gram-negative, facultatively anaerobic, catalase-positive, oxidase-negative, urease-negative, H_2S-negative, non-spore forming rod, and is motile with peritrichous flagella that can grow at temperatures ranging from 7–10 °C to 50 °C with an optimum at 37 °C (mesophilic). However, some ETEC strains can grow below 4 °C. In IMViC tests, most of the *E. coli* strains are indole and methyl red positive and VP and citrate negative. It is heat sensitive and so is killed by pasteurization process (62.8 °C for 30 min or 71.7 °C for 15 s) but can survive during refrigerated or frozen storage for extended times. The optimum pH for its growth is 7.0 but can grow at pH 4.4. It ferments lactose with producing gas. Since it normally lives in the intestines of humans, warm-blooded animals, and birds at a very high level, it has been used as an index organism of possible faecal contamination and the presence of enteric pathogens in water and food (Adams and Moss 2008; Lund 2008; Ray and Bhunia 2008).

E. coli is one of the major enteric species in the gut of humans, warm-blooded animal. As being an important part of a healthy human intestinal tract, most *E. coli* strains are harmless and in fact provide many health benefits such as preventing colonization of the gut by harmful pathogens and producing vitamin K_2. However, some *E. coli* strains, which are described as enterovirulent *E. coli*, diarrhoeagenic *E. coli*, or more commonly, pathogenic *E. coli*, have the potential to be pathogenic to humans and can cause severe diarrhoeal diseases. Currently, pathogenic *E. coli* strains are categorized into six recognized pathotypes which are enteroinvasive *E. coli* (EIEC), enteropathogenic *E. coli* (EPEC), enterotoxigenic *E. coli* (ETEC), enterohaemorrhagic *E. coli* (EHEC), enteroaggregative *E. coli* (EaggEC), and diffusely adherent *E. coli* (DAEC) (Adams and Moss 2008). Among them, EHEC is distinguished from others by the production of Shiga-like toxins (Stx) (also known as verotoxin). There are more than 200 serotypes of Shiga toxin-producing *E. coli* (STEC) or Vero (so-called because of its ability to kill Vero that is African Green Monkey Kidney cells) toxin-producing *E. coli* (VTEC), but not all have been involved in human illness. Thus, *EHEC* is a subset of the STEC that has been implicated in haemorrhagic colitis (HC), the potentially fatal haemolytic uremic syndrome (HUS) or thrombotic thrombocytopaenic purpura (TTP). However, EHEC not only produce the shiga toxin coded by the *Stx* gene but can also attach to the intestinal wall due to the protein intimin (an adhesin) coded by the *eae* gene, and also includes the *ehx*A gene that encodes the enterohemolysin. Although several serotypes, including O111:H8, O26:H11, O103:H2, O113:H2, O104:H21, have caused human illness and are recognized as EHEC, serotype O157:H7 is the most common and widely recognized EHEC serotype. *E. coli* are also categorized based on their serotype, defined by the O antigen determined by the polysaccharide portion of cell wall lipopolysaccharide (LPS) and the H antigen due to flagella protein. The following six STEC O serogroups (non-O157 STEC), in addition to O157, represent the greatest public health risk: O26, O111, O102, O121, O145, and O45 (Feng and Monday 2006). Besides, many of the virulence genes carried by pathogenic *E. coli* exist in mobile genetic elements and can be transferred. As an example, serotype O104:H4 which have been known to cause persistent diarrhoea in under-developed countries, but rarely have been associated with major food-borne incidents, caused a large outbreak in Germany in 2011 by producing Shiga toxin, a characteristic of EHEC, although it was genetically EAEC. Thus, the O104:H4 strain that caused this outbreak appeared to be an EAEC strain that had acquired the ability to produce Shiga toxin (FDA 2012).

Enteroaggregative *E. coli* (EAggEC)

EAggEC, which was formerly categorised as EPEC, differs from this group due to their aggregative adherence on Hep 2 cells and produce a heat-stable enterotoxin called EAST 1. EAggEC causes chronic diarrhoea in children (Duffy 2006).

Enterohaemorrhagic *E. coli* (EHEC)

The symptoms of infection from this group of organisms includes severe abdominal cramps, watery diarrhoea which may develop into bloody diarrhoea (haemorrhagic colitis-HC), HUS, and TTP that may result in a fatality rate of approximately 5 % in patients. HC lasts 4–10 days. HUS is characterized by the triad of haemolytic anaemia, thrombocytopaenia, and renal insufficiency and mostly affects children under 5 years of age. TTP is a serious condition describing the presentation with HUS, fever and neurologic symptoms. The infective dose for O157:H7 is estimated to be 10–100 cells (Duffy 2006; Adams and Moss 2008; Scannell 2011). Although O157:H7 is currently the predominant strain and accounts for approximately 75 % of the EHEC infections in the world, other non-O157 STEC are also emerging as a cause of food-borne illnesses (FDA 2012).

Many cases of O157:H7 infection have been related to consumption of raw or under-cooked ground beef and beef products and raw milk. O157:H7 can develop acid tolerance and so also isolated from yogurt, mayonnaise, fermented sausages, cheeses, and unpasteurized fruit juices. Various water sources, including potable, well, and recreational water, have also caused O157:H7 infections. The other food sources are salad vegetables (especially lettuce, spinach, and alfalfa sprouts contaminated with faeces and manure) (Duffy 2006; Adams and Moss 2008).

The most important control way is to avoid contamination of carcasses during slaughtering. Besides, treating of animal waste applied to salad vegetables, adopting a hazard analysis and critical control point (HACCP) approach in food manufacturing, processing, and service sectors; preventing cross-contamination from raw to cooked meat; not drinking raw milk; and cooking minced or ground meat thoroughly can also prevent and control the illness caused by this pathogen (Adams and Moss 2008; FDA 2012; Lund 2008; WHO 1997).

Enteroinvasive *E. coli* (EIEC)

Infection with EIEC results in an illness with the classical symptoms of invasive bacillary dysentery similar to that caused by *Shigella* species. Illness caused by EIEC usually occurs 24 h after ingestion and symptoms are fever, severe abdominal pains, malaise, and often a watery diarrhoea, followed by the passage of bloody stools (Ewing and Gravatti 1947; Duffy 2006; Adams and Moss 2008).

Enteropathogenic *E. coli* (EPEC)

While EPEC was a frequent causative agent of infantile diarrhoea in the USA in the 1940s and 1950s, its infections are less important in developed countries, but continue to be a major cause of diarrhoea in developing countries, especially in infants nowadays (FDA 2012). Symptoms generally appear about 12–36 h after ingestion and include vomiting, malaise, and diarrhoea with stool containing mucus, but rarely blood (Adams and Moss 2008). The major O groups within this group which are linked to human illness include O55, O86, O111, O119, O126, O127,

O128, and O142. Food and water are the vectors of this pathogen and cases are seen mostly in under-developed countries where there is inadequate sanitation and poor water quality (Duffy 2006).

Enterotoxigenic *E. coli* (ETEC)

ETEC, which is also a common cause of infantile diarrhoea and serious dehydration in developing countries, is best known as the causative agent of travellers' diarrhoea. It produces both heat-labile (LT) and heat-stable (ST) toxins and has several colonization-factor antigens (FDA 2012). Illness caused by ETEC usually occurs between 12 and 36 h after ingestion and symptoms can range from mild diarrhoea to a severe cholera-like illness with diarrhoea characterised by watery stools accompanied by high fever or vomiting and severe stomach pain. Serotypes which cause illness in humans include O6, O8, O15, O25, O78, O148, O159, and O167 (Duffy 2006; Adams and Moss 2008; FDA 2012). Foods implicated include turkey, mayonnaise, imported French cheese, and salad vegetables. Besides, food handlers with diarrhoea are also the vehicles for this illness (Lund 2008).

2.1.4 *Listeria monocytogenes*

L. monocytogenes, which was firstly described as *Bacterium monocytogenes* due to its causal of mononuclear leucocytosis in rabbits by Murray in 1926 (Murray et al. 1926), has been known as the causative agent of meningoencephalitis in sheeps and cattles, stillbirth, abortion, and septicaemia in monogastrics and young ruminants (Adams and Moss 2008). The role of *L. monocytogenes* as a causative agent in food-borne disease had not been confirmed until the large-scale outbreak involving Mexican-style soft cheese (Linnan et al. 1988).

The genus *Listeria which was formerly in the family of* Corynebacteriaceae *and has been in the family of Listeriaceae* since 2001, includes eight recognized species which are *L. monocytogenes*, *L. innocua*, *L. welshimeri*, *L. seeligeri*, *L. ivanovii*, *L. grayi*, *L. Marthii*, and *L. rocourtiae*, the latter two were described in 2009. However, *L. monocytogenes* which is closely related to *L. innocua* and *L. marthii*, and *L. ivanovii* which is closely related to *L. seeligeri* have been identified as the pathogens of warm-blooded hosts (den Bakker et al. 2010). *L. monocytogenes* has 13 serotypes, including 1/2a, 1/2b, 1/2c, 3a, 3b, 3c, 4a, 4ab, 4b, 4c, 4d, 4e, and 7 in which serotypes 1/2a, 1/2b, and 4b have been associated with the majority of food-borne infections (FDA 2012).

L. monocytogenes is a Gram-positive, facultatively anaerobic, catalase-positive, oxidase-negative, non-spore forming, motile rod that can grow at temperatures ranging from 0 to 45 °C with an optimum between 30 and 35 °C (mesophilic) and a pH range from 6.0 to 9.0, or in nutrient broth with up to 10 % (w/v) NaCl. It is an acid-tolerant pathogen owing to its glutamate decarboxylase (GAD) enzyme system. It cannot ferment xylose but ferments rhamnose and glucose but the latter one without producing any gas. It is sensitive to pasteurization temperature (Adams and Moss 2008; Ray and Bhunia 2008; FDA 2012; Wang and Orsi 2013).

Listeriosis is a generic term for a variety of syndromes caused by infection with *Listeria* spp. *L. monocytogenes* causes two types of diseases: invasive listeriosis, which occurs predominantly in the elderly, the pregnant women, the neonates, the cancer patient, and the immunocompromised people, has an infectious dose in the range of 10^2–10^4 cells and an incubation period ranging from 24 h to 91 days. However, more severe form of this type of listeriosis has a very long incubation period (e.g. 3 days to 3 months) ending with septicaemia, meningitis, encephalitis, endocarditis, liver abscess, foetal loss, and death with a fatality rate of 20–30 %. On the other hand, febrile gastroenteritis, which is mostly associated with healthy individuals, has an infectious dose in the range of 10^8–10^{10} cells and a relatively short incubation period (e.g. a few hours to 2 or 3 days) and its mild flu-like symptoms include slight fever, nausea, vomiting, and diarrhoea (Farber and Peterkin 2000; WHO/FDA 2004; Ray and Bhunia 2008; FDA 2012; Wang and Orsi 2013).

L. monocytogenes is widespread in the environment and so can be found in soil, water, silage, sewage, slaughterhouse waste, milk of both normal and mastitic cows, human and animal faeces, and foods. Almost all (99 %) of human listeriosis are food-borne (Scallan et al. 2011). Although food-borne listeriosis in humans is an opportunistic rare disease and is mostly sporadic, outbreaks were reported from the consumption of raw milk, inadequately pasteurized milk and dairy products (e.g. ice cream), soft cheeses, meat pate, turkey franks, cold cut meats, improperly cooked chicken, contaminated coleslaw, and smoked mussels. Besides, *L. monocytogenes* outbreaks are mainly associated with ready-to-eat (RTE) foods. It is quite resistant to curing ingredients and so can be found in delicatessen meats such as salami, ham, and corned beef. The infective dose of *L. monocytogenes* is estimated higher than 10^3 CFU/g food, but is believed to vary with the strain and the susceptibility of the host, and the food matrix (WHO/FDA 2004; Adams and Moss 2008; Lund 2008; Ray and Bhunia 2008; FDA 2012; Wang and Orsi 2013).

The ability of *L. monocytogenes* to colonize on food processing niches which acts for an important source of contamination in the finished food products and the ability to grow in many foods at refrigerated temperature due to its resistance to the freezing effect which helps the organism to reach to a level of infective dose during storage of refrigerated foods, make this organism an important problem for the food industry. On the other hand, it is inactivated in meats at 70 °C for 2 min, giving an estimated 7 log inactivation (Farber and Peterkin 2000). Since the majority of listeriosis result from the consumption of contaminated food, foods especially the RTE ones, are in the high risk food group. Because they can become contaminated after cooking and before packaging, or during fermentation and ripening of certain cheeses, in other words before the RTE food is eaten (Tompkin 2002). Therefore the most effective control way of this pathogen is to eliminate the post-processing contamination (PPC) by using lethal or post-lethality treatments and/or growth inhibitors, to prevent in-plant contamination by performing good hygiene and manufacturing practices, testing and sanitation of food contact surfaces, and pre- and post-packaging interventions, and to control measures in the processing facilities (e.g. storage temperatures and periods) (Wang and Orsi 2013).

2.1.5 *Salmonella* spp.

The genus *Salmonella*, which was created in 1900 by Ligniéres and named in honour of the American veterinary pathologist Dr. Salmon, belongs to the family of Enterobacteriaceae. It contains a single species, *S. enterica* (earlier named as *S. choleraesuis*), which consists of seven subspecies. However, subspecies I (*S. enterica* subsp. *enterica*), which accounts for more than 59 % of the 2,400 serovars known, is responsible for nearly all infections in humans and warm-blooded animals, while six other subspecies are isolated principally from cold-blooded animals. The taxonomic nomenclature of this genus is relatively different from that of other genera. Until 2005, most of the different serovars were named as if they were different species. However now, for the taxonomic integrity, the non-italicized serovar name is used after the species name so that *S. enteritidis* becomes *S. enterica* subsp. *enterica* ser. Enteritidis or, shortly, *Salmonella* Enteritidis. For the other subspecies, the serovar formula is used after the name of the subspecies, e.g. *S. fremantle* would be *S. enterica* subsp. *salamae* ser.42;g, t:- (Adams and Moss 2008; Desai et al. 2013).

Salmonellas are Gram-negative, facultatively anaerobic, catalase-positive, oxidase-negative, urease-negative, H_2S-positive, non-spore forming rods, and are generally motile with peritrichous flagella that can grow at temperatures ranging from 5 to 47 °C with an optimum at 37 °C (mesophilic). They are heat sensitive and so are killed by pasteurization process. The minimum pH for their growth changes with the acidulant from 5.4 with acetic acid to 4.05 with hydrochloric and citric acids. However, the optimum pH for their growth is 7.0 (Adams and Moss 2008).

Salmonellosis, which is described as a zoonotic infection, is the second ranking food-borne disease in humans after campylobacteriosis in most European countries. The most common diseases caused by *Salmonella* in human are enteritis, septicaemia, and abortion. The symptoms of *Salmonella* infection usually appear 12–72 h after infection and include fever, abdominal pain, cramps, chills, diarrhoea, nausea, and sometimes vomiting. The illness usually lasts 4–7 days, and most people recover without treatment. Although most infections cause mild to moderate self-limited gastroenteritis, serious infections, especially the ones in the very young and the elderly, and in cases when the bacteria enter the bloodstream, leading to death do occur. The mortality rate for the most of the serovars is usually low (<1 %) but it is 20–30 % for the host-adapted serovars such as *S. Dublin* and *S. Choleraesuis* (Clarke and Gyles 1993; Mølbak et al. 2006; Adams and Moss 2008; Hald 2013).

Salmonella species enjoy widespread occurrence in the environments, including a wide range of domestic and wild animals and a variety of foodstuffs of both animal and plant origin and cause salmonellosis in humans and animals. Transmission is by the faecal–oral route whereby the intestinal contents from an infected animal are ingested with the foodstuff or water. Human infections also result from eating raw or under-cooked foods, including meats, poultry and poultry products, eggs, and dairy products (Adams and Moss 2008; Hald 2013).

In spite of the efforts to prevent and control food-borne salmonellosis during the last decades, it continues to be one of the leading causes of human gastroenteritis. Preharvest food safety is the most important component of an effective prevention and control strategy for *S.* Enteritidis infection. HACCP plans, disinfection protocols, and preventing the recycling of offal and inedible raw materials into animal feeds may reduce direct horizontal transmission of *Salmonella* spp. within and between herds and flocks. In the EU, the Zoonosis Directive (92/117/EEC) starts an EU-wide control effort against *Salmonella*, especially in broiler and layer breeders (Hald 2013).

2.1.6 *Vibrio parahaemolyticus* and *Vibrio vulnificus*

Among the vibrios, *V. parahaemolyticus, V. vulnificus*, and *Vibrio cholerae* which belong to the family of Vibrionaceae, are responsible for most cases of the water-borne and food-borne illnesses (Oliver and Kaper 2001; Morris 2003, Nair et al. 2006; Ray and Bhunia 2008; Wright and Harwood 2013). The first two will be discussed in this section and the last one will be included in Sect. 2.2.

V. parahaemolyticus which was firstly described as the cause of an outbreak of food-borne illness in Japan in 1950 (Lund 2008), is a Gram-negative, facultatively anaerobic, catalase-positive, oxidase-positive, non-spore forming, motile curved rod that can grow at temperatures ranging from 5 to 42 °C with an optimum between 30 and 37 °C (mesophilic), or in media with 3–5 % (w/v) NaCl (halophilic). However, it cannot tolerate 10 % (w/v) NaCl. It is slowly inactivated at temperatures below 10 °C and so the cultures should never be stored in refrigerators. It hardly grows at pH 5.0 or below. It ferments glucose without producing gas, but cannot ferment lactose and sucrose. It is highly sensitive to pasteurization temperature, low pH, drying, and freezing. All *V. parahaemolytic* strains are not pathogenic; the pathogenic ones are haemolytic which can be detected on human blood agar plates, known as the Kanagawa positive (KP) since they produce thermal stable direct hemolysin (TDH) or TDH-related hemolysin (TRH). However, although the most environmental isolates do not produce TDH or TRH, they have also been associated with food-borne outbreaks. On the other hand; *V. vulnificus* is a lactose-positive and salicin-positive and a highly invasive bacterium that produces a haemolytic cytotoxin (Nair et al. 2006; Ray and Bhunia 2008; FDA 2012; Wright and Harwood 2013).

The infective dose of *V. parahaemolyticus* and *V. vulnificus* given in the literature varies from 10^5 to 10^7 organisms due to the type of contaminated food consumed and the general health of the exposed person. The most common symptoms of *V. parahaemolyticus* and *V. vulnificus* gastroenteritis are abdominal pain, nausea, vomiting, and rapid onset non-bloody diarrhoea, fever, chills which are much milder symptoms than of *V. cholera* and so recovery without treatment normally occurs after several days without any long-term consequences. However; *V. vulnificus* can infect the bloodstream, causing a severe and life-threatening illness characterized by decreased blood pressure (septic shock) in immunocompromised persons. In addition to these symptoms, septicaemia and severe necrotizing infections of soft tissues that typically result from exposure of open wounds to water harbouring *V. vulnificus*, occur with a very high fatality rate (40–60 %) (Ray and Bhunia 2008; Jones and Oliver 2009; Wright and Harwood 2013).

V. parahaemolyticus and *V. vulnificus* are ubiquitous in the natural flora of coastal marine, in estuarine and freshwater environments, especially being present in the highest numbers during the summer months when water gets warmer and have been isolated from water, sediments, and various seafood such as shrimps, squid, octopus, fish, crabs, oysters, lobsters, and clams (Ray and Bhunia 2008; Jones and Oliver 2009, FDA 2012). *V. parahaemolyticus* and *V. vulnificus* gastroenteritis is commonly attributed to the consumption of raw, under-cooked, or post-heat-contaminated seafood (primarily raw oysters) (Jones and Oliver 2009; Lipp et al. 2002; Morris 2011).

The most important control way is to harvest seafood only from approved waters free from faecal contamination and to chill promptly. In addition to this; avoiding exposure of vegetables to contaminated irrigation water, cooking seafood thoroughly before serving, applying proper refrigeration to raw and heated products, avoiding cross-contamination of cooked seafood and other foods with raw seafood, applying employee health, and hand washing policy are necessary for prevention and control of this bacteria (Lund 2008; FDA 2012).

2.1.7 *Yersinia* spp.

The genus *Yersinia*, which was firstly described as the causative agent of the bubonic plague by the French bacteriologist Alexandre Yersin in 1894, killed an estimated 25 % of the European population in the fourteenth century (Adams and Moss 2008). Among the 11 species within the genus, four are pathogenic but only *Y. enterocolitica* and *Y. pseudotuberculosis* are food-borne diarrhoeagens. Although *Y. pestis* is closely related with *Y. pseudotuberculosis* with a gene homology of almost 97 %, it infects humans by routes aside from food and causes plague (Scannell 2011; FDA 2012; Nesbakken 2013). *Y. enterocolitica* strains have been differentiated into biotypes and serotypes. Most of the pathogenic strains are included in Biotype 1B (O:4, 32; O:8; O:13a; O:13b; O:18; O:21), Biotype 2 (O:5,27; O:9), Biotype 3 (O:1, 2, 3; O:3; O:5,27), Biotype 4 (O:3), and Biotype 5 (O:2,3); however, among them Biotype 4/serotype O:3 is the most frequently isolated pathogen in the world, especially in Europe, Canada, Japan, and South Africa. In the USA, Biotype 1/serotype O:8 most commonly causes human yersiniosis. On the other hand; *Y. pseudotuberculosis* is divided into five serotypes (I to V) among which Serotype I is the most common serotype associated with human and animal infections in Europe (Robins-Browne 2001; Lund 2008; Nesbakken 2013).

Y. enterocolitica, which is a member of the family Enterobacteriaceae, is a Gram-negative, facultatively anaerobic, catalase-positive, oxidase-negative, urease-positive, H_2S-negative, non-spore forming rod (occasionally coccoid) that can grow at temperatures ranging from −1 to +40 °C, with an optimum around 29 °C (psychrotrophic) and over a pH range of 4–10, with an optimum 7.0–8.0, or at NaCl concentrations greater than 7 % (w/v). It is non-motile at 35–37 °C, but motile with peritrichous flagella at 22–25 °C. It ferments mannitol and glucose but the latter one without producing any gas. It multiplies more rapidly at 0–5 °C than any other food-borne pathogenic bacterium and can grow in a range of foods at refrigeration temperature. It can tolerate freezing and so can survive in frozen foods for extended time. It is sensitive to pasteurization (Bercovier and Mollaret 1984; Adams and Moss 2008; FDA 2012; Nesbakken 2013).

Yersiniosis, which is a generic term for gastroenteritis and other syndromes caused by infection with *Y. enterocolitica*, occurs most commonly in children less than 14 years of age, especially infants under 1 year being at greatest risk. It has an incubation period of 4–7 days and may last for 1–3 weeks or for several months. The symptoms of gastroenteritis which dominates in children and young people are low-grade fever, abdominal pain, watery or mucoid diarrhoea, bloody stools (rarely), and acute lower right quadrant abdominal pain that may be mistaken diagnosis of appendicitis. However; various forms of reactive arthritis and erythema nodosum (red skin lesion) are most common in adults, especially in adult females (Robins-Browne 2001; Sutherland and Varnam 2002; Adams and Moss 2008; FDA 2012).

Pigs are the main reservoir and chronic carriers for the human pathogenic types of *Y. enterocolitica*, however, pathogenic serotypes have also been found in cattle, sheep, deer, rodents, cats, and dogs (Adams and Moss 2008; Nesbakken 2013). *Y. enterocolitica* strains can be found in pork, beef, and lamb meats, oysters, fish, crabs, raw milk, and contaminated water. However, poor sanitation and improper sterilization techniques by food handlers, including improper storage offer many opportunities for *Yersinia* to enter the food supply (FDA 2012).

Most of the preventive and control measures against salmonellosis are also valid against this zoonotic infection, *yersiniosis*. Cooking meat products, particularly pork, poultry, and seafood thoroughly before serving; avoiding drinking unpasteurized milk, avoiding exposure of vegetables to contaminated irrigation water; following good hygienic practices; proper sanitation of food contact surfaces (avoiding cross-contamination); applying employee health and hand washing policy are necessary for prevention and control of this bacteria (Lund 2008).

2.2 *Bacterial Intoxications and Toxico-Infections*

2.2.1 *Bacillus cereus*

The taxonomy of the genus *Bacillus*, which is in the family of Bacillaceae, has undergone considerable revision in recent years; however it still contains about 80 species, including *B. cereus* which is best known as one of the causative agent of food-borne microbial diseases (Jay et al. 2005; Adams and Moss 2008).

B. cereus was first recognized as a food-borne pathogen in 1949 (Hauge 1950) and then in 1955 (Hauge 1955). It is a Gram-positive, facultatively anaerobic, catalase-positive, and endospore-forming rod that grows over a temperature range from 8 to 55 °C, optimally around 28–35 °C (mesophilic) and a pH range from 4.5 to 9.5, optimally around 6.0–7.0 (Ehling-Schulz et al. 2004; Adams and Moss 2008; Lund 2008; FDA 2012).

B. cereus is responsible for two different types of food-borne illnesses which are relatively late-onset toxico-infection, "diarrhoeal syndrome" and rapid-onset intoxication, "emetic syndrome (vomiting illness)". The symptoms of diarrhoeal syndrome which resemble those of *Clostridium perfringens* food poisoning, are profuse watery diarrhoea, abdominal cramps, and pain that last for 12–24 h. The emetic syndrome which resembles the illness caused by *Staphylococcus aureus*, is characterized by nausea, malaise, and vomiting that subside within 6–24 h (Kramer and Gilbert 1989; Granum and Baird-Parker 2000; Kotiranta et al. 2000; Adams and Moss 2008; Lund 2008).

B. cereus, like many other bacilli, is common in soil and as a result of its spore-forming ability, it is distributed in the environment thus can easily contaminate foodstuffs, especially the ones with plant origin. Since spores of *B. cereus* are resistant to drying and heat treatment, a wide variety of foods, including spaghetti, pasta, rice, dairy and dried milk products, spices, and other dried foodstuffs as well as meat, chicken, vegetables, fruits, grain, and seafood allow the survival of *B. cereus* cells and their spores (Kamat et al. 1989; FDA 2012). Also the emetic toxins stay active after a heat treatment of 100 °C for 150 min (Ehling-Schulz et al. 2004). The diarrhoeal syndrome is mainly associated with the consumption of milk products, vegetables, fish, and meats, while the emetic syndrome is transmitted by rice and pasta (FDA 2012). Most *B. cereus* food poisoning cases are related to cereal-based or protein-based foods, slowly cooled and stored between 10 and 50 °C which allows the spores to germinate and cause the illness (Adams and Moss 2008).

Cleaning and disinfection of food manufacturing and production equipments and devices are important for preventing a build-up of *B. cereus* cells and their spores. On the other hand, most heating and cooking treatments as well as, steaming under pressure, frying, grilling, and roasting, generally kill the vegetative cells, and probably the spores (ICMSF 1996). However, only canning can guarantee complete destruction of *B. cereus* spores. In addition to these, cooked foods should be consumed promptly, or kept above 63 °C for a short time, or cooled rapidly and kept below the temperatures of about 7–8 °C (ideally below 4 °C). Dried milk products and similar powdered foods should be used promptly after reconstitution and holding them at room temperature for several hours should be avoided (EFSA 2005).

2.2.2 *Clostridium botulinum*

Thirteen people became ill and six of them later died after consuming a type of sausage called Blunzen that was made by packing blood and other ingredients into a pig's stomach in Wildbad in 1793. However, the causative agent of this illness which was later named botulism (derived from the Latin word; *botulus*: sausage), was primarily isolated in 1897 by Dr. Van Ermengem, Professor of Bacteriology at the University of Ghent. He found that botulism resulted from the consumption of food containing a heat-labile toxin produced by an obligate anaerobic, spore-forming bacillus firstly named *Bacillus botulinum* (Adams and Moss 2008).

C. botulinum which is a member of the family Bacillaceae, is a Gram-positive, obligate anaerobic, catalase-negative, spore-forming straight or slightly curved rod that produces central or subterminal oval spores and is motile with peritrichous flagella. The vegetative cells are sensitive to low pH (<4.6), low A_w (0.93), and quite high salt concentrations (5.5 %) and can be killed by pasteurization, whereas their spores are highly heat resistant and thus can only be killed at the temperatures above 115 °C (Adams and Moss 2008; Ray and Bhunia 2008).

C. botulinum strains have been divided into seven types which are A, B, C, D, E, F, and G according to the type of toxin that they produce. However, only the types A, B, E, and F (rarely) strains are associated with human food-borne intoxications since types C and D strains cause botulism in animals and no outbreaks of type G strain have been reported till now. Types C and E (psychrotrophic) strains also cause botulism in birds. Type A strains are proteolytic and can grow between 10 and 48 °C, with the optimum at 35 °C (mesophilic). Type E strains are non-proteolytic and grow optimally at 30 °C (mesophilic), with a range between 3.3 and 45 °C. However, types B and F strains can be either proteolytic or non-proteolytic. Actually type C is recognized as C_1 and C_2 toxin (though C_2 is not a neurotoxin) (Adams and Moss 2008; Ray and Bhunia 2008; Hill et al. 2009; FDA 2012).

There are 5 types of botulism which are food-borne botulism, infant or infectious botulism, hidden botulism, wound botulism, and inadvertent botulism. Food-borne botulism which results from the ingestion of an exotoxin produced by *C. botulinum* growing in the food, usually occurs within 12–36 h of ingestion of toxin, but may occur within 6 h or after 10 days. The first effect of the toxin is often on neuromuscular junctions in the head and neck, causing symptoms such as double vision, failure in focusing, drooping eyelids (ptosis), dry mouth, failure in speaking clearly (dysphonia), and difficulty in swallowing (dysphagia) accompanying with some gastrointestinal disorders (e.g. nausea, vomiting, diarrhoea, constipation, and urine retention). Subsequently, paralysis occurs in the muscles of arms, legs, trunk, and heart which results in death. The mortality rate is high (20–50 %), but will depend on the type of toxin (type A usually produces a higher mortality than B or E), the amount of ingested toxin, the type of food and the speed of treatment (Adams and Moss 2008; Lund 2008; Ray and Bhunia 2008). Infant botulism has been caused by the ingestion of spores of *C. botulinum* that colonize and form neurotoxin in the intestinal tracts of infants, especially babies under the age of 1 year (i.e. intestinal toxaemia botulism) (Lund 2008; Glass and Marshall 2013). Constipation, lethargy, inability to suck, weak cry, and difficulty in swallowing are the initial symptoms continuing with paralysis. Honey and corn syrup have been associated with infant botulism cases (Lund 2008; Ray and Bhunia 2008). Hidden botulism is seen in adults suffering from chronic gastrointestinal disorders (i.e. adult variant of infant botulism) (Ray and Bhunia 2008). Wound botulism which is caused by a subcutaneous infection with *C. botulinum*, has been more commonly associated with accidental cut and intravenous drug usage (Adams and Moss 2008; Ray and Bhunia 2008; Glass and Marshall 2013) whereas inadvertent botulism has been associated with therapeutic or cosmetic usage of botulism neurotoxins (Ray and Bhunia 2008).

Almost any type of food that is not very acidic ($pH > 4.6$) can support growth and toxin production of *C. botulinum* (FDA 2012). In other words, both the vegetative cells and the spores can be isolated from a wide range of foods, including fish, meat, vegetables (e.g. green beans, corn, spinach, asparagus, pepper, and mushrooms), fruits (e.g. figs and peaches), honey, mushrooms, cheese, and nuts. The main reason of outbreaks is improper home canning of the contaminated foodstuffs (Adams and Moss 2008; Ray and Bhunia 2008; Glass and Marshall 2013).

C. botulinum spores usually enter the food chain via direct contamination of the animal or plant during production or via cross-contamination during harvest, handling, or manufacturing and processing. Foods that allow the growth of *C. botulinum* should either undergo a processing that will inactivate the spores, or that the composition and storage conditions are controlled so as to prevent growth. The most important control way is to use proper temperature and time in home canning of low-acid products Cooked foods in which spores may survive should be stored at low temperatures (≤3 °C) Prepared ready meals, vacuum-packed vegetables, pasteurized, chilled foods and herbs/spices/vegetables stored in oil, should be controlled. Honey should not be given to babies under the age of 1 year, or to adults with recent abdominal surgery or gastrointestinal abnormalities (Lund 2008; Glass and Marshall 2013).

2.2.3 *Clostridium perfringens*

C. perfringens (earlier named as *C. welchii*), which was firstly described by the American bacteriologist Welch in 1892, has been known as the causative agent of gas gangrene (Adams and Moss 2008). The role of *C. perfringens* as a causative agent in food-borne disease had not been confirmed until the large-scale outbreak of a food poisoning among schoolchildren in Leicester, England, due to the consumption of gravy contaminated by *C. perfringens* (Knox and MacDonald 1943).

C. perfringens which is a member of the family Bacillaceae, is a Gram-positive, anaerobic (but can tolerate some oxygen), catalase-negative, non-motile, blunt-ended, and oval subterminal spore-forming rod that can be characterized by the reduction of nitrate, the liquefaction of gelatin, the production of lecithinase (α-toxin), and the fermentation of lactose. It is intolerant to low temperatures and thus cannot grow below the temperature of 12 °C. It grows optimally around 43–45 °C (thermoduric) and continues growing up to 50 °C. It grows at a pH range from 5.0 to 9.0, optimally around 6.0–7.0, but cannot grow at $pH < 5.0$ and also in the presence of 6 % (w/v) NaCl. The vegetative cells can be killed by pasteurization, whereas their spores are extremely heat resistant, and may survive even boiling for several hours (de Jong et al. 2004; Adams and Moss 2008; Lund 2008; Ray and Bhunia 2008; Labbé and Juneja 2013).

C. perfringens strains have been classified into 5 types which are A, B, C, D and E according to the production of four major lethal enterotoxins, α, β, ε, and ι. However, Type A strains are mainly associated with food borne toxico-infections and gas gangrene and produce only the α-toxin which has lecithinase (phospholipase

C) activity. Type B strains produce α, β, and ε toxins, whereas Type C strains produce α and β toxins which cause more severe, but more rare enteric disease, necrotic enteritis (NE). Type D strains produce α and ε toxins and Type E strains produce α and ι toxins (de Jong et al. 2004; Adams and Moss 2008; Ray and Bhunia 2008).

Because of its almost ubiquitous distribution in the environment, including the intestinal tract of animals, soils, and retail foods, the common form of poisoning is *C. perfringens* Type A poisoning (CPTA) (Heikinheimo et al. 2006). The vegetative cells survive in the acidity of stomach while passing through it and then enter into the small intestine where they multiply, sporulate, and release their enterotoxins. The enterotoxin which is synthesized almost by the sporulating cells causes only gastroenteritis. Diarrhoea and lower abdominal cramps typically occur usually 8–24 h after the consumption of food containing large numbers of the vegetative cells of the organism (or 10^6 spores/g foods). Nausea, vomiting, and fever are less common. Mortality occurs mostly in elderly people as a result of dehydration. Symptoms lessen within 1–2 days, but cramps can continue a little longer (Ray and Bhunia 2008; FDA 2012; Labbé and Juneja 2013).

High-protein foods such as meat (cooked beef which is prepared with its gravy, meat pies, sauces, roasts, casseroles) and poultry are the most related vehicles involved in *C. perfringens* food-borne toxico-infection. In addition to this, temperature-abused meat and poultry products (i.e. inadequate holding temperature and cooking, slow cooling, or prolonged storage at room temperature after cooking) allow the spores to germinate and multiply rapidly (Kalinowski et al. 2003; Adams and Moss 2008; Ray and Bhunia 2008; Labbé and Juneja 2013).

The major factors in prevention of *C. perfringens* toxico-infection are rapid cooling, refrigerated storage of cooked foods, adequate reheating of cooked and cooled foods which inhibit germination of surviving spores and growth of vegetative bacteria. In other words, cooked foods should be eaten immediately or kept above 63 °C for a short time, or be cooled rapidly and uniformly and maintained below 7–8 °C (ideally below 4°) and reheated to at least 72 °C before consumption (Lund 2008).

2.2.4 *Staphylococcus aureus*

The genus *Staphylococcus*, which belongs to the family of Staphylococcaceae, includes over 40 species (Jay et al. 2005). Among them, *S. aureus* is the most important member of this genus since it is an enterotoxin-producing species. However; *S. intermedius* and *S. hyicus* have also been reported with the enterotoxin production (Adams and Moss 2008; Landgraf and Destro 2013).

The Staphylococci (derived from the Greek word: *staphyle* : bunch of grapes and *coccus*: a grain or berry) (Adams and Moss 2008) were firstly described by the Scottish surgeon, Sir Alexander Ogston during his observation of pyogenic abscesses in humans (Ogston 1882). In 1884, while the investigation of an outbreak in Michigan by Vaughan and Sternberg, the staphylococci had been found as the causative agents of a food poisoning regarding the consumption of cheddar cheese (Dack 1956).

In 1914, Barber found out that a toxin produced by staphylococci was responsible for staphylococcal food poisoning (SFP) (i.e. intoxication) (Bhunia 2008).

S. aureus is a Gram-positive, facultatively anaerobic, catalase-positive, oxidase-negative, non-spore forming, non-motile coccus that occurs in pairs, tetrads, short chains, or bunched in grape-like clusters and can grow at temperatures ranging from 7 to 48 °C with the optimum at 37 °C (mesophilic) and a pH range from 4.0 to 9.8–10.0 with the optimum 6.0–7.0. The optimum temperature for enterotoxin production is between 35 and 40 °C. It is one of the most salt-tolerant pathogenic microorganisms in foods, grows easily in media containing 5–7 % (w/v) NaCl and some strains can grow in media with up to 20 % (w/v) NaCl. Optimum growth of *S. aureus* occurs at A_w of >0.99. However, it can grow at a A_w as low as 0.83, depending on the temperature, pH, type of humectants, and other parameters. The minimum A_w for enterotoxin production is 0.86. It ferments glucose which is used for distinguishing it from the strictly aerobic genus *Micrococcus* (Bannerman and Peacock 2007; Adams and Moss 2008; Schelin et al. 2011; Landgraf and Destro 2013).

The genus *Staphylococcus* includes both coagulase-negative and coagulase-positive strains which can produce highly heat-stable enterotoxins [known as staphylococcal enterotoxins (SEs)] that cause gastroenteritis in humans. However, the bacterium itself can be destroyed by pasteurization. There are 21 different SEs or enterotoxin-like proteins (SE-*l*) described (from SEA to SEV, except SEF which was the original name of toxic shock syndrome toxinTSST-1), excluding molecular variants. The first five (SEA; SEB; SEC1,2,3; SED; and SEE) have emetic activity and are called classical SEs which cause 95 % of human SFP. Among them, SEA is the most often reported toxin involved in SFP followed by SEB, SEC, or SED, depending on the district of the world. There is a high correlation between the coagulase activity and SEs. In other words; the detection of coagulase activity is important in distinguishing *S. aureus*-related food-borne illness from other strains. However other species that are coagulase-negative can also produce SEs. SEs are also resistant to proteolytic enzymes, such as trypsin and pepsin, which allows them to pass through the digestive tract without any denaturation. In addition to SEs, *S. aureus* produces a variety of extracellular products such as adhesion proteins, coagulase, superantigens, thermostable nuclease (TNase), ADP-ribosylating toxins, and hemolysins, many of which play a role as virulence factor (Bhunia 2008; Ono et al. 2008; RůŽičkova et al. 2008; FDA 2012; Hennekinne et al. 2012; Landgraf and Destro 2013).

CDC estimates that SFP causes approximately 241,188 illnesses, 1,064 hospitalizations, and 6 deaths each year in the USA (FDA 2012). Nausea, stomach cramps, retching, vomiting, and diarrhoea, which are the most common symptoms of SFP, usually occur within 2–4 h of ingestion because of fast-acting ability of SEs, depending on the amount of toxin ingested and host-specific factors. Breathing of SEs may cause rapid onset of fever, chills, cough, and difficulty in breathing. However; headache, muscle cramping, dehydration, temporary changes in blood pressure, and prostration can also be experienced in more severe cases. Ingested bacteria do not produce toxin, and the symptoms therefore normally subside after 24 h. Besides, infections caused by methicillin-resistant (MRSA) and vancomycin-resistant strains may be fatal because of lack of suitable antibiotics. The intoxication dose of SE is less than 1.0 μg which is barely reached when there are more than 10^5cells of *S. aureus* in 1 g of foodstuff (Scannell 2011; Schelin et al. 2011; FDA 2012).

S. aureus is natural inhabitant of human and animal skin so it occurs commonly on the skin, nostrils, and mucous membranes of warm-blooded animals. Staphylococcal contamination may be introduced into foods by direct contact such as through the hands of the workers, or indirectly such as through skin fragments and also by coughing and sneezing which is seen when there is a respiratory infection. Creamy food prepared with milk, custard (pudding), mashed potato made with raw milk, deli foods, salad dressings, meats, hams, fish, shellfish, raw milk, and cheeses made from unpasteurized milk are the most related foods involved in SFP. It is better to remind that the foodstuffs may not smell terrible or look spoiled in order to produce SEs (Bhunia 2008; Landgraf and Destro 2013).

The most important control way is to avoid time and temperature abuse of food products, inadequate refrigeration, inadequate heating or cooking, and preparation or serving of food by a worker who have wounds or skin infections on his hands or wrists (poor personal hygiene). Hygienic practices are crucial in preventing SFP (Bhunia 2008; Landgraf and Destro 2013).

2.2.5 *Vibrio cholerae*

V. cholerae which is a member of the family Vibrionaceae, was firstly isolated in pure culture by Robert Koch in 1883. It is classified on the basis of its somatic (O) antigens into serovars or serogroups, and there are at least 206 serogroups. *V. cholerae* serogroup O1 and O139 include all the strains responsible for epidemic and endemic cholera. Serogroup O1 has two major serotypes, Ogawa and Inaba, and rarely reported serotype Hikojima. These serotypes can be further divided into two biotypes named as "Classical" and "El Tor", based on antibiotic resistance and hemolysin expression. All strains that were identified as *V. cholerae* but cannot agglutinate with O1 antiserum belonged to the non-O1*V. cholerae* until 1993 in which a new bacterium was determined as the causative agent of the epidemic cholera-like disease in Bangladesh (8th great pandemic). This bacterium did not agglutinate with O1 antiserum and therefore did not belong to any of the O serogroups previously described for *V. cholerae* but to a new serogroup O139 (Ramamurthy et al. 1993; Shimada et al. 1993; Faruque et al. 1998; Maheshwari et al. 2011; Wright and Harwood 2013).

V. cholerae is a Gram-negative, facultatively anaerobic, catalase-positive, oxidase-positive, non-spore forming, motile curved rod that is capable of respiratory and fermentative metabolism (*V. cholerae* serogroup O1 ferment sucrose and mannose but not arabinose and they produce acid but not gas) and can grow at temperatures ranging from 5 to 42 °C with an optimum between 30 and 37 °C (mesophilic), or in media with 3–5 % (w/v) NaCl (halophilic). It tolerates alkaline media that kill most of the intestinal commensals, but it is sensitive to acid and so dies rapidly in solutions below pH 6. It is killed by pasteurization but can persist in raw milk as long as 4 weeks, even if refrigerated (Maheshwari et al. 2011).

Worldwide, between the year 1817 and the present time eight great pandemics caused by toxigenic *V. cholerae* have been recorded (Lund 2008). Water from public supplies was implicated in the first six pandemics. *V. cholerae* causes a water-borne disease called cholera that is characterized by a harmful watery diarrhoea which leads to rapid dehydration, and death occurs in 50–70 % of untreated people (Faruque et al. 1998). Infection due to *V. cholerae* begins with the ingestion of contaminated water or food contaminated directly or indirectly with faeces of infected individual. Infected people excrete between 10^7 and 10^8 cells of *V. cholerae* per gram of stool and total output of *V. cholerae* by a patient can be in the range of 10^{11}–10^{13} CFU (Nair et al. 2006). After its passage through the acid barrier of the stomach, *V. cholera* colonizes in the small intestine and produces an enterotoxin called cholera toxin (CT) which has been shown to have biological similarities with the *Escherichia coli* enterotoxin (LT). In addition to CT, several other toxins such as neuraminidase, disulphide isomerase, protease, haemolysin-cytolysin toxin, ZO toxin, accessory cholera enterotoxin, Shiga-like toxin, and thermostable direct hemolysin (TDH) can be produced by the *V. cholera* (Faruque et al. 1998; Maheshwari et al. 2011).

Likewise *V. parahaemolyticus* and *V. vulnificus*, *V. cholerae* is found in coastal waters, especially is widely distributed in temperate and tropical aquatic environments in association with a wide range of aquatic life, including cyanobacteria, diatoms, oysters, water hyacinths, and blue crab. Faecal–oral spread is the primary mode of cholera transmission. It can be isolated from areas where poor environmental sanitation is coupled with poor personal hygiene. Using polluted water for irrigation and inadequately treated sewage sludge make it spread to foods such as leafy green vegetables. However, cholera is commonly attributed to the consumption of faecally contaminated water followed by human-to-human transmission and can be rapidly fatal due to massive dehydration (Lipp et al. 2002; Ray and Bhunia 2008; Maheshwari et al. 2011; Morris 2011; FDA 2012; Wright and Harwood 2013).

The application of main sanitary principles and ensuring safe drinking-water would go a long way toward controlling the disease cholerae. A number of approaches to the control of *V. cholera* have been proposed which are generally similar to those proposed for the control of *V. parahaemolyticus* and *V. vulnificus*, such as avoiding exposure of vegetables to contaminated irrigation water, cooking seafood thoroughly before serving, applying proper refrigeration to raw and heated products, avoiding cross-contamination of cooked seafood and other foods with raw seafood, applying employee health and hand washing policy (Lund 2008; FDA 2012).

3 Toxigenic Fungi

Fungi are organisms made up of eukaryotic cells that have cell walls containing chitin and are heterotrophs that obtain their nutrients by extracellular digestion based on the activity of secreted enzymes, followed by absorption of the solubilized breakdown products, which all distinguish them from animals. They are not capable of forming true tissues like complex plants and animals and they can reproduce both sexually and asexually (Moss 2000; Brock 2006; Webster and Weber 2007).

Fungi are ubiquitous in nature and there are probably over 1.5 million species of fungi in which about 80,000 to 120,000 species of fungi have been identified up to date (Hawksworth 2001; Brock 2006). They have colonized in a wide range of ecosystems and they have the ability to produce many extracellular chemicals that are known as secondary metabolites, in which some may be very useful for pharmaceutical usages, and the others are toxic and so called mycotoxins (Magan and Aldred 2007).

Mycotoxins which are produced by filamentous fungi (molds) and contaminate agricultural commodities pre- or postharvest and so foods and foodstuffs, pose important food safety risks and public health hazards, and result in economic losses by reducing the commercial value of crops and so limiting the marketability of grain supplies (Atanda et al. 2011; Gnonlonfin et al. 2013; Woloshuk and Shim 2013). Although hundreds of mycotoxins have been identified up to now, only few are known to impact global agriculture (Bennett and Klich 2003). Aflatoxins, ochratoxins, trichothecenes, zearelenone, fumonisins, tremorgenic toxins, and ergot alkaloids are the mycotoxins that have the greatest agro-economic importance (Atanda et al. 2011) (Table 1). The toxicity and acute and chronic disorders caused by these mycotoxins in humans and animals have been comprehensively documented many times before (Bennett and Klich 2003; Richard 2007).

Although a very diverse range of fungi can produce mycotoxins, some of them are the most carcinogenic compounds in nature and are toxic to vertebrates. Besides, there are three genera of mycotoxigenic moulds that are especially important in foods: *Aspergillus*, *Penicillium* and *Fusarium* (Moss 2000). Therefore, the following section will deal only with these major fungi and their characteristics.

3.1 **Aspergillus** *and Related Teleomorphs*

The genus *Aspergillus* and related 12 teleomorph genera (*Chaetosartorya*, *Dichotomomyces*, *Emericella*, *Eurotium*, *Fennellia*, *Neocarpenteles*, *Neopetromyces*, *Neosartorya*, *Penicilliopsis*, *Petromyces*, *Sclerocleista*, and *Warcupiella*) belong to the family Trichocomaceae of the class Euascomycetes in the phylum Acsomycota. Among the teleomorps, *Eurotium*, *Neosartorya*, and *Emericella* are significant in foods (Pitt and Hocking 2009; Samson and Varga 2010; Public Health Agency of Canada 2013). The genus *Aspergillus* includes eight subgenus (subgenus *Aspergillus* with the sections *Aspergillus* and *Restricti*; subgenus *Fumigati* with the sections *Fumigati*, *Clavati*, and *Cervini*; subgenus *Circumdati* with the sections *Circumdati*, *Nigri*, *Flavi*, and *Cremei*; subgenus *Candidi* with the section *Candidi*; subgenus *Terrei* with the sections *Terrei* and *Flavipedes*; *subgenus Nidulantes* with the sections *Nidulantes*, *Usti* and *Sparsi*; subgenus *Warcupi* with the sections *Warcupi* and *Zonati* and subgenus *Ornati* with the section *Ornati*), each containing several species (Samson and Varga 2010).

Table 1 Some of the mycotoxins and their producers [Adapted from Moss (2000), Frisvad et al. (2007)]

Mycotoxin name	Producer name
Aflatoxin	*Aspergillus bombycis, A. flavus, A. nomius, A. ochraceoroseus, A. parasiticus, A. parvisclerotigenus, A. pseudotamarii, A. rambellii, A. toxicarius, Emericella astellata, E. olivicola, E. venezuelensis*
Citreoviridin	*A. terreus, Eupenicillium cinnamopurpureum, Penicillium citreonigrum, P. manginii, P. miczynskii, P. smithii*
Citrinin	*A. terreus chemotype II, A. carneus, A. niveus, Blennoria sp., Clavariopsis aquatica, Monascus ruber, P. chrzaczszii, P. citrinum, P. expansum, P. manginii, P. odoratum, P. radicicola, P. verrucosum, P. westlingii*
Cyclopiazonic acid	*A. flavus, A. lentulus, A. oryzae, A. parvisclerotigenus, A. pseudotamarii, A. tamarii, P. camemberti, P. commune, P. dipodomyicola, P. griseofulvum, P. palitans*
Ergot alkaloids	*Claviceps fusiformis, C. paspali, C. purpurea*
Fumonisins	*Fusarium anthophilum, F. dlamini, F. napiforme, F. nygamai, F. proliferatum, F. thapsinum, F. verticillioides*
Moniliformin	*F. avenaceum, F. napiforme, F. nygamai, F. oxysporum, F. proliferatum, F. subglutinans, F. tricinctum, F. thapsinum, F. verticillioides*
Ochratoxin A	*A. carbonarius, A. cretensis, A. flocculosus, A. lacticoffeatus, A. niger, A. ochraceus, A. pseudoelegans, A. roseoglobulosum, A. sclerotioniger, A. sclerotiorum, A. steynii, A. sulphureus, A. westerdijkiae, Neopetromyces muricatus, P. nordicum, P. verrucosum, Petromyces albertensis, Petromyces alliaceus*
Patulin	*A. clavatonanica, A. clavatus, A. giganteus, A. longivesica, Byssochlamys nivea, P. carneum, P. clavigerum, P. concentricum, P. coprobium, P. dipodomyicola, P. expansum, P. formosanum, P. gladioli, P. glandicola, P. griseofulvum, P. marinum, P. paneum, P. sclerotigenum, P. vulpinum*
Penicillic acid	*A. auricomus, A. bridgeri, A. cretensis, A. flocculosus, A. insulicola, A. melleus, A. neobridgeri, A. ochraceus, A. ostianus, A. persii, A. petrakii, A. pseudoelegans, A. roseoglobulosus, A. sclerotiorum, A. sulphureus, A. westerdijkiae, Neopetromyces muricatus, P. aurantiogriseum, P. brasilianum, P. carneum, P. cyclopium, P. fennelliae, P. freii, P. matriti, P. polonicum, P. radicicola, P. tulipae, P. viridicatum*
Penitrem A	*P. clavigerum, P. crustosum, P. glandicola, P. janczewskii, P. melanoconidium, P. tulipae*
Sterigmatocystin	*A. multicolor, A. ochraceoroseus, A.rambellii, A. versicolor, Chaetomium thielavioideum, Chaetomium spp., Emericella nidulans, Emericella spp., Humicola fuscoatra, Monocillium nordinii*
Tenuazonic acid	*Alternaria citri, Alternaria japonica, Alternaria kikuchiana, Alternaria longipes, Alternaria mali, Alternaria oryzae, Alternaria solani, Alternaria tenuissima, Phoma sorghina*
Trichothecenes	*F. crookwellense, F. culmorum, F.equiseti, F. graminearum, F. langsethiae, F. poae, F. pseudograminearum, F. sambucinum, F. sporotrichioides, F. venenatum*
Zearalenone	*F. crookwellense, F. culmorum, F. equiseti, F. graminearum*

The genus *Aspergillus* is characterized by unbranched, aseptate conidiophores with usually swollen spherical vesicles which are covered with phialides, or metulae and phialides that are borne simultaneously. This character definitely distinguishes the genus *Aspergillus* from the genus *Penicillium.* Compared to *Penicillium, Aspergillus* spp. have ability to grow at higher temperatures and/or lower water activities and usually grow more rapidly although they take longer to sporulate. In addition to these, their spores are more resistant to light and chemicals (Pitt and Hocking 2009). However, *Aspergillus* spp. are susceptible to sodium hypochlorite solutions. Although conidia of *A. fumigates (*the most pathogenic species) are generally heat-resistant, the conidia of *A. niger* and *A. flavus* are easily inactivated at 60 °C for 45 min (Public Health Agency of Canada 2013).

Aspergillus spp. contain approximately 184 species, 40 of which have been implicated in human or animal infections, a group of diseases termed as aspergillosis. Aspergillosis which is mostly caused by *A. fumigatus*, *A. Flavus*, and *A. niger*, include illnesses that usually affect the respiratory system: clinical allergies [allergic bronchopulmonary aspergillosis (ABPA), rhinitis, Farmers's lung], superficial and local infections (cutaneous infections, otomycosis, tracheobronchitis), damaged tissue infections (aspergilloma and osteomyelitis), and chronic invasive pulmonary infections (CPA) (Public Health Agency of Canada 2013). These illnesses are common among people who work in the farming industry, and are considered an occupational hazard.

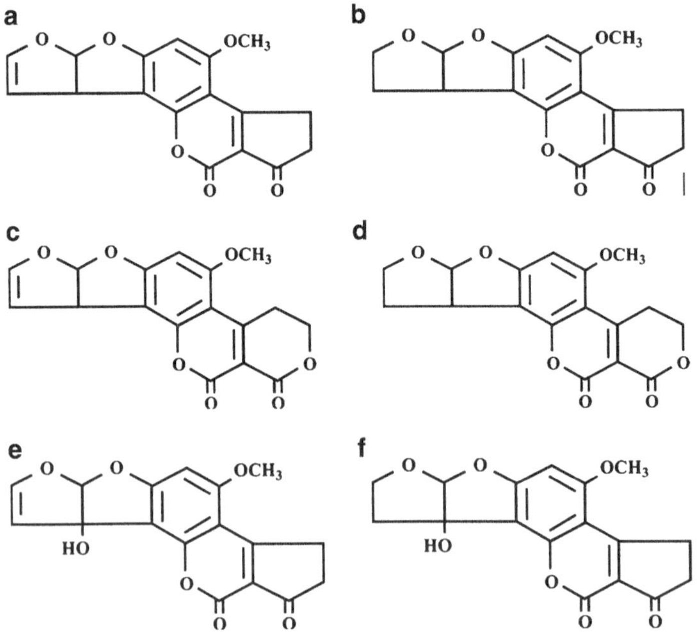

Fig. 1 Structures of (**a**) AFB_1, (**b**) AFB_2, (**c**) AFG_1, (**d**) AFG_2, (**e**) AFM_1, (**f**) AFM_2

Although *Aspergillus* spp. can produce several mycotoxins (Table 1), we deem giving detailed information on aflatoxins more appropriate in the following lines. The structural formulas of all some types of aflatoxins are given in Fig. 1. The genus *Aspergillus* includes notorious pathogens that produce aflatoxins B_1, B_2, G_1, or G_2, which are one of the most effective naturally produced mycotoxins in the world "B" and "G" refer to the blue and green fluorescent colours, respectively, produced under UV light on thin layer chromatography (TLC) plates and the subscript numbers 1 and 2 indicate major and minor compounds, respectively (Pitt and Hocking 2009).

AFM_1 which is the major metabolite of AFB_1 (4-hydroxy derivative of AFB_1) in mammals can partially excreted into milk. The carryover of AFB_1 in feeds as AFM_1 in milk varies from animal to animal and day to day changing 1–2 % (Moss 2000). The maximum permitted concentration of AFM_1 in cows' milk is 0.05 μg/kg in the European Union (EU) (Britzi et al. 2013). Consequently limiting AFB_1 in animal feeds is the most effective way of controlling AFM_1 in milk (FAO 2003).

Aspergillus spp. are widely distributed in the environment, including soil, plant debris, wood, and air. Water and inadequately harvested and stored grains and other foodstuffs such as peanuts, almonds, rye, wheat, and maize kernels, or whole-meal products are also act as reservoirs of transmission of *Aspergillus* spp. (Public Health Agency of Canada 2013). Even non-mouldy foods may contain aflatoxins since aflatoxins excreted by the mycelium can penetrate several centimetres into foods during a few days. Relative humidity <65 %, temperature <10 °C and moisture content <12 % are found as the conditions for preventing mould from growing and toxin production in the stored grains and other foodstuffs (http://pac.iupac.org/publications/pac/pdf/1973/pdf/3503x0239.pdf). The usage of stress and aflatoxin-tolerant cultivars, irrigation practices and management of insect pests in the field before harvest; avoiding direct contact of grains with soil after harvest; rapid grain drying and moisture control during storage may also given as effective ways to control aflatoxins from field to fork (http://www.aflasafe.com/aflatoxins).

3.2 Penicillium *and Related Teleomorphs*

The genus *Penicillium* which is known as opportunistic saprophytes is more diverse than the genus *Aspergillus*. Although most of the species in this genus are soil fungi, some species are directly related with food spoilage. Pitt and Hocking (2009) stated the taxonomy of this genus as difficult due to the similar colour and general colony appearance of the species occurring in foods.

The genus *Penicillium* has branched, septate conidiophores with a relatively slight, apparent stipe ending with a penicillus which helps to determine the subgenera of this genus. In other words, the number of branch points between phialide or conidial chain and stipe down the penicillus helps to find out the subgenera belonging to this genus. *Aspergilloides* is the simplest subgenus with only one branch point between conidial chain and stipe. Species with two branch points are classified in one of two subgenera named *Furcatum* or *Biverticillium*, while species with three or four branch points are classified in subgenus *Penicillium* (Pitt and Hocking 2009).

The genus *Penicillium* is associated with two teleomorph genera; *Eupenicillium* and *Talaromyces*. *Penicillium* species associated with *Talaromyces* teleomorphs are all classified in the *Penicillium* subgenera *Biverticillium* while the species associated with *Eupenicillium* teleomorphs are classified in the *Penicillium* other subgenera.

Penicillium spp. can produce several mycotoxins (Table 1). The growth of *P. citreonigrum* in rice resulted in acute cardiac beriberi due to the production of a neurotoxin named as citreoviridin; *P. citrinum*'s and *P.verrucosum's* production of citrinin in various cereals which is a renal toxin to domestic animals, causes kidney degeneration in chickens and turkeys, teratogenic effects in rats and adverse effects on human T cells; *P. crustosum's* production of penitrem A which is a powerful neurotoxin, causes tremor syndrome in humans; *P. expansum*'s production of patulin which is also an indicator of the use of poor quality raw materials in fruit juice manufacturing and processing plants, causes damages in the DNA of mammalian cells; *P. verrucosum*'s production of ochratoxin A which is a chronic nephrotoxin and also produced by several *Aspergillus* species, has immunosuppressive, teratogenic, genotoxic, and carcinogenic effects in animals and in humans (Frisvad et al. 2007; Pitt and Hocking 2009). This list can be extended.

Although there is not any data on human toxicity of patulin, an upper limit as 50 mg patulin in 1 L of fruit juices, especially in apple juice (FAO 2003) and a limit as 10 mg patulin in 1 kg food products produced for babies and infants, have been approved in some countries (EU 2004). The maximum level of ochratoxin A is set to 5 mg/kg for raw cereal grains whereas it is set to 3 mg/kg for cereal products and cereal grains intended for direct human consumption. In addition to these, its level is set to 10 mg/kg for currants and raisins (Adams and Moss 2008).

3.3 **Fusarium** ***and Related Teleomorphs***

The genus *Fusarium* is well known for taxonomic difficulties, comprising numerous members which can be recovered from plants, plant debris, and cultivated soils worldwide as pathogens, endophytes and harmless saprobes. Therefore, the *Fusarium* species are mostly regarded as soil-borne fungi due to their abundance in soil and plant roots (Nelson et al. 1994). It can be characterized by the production of macroconidia which are in crescent shapes, microconidia which are smaller one to two-celled conidia in various shapes (oval, ampulliform, ovoid, clavate, falcate, apiculate, reniform, and etc.), and chlamydospores which are thick-walled spores (formed singly, doubles, in clumps, or in chains) filled with lipid-like material that serves to carry the fungi over winter in soil when a suitable host is not available. Some species produce all these three types of spores although other species do not (Nelson et al. 1994; Pitt and Hocking 2009).

A number of *Fusarium* species have teleomorphs belonging to the genera *Gibberella*, *Haematonectria* (which includes *F. solani complex*), and *Albonectria* (which includes *F. decemcellulare*). Among these telemorph genera, *Gibberella* includes most of the significant pathogens such as *F. graminearum* (*G. zeae*) which commonly infects barley if there is rain late in the season and *F. verticillioides* (*G. moniliformis*) (Summerell and Leslie 2011).

Fusarium spp. are opportunistic fungi that can affect humans mainly in immunocompromised hosts, by producing either mycotoxicosis or invasive diseases. The main toxins produced by these *Fusarium* species are fumonisins and trichothecenes (a family of sesquiterpenes). During the closing years of World War II, *F. sporotrichoide* and *F. poae*-contaminated over-wintered wheat (wheat which had been left in the fields throughout the winter) the flour of which was baked into bread caused alimentary toxic aleukia (ATA) with the deaths of hundreds of thousands of people in the USSR (Nelson et al. 1994; Pitt and Hocking 2009). The active ingredient was found to be trichothecene T-2 mycotoxin. Two to three weeks after the consumption of toxic grain, ATA appears with an abdominal pain, diarrhoea, vomiting and prostration, and continues with fever, chilling, bloody diarrhoea, myalgias, and bone-marrow depression with granulocytopenia and secondary sepsis within a few days.

In Japan, *F. graminearum* or *F. sporotrichioides*-infected cereal grains (barley, oats, rye, rice, and wheat) are commonly associated with outbreaks of akakabi-byo (red mould disease or scabby grain intoxication) which is characterized by symptoms such as abdominal pain, diarrhoea, nausea, vomiting, headache, chills, giddiness, and anorexia. *Fusarium* isolates isolated from these scabby cereal grains are found to produce the trichothecenes: deoxynivalenol, nivalenol, fusarenon-X, diacetoxyscirpenol, neosolaniol, and T-2 mycotoxins in culture (Nelson et al. 1994).

F. moniliforme is one of the most prevalent fungi associated with mouldy corn toxicosis which causes human oesophageal cancer.

As an example, the highest rate of human oesophageal cancer occurs in the southwestern districts of the Transkei in southern Africa, where corn is consumed as the main foodstuff. The ingestion of *F. moniliforme*-contaminated corns gave rise to outbreaks of a neurotoxic disease called equine leukoencephalomalacia (LE) in horses, donkeys, and mules and pulmonary oedema (PPE) in swines (Nelson et al. 1994).

Zearalenone, which is not a true mycotoxin but is an oestrogen analogue, is mostly produced by *F. graminearum* and *F. culmorum* in cereals. Limits for zearalenone in maize and other cereals vary from 50 to 1,000 mg/kg (FAO 2003).

4 Non-bacterial and Non-fungal Food-borne Microbial Diseases

Viruses, protozoa, toxigenic algae, nematodes, and helminths are also the causative agents of some other food-borne microbial diseases. However, the following section will not explain them as in the same detail as food-borne bacterial pathogens and toxigenic fungi but it will give the readers a brief significance and occurrence of important food-borne viruses and protozoa.

4.1 Food-borne Viruses

Viruses (derived from the Latin word; *poisons*) are non-cellular, strict intracellular parasites that have only one type of nucleic acid (either DNA or RNA) wrapped in a protein coat or capsid. They require a live host for replication and cannot multiply in foods, and so foods only play passive roles in the transmission of viral infection. Sensory characteristics of foods containing these viral pathogens and those of non-contaminated foods are identical (Koopmans and Duizer 2004). Food-borne viral infections can occur only from human enteric pathogenic viruses due to contamination of the fresh produce or processed foods by virus-containing faecal material. Almost more than 100 of them have been identified as the causative agents of food-borne diseases in humans. They enter the body via the gut but they differ in their target tissues. As an example; gastroenteritis viruses stay, multiply, and cause illnesses in gut while polio and hepatitis viruses cause illnesses when they have migrated to and multiply in the other organs of the host (Adams and Moss 2008; Scallan et al. 2011).

Poliovirus, which is in the genus *Enterovirus* of the family Picornaviridae, is a single-stranded RNA virus that causes an illness characterized by headache, fever, and sore throat. When it spreads to neurons of the spinal cord, cell destruction, and paralysis occur. Before the 1940s, contaminated milk had been the main source of it; however, this problem is overcome by the application of good sanitation and personal hygiene habits (Adams and Moss 2008; Ray and Bhunia 2008).

Several types of hepatitis viruses exist but only hepatitis A (mostly) which is in the genus *Hepatovirus* of the family Picornaviridae, and hepatitis E (very rare) which is classified in a separate genus *Hepevirus* of the family Hepeviridae are generally associated with food-borne illnesses and are major public health concerns (Schlauder and Mushahwar 2001; Vasickova et al. 2005). Hepatitis A virus (HAV) is a single-stranded RNA virus that causes an illness characterized by anorexia, fever, malaise, nausea, and vomiting, followed after a few days by abdominal discomfort, inflammation of liver, and jaundice. It causes the vast majority of mortality associated with food-borne viral disease. In 1988, the largest outbreak (approximately 300,000 people) of HAV, which was resulted with 47 deaths, occurred in Shanghai, China due to consumption of contaminated dairy clams (Bhunia 2008). On the other hand, Hepatitis E virus (HEV) can cause life-threatening infections in women in the later stages of pregnancy. HEV has been isolated in swine and rats, and in pig livers sold in the local grocery stores in the USA. Under-cooked pork and deer meat were the sources of infection in the major HEV outbreaks in Japan (Feagins et al. 2007). Faecal–oral spread is the primary mode of HAV transmission. RTE foods which are contaminated by infected food handlers and workers are the main source of it however; sewage contaminated-drinking water, milk, fruits (especially strawberries and raspberries when contaminated water is used to rinse them), salad vegetables contaminated with polluted water and raw or improper-cooked shellfish are also the other sources of it. It survives in water and sewage for months and survives in foods for several days, even at refrigeration temperatures but can be inactivated by boiling or cooking to 85 °C. Besides, only HAV has available vaccines to prevent illness associated with food-borne viruses (Atreya 2004; Adams and Moss 2008; Ray and Bhunia 2008).

Norovirus (NoV) [earlier named as Norwalk-like viruses (NLV) due to the virus which caused an outbreak of gastroenteritis in elementary school children in Norwalk, Ohio, in the USA in 1968 (Adler and Zickl 1969)] or small round structures viruses (SRSVs) which is in the family Caliciviridae and is a plus-strand RNA virus, causes gastroenteritis characterized by vomiting, diarrhoea, and abdominal pain. They are estimated to be responsible for over 95 % of non-bacterial epidemic gastroenteritis outbreaks, and 50 % of all gastroenteritis outbreaks, worldwide (Jones and Karst 2013). However, asymptomatic infections are common and may contribute to the spread of the infection. Large numbers of virus particles are excreted by diarrhoeal stools during the illness since it multiplies in the gut. It is spread by the faecal–oral route. Foods can be contaminated with NoVs at the source (where the food is produced) by contaminated irrigation water or during handling and preparation processes. RTE foods and shellfish are the common sources of NoVs (Adams and Moss 2008; Glass et al. 2009; Jones and Karst 2013).

Norovirus is not the only virus that causes diarrhoeal disease. Astrovirus (AstV), rotavirus, sapovirus, and parvovirus are the other viruses that cause gastroenteritis. Among them, AstV and rotavirus which affect primarily the infants have more importance in food-borne viral infections. AstV [named for its star-like appearance (derived from the Greek word *astron*: star)] which is a single-stranded RNA virus including the only members of the family Astroviridae, and rotavirus [named for its wheel-like appearance (derived from the Latin word *rota* : wheel)] which is a double-stranded RNA virus belonging to the family Reoviridae, show the same symptoms such as vomiting for up to 48 h which is followed by 24 h of diarrhoea. If dehydration and electrolyte imbalances occur, death may occur. Poor water quality (using polluted water for irrigation), inadequately treated sewage sludge, and oysters, clams, mussles, and cockles, lettuce, and other vegetables are the main sources of AstVs and rotaviruses (Vasickova et al. 2005; Gastañaduy et al. 2013; Karlsson and Schultz-Cherry 2013).

Consequently, good agriculture practice (GAP), good manufacturing practice (GMP) and HACCP should be applied in order to avoid introduction of viruses onto the raw food material and into the food-manufacturing and processing environment. Good personal hygiene, improved sanitation and provision of clean drinking water should be provided. Since viruses remain infectious in most foods for several days or weeks, especially if kept cooled (at 4 °C), more attention should be given to personal hygiene during handling, preparation and service of foods (Koopmans and Duizer 2004).

4.2 Food-borne Protozoa

Amongst the protozoa, the flagellate *Giardia lamblia*, the amoeboid *Entamoeba histolytica*, and three sporozoid (*Cryptosporidium parvum, Toxoplasma gondii* and *Sarcocystis*) are of special concern from the point of view of food-borne microbial diseases.

Giardia lamblia which is also known as *G. intestinalis*, *G. lamblia*, or *G. duodenalis*, survives as cysts in soil, food, water, or on surfaces that has been contaminated with faecal material from infected humans or wild animals, such as beavers, for weeks or months although it is a parasite that feeds off a live host to survive. After the ingestion of its cysts, excystation occurs and so two active flagellate protozoa known as trophozoites which are characterized by eight flagella, two nuclei and tumbling motility, are released from each of the cyst into the small intestine. The main food sources are contaminated water, salad vegetables and fruits (especially lettuce in sandwiches, strawberries and raspberries when contaminated water is used to rinse them), improperly handled (by infected persons not observing good hygiene practices—poor personal hygiene) and improperly cooked foods. However, swallowing water during swimming in water where *Giardia* may live and having contact with a person who is ill also cause a disease called giardiasis which is characterized by recurrent abdominal cramps, nausea, acute or chronic dysentery type diarrhoea, with malabsorption and failure of children to thrive, greasy stools that tend to float and dehydration. The cysts are resistant to chlorine but are killed by cooking (Adams and Moss 2008; Bhunia 2008; CDC 2012).

Entamoeba histolytica cysts enjoy widespread occurrence in the environments wherever there is poor hygiene and inadequate sanitation facilities (endemic especially in the tropical areas with poor sanitary conditions), including contaminated water, sewage, insects, a wide range of domestic and wild animals, and a variety of foodstuffs of both animal and plant origin, and cause amoebiasis which is characterized by stomach pain, stomach cramping, and loose faeces in humans and animals. In severe forms, amoebic dysentery associated with bloody stools and fever occurs. However, asymptomatic infections are common and may contribute to the spread of the infection. Transmission is by the faecal–oral route. A person with amoebic dysentery may pass up to 50 million cysts per day. Following the ingestion of its cysts, excystation occurs and each cyst gives rise to eight daughter trophozoites. The cysts can survive outside the host for several weeks to months although it is a parasite that feeds off a live host to survive. The cysts are sensitive to temperature below −5 °C or over 40 °C (Adams and Moss 2008; Bhunia 2008; Nagata et al. 2012).

Cryptosporidium parvum which causes a waterborne diarrhoeal disease called cryptosporidiosis that is characterized by a cholera-like illness with stomach and muscle pain, nausea, vomiting, low-grade fever, dehydration, weight loss, and anorexia. Some people have no symptoms at all. After the ingestion of its oocysts, excystation occurs and so four sporozoites which parasitize epithelial cells of the gastrointestinal or respiratory tract, are released from each of the cyst into the small intestine. The oocysts complete their life cycles in one host (homoxenous). Vegetables or foods which are exposed to contaminated water, serve as the major sources of it. The oocysts are resistant to disinfectants usually used to treat water. In the midst of 1990, a large outbreak of *Cryptosporidium* occurred among HIV patients in Wisconsin, the USA due to the consumption of water from municipal water supply, which was contaminated with cattle manure (Adams and Moss 2008; Bhunia 2008). Between October and December 2011, an outbreak of 26 cases (24 children under 2 years of age and two caregivers) of cryptosporidiosis occurred in a day-care centre in Gipuzkoa, Spain (Artieda et al. 2012).

Toxoplasma gondii which causes a zoonotic disease called toxoplasmosis is mostly transmitted by the members of the family Felidae (domestic cats and their relatives). It not transmitted by person-to-person contact, except in cases of mother-to-child (congenital) transmission. After the ingestion of its oocysts, they dissolve in the gut and tachyzoites are released which penetrate the intestinal epithelial cells, reach to blood circulation and invade muscle tissues, and develop into tissue cyst bradyzoites. Although food-borne infection in humans is uncommon, it could occur by drinking water contaminated with cat faeces and consumption of raw or under-cooked meat, especially mutton or pork. It is normally symptomless or associated with a mild influenza-like illness (fatigue, joint, and muscle pain) in healthy humans. However, spontaneous abortion or stillborn child usually occurs in pregnant women (Montoya and Liesenfeld 2004; Adams and Moss 2008; Bhunia 2008).

On the other hand, *Sarcocystis* (derived from the Greek words; *sarx*: flesh and *kystis*: bladder) species are intracellular protozoan parasites with a requisite two-host life cycle based on a prey–predator (intermediate-definitive) host relationship: an intermediate host such as cattle, sheep, or pigs in the tissues of which the asexual cysts are formed, and the definitive host such as cats, dogs, or humans, in which sexual reproduction of this parasite takes place. In other words, sarcocysts in meat eaten by humans initiate sexual stages in the intestine which terminate in oocysts excreted in the faeces. *S. hominis* and *S. suihominis* can infect humans who eat raw or under-cooked meat from cattle and pigs containing mature sarcocysts, respectively. The most likely source of sarcocysts is water contaminated with faeces or foods washed or irrigated with contaminated water. They usually cause mild illness starting with nausea and diarrhoea. They can be prevented by thoroughly cooking or freezing meat to kill bradyzoites (spindle- or crescent-shaped bodies) in the sarcocysts (Fayer 2004; Adams and Moss 2008).

5 Conclusion

A significant amount of foodstuff in our daily life goes to the garbage since it is spoiled by different microorganisms. Among these microorganisms, pathogens have received abundant attention because of the illnesses and diseases that they or their toxins cause. Although there are certainly many more unrecognized microbial pathogens awaiting identification, a great deal is known about pathogens. However; we are still not able to control them and thus food-borne diseases caused by pathogens still occur at unacceptably high rates even in industrialized and developing countries.

Food-borne microbial diseases are important causes of personal suffering, illness, death, and economic burden. Increased public awareness of the health-related and economic impact of food-borne microbial contamination and disease has resulted in greater efforts to develop of new methods of food manufacturing, processing and preservation in order to prevent and control the contamination during the flow of food from the farm to table. This technological improvement together with the good hygienic and good agricultural and manufacturing practices and the knowledge on these pathogens, supported by research, help us control the contamination and the incidence of food-borne microbial pathogens.

References

Adams MR, Moss MO (2008) Food microbiology, 3rd edn. The Royal Society of Chemistry, Cambridge, p 463

Adler JL, Zickl R (1969) Winter vomiting disease. J Infect Dis 119:668–673

Allos M (2001) *Campylobacter jejuni* Infections: update on emerging issues and trends. Food Saf 32:1291–1296

Artieda J, Basterrechea M, Arriola M, Yagüe M, Albisua E, Arostegui N, Astigarraga U, Botello R, Manterola JM (2012) Outbreak of cryptosporidiosis in a child day-care centre. Gipuzkoa, Spain, Oct–Dec 2011. Euro Surveill 17:pii=20070

Atanda SA, Pessu PO, Agoda S, Isong IU, Adekalu OA, Echendu MA, Falade TC (2011) Fungi and mycotoxins in stored foods. Afr J Microbiol Res 5:4373–4382

Atreya CD (2004) Major foodborne illness causing viruses and current status of vaccines against the diseases. Foodborne Pathog Dis 1:89–96

Bannerman TL, Peacock SJ (2007) *Staphylococcus*, micrococcus, and other catalase-positive cocci. In: Murray PR, Baron EJ, Jorgenseni JH, Landry ML, Pfaller MA (eds) Manual of clinical microbiology, 9th edn. American Society for Microbiology Press, Washington, DC, pp 390–411

Bennett JW, Klich M (2003) Mycotoxins. Clin Microbiol Rev 16:497–516

Bercovier H, Mollaret HH (1984) Genus XIV. *Yersinia*. In: Krieg NR (ed) Bergey's manual of systematic bacteriology, vol 1. Williams & Wilkins, Baltimore, MD, pp 498–506

Bhunia AK (2008) Foodborne microbial pathogens: mechanisms and pathogenesis. Springer, New York, p 277

Black RE, Levine MM, Clements ML (1988) Experimental *Campylobacter jejuni* infections in humans. J Infect Dis 157:472–479

Britzi M, Friedman S, Miron J, Solomon R, Cuneah O, Shimshoni JA, Soback S, Ashkenazi R, Armer S, Shlosberg A (2013) Carry-over of aflatoxin B_1 to aflatoxin M_1 in high yielding Israeli cows in mid- and late-lactation. Toxins (Basel) 16:173–183

Brock DL (2006) Deadly diseases and epidemics: infectious fungi. Infobase, New York, p 126

Centers for Disease Control (CDC) (2012) Giardiasis surveillance-the United States, 2009–2010. 61(SS05):13–23

Clarke RC, Gyles CL (1993) *Salmonella*. In: Gyles CL, Thoen CO (eds) Pathogenesis of bacterial infections in animals, 2nd edn. Iowa State University Press, Ames, pp 133–153

Dack GM (1956) Food poisoning, 3rd edn. University of Chicago Press, Chicago, p 251

de Jong AEI, Rombouts FM, Beumer RR (2004) Behaviour of *Clostridium perfringens* at low temperatures. Int J Food Microbiol 97:71–80

den Bakker H, Cummings CA, Ferreira V, Vatta P, Orsi RH, Degoricija L, Barker M, Petrauskene P, Furtado MR, Wiedmann W (2010) Comparative genomics of the bacterial genus *Listeria*: genome evolution is characterized by limited gene acquisition and limited gene loss. BMC Genomics 11:688

Desai PT, Porwollik S, Long F, Cheng P, Clifton SW, Weinstock GM, McClelland M (2013) Evolutionary genomics of *Salmonella enterica* subspecies. mBio 4. doi:10.1128/mBio.00579-12

Dhama K, Rajagunalan S, Chakraborty S, Verma AK, Kumar A, Tiwari R, Kapoor S (2013) Food-borne pathogens of animal origin-diagnosis, prevention, control and their zoonotic significance: a review. Pak J Biol Sci 16:1076–1085

Duffy G (2006) Emerging pathogenic *E. coli*. In: Motarjemi Y, Adams M (eds) Emerging foodborne pathogens. Woodhead, Cambridge, pp 253–281

EFSA (European Food Safety Authority) (2005) Opinion of the scientific panel of biological hazards on *Bacillus cereus* and other *Bacillus spp.* in foodstuffs. EFSA J 175:1–48

Ehling-Schulz M, Fricker M, Scherer S (2004) *Bacillus cereus*, the causative agent of an emetic type of foodborne illness. Mol Nutr Food Res 48:479–487

European Union (EU) (2004) Commission regulation (EC) no. 455/2004 of 11 March 2004 amending regulation (EC) no. 466/2001 as regards patulin. Off J Eur Union L74/11

Ewing WH, Gravatti JL (1947) *Shigella* types encountered in the mediterranean area. J Bacteriol 53:191–195

Farber JM, Peterkin PI (2000) *Listeria monocytogenes*. In: Lund BM, Baird-Parker TC, Gould GW (eds) The microbiological safety and quality of food. Aspen, Maryland, pp 1178–1232

Faruque SM, Albert MJ, Mekalanos JJ (1998) Epidemiology penetics, and ecology of toxipenic *vibrio cholerae*. Microbiol Mol Biol Rev 62:1301–1314

Fayer R (2004) *Sarcocystis* spp. in human infections. Clin Microbiol Rev 17:894–902

Feagins AR, Opriessnig T, Guenette DK, Halbur PG, Mengl X (2007) Detection and characterization of infectious Hepatitis E virus from commercial pig livers sold in local grocery stores in the USA. J Gen Virol 88:912–917

Feng PCH, Monday SR (2006) Multiplex PCR for specific identification of enterohemorrhagic *Escherichia coli* strains in the O157:H7 complex. In: Adley CC (ed) Foodborne pathogens methods and protocols. Humana, New Jersey, pp 47–55

Food and Agriculture Organization (FAO) (2003) Worldwide regulations for mycotoxins in food and feed. http://www.fao.org/docrep/007/y5499e/y5499e07.htm. Accessed on 3 Sept 2013

Food and Drug Administration (FDA) (2012) Bad bug book: handbook of foodborne pathogenic microorganisms and natural toxins, 2nd edn. http://www.fda.gov/downloads/Food/FoodborneIllnessContaminants/UCM297627.pdf. Accessed on 12 Sept 2013

Frisvad JC, Thrane U, Samson RA (2007) Mycotoxin producers. In: Dijksterhuis J, Samson RA (eds) Food mycology: a multifaceted approach to fungi and food. CRC Press, Boca Raton, pp 135–159

Gastañaduy PA, Hall AJ, Parashar UD (2013) Rotavirus. In: Morris JG Jr, Potter ME (eds) Foodborne infections and intoxications, 4th edn. Academic, Boston, pp 303–312

Giovannini M, Verduci E, Ghisleni D, Salvatici E, Riva E, Agostoni C (2008) *Enterobacter sakazakii*: an emerging problem in paediatric nutrition. J Int Med Res 36:394–399

Glass K, Marshall K (2013) *Clostridium botulinum*. In: Morris JG Jr, Potter ME (eds) Foodborne infections and intoxications, 4th edn. Academic, Boston, pp 371–388

Glass RI, Parashar UD, Estes MK (2009) Norovirus gastroenteritis. J N Engl Med 361: 1776–1785

Gnonlonfin GJB, Hell K, Adjovi Y, Fandohan P, Koudanda DO, Mensah GA, Sanni A, Brimer L (2013) A review on aflatoxin contamination and its implications in the developing world: a sub-saharan African perspective. Crit Rev Food Sci 53:349–365

Granum PE, Baird-Parker TC (2000) *Bacillus* species. In: Lund BM, Baird-Parker TC, Gould GW (eds) The microbiological safety and quality of food. Aspen, Maryland, pp 1029–1039

Hald T (2013) *Salmonella*. In: Morris JG Jr, Potter ME (eds) Foodborne infections and intoxications, 4th edn. Academic, Boston, pp 67–98

Hauge S (1950) Matforgiftninger fremkalt av *Bacillus cereus*. Forel øpig meddelelse. Nord Hyg Tidskr 31:189–206

Hauge S (1955) Food poisoning caused by aerobic spore forming bacilli. J Appl Bacteriol 18:591–595

Hawksworth DL (2001) The magnitude of fungal diversity: the 1.5 million species estimate revisited. Mycol Res 105:1422–1432

Healy B, Cooney S, O'Brien S, Iversen C, Whyte P, Nally J, Callanan JJ, Fanning S (2010) *Cronobacter (Enterobacter sakazakii)*: an opportunistic foodborne pathogen. Foodborne Pathog Dis 7:339–350

Heikinheimo A, Lindstrom M, Granum PE, Korkeala H (2006) Humans as reservoir for enterotoxin gene-carrying *Clostridium perfringens* type A. Emerg Infect Dis 12:1724–1729

Hennekinne JA, De Buyser ML, Dragacci S (2012) *Staphylococcus aureus* and its food poisoning toxins: characterization and outbreak investigation. FEMS Microbiol Rev 36:815–836

Hill KK, Xie G, Foley BT, Smith TJ, Munk AC, Bruce D, Smith LA, Brettin TS, Detter JC (2009) Recombination and insertion events involving the botulinum neurotoxin complex genes in *Clostridium botulinum* types A, B, E and F and *Clostridium butyricum* type E strains. BMC Biol 7:66

Horrocks SM, Anderson RC, Nisbet DJ, Ricke SC (2009) Incidence and ecology of *Campylobacter jejuni* and *coli* in animals. Anaerobe 15:18–25

http://pac.iupac.org/publications/pac/pdf/1973/pdf/3503x0239.pdf. Accessed on 3 Sept 2013

http://www.aflasafe.com/aflatoxins. Accessed on 3 Sept 2013

http://www.efsa.europa.eu/en/efsajournal/doc/175.pdf. Accessed on 13 Sept 2013

ICMSF (International Commission on Microbiological Specifications for Foods) (1996) *Bacillus cereus*. In: Roberts TA, Baird-Parker AC, Tompkin RB (eds) Microorganisms in foods 5. Microbiological specifications of food pathogens. Blackie Academic, London, pp 20–35

Jones M, Karst SM (2013) Noroviruses. In: Morris JG Jr, Potter ME (eds) Foodborne infections and intoxications, 4th edn. Academic, Boston, pp 261–278

Jones MK, Oliver JD (2009) *Vibrio vulnificus*: disease and pathogenesis. Infect Immun 77:1723–1733

Jore S, Viljugrein H, Brun E, Heier BT, Borck B, Ethelberg S, Hakkinen M, Kuusi M, Reiersen J, Hansson I, Engvall EO, Løfdahl M, Wagenaar JA, van Pelt W, Hofshagen M (2010) Trends in *Campylobacter* incidence in broilers and humans in six European countries, 1997–2007. Prev Vet Med 93:33–41

Kalinowski RM, Tompkin RB, Bodnaruk PW, Pruett WP Jr (2003) Impact of cooking, cooling, and subsequent refrigeration on the growth of survival of *Clostridium perfringens* in cooked meat and poultry products. J Food Protect 66:1227–1232

Kamat AS, Nerkar DP, Nair PM (1989) *Bacillus cereus* in some indian foods, incidence and antibiotic, heat and radiation resistance. J Food Saf 10:31–41

Karlsson EA, Schultz-Cherry S (2013) Astroviruses as foodborne infections. In: Morris JG Jr, Potter ME (eds) Foodborne infections and intoxications, 4th edn. Academic, Boston, pp 293–302

Knox R, MacDonald EK (1943) Ourbreaks of food poisoning in certain Leicester institutions. Med Off 69:21–22

Koopmans M, Duizer E (2004) Foodborne viruses: an emerging problem. Int J Food Microbiol 90:23–41

Kotiranta A, Lounatmaa K, Haapasalob M (2000) Epidemiology and pathogenesis of *Bacillus cereus* infections. Microbes Infect 2:189–198

Kramer JM, Gilbert RJ (1989) *Bacillus cereus* and other *Bacillus* species. Foodborne bacterial pathogens. Marcel Dekker, New York

Labbé RG, Juneja VK (2013) *Clostridium perfringens* gastroenteritis. In: Morris JG Jr, Potter ME (eds) Foodborne infections and intoxications, 4th edn. Academic, Boston, pp 99–112

Lai KK (2001) *Enterobacter sakazakii* infections among neonates, infants, children, and adults. Case reports and a review of the literature. Medicine 80:113–122

Lambert JW, Bidlas E (2007) A study of the gamma hypothesis: predictive modelling of the growth and inhibition of the *Enterobacter sakazakii*. Int J Food Microbiol 115:204–213

Landgraf M, Destro MT (2013) Staphylococcal food poisoning. In: Morris JG Jr, Potter ME (eds) Foodborne infections and intoxications, 4th edn. Academic, Boston, pp 389–401

Linnan MJ, Mascola L, Lou XD, Goulet V, May S, Salminen C, Hird DW, Yonekura ML, Hayes P, Weaver R et al (1988) Epidemic listeriosis associated with Mexican-style cheese. J N Engl Med 319:823–828

Lipp EK, Huq A, Colwell RR (2002) Effects of global climate on infectious disease: the cholera model. Clin Microbiol Rev 15:757–770

Lund BM (2008) Properties of microorganisms that cause foodborne disease. In: Lund BM, Hunter PR (eds) The microbiological safety of food in healthcare settings. Blackwell, Oxford, pp 12–233
Magan N, Aldred D (2007) Why do fungi produce mycotoxins? In: Dijksterhuis J, Samson RA (eds) Food mycology: a multifaceted approach to fungi and food. CRC Press, Boca Raton, pp 121–133
Maheshwari M, Nelapati K, Kiranmayi B (2011) *Vibrio cholerae* - a review. Vet World 4:423–428
Mead PS, Slutsker L, Dietz V, McCaig LF, Bresee JS, Shapiro C, Griffin PM, Tauxe RV (1999) Food-related illness and death in the United States. Emerg Infect Dis 5:607–625
Mølbak K, Olsen JE, Wegener HC (2006) *Salmonella* infections. In: Rieman HP, Cliver DO (eds) Foodborne infections and intoxications, 3rd edn. Academic, Boston, pp 57–136
Montoya JG, Liesenfeld O (2004) Toxoplasmosis. Lancet 363:1965–1976
Morris JG Jr (2003) Cholera and other types of vibriosis: a story of human pandemics and oysters on the half shell. Clin Infect Dis 37:272–280
Morris JG Jr (2011) Cholera-modern pandemic disease of ancient lineage. Emerg Infect Dis 17:2099–2104
Moss MO (2000) Toxigenic fungi and mycotoxins. In: Lund BM, Baird-Parker TC, Gould GW (eds) The microbiological safety and quality of food. Aspen, Maryland, pp 1491–1517
Murray EGD, Webb RA, Swarm MBR (1926) A disease of rabbits characterized by a large mononuclear leucocytosis, caused by a hitherto undescribed bacillus *Bacterium monocytogenes* (n.sp). J Pathol Bacteriol 29:437–439
Nachamkin I (2001) *Campylobacter jejuni*. In: Doyle MP, Beuchat LR, Montville TJ (eds) Food microbiology: fundamentals and frontiers, 2nd edn. ASM Press, Washington, DC, pp 179–193
Nagata N, Shimbo T, Akiyama J, Nakashima R, Nishimura S, Yada T, Watanabe K, Oka S, Uemura N (2012) Risk factors for intestinal invasive amebiasis in Japan, 2003-2009. Emerg Infect Dis 18:717–724
Nair GB, Faruque SM, Sack DA (2006) Vibrios. In: Motarjemi Y, Adams M (eds) Emerging foodborne pathogens. Woodhead, Cambridge, pp 332–372
Nelson PE, Dignani MC, Anaissie EJ (1994) Taxonomy, biology, and clinical aspects of *Fusarium* species. Clin Microbiol Rev 7:479–504
Nesbakken T (2013) *Yersinia*. In: Morris JG Jr, Potter ME (eds) Foodborne infections and intoxications, 4th edn. Academic, Boston, pp 187–198
Nguyen VT, Turner MS, Dykes GA (2010) Effect of temperature and contact time on the attachment to, and probability of detachment from, stainless steel. J Food Protect 73:832–838
Ogston A (1882) Micrococcus poisoning. J Anat 17:24–58
Oliver JD, Kaper JB (2001) Vibrio species. In: Doyle MP, Beuchat LR, Montville TJ (eds) Food microbiology: fundamentals and frontiers, 2nd edn. ASM Press, Washington, DC, pp 263–300
Ono HK, Omoe K, Imanishi K, Iwakabe Y, Hu DL, Kato H, Saito N, Nakane A, Uchiyama T, Shinagawa K (2008) Identification and characterization of two novel staphylococcal enterotoxins, types S and T. Infect Immun 76:4999–5005
Pitkänen T (2013) Review of *Campylobacter* spp. in drinking and environmental waters. J Microbiol Methods 95:39–47
Pitt JI, Hocking AD (2009) Fungi and food spoilage, 3rd edn. Springer, New York, p 519
Public Health Agency of Canada (2013) *Aspergillus* spp.: pathogen safety data sheet-infectious substances. http://www.phac-aspc.gc.ca/lab-bio/res/psds-ftss/aspergillus-spp-eng.php. Accessed on 1 Sept 2013
Ramamurthy T, Garg R, Sharma R, Bhattacharya SK, Nair GB, Shimada T, Takeda T, Karasawa T, Kurazano H, Pal A, Takeda Y (1993) Emergence of novel strains of *Vibrio cholerae* with epidemic potential in southern and eastern India. Lancet 314:703–704
Ray B, Bhunia A (2008) Fundamental food microbiology, 4th edn. CRC Press; Taylor & Francis, Boca Raton, p 492
Richard JL (2007) Some major mycotoxins and their mycotoxicoses – an overview. Int J Food Microbiol 119:3–10
Robins-Browne R (2001) *Yersinia enterocolitica*. In: Doyle MP, Beuchat LR, Montville TJ (eds) Food microbiology: fundamentals and frontiers, 2nd edn. ASM Press, Washington, DC, pp 125–247

Robinson DA (1981) Infective dose of *Campylobacter jejuni* in milk. Brit Med J 282:1584
RůŽičkova V, Karpíšková R, Pantůček R, Markéta Pospíšilová M, Černíková P, Doškař J (2008) Genotype analysis of enterotoxin H-positive *Staphylococcus aureus* strains isolated from food samples in the Czech Republic. Int J Food Microbiol 121:60–65
Samson RA, Varga J (2010) Molecular systematics of *Aspergillus* and its teleomorphs. In: Machida M, Gomi K (eds) *Aspergillus*: molecular biology and genomics. Caister Academic, Norfolk, pp 19–40
Scallan E, Hoekstra RM, Angulo FJ, Tauxe RV, Widdowson MA, Roy SL, Jones JL, Griffin PM (2011) Foodborne illness acquired in the United States—major pathogens. Emerg Infect Dis 17:7–15
Scallan E, Kirk M, Griffin PM (2013) Estimates of disease burden associated with contaminated food in the United States and globally. In: Morris JG Jr, Potter ME (eds) Foodborne infections and intoxications, 4th edn. Academic, Boston, pp 3–18
Scannell AGM (2011) Overview of foodborne pathogens. In: Sun D-W (ed) Handbook of food safety engineering. Blackwell, Oxford, UK, pp 18–56
Schelin J, Wallin-Carlquist N, Cohn MT, Lindqvist R, Barker GC, Rådström P (2011) The formation of *Staphylococcus aureus* enterotoxin in food environments and advances in risk assessment. Virulence 2:580–592
Schlauder GG, Mushahwar IK (2001) Genetic heterogeneity of hepatitis E virus. J Med Virol 65:282–292
Schnider A, Overesch G, Korczak BM, Kuhnert P (2010) Comparison of real-time PCR assays for detection, quantification and differentiation of *Campylobacter jejuni* and *Campylobacter coli* in broiler neck skin samples. J Food Protect 73:1057–1063
Shimada T, Nair GB, Deb BC, Albert MJ, Sack RB, Takeda Y (1993) Outbreaks of *Vibrio cholerae* non-O1 in India and Bangladesh. Lancet 341:1347
Summerell BA, Leslie JF (2011) Fifty years of Fusarium: how could nine species have ever been enough? Fungal Divers 50:135–144
Sutherland J, Varnam A (2002) Enterotoxin-producing *Staphylococcus*, *Shigella*, *Yersinia*, *Vibrio*, *Aeromonas* and *Plesiomonas*. In: Blackburn Cde W, McClure PJ (eds) Foodborne pathogens: hazards, risk analysis and control. CRC Press, Boca Raton, pp 385–415
Tall BD, Grim CJ, Franco AA, Jarvis KG, Hu L, Kothary MH, Sathyamoorthy V, Gopinath G, Fanning S (2013) *Cronobacter* species (formerly *Enterobacter sakazakii*). In: Morris Jr, Potter ME (eds) Foodborne infections and intoxications, 4th edn. Academic, Boston, pp 251–260
Tompkin RB (2002) Control of *Listeria monocytogenes* in the food-processing environment. J Food Protect 65:709–725
Vasickova P, Dvorska L, Lorencova A, Pavlik I (2005) Viruses as a cause of foodborne diseases: a review of the literature. Vet Med-Czech 50:89–104
Wang S, Orsi RH (2013) *Listeria*. In: Morris JG Jr, Potter ME (eds) Foodborne infections and intoxications, 4th edn. Academic, Boston, pp 199–216
Webster J, Weber R (2007) Introduction to fungi, 3rd edn. Cambridge University Press, Cambridge, p 841
Woloshuk CP, Shim W-B (2013) Aflatoxins, fumonisins, and trichothecenes: a convergence of knowledge. FEMS Microbiol Rev 37:94–109
World Health Organization (WHO) (1997) Prevention and control of enterohaemorrhagic *Escherichia coli* (EHEC) infections. Report of a WHO consultation. Document WHO/FSF/FOS/97.6
World Health Organization (WHO) (2007) Food safety and foodborne illness. WHO Media Centre, No 237. http://www.who.int/mediacentre/factsheets/fs237/en/. Accessed on 28 Jan 2008
World Health Organization/Food and Drug Admistration (WHO/FAO) (2004) Risk assessment of *Listeria monocytogenes* in ready-to-eat foods, interpretative summary. Microbiological risk assessment series 4. http://www.who.int/foodsafety/publications/micro/en/mra4.pdf. Accessed on 18 Sept 2013
Wright AC, Harwood WJ (2013) Vibrios. In: Morris Jr, Potter ME (eds) Foodborne infections and intoxications, 4th edn. Academic, Boston, pp 113–128

Microbial Metabolites as Biological Control Agents in Food Safety

Zerrin Erginkaya, Emel Ünal, and Selin Kalkan

1 Introduction

Ensuring food safety has resulted in increased interest in alternative preservation techniques for inactivating microorganisms and enzymes in foods. This increasing demand has opened new dimensions for the use of natural preservatives derived from plants, animals, and microflora. Biopreservation has been explored as a means of increasing positive characteristics. In biopreservation, storage life is extended, and/or safety of food products is enhanced by using natural or controlled microflora, mainly lactic acid bacteria (LAB) and/or their antibacterial products such as lactic acid, bacteriocins, and others (Tiwari et al. 2009). Some of the desirable characteristics of biopreservative agents are listed : (1) safe (i.e., do not contribute to health risk of the food), (2) stable (i.e., maintains inhibitory activity during storage), (3) effective (i.e., broadly active against all major infective/spoilage bacteria and fungi), resistant to selection (i.e., resistant strains of target bacteria not readily selected for), (4) complementary (i.e., not significantly neutralized by antagonistic activity of the food environment (pH, fats, etc.) and compatible with the physiological and chemical characteristics of the food material), and (5) lethal (i.e., bacteri-/fungi-/sporicidal in preference to static).

Z. Erginkaya (✉) • E. Ünal
Faculty of Agriculture, Department of Food Engineering, Cukurova University, Adana 01330, Turkey
e-mail: zerriner@cu.edu.tr; unale@cu.edu.tr

S. Kalkan
Bahce Vocational School of Higher Education, Department of Food Technology, Osmaniye Korkut Ata University, Osmaniye 80500, Turkey
e-mail: selinkalkan@osmaniye.edu.tr

A. Malik et al. (eds.), *Food Processing: Strategies for Quality Assessment*, Food Engineering Series, DOI 10.1007/978-1-4939-1378-7_9

Naturally derived antimicrobial systems from microbial origin have been used as biocontrol agents in food safety. It tends to be fermentative organisms that are most typically evaluated for their potential in food biopreservation. For the preservation of foods where minimal flavor and textural changes are desired, LAB strains may provide a greater range of seeding candidates than heterofermentative strains due to their production of relatively innocuous (from a sensory point of view) antimicrobial compounds, such as lactic acid, as a dominant proportion of their fermentation end-products. In addition to LAB, other microorganisms like molds, yeast, and even pathogenic bacteria can produce antimicrobial metabolite and these metabolites may utilize for preservation of food (McIntyre et al. 2007; Lacroix 2011; Ivey et al. 2013)

Much of the preservative effect conferred on fermented food materials is attributable to its content of acids (especially lactic and acetic), resulting in a reduction of pH and the antimicrobial activity of the un-dissociated acid molecules. A wide variety of small inhibitory molecules including hydrogen peroxide, diacetyl, hypothiocyanate, exopolysaccharides, biofilm, reuterin and bacteriocins, sometimes powerfully active against pathogens. Both Gram-negative and Gram-positive bacteria produce bacteriocins. Bacteriocins are proteinaceous antibacterial compounds, which constitute a heterologous subgroup of ribosomally synthesized antimicrobial peptides. Bacteriocin production can be exploited by food processors to provide an additional barrier to undesirable bacterial growth in foods (Ivey et al. 2013).

2 Antimicrobial Metabolites

Antimicrobial is a general term used broadly to refer to any compound, including antibiotics, food antimicrobial agents, sanitizers, disinfectants, and other substances, that acts against microorganisms. The definitions and use of each of these terms differ among various groups. Antimicrobial metabolite defined as compounds that and has antimicrobial affect on pathogens/spoilage microorganisms and were produce by microorganisms (Hawke 2006).

The preservative action of starter strains in food and beverage systems is attributed to the combined action of a range of antimicrobial metabolites produced during the fermentation process. Bacteria produce many compounds that are active against other bacteria, which can be harnessed to inhibit the growth of potential spoilage or pathogenic microorganisms. These include fermentation end products such as organic acids, hydrogen peroxide, EPS and biofilms, carbon dioxide, and diacetyl, in addition to bacteriocins and other antagonistic compounds such as reuterin. Preservation of foods by fermentation depends on the principle of oxidation of carbohydrates and related derivatives to generate end-products. These end-products control the growth of food spoilage microorganisms and because the oxidation is only partial, the food retains sufficient energy potential to be of nutritional benefit to the consumer (Erginkaya et al. 2011; Stoyanova et al. 2012; Ivey et al. 2013).

2.1 Bacteriocin

Currently, natural antimicrobials such as bacteriocin have become increasingly attractive for application in food products. Bacteriocins are ribosomally synthesized peptides or proteins with antimicrobial activity, produced by different groups of bacteria (Galvez et al. 2007). Bacteriocin was firstly identified by Gratia (1925) as an antimicrobial protein produced by *Escherichia coli* and named colicin (Balciunas et al. 2013). Although bacteriocins are produced by many Gram-positive and Gram-negative species, those produced by the LAB are of particular interest to the food industry as potential novel preservatives. The bacteriocins produced by gram-positive bacteria are often proteins smaller than 6 kDa. In contrast most of bacteriocins produced by gram negative are often produced peptide bacteriocins larger than 20 kDa. Studies on bacteriocins produced by LAB began in the 1930s, starting from investigations of nisin produced by the lactococcus *L. lactis* subsp. *lactis* (Naidu 2000). The group of bacteriocin producing LAB includes representatives of different genera: *Lactococcus, Lactobacillus, Leuconostoc, Pediococcus*, and *Streptococcus* (Deegan et al. 2006).

LAB are an important group of industrial microorganisms which are involved in various food fermentation processes and contribute to the enhancement of organoleptic attributes of the food, as well as to the preservation and microbial safety of the end-product. One important attribute of LAB is their ability to produce antimicrobial compounds called bacteriocin. They are able to inhibit the growth of many undesirable microbes because of the production of bacteriocins (Brzozowski et al. 2009; Ibrahim and Desouky 2009; Pyar et al. 2013). Bacteriocins produced by LAB are low-molecular-mass peptides or proteins with an antibacterial mode of action towards sensitive species that are closely related to the producer cell. They may cause the inhibition of spoilage bacteria and food-borne pathogens (Davidson et al. 2005; Rattanachaikunsopon and Phumkhachorn 2010; Lacroix 2011).

It is known that bacteriocins are bactericidal, antibiotic-like substances, apparently protein in nature, which are produced by many bacteria and have a killing action on strains of the same or closely related species (Daw and Falkiner 1996). So that, bacteriocins are often confused in the literature with antibiotics. There is distinction between bacteriocins and antibiotics. This difference between bacteriocins and antibiotics are on the basis of synthesis, mode of action, antimicrobial spectrum, toxicity, and resistance mechanisms (Cleveland et al. 2001). Bacteriocins can be distinguished from peptide antibiotics in that they are ribosomally synthesized rather than secondary metabolites and the genes responsible for production and immunity are generally found clustered in operons (Deegan et al. 2006). Bacteriocins should be safely and effectively used to control the growth of target pathogens in foods (Cleveland et al. 2001).

Bacteriocins can be incorporated directly into fermented foods by using a bacteriocin producer as a starter or adjunct culture. Alternatively, the producer can be used to make a food-grade fermentate, which can be dried to make a powdered ingredient. This powder can be then incorporated into either fermented or non-fermented foods (Deegan et al. 2006).

2.1.1 Bacteriocins from Lactic Acid Bacteria

Bacteriocins are ribosomally synthesized antimicrobial compounds that are produced by many different bacterial species including many members of the LAB. Bacteriocins produced from LAB were called as lantibiotics (Abee et al. 1995). Some bacteriocins produced by LAB, such as nisin, inhibit not only closely related species but are also effective against food-borne pathogens and many other gram-positive spoilage microorganisms. For this reason, bacteriocins have attracted considerable interest for use as natural food preservatives in recent years, which has led to the discovery of an ever increasing potential source of these protein inhibitors (Galvez et al. 2007).

The best studied member of this group is Nisin. Nisin is the only bacteriocin with GRAS (Generally regarded as safe) status for use in specific foods and this was awarded as a result of a history of 25 years of safe use in many European countries and was further supported by the accumulated data indicating its nontoxic, nonallergenic nature. Other bacteriocins without GRAS status will require premarket approval (Suskovic et al. 2010; Balciunas et al. 2013).

An advantage of bacteriocins over classical antibiotics is that digestive enzymes destroy them. Bacteriocin producing strains can be used as part of, or adjuncts to starter cultures for fermented foods in order to improve safety and quality (Suskovic et al. 2010). Bacteriocins have a broad inhibitory spectrum, including many Gram-positive and Gram-negative bacteria. The best known bacteriocin from LAB is nisin, produced by *L. lactis* reported that *L. reuteri* produces the bacteriocin reuterin which affects Gram-negative (*Salmonella* and *Shigella*) and Gram-positive (*Clostridia* and *Listeria*) bacteria (Naidu 2000).

Bacteriocins have been mostly isolated from a variety of LAB. The bacteriocins produced by LAB were classified into four major classes: (1) lantibiotics (<5 kDa), (2) small thermostable peptides (nonlantibiotics) (<10 kDa) (3) high molecular weight (>30 kDa) thermolabile peptides, and (4) large peptides complexed with carbohydrates or lipids (Garneau et al. 2002; Stoyanova et al. 2012; Balciunas et al. 2013).

Classification of bacteriocins is on the basis of their mode of action, activity of spectrum, molecular weight, biochemical properties, and genetic origin (Klaenhammer et al. 1993; Abee et al. 1995; Naidu 2000; Papagianni 2003; Stoyanova et al. 2012)

Class I Lantibiotics, bacteriocins that contain the unusual amino acids lanthionine and 3- methyl-lanthionine (e.g., nisin).

Class II Small, heat stable, non-lantibiotic peptides, subdivided as:

- Pediocin-like, one-peptide bacteriocins with a double-glycine leader peptide and a dedicated secretion and processing machinery. Consist of a relatively well-conserved, hydrophilic N-terminal sequence, containing the distinctive YGNGVXC motif (X means any amino acid residue), and a more diverse, relatively hydrophobic C-terminal half. Display inhibition against *Listeria* (e.g., pediocin PA-1).

- Bacteriocins as in a) but require combination of two polypeptide chains for full activity (e.g., lactococcin G).
- Bacteriocins with a *sec*-dependent signal sequence (e.g., divergicin A).

Class III Large, heat labile protein bacteriocins (e.g., helveticin J).

Class IV Complex bacteriocins with carbohydrate or lipid moiety (not fully accepted).

Bacteriocins of LAB, according to the classification procedure proposed are divided into four classes. The majority of bacteriocins produced by bacteria associated with food belong to classes I and II (O'Sullivan et al. 2002). Some of bacteriocins isolated from lactobacilli are listed in Table 1.

The bacteriocins of LAB are generally ineffective against Gram-negative bacteria due to the possession by such organisms of an outer membrane. Some bacteriocinogenic LAB have also been found to have limited direct inhibitory activity against Gram-negative bacteria. The production of a bacteriocin is a relatively simple inhibitory mechanism whereby the proliferation of one organism is restricted by the generation by another of a proteinaceous inhibitory molecule (Naidu 2000).

Nisin

Nisin is a polypeptide antibacterial substance or bacteriocin of lantibiotic family produced by the fermentation of a suitable substrate by certain strains of *Lactococcus lactis* subsp. *lactis*. The peptide has been employed as a food preservative as it has a high antibacterial activity and a relatively low toxicity for humans. Accumulating evidence shows that nisin is active against Gram-positive bacteria but has little or no effect against Gram-negative bacteria, yeasts, and molds. The insensitivity of Gram-negative bacteria to nisin could be due to the large size (1.8–4.6 kDa) of nisin, which restricts its passage across the outer membrane of Gram-negative bacteria (Guiotto et al. 2003; Arques et al. 2004; Kuwano et al. 2005).

Development as a food preservative began in the 1950s. The first report of nisin used as a food preservative was the use of a nisin producing starter to prevent clostridial spoilage of Swiss cheese successfully used nisin-producing cultures to inhibit the development of clostridial spores in Gruyere cheese, but problems with inhibition of cheese starter cultures hampered such use (Lacroix 2011). Early uses of nisin were for prevention of clostridial spoilage of processed cheese but since then numerous other applications have been identified and its use is now approved in over 50 countries for a variety of applications. The Nisaplin® product is still in use today, but is now manufactured by Danisco who acquired Aplin and Barrett in 1999. Other nisin preparations apart from Nisaplin® are now commercially available: brand names include Chrisin® (Chr. Hansens, Denmark), Delvoplus® (DSM, Holland) and Silver Elephant Nisin made by Zheijiang Silver Elephant Bio-Engineering in China. There are also four or five other smaller manufacturers in China. All these preparations have a similar potency and contain 1,000,000 international units (IU) per gram or approximately 2.5 % nisin. It is the only bacteriocin that has been

Table 1 Bacteriocins from LAB (Suskovic et al. 2010; Rattanachaikunsopon and Phumkhachorn 2010; Ivey et al. 2013)

Producing LAB	Produced bacteriocin	Affected microorganism
Nisin	*L. lactis*	*Listeria monocytogenes*, *Brochothrix thermosphacta*
AcH Pediocin	*Lactobacillus plantarum*	*L. monocytogenes*
Enterocin	*Enterococcus faecium*, *Enterococcus faecalis*	*L. monocytogenes*, *Staphylococcus aureus*
Nisin Z	*Lactococcus lactis lactis*	*S. aureus*
Nisin	*Lactococcus lactis*	Gram-positive bacteria
Lacticin 481	*L. lactis*	Gram-positive bacteria
Lactocin	*Lactobacillus helveticus*	*Lactobacillus acidophilus*, *L. helveticus*
Leucocin B	*Leuconostoc mesenteroides*	LAB
Lactacin B	*L. acidophilus*	*Lactobacillus leichmannii*, *L. helveticus*, *Lactobacillus bulgaricus*, *L. lactis*
Pediocin PA-1/AcH	*Pediococcus acidilactic*	Gram-positive bacteria, *Listeria monocytogenes*
Mundticin	*Enterococcus mundtii*	*Listeria monocytogenes*, *Clostridium botulinum*, LAB
Enterocin A	*Enterococcus faecium*	*Listeria monocytogenes*
Enterolysin A	*Enterococcus faecalis* LMG 2333	Enterococci, pediococci, lactococci, and lactobacilli
Helveticin J	*L. helveticus* 481	*L. bulgaricus*, *L. lactis*
Reuterin	*Lactobacillus reuteri*	Fungi, protozoa, Gram-positive and Gram-negative bacteria
Reutericyclin	*Lactobacillus reuteri*	Gram-positive bacteria
Lactocin S, Sakacin P, Sakadin A	*L. sake*	*Lactobacillus* spp. *Leuconostoc* spp. *Pediococcus* spp., *Carnobacterium* spp., *Carnobacterium piscicola* *Enterococcus* spp. *Listeria monocytogenes*
Bavaricin A	*L. bavaricus*	*Lactobacillus* spp. *Lactococcus* spp. *Pediococcus* spp. *Enterococcus* spp. *Listeria monocytogenes*
Lactacin F	*L. acidophilus*	*Lactobacillus* spp. *Enterococcus faecalis*
Curvacin A	*L. curvatus*	*Lactobacillus* spp. *Carnobacterium* spp. *Listeria monocytogenes*
Helveticin J	*L. helveticus*	*L. helveticus* *L. bulgaricus* *L. lactis*

approved as a food additive in Europe and which has achieved GRAS status in the USA (Naidu 2000; Schillinger et al. 2001; Lacroix 2011).

One type of LAB, and even one strain, can produce bacteriocins of different classes. For example, *Lactococcus lactis* subsp. *lactis* produces the following lantibiotics: nisins A, B, C, L, Z, Q, and F; lacticins 481, and two peptide lantibiotic lacticin 3147. Lactococci produce lactococcin 972 and lacticin QU 5 out of second class bacteriocins. Lactocyclin Q is a recently discovered circular bacteriocin produced by *Lactococcus* sp. QU 12 (Stoyanova et al. 2012).

Two natural variants of nisin are known: nisin A and nisin Z and particular lactococcus producer strains will only make one nisin variant. Nisin is a cationic molecule due to the combination of three lysine residues and one or more histidine residues (depending on the nisin variant) together with a lack of glutamate and aspartate (Naidu 2000).

Nisin A is a polypeptide consisting of 34 amino acids with a molecular weight of 3,510 Da. It is an atypical protein in that it contains unusual amino acids and lanthionine rings. The presence of lanthionine is now known to be characteristic of a larger group of bacteriocins produced by different Gram-positive bacteria and collectively known as "lantibiotics." Various natural nisin variants have been discovered. Only nisin A and Z are used in commercial preparations. Most published scientific information pertains to nisin A. Solubility of nisin A is pH dependent (Lacroix 2011).

As known in literature, nisin does not significantly inhibit Gram-negative bacteria, yeasts and molds. Among Gram-positive bacteria that are sensitive to nisin are members of the mesophilic spore forming genera *Bacillus*, *Alicyclobacillus*, *Clostridium*, *Desulfomaculum*, and the thermophilic spore-forming genera *Geobacillus* and *Themoanaerobacterium* (Lacroix 2011). Nisin, however, is active against a broad spectrum of Gram-positive bacteria including *Listeria monocytogenes* (Schillinger et al. 2001) and gram-negative bacteria such as *E. coli* O 157:H7, *Salmonella* Enteritidis. Nisin also shows activity against many types of LAB. As such bacteria are capable of growth at low pH, nisin can be used as a preservative in low pH foods and beverages that are not heat processed, such as salad dressings, acidified cheese, and alcoholic beverages. The fact that yeasts are insensitive to nisin means that nisin can be used in fermentations alongside yeasts to control the growth of LAB with no effect on the yeast (Guiotto et al. 2003; Kuwano et al. 2005; Lacroix 2011).

Numerous toxicological studies have been carried out and it should be noted that all these have been confined to nisin A preparation. No toxicological study has been carried out with nisin Z or any other nisin variant. The studies carried out with nisin A preparation confirm that nisin A is non toxic at levels much higher than those used in food. In 1969 the FAO/WHO Expert Committee decided from the available evidence that a suitable acceptable daily intake (ADI) was 33,000IU (0.825 mg nisin A)/kg of body weight/day (Lacroix 2011).

Nisin activity in a food matrix may depend on several factors. These include: (1) changes in nisin solubility and charge, (2) binding of nisin to food components (e.g., meat phospholipids), (3) inactivation of nisin (by proteases or food ingredients), and (4) changes in the cell envelope of the target organisms in response to environmental factors. The activity of nisin produced in situ may also be affected by food components (Naidu 2000).

Nisin is suitable for use in a wide range of foods—liquid or solid, canned or packaged, chill or warm ambient storage. Based on target organisms, its usage falls into three broad categories: (1) to prevent spoilage by Gram-positive endospore formers (particularly in heat processed food) (2) to prevent spoilage by LAB and similar organisms such as *Brocothrix thermosphacta*, (3) to kill or inhibit Gram-positive pathogens, e.g., *L. monocytogenes*, *B. cereus*, and *Cl. botulinum* (Naidu 2000).

Nisin serves many purposes in food. It can extend shelf life of both chilled and ambient stored food. Typical examples of this are canned products stored at ambient temperatures in warm climates. Nisin also protects products, particularly chilled foods, from storage temperature abuse (Naidu 2000).

Natamycin

Natamycin, known as pimaracin or tennectin, is a polyene (tetraene) macrolide antimycotic produced by the actinomycete *Streptomyces natalensis* and other closely related *Streptomyces* spp.. Natamycin is active against yeasts and molds and shows no activity against bacteria. Natamycin belongs to a group of antifungals known as polyene macrolides. This antibiotic has strong antifungal activity against various types of fungal pathogens, such as *Candida, Aspergillus, Penicillium,* and *Trichoderma* Natamycin has the chemical formula $C33H47NO13$ and a molecular weight of 665.7 (Hawke 2006; Elsayed et al. 2013). Natamycin is a white/cream-colored crystalline powder with no taste and little odor. It is stable in powder form if stored at room temperature but in aqueous solutions is less stable particularly if exposed to acidic conditions, light, certain oxidants and heavy metals. Natamycin has low solubility in water (approximately 40 μg/ml), but this low solubility is an advantage in the surface treatment of foods because it ensures that the preservative remains on the surface of the food where it is needed, rather than migrating into the foods. Increased solubility occurs with a range of solvents (Lacroix 2011; Jiang et al. 2013).

Natamycin is used at concentrations between 1 and 20 ppm. In general yeasts are more sensitive (minimum inhibitory concentration (MIC) below 5 ppm) than molds (MIC above 10 ppm). Natamycin is permitted as an antimycotic in surface and cheese treatments in 32 countries but its use as a general food additive is more limited. In the European Union, natamycin (designated as preservative E235) is allowed for surface treatment of hard, semi-hard, and semi-soft cheeses as well as dry sausages (Fajardo et al. 2010; Lopez et al. 2012a, 2012b; Kallinteri et al. 2013; Tsiraki and Savvaidis 2014).

Natamycin as fungicide is commonly employed in dairy-based food products to control spoilage by fungi, especially in cheese. It has a broad spectrum of activity against spoilage fungi and is considered to be a very stable product with efficacy against *Aspergillus flavus* and aflatoxin production, *Aspergillus carbonarius* and OTA production in grape juice-based medium, *Aspergillus niger, Aspergillus versicolor, Penicillium chrysogenum, Penicillium glabrum, Penicillium commune, Penicillium verrucosum, Byssochlamys nivea*, and others (Hondrodimou et al. 2011).

The mode of action of natamycin involves an interaction between natamycin and ergosterol, an essential component of membranes of yeasts and molds. Natamycin inhibits yeasts by specifically binding to ergosterol but without permeabilizing the plasma membrane. Therefore, it is active against yeasts and molds but not against bacteria, viruses, and protozoa. Originally it was proposed that this interaction resulted in increased membrane permeability efflux of cellular material. The uses of natamycin as a preservative in foods and beverages are listed as cheese, dry sausage, yoghurt, bakery products, tomato puree/paste, olives, fruit juice, malt beverage, and wine. The main applications are for the surface treatment of cheeses (Lacroix 2011; Resa et al. 2014).

Due to its low toxicity, natamycin is one of the few antibiotics that is still regarded as a GRAS chemical compound by Food and Drug Administration (FDA) guidelines. Therefore, natamycin applications are not limited to medical fields but are also extended to the food sector as a potentially safe preservative agent in the cheese, olive, meat, and fruit processing industries (Elsayed et al. 2013; Resa et al. 2014). Natamycin was last extensively reviewed in 2003 by JECFA who confirmed that the previously established ADI of 0–0.3 mg/kg body weight was satisfactory and that consumption of treated cheese and meats would not exceed this ADI (Lacroix 2011).

Lacticin

Lacticin 3147 is a plasmid-encoded bacteriocin produced by *Lactococcus lactis* subsp. *lactis* DPC3147, a strain isolated from an Irish Kefir grain during a screening of natural sources for food-grade producers of antimicrobial compounds. Lacticin 3147 is a two-component antibiotic (consisting of LtnA1 2 and LtnA2 3), that binds lipid II and exhibits a similar antimicrobial activity spectrum as nisin A (Silkin et al. 2008). These, and other lactococci, are considered GRAS organisms, since they have been isolated from natural food sources, and, more importantly, because lactococci from dairy products have a long history of use and consumption by humans (Lacroix 2011).

An alternative to nisin is the two-peptide lantibiotic lacticin 3147, which has been shown to exhibit antimicrobial activity against a wide range of food pathogenic and food spoilage bacteria in addition to other LAB. With its broad bactericidal action, lacticin 3147 offers considerable potential for improving food safety and quality. In this regard, lacticin 3147- producing strains have previously been used to control non-starter LAB (NSLAB) during cheese ripening and to inhibit growth of *Listeria monocytogenes* in cheeses. Lacticin 3147 has commonly the inhibitory effect on Gram positive (*Listeria, Clostridium, Staphylococcus*, and *Streptococcus* species, methicillin-resistant *Staphylococcus aureus* (MRSA), vancomycin resistant *Enterococcus faecalis* (VRE), penicillin-resistant *Pneumococcus* (PRP), *Propionibacterium acne*, and *Streptococcus mutans*) (Martinez-Cuesta et al. 2010).

Lacticin 481 is a narrow-spectrum lantibiotic bacteriocin produced by strains of *L. lactis*. Also named lactococcin and lactococcin DR, the bacteriocin was first isolated from *L. lactis* subsp. *lactis* CNRZ 481 during a screening for bacteriocin producers. Considering the GRAS status of *L. lactis* strains, lacticin 481 may be regarded as food-grade and therefore the use of ex situ or in situ produced lacticin 481 for food preservation do not pose any legislative issues. Lacticin 481 exerted a bactericidal effect on all species of *Lactococcus*, some lactobacilli, and leuconostoc. Of particular interest is the sensitivity of *Clostridium tyrobutyricum* to lacticin 481, as this spoilage organism is responsible for butyric acid formation and late swelling in Emmental-type cheese. During trials for food preservation, in-situ produced lacticin 481 was also found to control growth of *L. monocytogenes* (Lacroix 2011).

Enterocin

Enterocins are most frequently produced by *Enterococcus faecium* strains, however, other species of enterococci have also been found to produce bacteriocins, e.g., *E. faecalis, E. hirae, E. mundtii, E. durans, E. avium, E. gallinarum, E. Casseliflavus*, and *E.columbae*. Although, the efficiency of enterocins against food spoilage and pathogenic bacteria (*Listeria* sp., *Staphylococcus* sp., *Bacillus* sp., *Clostridium* sp., *Escherichia coli, Salmonella* sp., *Campylobacter* sp.) in various food systems is well demonstrated, only little information is available on the role of bacteriocins in the animal ecosystem. They generally belong to class II bacteriocins and have the potential to inhibit the growth of a narrow range of strains closely related to the producer microorganism; some are also active against Gram-positive food-borne pathogens and spoilage bacteria and Gram-negative species (Sabia et al. 2002).

Enterocin EJ97 is a low-molecular weight cationic peptide with antilisteria activity produced by *E. faecalis* EJ97. This bacteriocin has been purified to homogeneity, and its N-terminal amino-acid sequence indicates that it is different from other bacteriocins with known sequences. The plasmid-borne genetic determinants of enterocin EJ97 and its complete amino-acid sequence have been reported recently. Enterocin EJ97 shows antimicrobial activity against bacteria involved in food spoilage (*B. coagulans, B. stearothermophilus*) and food-poisoning (*Listeria monocytogenes, Staphylococcus aureus*) (Garcia et al. 2003, 2004).

Enterocin AS-48 is a cationic cyclic antimicrobial peptide produced by *E. faecalis* S-48 and *E. faecium*, with high stability in a wide range of temperature and pH values, sensitivity to digestive proteases and broad bactericidal activity against most of the Gram-positive bacteria and some Gram-negative bacteria. They have been ascribed beneficial roles such as producing antilisterial bacteriocins, contributing to ripening and flavor development in some artisanal cheeses, and as probiotics to improve the microbial balance of the intestine and for treating gastroenteritis in humans and animals (Sabia et al. 2002; Ananou et al. 2010a, 2010b; Gomez et al. 2013).

The inhibitory spectrum of AS-48 is remarkably broad, being highly active against most of Gram-positive and some Gram-negative bacteria. Enterocin

AS-48 is stable to pH and heat, sensitive to digestive proteases, non-toxic to eukaryotic cells, and offers a good potential for application in food preservation. Recently, satisfactory results on application of enterocin AS-48 have been reported for dairy products, meat, and fruit juices including pathogenic and/or spoilage bacteria like *Staphylococcus aureus, Listeria monocytogenes, Bacillus cereus*, and *Alicyclobacillus* sp. They have broad bactericidal activity against most Gram-positive bacteria, including several pathogens such as *Listeria monocytogenes, Staphylococcus aureus, Mycobacterium spp., Bacillus cereus,* and some Gram-negative bacteria (Grande et al. 2006; Banos et al. 2012).

Enterocin AS-48 has significant potential as a biopreservative in a large variety of food systems, although its efficacy is noticeably decreased in the food environment compared with laboratory media. This has been attributed to the interaction of AS-48 molecules with food components which may result in a higher retention, or slower diffusion, or irregular distribution of the bacteriocin molecules in the food. Amount of AS-48 (175 AU/g) reduced viable counts of *S. aureus* by only 1.8 log (Lacroix 2011).

Sakacin

Sakacins are class IIa bacteriocins produced by certain strains of *Lactobacillus sakei*. As all bacteriocins of this subclass, they are small, cationic and hydrophobic peptides that contain an N-terminal leader sequence of the double-glycine type, and possess strong antilisterial activity (Lacroix 2011).

Sakacin C2 is a novel bacteriocin with a broad inhibitory spectrum secreted by *Lactobacillus sake* C2 isolated from Chinese traditional fermented cabbage. Sakacin C2 displayed the activity against some food-borne spoilage and pathogenic bacteria including not only Gram-positive bacteria such as *Staphylococcus aureus, Listeria innocua, Streptococcus thermophilus,* and *Bacillus cereus* but also Gram-negative bacteria such as *E. coli* and *Salmonella typhimurium*. Bacteriocins produced by LAB generally have inhibitory activity against Gram-positive bacteria. Sakacin C2 has inhibitory activity against not only many Gram-positive bacteria but also many Gram-negative bacteria (Aasen et al. 2003; Trinetta et al. 2008; Gao et al. 2011, 2013)

Sakacin P produced by *Lactobacillus sakei* is very potent against food pathogen (*L. monocytogenes* and *E. faecalis*) and spoilage (*Carnobacterium*) species. The mode of action of this and other pediocin-like bacteriocins has not been fully elucidated to date. Because of the combination of high anti-listerial activity and a narrow inhibitory spectrum, sakacin P is one of the most promising bacteriocins for preservation of foods in which contamination with listeria is a problem (Katla et al. 2001; Lacroix 2011)

Lactobacillus sakei CTC494, producing sakacin K, a pediocin-like bacteriocin is a competitive starter culture in sausage fermentation being specially active in inhibiting *L.monocytogenes* in in situ studies. The inhibitory effect of *Lact. sakei* CTC494 against *L. monocytogenes* in fermented sausages is influenced by the ingredients used in the formulation of sausages (Hugas et al. 2002).

Reuterin

Reuterin is an antimicrobial compound consisting of hydrated, non-hydrated, and dimeric forms of 3-hydroxypropionaldehyde (3-HPA). Reuterin (h-hydroxypropionaldehyde) is a broad spectrum antimicrobial compound produced by some strains of *Lactobacillus reuteri* during anaerobic fermentation of glycerol. The ability to produce reuterin is only reported for strains of *L. reuteri* isolated from human and animal intestine. Reuterin is soluble in water, resistant to heat, pH, proteolytic and lipolytic enzymes, and stable over a wide range of pH values (Arques et al. 2004, Arques et al. 2011; Langa et al. 2012; Stevens et al. 2013).

The use of reuterin to control Gram-positive and Gram-negative pathogens has been investigated in milk and dairy products and in meat products. Reuterin has been proposed as a potential food additive to prevent the growth of pathogenic and spoilage microorganisms. Reuterin control food-borne pathogens such *Salmonella* sp., *Escherichia coli* O157:H7, *Listeria monocytogenes* and *Staphylococcus aureus* has been investigated in milk and dairy products (Langa et al. 2013). Since reuterin is also active against *Escherichia coli*, it is a potential biopreservative agent against *E. coli* O157:H7, which is an increasing problem in the food industry (Rasch 2002). Reuterin exhibits a broad spectrum of antimicrobial activity against certain Gram-positive and Gram-negative bacteria, yeast, fungi, and protozoa (Suskovic et al. 2010). Spoilage organisms sensitive to reuterin include species of *Salmonella, Shigella, Clostridium, Staphylococcus, Listeria, Candida*, and *Trypanosoma* (Davidson et al. 2005).

Reuterin has been shown to inhibit ribonucleotide reductase activity in vitro, and since ribonucleotide reductase is catalyzing the first step in DNA synthesis, this is likely to be the reason for the broad antimicrobial spectrum. The possible use of reuterin for biopreservation has been investigated in meat milk and cottage cheese (Rasch 2002).

The reuterin producer organism *L. reuteri* is considered GRAS and this GRAS status simplifies the acceptation of reuterin as a food preservative by the authorities. Therefore, in the USA the use of *L. reuteri* cells and glycerol to produce reuterin in situ should be acceptable according to the Code of Federal Regulations (U.S.G.P. Office 1990), although the FDA may require justification of the GRAS status (Lacroix 2011).

Finally, the first patent about the use of the dimer form of reuterin as an antibiotic compound was filed in 1988 and legislation in most countries regulates patent-protection for only a limited time, free use of reuterin can be possible within the near future (Lacroix 2011).

Pediocin

Pediocin PA-1 is a representative 44 amino acid class IIa bacteriocin which is small and heat-stable and has a consensus motif, –YGNGV–, especially (Moon et al. 2006; Renye et al. 2011). Pediocin PA-1 produced by *Pediococcus acidilactici* PAC1.0 is a bacteriocin encoded by the *pedABCD* gene operon located on the 9.4 kb plasmid pSRQ11. Pediocin PA-1 shows a particularly strong activity against pathogenic *Listeria monocytogenes*. This pathogen is a major concern to food industries since its presence in contaminated dairy (cheese, ice cream) and meat (sausages) products may be responsible for the outbreak of foodborne illnesses (Beaulieu et al. 2005). Pediocin is likely to have higher activity and acts more specifically against *L. monocytogenes* than nisin (Woraprayote et al. 2013).

Pediocin PA-1 producing cultures and pediocin PA-1 containing fermentates have found numerous applications in food industry either to control the microbial succession during fermentation, or to inhibit the growth of spoilage microorganisms during storage (Diez et al. 2012).

A number of bacteriocins have been described for pediococci, viz. pediocin A, produced by different strains of *Pediococcus pentosaceus* isolated from cucumber brine; pediocin AcH (also referred to as pediocin H), pediocin PA-1, pediocin JD1 and pediocin SJ-1, produced by *Pediococcus acidilactici* isolated from fermented meat; and pediocin PD-1, produced by *Pediococcus damnosus* NCFB 1832, originally isolated from spoiled beer. Pediocins have also been reported for *P. pentosaceus* strains isolated from pepperoni and strains of *P. acidilactici* isolated from human clinical sources (Todorov and Dicks 2005).

Compared to other bacteriocins in the pediocin-family and to some lantibiotics, pediocin A of *Ped. pentosaceus* FBB61 and pediocins from *Ped. acidilactici* strains have a relatively wider bactericidal spectrum against Gram-positive bacteria. Pediocin A inhibits growth of several strains of *P. pentosaceus, P. acidilactici, S. aureus, L. lactis, Lactobacillus* spp., *Cl. botulinum, Cl. Perfringens*, and *Cl. sporogenes*. It has already been mentioned that pediocin A from *P. pentosaceus* FBB61 and pediocin PA-1/AcH from *P. acidilactici* strains as well as from *Ped. parvulus* strains and *Lab. plantarum* WHE92 are bactericidal against many normal as well as injured foodborne spoilage and pathogenic bacteria. Thus, these bacteriocins can be used to control these bacteria in foods (Naidu 2000).

2.1.2 Bacteriocin from Other Microorganisms

Although several bacteria produce bacteriocins, those of LAB have been extensively studied with the perspective of use as natural biopreservatives for the food industry. In addition to LAB bacteriocins, other bacteria even pathogen bacteria such as *Bacillus* spp., *Staphylococcus* spp., and *Listeria* spp. produce bacteriocin. *Bacillus* spp., *Staphylococcus* spp., and *Listeria* spp. are interesting genus to investigate for antimicrobial activity (Deegan et al. 2006; Balciunas et al. 2013)

Non-LAB bacteriocin production is also common among many Gram-positive bacteria especially soil bacteria of the genus *Bacillus*. Like LAB, the genus *Bacillus* includes a variety of industrially species which has been GRAS status by FDA, USA. *Bacillus* spp. produce a large number of peptide antibiotics, include a variety of industrially important species, and has a history of safe use in both food and industry. The production of bacteriocins or bacteriocin-like substances has been already described for many *Bacillus* species such as *Bacillus cereus, B. subtilis, B. megate-*

rium, *B. coagulans*, and *B. thuringiensis*. The best studied are subtilin and coagulin (Olivera et al. 2004; He et al. 2006; Cherif et al. 2008; Kaewklom et al. 2013; Compaoré et al. 2013). Staphylococcins are bacteriocins produced by strains belonging to the genus *Staphylococcus*. Most are either lantibiotics (Pep5, epicidin 280, epidermin, epilancin K7, nukacin ISK-1, and staphylococcin C55/BacR1) or class II bacteriocins (aureocin A70 and aureocin A53). Only one staphylococcin belonging to class III, lysostaphin, has been described thus far (Nascimento et al. 2005).

Staphylococcus aureus, an important pathogen with an alarming proportion of multi-resistant isolates, is another prominent producer of bacteriocins such as the two-peptide lantibiotic staphylococcin C55 and the closely related staphylococcin BacR1, independently isolated and characterized. It have been recently identified a new bacteriocin, aureocin A70, produced by an *S. aureus* strain isolated from pasteurized milk. This highly unusual bacteriocin possessed anti-microbial activity resulting from four unmodified peptides, AurA, AurB, AurC, and AurD, a hitherto unique composition. In previous work, it was identified another broad-spectrum bacteriocin, aureocin A53, from *S. aureus* A53, a strain also isolated from commercial milk. This bacteriocin shows bactericidal activity against a broad range of LAB, *L. monocytogenes* and many epidemiologically unrelated strains of *S. aureus* involved in bovine mastitis (Netz et al. 2001, 2002; Nascimento et al. 2002; Ceotto et al. 2009).

Bacteriocin-producer strains of *S. aureus* have been reported in the literature and in most cases this phenotype was associated with plasmids. Bac plasmids found in strains of *S. aureus* could be divided into two major groups: plasmids larger than 40 kb and small plasmids, ranging in size from 8.0 to 10.4 kb. These small Bac plasmids, whose prototypes are pRJ6 and pRJ9 (with 8.0 and 10.4 kb, respectively), carry genes involved in production of bacteriocins and immunity to bacteriocin action. pRJ6 codes for aureocin A70 and pRJ9 encodes aureocin A53. Analyses of the spectrum of action have shown that both aureocins have a broad spectrum of activity, inhibiting different species of bacteria, including important human and animal pathogens (Netz et al. 2001, 2002; Nascimento et al. 2002; Ceotto et al. 2009).

Variacin, a lanthionine-containing non-LAB bacteriocin produced by *Kocuria varians*. Variacin has been used to produce a milk-based fermented ingredient and was used to evaluate its effectiveness against *B. cereus* in chilled dairy products, vanilla, and chocolate desserts. (O'Mahony et al. 2001; Galvez et al. 2007, 2010; Settanni and Corsetti 2008; Balciunas et al. 2013).

Variacin is resistant to heat (100 °C for 15, 30, and 45 min) and pH values from 2 to 10. Total aminoacid composition of variacin was determined by means of peptide sequencing and mass spectrometry analysis of the FPLC active fractions, returning a peptide molecular weight of 2,658.61 Da. Variacin exhibited a wide spectrum of activity inhibiting all Gram-positive bacteria tested, including pathogenic and spoilage organisms such as listeriae, staphylococci, and the vegetative cells and spores of clostridia and bacilli. Common to most known bacteriocins produced by Gram positives, variacin did not exhibit inhibitory activity against Gram-negative bacteria (Lacroix 2011).

As some bacteriocins have a rather narrow inhibitory spectrum directed towards closely related species, *L. innocua* may produce an inhibitory factor such as bacteriocin inhibiting the growth of *L. monocytogenes* (Mollerach et al. 1988; Kalmokoff et al. 2001). It was shown bacteriocin from other microorganisms and inhibition effects in Table 2.

2.2 Organic Acids and Others

The antimicrobial effects of these acids are attributed to the reduction of pH to a level below the range of growth and metabolic inhibition by non-dissociated organic acid molecules (Crowley et al. 2013).

Fermentation by microorganisms is characterized by the accumulation of organic acids and the accompanying reduction in pH. The levels and types of organic acids produced during the fermentation process depend on the species of organisms, culture composition, and growth conditions. The antimicrobial effect of organic acids lies in the reduction of pH, as well as the undissociated form of the molecules. It has been proposed that the low external pH causes acidification of the cell cytoplasm, while the undissociated acid, being lipophilic, can diffuse passively across the membrane. The undissociated acid acts by collapsing the electrochemical proton gradient, or by altering the cell membrane permeability which results in disruption of substrate transport systems (Davidson et al. 2005).

Fermentation involving mostly LAB results in accumulation of organic acids, primarily lactic acid, as a major end-product of carbohydrate metabolism, generated from pyruvate by lactic acid dehydrogenase. The antimicrobial effect of LAB is mainly related to the production of lactic- and acetic acids, as well as propionic-, sorbic-, benzoic-acids, hydrogen peroxide, diacetyl, ethanol, phenolic- and proteinaceous compounds (Cizeikiene et al. 2013).

2.2.1 Lactic acid and Acetic Acid

Lactic and acetic acid have been used individually as mold inhibitors, both hold a GRAS status; their maximum concentration is given by sensorial parameters, they are present in fermented foods and are also easy to obtain commercially. Their inhibitory effect is based on the “weak acid preservative theory” and previous publications have reported synergistic antimicrobial effects between acetic and lactic acid (Peláez et al. 2012).

Weak organic acids have been widely used in food to maintain microbial stability of low pH products and among the most commonly used acidulants are acetic and lactic acid . Extensive research has been carried out to characterize andelucidate the inhibitive activities of these two acids on the growth of *Z. bailii*. According to previous result, acetic acid is more lethal to microorganisms than lactic acid. Nevertheless,

Table 2 Bacteriocin from other microorganisms and inhibition effects

Other microorganisms as bacteriocin producer	Produced bacteriocin	Inhibition effect	References
Bacillus licheniformis	Bacteriocin-like peptides	Some pathogenic and spoilage microorganisms such as *S. aureus*, *M. flavus*, and some plant pathogens	Olivera et al.(2004); He et al. (2006))
Bacillus amyloliquefaciens	Amysin	*L. monocytogenes*, *Salmonella* sp., and *Shigella*	Kaewklom et al. (2013)
Bacillus thuringiensis subsp. Entomocidus HD110	Entomocin 110	*Bacillus thuringiensis, Bacillus cereus, Bacillus mycoides, Bacillus pseudomycoides,Bacillus weihenstephanensis, Bacillus coagulans, Bacillus megaterium, Paenibacillus alvei, Paenibacillus polymyxa, Paenibacillus larvae, Listeria monocytogenes, Lactococcus lactis*	Cherif et al. (2008)
Bacillus subtilis subsp. *subtilis* H4	Subtilosin and subtilin	*B. cereus, Salmonella spp., L.monocytogenes, Y. enterocolitica, M. luteus, and S. aureus*	Compaoré et al. (2013)
Bacillus licheniformis AnBa9	Antibacterial peptide	*Staphylococcus aureus GCS1, Bacillus cereus GCS2, Staphylococcus epidermidis GCS4, Kurthia gibsonii GCS6, Micrococcus luteus GCS7, Streptococcus mitis GCS9, Bacillus subtilis B-4219, L. lactis B-1821, Staphylococcus epidermidis B-4268, Bacilus smithii NRS-173, Lactobacillus acidophilus B-4495, Micrococcus luteus B-287, Pediococcus acidilactici, B-14958, and Leuconostoc mesenteriodes B-1118*	Anthony et al. (2009)
Bacillus subtilis subsp. *subtilis*	Surfactin	*Listeria monocytogenes*	Sabaté and Audisio (2013)
Bacillus licheniformis 26 L-10/3RA	Lichenin	*Streptococcus bovis* SB3	Pattnaik et al. (2005)
Staphylococcus aureus	Aureocin A70, A53	*Bacillus cereus, Listeria monocytogenes,* and *Staphylococcus* spp.	Netz et al. (2002), Nascimento et al. (2002), Netz et al. (2001),Ceotto et al. (2009)
coagulase-negative Staphylococcus (CNS	Aureocin A70	*Listeria monocytogenes, Streptococcus agalactiae*(streptococcal mastitis)	Nascimento et al. (2005)
Staphylococcus hyicus 3682	Hyicin 3682	*Bacillus cereus, Listeria monocytogenes,* and *Staphylococcus aureus*	Fagundes et al. (2011)
Staphylococcus simulans 3299	Nukacin 3299	in bovine mastitis: staphylococci and streptococci	Ceotto et al. (2010)
Listeria innocua	Bacteriocin-like-substance	*Listeria monocytogenes*	Yokoyama et al. (1998)
Listeria innocua 743	New plasmid-encodedsec-dependent bacteriocin	*L. monocytogenes*	Kalmokoff et al. (2001)
Listeria innocua	Linnocuicina 819	–	Mollerach et al. (1988)
Bacillus licheniformis ZJU12	Bacteriocin-like peptides	*Bacillius subtilis, Enterococcus faecium, Micrococcus flavus, Staphylococcus aureus, Staphylococcus epidermidis, Xanthomonas oryzae pv.oryzae* Zhe 173, *Alternaria brassicae, Fusarium oxysporum, Guignardia* sp., *Pyricularia grisea, Rhizoctonia solani*	He et al. (2006)
Leuconostoc mesenteroides E131	Mesenterocin E131	*L. monocytogenes*	Xiraphi et al. (2008)
Ruminococcus gnavus E1	Ruminococcin C	*Clostridium perfringens*	Crost et al. (2011)

due to its pungent odour and taste, the level of acetic acid in foods is limited. Conversely, lactic acid has a very smooth and mild taste which allows it to be applied at substantial amounts in fermented meat, dairy products, sauces, pickled vegetables and salad dressings, etc. Nowadays, in order to meet the increasing consumers' demand for less "acid" products and to improve the microbial stability of their products, food manufacturers have used acetic–lactic acid mixtures as substitutes for high acetic acid concentrations. The use of weak organic acids, in particular acetic and lactic acid, as antimicrobial agents is quite common in the food industry, especially to maintain the microbial stability of low pH products. Moreover, the synergistic inhibitive effect of these two acids on yeasts has been reported (Dang et al. 2009).

Lactic acid is the major metabolite of LAB fermentation where it is in equilibrium with its undissociated and dissociated forms, and the extent of the dissociation depends on pH (Cizeikiene et al. 2013).

Lactic acid is also found in a wide array of microorganisms. Several bacterial species produce large quantities of lactic acid, among which the best known are *Lactobacillus, Sporolactobacillus, Enterococcus, Lactococcus, Bacillus, Streptococcus, Pediococcus, Leuconostoc*, and bifidobacteria (Naidu 2000).

Lactic acid elicits antimicrobial effects on a variety of microorganisms (*Mycobacterium tuberculosis, Bacillus coagulans, Yersinia enterocolitica, Aeromonas hydrophila, Cl. botulinum* A, B, & E, *Clostridium botulinum, Enterobacteriaceae, Lactobacillaceae, Asperigillus* sp., *Pseudomonas fragi, Clostridium sporogenes, Listeria monocytogenes, Vibrio vulnificus, Helicobacter pylori, Escherichia coli O157:H7, Pseudomonas* sp., *Salmonella typhimurium*) (Naidu 2000).

This antimicrobial activity is based on the ability of lactic acid to reduce pH of the milieu. Thus, in fermented foods, lactic acid in combination with other antimicrobial factors such as bacteriocins produced by LAB inhibits the competing microorganisms (Naidu 2000).

Finally, lactic acid and its derivatives elicit a broad spectrum of antimicrobial activity against Gram-positive bacteria including spore-forming *Clostridia* and *Bacillus* sp., as well as Gram-negative pathogens such as *E.coli* O157:H7 and *Salmonella* sp. Lactate compounds also demonstrate antifungal activity against aflatoxin-producing *Aspergillus* sp. Considering these attributes, lactic acid and lactates are widely used in the food industry for decontamination of meat foods such as beef, poultry, and pork during processing and packaging. They are also used in shelf life enhancement of fresh and semi-processed foods. Consumer acceptance of such processed foods, however, is largely based on visual and organoleptic qualities. Currently, the worldwide utility of lactic acid and its derivatives amounts to 100,000 metric tons per year (Naidu 2000).

At low pH, a large amount of lactic acid is in the undissociated form, and it is toxic to many bacteria, fungi and yeasts. However, different microorganisms vary considerably in their sensitivity to lactic acid. At pH 5.0 lactic acid was inhibitory towards spore-forming bacteria but was ineffective against yeasts and molds. It was possible to grow *Aspergillus parasiticus* NRRL 2999 in a medium containing 0.5 or 0.75 % lactic acid at pH 3.5 or 4.5. Lindgren and Dobrogosz (1990) showed that at different pH ranges the MIC of the undissociated lactic acid was different against

Clostridium tyrobutyricum, *Enterobacter* sp., and *Propionibacterium freudenreichii* ssp. *shermanii*. In addition, the stereoisomers of lactic acid also differ in antimicrobial activity, L-lactic acid being more inhibitory than the D-isomer (Yang 2000).

For lactic acid, the two main antimicrobial mechanisms are (a) the capacity of lactic acid to reduce pH of the medium and the lipophilic characteristic of its protonized (undissociated) form, which facilitates its penetration through the cell wall and (b) the effect of lactates (mainly sodium lactate) on water activity in foods with intermediate humidity (Lopez et al. 2012b).

Acetic acid, historically diluted in the form of vinegar, has been utilized perhaps longer than any other preservative for its antimicrobial effect that influences food keeping-quality, wholesomeness, and safety. Because acetic acid is the predominant flavoring and antimicrobial component in vinegars, studies about acetic acid will focus on the importance of this acid as a direct food additive or more recently as a food processing aid, to decontaminate foods prior to distribution and consumption. When acetic acid is dissolved in solution, it dissociates to release free protons, which decrease solution pH. The increased number of protons on the outer surfaces of microorganisms can disrupt membrane function by denaturing enzymes and by altering permeability leading to membrane destabilization. Undissociated acetic acid can also traverse the lipid bilayer of bacteria and yeasts and release protons into the cytoplasm (Naidu 2000).

Acetic acid caused inhibition by lowering the cytoplasmic pH to an inhibitory level, by direct tests of cytoplasmic acidification (pHi), when applied acetic acid to *A. niger* conidia. Most microorganisms are susceptible to antimicrobial effects in the presence of organic acids and such inhibition increases by lowering pH of the media. *A. flavus* is more sensitive to acetic acid than lactic acid. (Peláez et al. 2012).

Acetic acid is a monocarboxylic acid with a pungent odor and taste, which limits its use in foods. It is the principal component of vinegars and as such is primarily used for its flavoring abilities. Acetic acid is GRAS for miscellaneous and general-purpose usage (21 CFR 184.1005). Acetic acid has an antimicrobial effect on many microorganisms such as *Salmonella typhimurium*, *Enterobacteriaceae*, *Escherichia coli*, *Salmonella bareilly*, and *Listeria monocytogenes*. Many studies have demonstrated the ability of acetic acid to inhibit a variety of microorganisms. Acetic acid is more lethal than other organic acids such as lactic acid, formic acid, citric acid, and sulfuric acids on some microorganisms (*Helicobacter pylori, Salmonella* spp., *Yersinia enterocolitica, E. coli* O157:H7, *Bacillus cereus, B. subtilis, B. licheniformis, L. monocytogenes, Campylobacter jejuni, Staphylococcus aureus, Salmonella enteritidis, and Aspergillus parasiticus* (Naidu 2000; Lopez et al. 2012a,b).

Several reports probed the acetic acid efficiency to decontaminate meat and meat products (beef, pork, and poultry. Several derivatives of acetic acid are currently in use as antimicrobial agents. The salt form, however, requires different handling and utilization procedures than the acid. The sodium and calcium salts are sometimes used in foods and would be expected to have the same antimicrobial properties as acetic acid at the same pH values used a dip of sodium acetate in combination with potassium sorbate (10 %) and phosphates (10 %) to extend the shelf life of pork chops. Sodium diacetate reduced the pH of the meat slightly. A surface application of 1–3 mg/cm^2 (2–6 g/chicken) sodium diacetate powder extended the shelf life of chickens about 4 days when held at 2 °C (Lopez et al. 2012a, 2012b)

As a result, Acetic acid and its GRAS salts have a long history of functional use in foods. The ability to lower pH with acetic acid or vinegar provides pickled foods with unsurpassed keeping quality, flavor, and safety (Naidu 2000).

2.2.2 Citric Acid

There are two organisms that are widely used in the production of citric acid, i.e., yeast and *Aspergillus niger*. Many other fungi are also able to produce citric acid such as *A. clavatus, A fumigatus, A. citrinum* and *Candida* species. Yeast has certain advantages over *Aspergillus* due to its independence with trace metals and higher fermentation rates. It also has the potential for higher sugar concentrations using osmophilic variants. Citric acid, like many other organic acids, elicits antimicrobial activity by various mechanisms. The antimicrobial outcome is influenced by various factors including pH, concentration, temperature, chain length, etc. In addition to the antimicrobial property, the choice of the acidulant may depend upon secondary activity. Although citric acid is not used solely for its antimicrobial activity, it is shown to exhibit certain antimicrobial properties against molds and bacteria. In 1950, Murdock reported that some of the organisms isolated from tomato juice were inhibited by citric acid. Citric acid was more inhibitory to thermophilic bacteria than acetic and lactic acids. It was reported that citric acid has some antimicrobial effect on *Salmonella* spp., *Listeria monocytogenes*, and *Staphylococcus aureus* (Naidu 2000).

Combinations of alkaline-electrolyzed water and citric acid showed a strong synergistic antimicrobial effect that reduced background flora and food-borne pathogens on fresh-cut produce and cereal grains (Rahman et al. 2011)

Currently, ASC (acidified sodium chlorite) is commercially supplied as a kit containing citric acid (CA) and SC (sodium chlorite). These chemicals when combined produce active chlorine dioxide (ClO_2), which is more soluble than sodium hypochlorite (NaOCl) in water and has about 2.5 times greater oxidizing capacity than hypochlorous acid (HOCl). A number of reports have described the strong efficacy of ASC in the FDA approved application concentration range of 0.5–1.2 g/L on inactivation of pathogens, including *E. coli* O157:H7 and *Salmonella* spp.. However, a negative impact on organoleptic quality of red meat and shredded carrots occurred when ASC was used within the approved concentration range. Therefore, it is critical to find the concentration of ASC that will optimize microbial safety while maintaining quality of fresh-cut cilantro (Allende et al. 2009)

In previous studies, it was reported the antimicrobial activity of citric acid against *E. faecalis* (Krause et al. 2007; Moliz et al. 2008). In the study of Eswaranandam et al. (2004) Lactic acid-incorporated films showed inhibition of *E. coli* O157: H7 with (7.0 log number CFU/mL) and without nisin (7.3 log number CFU/mL).

In our study, citric acid-incorporated films and tartaric acid-incorporated films showed less killing effect against *L. monocytogenes, S. gaminara*, and *E. coli* O157:H7. Citric acid was mainly tested for antimicrobial activity of organic acids and reported to have less antimicrobial activity.

Sodium citrate could elicit bacteriostatic activity. The antimicrobial activity of citrates has been attributed to its ability to chelate metal ions. *Lactobacillus casei* is inhibited by sodium citrate induced metal-chelation at 12–18 μM concentrations. Chelation effect also appears to be an influencing factor in the growth-inhibition of *Staphylococcus aureus* and *Arthrobacter pascens*. A concentration of 0.75 % citric acid was sufficient to reduce inoculated populations of *Salmonella typhimurium, Yersinia enterocolitica, Escherichia coli*, and *S. aureus*. Citric acid solution was found to exhibit antibacterial effects on the bacteria used (Naidu 2000).

2.2.3 Diacetyl

Diacetyl is a product of citrate metabolism and is responsible for the aroma and flavor of butter and some other fermented milk products. Diacetyl is produced by strains within all genera of LAB by citrate fermentation. Many LAB including strains of *Leuconostoc*, *Lactococcus*, *Pediococcus*, and *Lac*tobacillus may produce diacetyl although production is repressed by the fermentation of hexoses. The antimicrobial effect of diacetyl has been known since the 1930s. It inhibits the growth of Gram-negative bacteria by reacting with the arginine-binding protein, thus affecting the arginine utilization. Gram-negative bacteria, yeasts, and molds are more sensitive to diacetyl than gram-positive bacteria and its mode of action is believed to be due to interference with the utilization of arginine. Diacetyl is rarely present in food fermentations at sufficient levels to make a major contribution to antibacterial activity (Caplice and Fitzgerald 1999; Yang 2000).

Among the substances of microbial origin, diacetyl seems to be a good candidate due to its selective antimicrobial activity and absence of toxicity against humans and animals. The antimicrobial activity of diacetyl has been documented since 1927. This author screened several microbial species for sensitivity to diacetyl and concluded that this molecule is characterized by a broad antimicrobial activity at concentrations ranging between 200 and 1,000 ppm. Gram-positive bacteria were more resistant, while Gram negative and yeasts exhibited a higher sensitivity to this molecule. More recently, An exposure to 50 ppm of diacetyl for 24 h was sufficient to exhibit a lethal effect on *Vibrio vulnificus*. The addition of same concentration was able to strongly inhibit *Escherichia coli* O157:H7 and *Salmonella typhimurium* in the presence of a starter culture in a laboratory medium. These latter authors, in light of their results, proposed the use of diacetyl as food ingredient during meat fermentation to control the growth of *S. typhimurium* and *E. coli* O157:H7 without compromising effects on the growth and acid production by *Pediococcus acidilactici* (Lancitti et al. 2002).

Gram-negative bacteria were more sensitive to diacetyl than Gram-positive bacteria; the former was inhibited by diacetyl at 200 μg/mL and the latter at 300 μg/mL.

Diacetyl at 344 μg/mL inhibited strains of *Listeria*, *Salmonella*, *Yersinia*, *Escherichia coli*, and *Aeromonas*. Since the production of diacetyl during lactic fermentation is low, e.g., 4 μg/mL produced by *Lc. lactis* ssp. *diacetylactis*, and the acceptable sensory levels of diacetyl are at 2–7 μg/mL, its practical use as a food preservative is limited. However, diacetyl may act synergistically with other antimicrobial factors and contribute to combined preservation systems in fermented foods (Yang 2000).

Diacetyl (2,3-butanedione) is an end product of pyruvate metabolism by citrate fermenting LAB. Diacetyl elicits a potent antimicrobial activity against various food-borne pathogens and spoilage microorganisms. Diacetyl is more effective against Gram-negative bacteria, yeasts, and molds, than against Gram-positive organisms. Diacetyl interferes with arginine utilization by reacting with the arginine-binding protein of Gram-negative bacteria. High concentration of diacetyl (0.4 mg/ml) is required for antimicrobial effect against most organisms. However, *E. coli* is susceptible to diacetyl at extremely low concentration. Dose-dependent inhibition experiments established that concentrations of 0.2 and 0.3 mg/ml are required for eliciting antimicrobial activity against yeasts/Gram-negative bacteria, and non-lactic Gram-positive bacteria, respectively. It was reported bactericidal effect for diacetyl (344 ppm) against strains of *Yersinia enterocolitica, Aeromonas hydrophila, E. coli,* and *Salmonella anatum*, but not against *Listeria* (Naidu 2000).

It is the component responsible for the characteristic aroma in butter. It is formed during transformation of citrate via pyruvate. The maximum formation of diacetyl is observed under slightly acidic pH. It is mainly active against Gram-negative bacteria, belonging to the genera *Salmonella*, *Yersinia*, *Escherichia*, and *Aeromonas*, and also against Gram-positive bacteria belonging to the genus *Bacillus* (Yang 2000).

2.2.4 Acetaldehyde

Acetaldehyde is a reactive, low molecular weight, flavor active compound found in a variety of foods and beverages such as cheese, yogurt, beer, and wine. Acetaldehyde formed during carbohydrate metabolism of heterofermentative LAB is reduced to ethanol by re-oxidation of pyridine nucleotides, catalyzed by an NAD dependent alcohol dehydrogenase. Acetaldehyde imparts the typical aroma of yogurt. Acetaldehyde is produced by *L. delbrueckii* ssp. *bulgaricus* by the action of a threonine aldolase, which cleaves threonine into acetaldehyde and glycine. Since *L. delbrueckii* ssp. *Bulgaricus,* and *S. thermophilus* in yoghurt cannot metabolize acetaldehyde, it accumulates in the product at a concentration of about 25 ppm. Acetaldehyde at 10–100 ppm inhibits the growth of *Staphylococcus aureus, Salmonella typhimurium,* and *E. coli* in dairy products (Yang 2000).

The mechanism of the acetaldehyde effect is unclear, but has been suggested to derive from the ability of added acetaldehyde to replace intracellular acetaldehyde lost from the cell when the permeability of the plasma membrane is disturbed by ethanol (Barber et al. 2002).

Pure acetaldehyde and benzaldehyde were shown to be inhibitory to *L. monocytogenes* and *S. typhimurium.* there are indications that *P. roqueforti* might produce metabolites inhibitory to, e.g., *L. monocytogenes*. Several metabolites such as aldehydes, alcohols, and organic acid in small amounts may all contribute to the inhibitory activity. Acetaldehyde and benzaldehyde may still been important part of the inhibitory activity detected as these compounds are known to be inhibitory to both fungal and bacterial strains. Acetaldehyde may account for some of the inhibition observed but more research is required to elucidate this (Larsen and Knochel 1997).

2.2.5 Hydrogen Peroxide

In the presence of oxygen, LAB produces hydrogen peroxide (H_2O_2) through electron transport via flavin enzymes. In the presence of H_2O_2, superoxide anions form destructive hydroxy radicals. This process may lead to peroxidation of membrane lipids and increased membrane permeability. The resulting bactericidal effect of these oxygen metabolites has been attributed to their strong oxidizing effect on the bacterial cell as well as destruction of nucleic acids and cell proteins. In addition, H_2O_2 could react with other cellular and milieu components to form additional inhibitory substances (Naidu 2000).

H_2O_2 formation by LAB and its effect on various microorganisms is well documented for years.

Fitzsimmons and Berry examined the inhibitory effect of H_2O_2 producing lactobacilli (LB+) on *Candida albicans*. A range of *Lactobacillus acidophilus* strains isolated from patients using oral, vaginal, and endocervical swabs were investigated for their ability to (a) inhibit the growth of *C. albicans*, and (b) generate peroxidase, H_2O_2, and hypothiocyanite. Inhibition of *C. albicans* and H_2O_2 production was detected in nine out of twelve strains, whereas peroxidase production was only detected in three out of twelve strains, all from oral swabs. Hypothiocyanite production was detected in two strains and it was only detected in these strains after growth in MRS medium in aerobic conditions (Naidu 2000).

Hydrogen peroxide is produced by LAB in the presence of oxygen as a result of the action of flavoprotein oxidases or nicotinamide adenine hydroxy dinucleotide (NADH) peroxidase. The antimicrobial effect of H_2O_2 may result from the oxidation of sulfhydryl groups causing denaturing of a number of enzymes, and from the peroxidation of membrane lipids thus the increased membrane permeability. H_2O_2 may also be as a precursor for the production of bactericidal free radicals such as superoxide (O2-) and hydroxyl (OH.) radicals which can damage DNA. It has been reported that the production of H_2O_2 by *Lactobacillus* and *Lactococcus* strains inhibited *Staphylococcus aureus*, *Pseudomonas* sp., and various psychotrophic microorganisms in foods. In raw milk, H_2O_2 activates the lactoperoxidase system, producing hypothiocyanate (OSCN-), higher oxyacids (O_2SCN- and O_3SCN-), and intermediate oxidation products that are inhibitory to a wide spectrum of Gram-positive and Gram-negative bacteria (Yang 2000).

In the presence of oxygen, LAB can also produce H_2O_2 when subjected to NADH oxidase and superoxide dismutase activities. When heme is absent from the environment, LAB does not produce catalase; this results in peroxide accumulation. The peroxide effect can be amplified in the presence of lactoperoxidase and thiocyanate, which are present in natural LAB habitats, such as milk. The antimicrobial activity of hydrogen peroxide is linked to the strong oxidizing effect. The observed growth inhibition of *Lactococcus* and Gram-negative *Pseudomonas* spp., responsible for food contamination, was due to the accumulation of peroxides by *Lactococcus* and *Lactobacillus* (Stoyanova et al. 2012).

The lactoperoxidase system, found in milk, has profound antimicrobial effects against both bacteria and fungi. A wide range of both Gram-negative bacteria and Gram-positive bacteria are inhibited by the lactoperoxidase system. Hydrogen peroxide on its own is also known to be bactericidal depending on the concentrations applied and on environmental factors such as pH and temperature. While the mechanism by which hydrogen peroxide kills spores is not known, killing of vegetative bacteria and fungi is known to involve DNA damage (Brul and Coote 1999).

2.2.6 Carbon Dioxide

Carbon dioxide (CO_2) is a major end product of hexose fermentation by heterofermentative LAB. A number of LAB are capable of CO_2 production from malate and citrate and also by metabolizing arginine via the arginine deaminase pathway. Finally, decarboxylation of amino acids (histidine, tyrosine) can also result in CO_2 formation (Naidu 2000).

The CO_2 also contributes to the antimicrobial activity of LAB. Its role in creating anaerobic environment by replacing existent molecular oxygen, its extra- and intracellular capability to decrease pH, and its destructive effects on cell membranes make CO_2 a potent inhibitory system against a wide variety of microorganisms. This protective role of CO_2 is critical, particularly in the fermentation of vegetables and silages to prevent growth of molds (Naidu 2000).

CO_2 is mainly produced by heterofermentative LAB. The precise mechanism of its antimicrobial action is still unknown. However, CO_2 may play a role in creating an anaerobic environment which inhibits enzymatic decarboxylations, and the accumulation of CO_2 in the membrane lipid bilayer may cause a dysfunction in permeability (Yang 2000).

CO_2 can effectively inhibit the growth of many food spoilage microorganisms, especially Gram-negative psychrotrophic bacteria. The degree of inhibition by CO_2 varies considerably between the organisms. CO_2 at 10 % could lower the total bacterial counts by 50 % and at 20–50 % it had a strong antifungal activity (Yang 2000).

2.2.7 Others

Under certain conditions, some lactobacilli and lactococci possessing lipolytic activities may produce significant amounts of fatty acids, e.g., in dry fermented sausage and fermented milk. The antimicrobial activity of fatty acids has been recognized for many years. The unsaturated fatty acids are active against Gram-positive bacteria, and the antifungal activity of fatty acids is dependent on chain length, concentration, and pH of the medium. The antimicrobial action of fatty acids has been thought to be due to the undissociated molecule, not the anion, since pH had profound effects on their activity, with a more rapid killing effect at lower pH (Yang 2000).

Pyrrolidone 5 carboxylic acid is only produced by certain types of LAB, such as *Lactobacillus casei* ssp. *casei* and *L. casei* ssp. *pseudoplantarum*, and has bactericidal activity against *Bacillus subtilis* and *Enterobacter cloacae* (Stoyanova et al. 2012).

Diketopiperazines are products of protein degradation. The mechanism of their formation still remains unclear. Nevertheless, it was found that the synthesis occurs via a nonribosomal pathway that uses a multifunctional enzyme. Diketopiperazines can also be formed from peptides in alkaline or acidic environments some LAB strains. Recently, a new compound with antifungal activity has been separated from *Lactobacillus plantarum* AF1. It was identified as 3,6-bis(2-methylpropyl)-2,5-piperazinedione. It was determined the antifungal activity of LAB, which resulted from the activity of a cyclic compound (Leu–Leu) that belongs to 2,5-diketopiperazines (Stoyanova et al. 2012).

Some LAB produce 2-hydroxy-hexane and 3-hydroxy-heptadecane carboxylic acids, which belong to this class of compounds. The MiLABH strain of the genus *L. plantarum* produces several hydroxylated fatty acids with strong antifungal activity: 3-hydroxydecanoic acid, 3-hydroxydodecanoic acid, 3-hydroxy-tetradecanoic, and 3-hydroxy-5-cis-dodecanoic acids. LAB produce hydroxy derivatives of fatty acids from their unsaturated counterparts. All of the above unsaturated fatty acids exhibit antibiotic activity against a broad range of yeasts and mold. The total inhibitory activity of hydroxy fatty acids ranges from 10 to 100 mg/ml due to their poor solubility in aqueous solutions (Stoyanova et al. 2012).

The compound is a metabolite in phenylalanine metabolism and can be formed in LAB cells from p-hydroxyphenylpyruvic acid. This end metabolite exhibits antibiotic activity against Gram-positive and Gram-negative bacteria and also affects a wide range of microscopic fungi. According to literature, *L. plantarum* can produce several related compounds: phenyllactic acid, 4-hydroxy-phenyllac tic acid, and 3-hydroxy-phenyllactic acid; while *L. coryniformis*, *L. sakei*, and *Pediococcus pentosaceus* only produce phenyllactic acid. Apart from lactobacilli, propionibacteria also synthesize phenyllactic acid. There are only a few literature references that contain information on lactococci possessing antifungal activity. Specifically, the *Lactococcus lactis* LI4 strain that inhibits the growth of *Candida albicans* DMST 5239 was isolated from cultured dairy products. The activity of this strain was sustained in the pH range 2.0–4.0 and retained even after autoclaving. The inhibition of growth and aflatoxin production by *Aspergillus flavus* fungi at cocultivation with lactococci were described (Stoyanova et al. 2012).

In this case, the active component was a phosphoglycolipid with a low molecular weight of less than 500 Da and containing an aromatic ring. A thermostable compound with a low molecular weight, which loses activity during prolonged storage, was responsible for the inhibition of aflatoxin production (Stoyanova et al. 2012).

The lactococci strains identified as *Lactococcus lactis* subsp. *lactis* synthesized alkyl ketones. These compounds defined the antifungal activity of those strains, suggesting that these lactococci can be potentially used to prevent spoilage of fruits and vegetables due to contamination by fungi and yeast (Stoyanova et al. 2012).

2.3 *Biofilms and Exopolysaccharides*

A biofilm is a multicellular layer of adherent bacteria surrounded by a matrix of extracellular polysaccharides with growth and survival advantages over planktonic cells, such as a documented increased resistance to antimicrobial compounds and thermal stress. It is demonstrated that numerous bacteria are able to attach to surfaces of equipment used for food handling or processing and remain viable even after cleaning and disinfection, becoming a chronic source of microbial contamination which may compromise food quality and represent a significant health hazard. It is worth noting that not all biofilms cause problems and there are some successful examples of their positive use, even if this aspect remains little studied. At the best of our knowledge, there are no studies about a potential use of microbial biofilms as a means to guarantee food safety. Moreover, the major emphasis on biofilm research is food-borne pathogens whereas biofilm formation by NSLAB has received little attention (Erginkaya et al. 2011; Unal 2013).

In recent years, it has been noted that biofilm produced LAB has antibacterial effect especially on some pathogen formed regard food safety in meat and dairy products. Researches were especially focused on antilisterial effect of biofilm formed LAB. Also, it was reported that biofilms produced by LAB are effective on fungus. But, it should be studied more about biofilms and focused on use of LAB biofilms as biocontrol agents with new studies in food industry (Trias et al. 2008; Guerrieri et al. 2009).

Controlling contamination and growth of *L. monocytogenes* during cheese manufacture, ripening, and storage is an important safety concern and consumer demand. This is a preliminary study, mainly initiated to (a) determine whether NSLAB can form biofilms on stainless steel, (b) evaluate whether NSLAB biofilms can be considered beneficial used as a mean to delay the growth of L. monocytogenes in soft cheeses (Speranza et al. 2009). The decrease in *L. monocytogenes* viable counts in the mixed species biofilm might be related with the enhanced acidification by *L. plantarum* of the medium containing glucose, which reached approximately pH3.4 after 48–72 h. In contrast, acidification during *L. monocytogenes* single species biofilm formation in medium containing glucose stopped at approximately pH4.3. Single and mixed species biofilm formation in BHI and BHI containing manganese sulfate resulted in a final pH of approximately 5.3–5.5.(Veen and Abee 2011).

The growth and spatial localization of *L. monocytogenes* with *L. lactis*, a competing model resident flora, in dual-species biofilms under constant nutrient renewal were investigated. *L. monocytogenes* cells were inhibited by the presence of *L. lactis* and were localized in the bottom biofilm layers in contact with the substratum. By using a simplified IBM framework, it was demonstrated that the initial disparity in generation times between *L. monocytogenes* and *L. lactis* most likely explained the species spatialization observed within dual-species biofilms, and hence the inhibition of the growth of the pathogen (Habimana et al. 2011).

Many adherent bacteria occur in natural environments as surface attached biofilms where they are contained within a self-produced extracellular matrix that protects them against hostile environmental conditions. Biofilms also play a role in the intimate relationship between the human body and its resident microbes for example in the gut (Lebeer et al. 2007).

Biofilms or adherent structured microbial communities in the oral cavity and respiratory tract are well-characterized and are associated with respiratory infections, dental caries, and periodontitis. In contrast, biofilm-like communities of the gastrointestinal and female urogenital tracts containing beneficial lactobacilli may have a protective role. In bacterial vaginosis, indigenous lactobacilli are replaced with pathogenic biofilms consisting of *Gardnerella vaginalis* and other bacteria. Probiotic *L. reuteri* can displace *G. vaginalis* biofilms and could potentially re-establish protective biofilms in the female urogenital tract (Jones and Versalovic 2009).

As mentioned before, a biofilm is an aggregate of microorganisms in which cells adhere to each other on a surface. These adherent cells are frequently embedded within a self-produced matrix of extracellular polymeric substance or exopolysaccharide (EPS). Biofilms are composed primarily of microbial cells and EPS. EPS may account for 50–90 % of the total organic carbon of biofilms and can be considered the primary matrix material of the biofilm. EPS may vary in chemical and physical properties, but it is primarily composed of polysaccharides. EPS may also contribute to the antimicrobial resistance properties of biofilms by impeding the mass transport of antibiotics through the biofilm, probably by binding directly to these agents (Kumar and Anand 1998).

EPS play an important role on bacteria resistance and form much level of biofilm. Many LAB are able to produce EPSs. Most LAB that have probiotic characteristics and produce biofilm form protective effect against pathogenic bacteria (Kim et al. 2009).

The ability of lactobacilli to produce EPSs has been recognized for many years. In 1968, Kooiman first reported the structure of a heteropolysaccharide produced by a *Lactobacillus brevis* strain isolated from kefir grains. This polysaccharide consists of a hexasaccharide repeating unit with D-galactose and D-glucose in the molar ratio 1:1. *L. helveticus* strains produce several EPSs with varying repeating units, though all containing galactose and glucose. The EPS produced by *L. helveticus* 776 has hexasaccharide repeating units containing D-galactose and D-glucose. The EPS produced by *L. helveticus* TY1-2 consists of heptasaccharide repeating units with D-galactopyranosyl and D-glucopyranosyl, and 2-acetamido-2-deoxy-D-glucopyranosyl residues. The EPS produced by *L. helveticus* Lh59 had an identical primary molecular structure as the one produced by *L. helveticus* TN-4, a presumed spontaneous mutant of the strain TY1-2.

EPSs produced by LAB have gained increasing attention due to their potential health benefits. LAB are food-grade organisms which are GRAS, and can produce EPSs that are potentially useful as additives to improve texture and viscosity of natural fermented milk products and to prevent syneresis. It has also been suggested that some EPSs produced by LAB may confer health benefits to the consumer. The formation of EPSs occurs in two forms depending on location: as a capsule (capsular polysaccharides) where the polymer is closely associated with the bacterial surface; and as slime polysaccharide loosely associated with the bacterial surface. Such a distinction may be difficult since some strains release capsular polysaccharide material at the periphery. For bacteria, EPSs are thought to play a role in protection against desiccation, toxic compounds, bacteriophages, osmotic stress, and to permit adhesion to solid surfaces and biofilm formation (Wu et al. 2010; Nikolic et al. 2012).

EPS, one of the primary metabolic products of LAB, have received an increasing amount of attention in recent years, and EPS have been attributed to positive health effects. However, the functional role that EPS plays in bacterial ecology still remains uncertain. Moreover, a recent study reported that a number of EPS isolated from commercial fermented milk "villi" were capable of interfering with the adhesion of several enteric pathogens. However, to date, there have been only a few reports regarding the inhibition of biofilm formation by EPS produced by probiotic bacteria (Kim et al. 2009).

EPSs are long-chain polysaccharides that are secreted mainly by bacteria and microalgae into their surroundings during growth and that are not permanently attached to the surface of the microbial cell. The physical characteristics of EPSs are responsible for the slime-forming or mucoid trait of many microorganisms. A second group of polysaccharides that are structurally similar but that are permanently attached to the cell surface are classified as capsular polysaccharides (Laws et al. 2001).

The formation of dental plaques is related to EPS synthesis by LAB. The EPSs from LAB are responsible for biofilm formation that can lead to biofouling (Otto 2006).

Microbial EPSs can be divided into two groups: homopolysaccharides (e.g., cellulose, dextran, pullulan, levan, and curdlan) and heteropolysaccharides (e.g., gellan and xanthan). Homopolysaccharides are constructed from monosaccharides joined by either a single linkage type (e.g., 1–2 or 1–4) or by a combination of a limited number of linkage types (e.g., 1–2 and 1–4). The current review will focus on recent work related to the synthesis and structural characterization of heteropolysaccharides from LAB. Heteropolysaccharides are constructed from multiple copies of an oligosaccharide. The oligosaccharide can contain between three and seven residues, it possesses a variety of two or more different types of monosaccharides and frequently has a range of different linkage patterns (Laws et al. 2001).

Studies about EPS may provide insight into the antimicrobial activity of EPS against pathogen and food-spoilage bacteria. The antibacterial activity of EPS was examined using seven different bacterial pathogens and food-spoilage bacteria, including the following 4 g negative and 3 g positive species: *E. coli* BCRC10239, *S. typhimurium* CRC10747 (ATCC14028), *P. aeruginosa* BCRC10261 (IFO3898), *V. parahaemolyticus* ATCC17802, *S. aureus* BCRC10451 (ATCC6538P), *B. subtilis* BCRC10029, and *B. cereus* ATCC10361 (Wu et al. 2010).

2.4 Antibiotics as Secondary Metabolite

Secondary metabolites with antimicrobial properties have been used for centuries. Microorganisms, and more specifically their products, were used in food conservation and production of wine, cheeses, and bread. For example, as a method of preservation, milk was converted to lactic acid to make yoghurt. Many of these antimicrobial compounds kill microorganisms by causing membrane permeabilization, although not necessary as their sole mode of action (Donadio et al. 2002).

Microbial secondary metabolites represent a large source of compounds endowed with ingenious structures and potent biological activities. Many of the products currently used for human or animal therapy, in animal husbandry and in agriculture are produced by microbial fermentation, or are derived from chemical modification of a microbial product. Different strains generally produce different compounds. Thus, new bioactive metabolites continue to be identified from microbial sources, thanks to the large variety of existing strains. In particular, the ability to produce a large number of chemically different secondary metabolites is associated mostly with the filamentous actinomycetes, the myxobacteria, the pseudomonads, and the cyanobacteria within the prokaryotic world, and mostly to the filamentous fungi for the eukaryotic microbes (Donadio et al. 2002; Gonzalez et al. 2003). Microbial secondary metabolites are compounds produced by actinomycetes and fungi. Antibiotics are the best known secondary metabolites. Antibiotics from secondary metabolite are defined as low molecular weight organic natural products made by microorganisms which are active against other microorganisms at low concentration. This activity develops through a limited number of mechanisms; antimicrobials interfere with cell wall synthesis, cell membrane integrity, protein synthesis, DNA replication and repair, transcription, and intermediate metabolism. In fact, secondary metabolites are accepted to be essential for the producing cell as inhibitors of other organisms that compete for the same food supply or as regulators of cellular differentiation processes. Microbial cells are the most importance source of this type of secondary metabolites. Indeed, from the known antibiotics 55 % are produced by filamentous bacteria of the genus actinomyces, 11 % from other actinomyces, 12 % from non-filamentous bacteria, and 22 % from filamentous fungi. Most of the work performed in this field has been focused in antibiotics produced by fungi and actinomycetes (Donadio et al. 2002).

The term antibiotic is used in this report to refer to drugs used to treat infectious disease in humans, animals, or plants, by inhibiting the growth of or destroying microorganisms; such substances may be naturally occurring, semisynthetic, or synthetic. Antibiotics are also used in food animals to prevent infectious disease and improve the efficiency of feed utilization. Antimicrobial drugs and antibiotics, by major class, approved in the USA for animal, plant, or human use are these: Aminoglycosides (gentamycin, neomycin, streptomycin), Beta-lactams (penicillins, cephalosporins), Chloramphenicol (Florfenicol), Cycloserines (cycloserine), Glycopeptides (vancomycin), Ionophores (monensin, salinomycin, semduramicin, lasalocid), Lincosamides (lincomycin), Macrolides (tylosin, tilmicosin erythromycin), Monobactrams (aztreonam), Polypeptides (bacitracin), Fluoroquinolones (enrofloxacin, danofloxacin), Streptogramins (virginiamycin), Sulfonamides (sulfadimethoxine, sulfamethazine, sulfisoxazole), Tetracyclines (chlortetracycline oxytetracycline tetracycline), and Others (Bambermycin, Carbadox, Novobiocin, Spectinomycin) (Hawke 2006).

Natural organic compounds produced by microorganisms are an important screening target for a variety of bioactive substances. It can be purified from microbial fermentation and modified chemically or enzymatic ally for either chemical use or for fundamental studies (Palanivel et al. 2012).

Bioactive natural products produced by microbes have almost limitless potential in pharmaceutical applications. Microbial natural products have been the source of most of the antibiotics in current use for the treatment of various infectious diseases. After the discovery of penicillin in 1928, many other drugs including chlortetracycline, chloramphenicol, streptomycin, erythromycin, rifamycin, lincomycin, cephalosporin C, vancomycin, erythromycin, nalidixic acid, amphotericin B, nystatin, and daunorubicin the antitumor agent were discovered from microorganisms (Sunazuka et al. 2006; Tawiah et al. 2012).

Extensive use of antibiotics has led to growing resistance and the spread of many bacterial pathogens, which now constitutes a serious medical problem. For this reason, the number of studies aimed at developing new analogs of known antibiotics, e.g., oxazolidinones, glycopeptides, quinolones, aminoglycosides, tetracyclines and ketolides, and at identifying novel antibacterial therapeutics and strategies, is growing exponentially. Due to their narrow action spectrum and toxicity, bacteriocins were replaced by antibiotics in clinical use and are now extensively used in food preservation. Bacteriocin resistance may be countered by the use of these compounds in combination with novel antimicrobials. Such a strategy might restore the potential of bacteriocin (e.g., nisin) to eliminate pathogenic bacteria, like *S. aureus*, from food. Currently, novel antimicrobials cannot replace antibiotics, but they may become valuable antibiotic complements. In order to exploit these new antimicrobials effectively in synergistic combination therapy, it will be necessary to determine the optimal ratio and dosing regimen, and to fully characterize the mechanisms of their activities by employing genomic, proteomic, and metabolomic technologies (Wolska et al. 2012).

The secondary metabolites exhibits either antimicrobial (antibacterial, antifungal, antiprotozoal), antitumor and/or antiviral activities, used to be called as antibiotics. The practical importance of antibiotics and other secondary metabolites is tremendous. They are widely used in the human therapy, veterinary, agriculture, scientific research, and in countless other areas. In general, natural products including the microbial metabolites may be practically utilized in three different ways: (1) Applying the natural/fermentation product directly in the medicine, agriculture, or in any other fields; (2) using as starting material for subsequent chemical or microbiological modification (derivatization); and (3) they can be used as lead compounds for chemical synthesis of new analogs or as templates in the rational drug design (RDD) studies.

References

Aasen IM, Markussen S, Moretro T, Katla T, Axelsson L, Naterstad K (2003) Interactions of the bacteriocins sakacin P and nisin with food constituents. Int J Food Microbiol 87:35–43

Abee T, Krockel L, Hill C (1995) Bacteriocins: modes of action and potentials in food preservation and control of food poisoning. Int J Food Microbiol 28:169–185

Allende A, McEvoy J, Tao Y, Luo Y (2009) Antimicrobial effect of acidified sodium chlorite, sodium chlorite, sodium hypochlorite, and citric acid on Escherichia coli O157:H7 and natural micro ora of fresh-cut cilantro. Food Control 20:230–234

Ananou S, Banos A, Maqueda M, Bueno MM, Galvez A, Valdivia E (2010a) Effects of combined physico-chemical treatments based on enterocin AS-48 on the control of *Listeria monocytogenes* and *Staphylococcus aureus* in a model cooked ham. Food Control 21:478–486

Ananou S, Garriga M, Jofre A, Aymerich T, Galvez A, Maqueda M, Bueno MM, Valdivia E (2010b) Combined effect of enterocin AS-48 and high hydrostatic pressure to control food-borne pathogens inoculated in low acid fermented sausages. Meat Sci 84:594–600

Anthony T, Rajesh T, Kayalvizhi N, Gunasekaran P (2009) Influence of medium components and fermentation conditions on the production of bacteriocin(s) by *Bacillus licheniformis* AnBa9. Bioresour Technol 100:872–877

Arques JL, Fernandez J, Gaya P, Nunez M, Rodriguez E, Medina M (2004) Antimicrobial activity of reuterin in combination with nisin against food-borne pathogens. Int J Food Microbiol 95:225–229

Arques JL, Rodriguez E, Nunez M, Medina M (2011) Combined effect of reuterin and lacticacid bacteria bacteriocins on the inactivation of food-borne pathogens in milk. Food Control 22:457–461

Balciunas EM, Martinez FAC, Todorov SD, Franco BDGM, Converti A, Oliveira RPS (2013) Novel biotechnological applications of bacteriocins: a review. Food Control 32:134–142

Banos A, Ananou S, Bueno MM, Galvez A, Maqueda M, Valdivia E (2012) Prevention of spoilage by enterocin AS-48 combined with chemical preservatives, under vacuum, or modified atmosphere in a cooked ham model. Food Control 24:15–22

Barber RA, Vriesekoop F, Pamment BN (2002) Effects of acetaldehyde on Saccharomyces cerevisize exposed to a range of chemical and environmental stresses. Enzyme Microb Technol 30:240–250

Beaulieu L, Groleau D, Miguez BC, Jette FJ, Aomari H, Subirade M (2005) Production of pediocin PA-1 in the methylotrophic yeast Pichia pastoris reveals unexpected inhibition of its biological activity due to the presence of collagen-like material. Protein Expr Purif 43:111–125

Brul S, Coote P (1999) Review preservative agents in foods Mode of action and microbial resistance mechanisms. Int J Food Microbiol 50:1–17

Brzozowski B, Bednarski W, Dziuba B (2009) Functional properties of *Lactobacillus acidophilus* metabolites. J Sci Food Agric 89:2467–2476

Caplice E, Fitzgerald GF (1999) Food fermentations: role of microorganisms in food production and preservation. Int J Food Microbiol 50:131–149

Ceotto H, Nascimento JS, Brito MAVP, Freire Bastos MCF (2009) Bacteriocin production by *Staphylococcus aureus* involved in bovine mastitis in Brazil. Res Microbiol 160:592–599

Ceotto H, Holo H, Costa KFS, Nascimento JS, Salehian Z, Nes IF, Bastos MCF (2010) Nukacin 3299, a lantibiotic produced by *Staphylococcus simulans* 3299 identical to nukacin ISK-1. Vet Microbiol 146:124–131

Cherif A, Rezgui W, Raddadi N, Daffonchio D, Boudabous A (2008) Characterization and partial purification of entomocin 110, a newly identified bacteriocin from Bacillus thuringiensis subsp. Entomocidus HD110. Microbiol Res 163:684–692

Cizeikiene D, Juodeikiene G, Paskevicius A, Bartkiene E (2013) Antimicrobial activity of lactic acid bacteria against pathogenic and spoilage microorganism isolated from food and their control in wheat bread. Food Control 31:539–545

Cleveland J, Montville TJ, Nes IF, Chikindas ML (2001) Bacteriocins: safe, natural antimicrobials for food preservation. Int J Food Microbiol 71:1–20

Compaoré CS, Nielsen DS, Ouoba LII, Berner TS, Nielsen KF, Lingani HS, Diawara B, Ouédraogo GA, Jakobsen M, Thorsen L (2013) Co-production of surfactin and a novel bacteriocin by *Bacillus subtilis* subsp. *subtilis* H4 isolated from Bikalga, an African alkaline Hibiscus sabdariffa seed fermented condiment. Int J Food Microbiol 162:297–307

Crost EH, Ajandouz EH, Villard C, Geraert PA, Puigserver A, Fons M (2011) Ruminococcin C, a new anti-*Clostridium perfringens* bacteriocin produced in the gut by the commensal bacterium *Ruminococcus gnavus* E1. Biochimie 93:1487–1494

Crowley S, Mahony J, Sinderen DV (2013) Current perspectives on antifungal lactic acid bacteria as natural bio-preservatives. Trends Food Sci Technol 1–17

Dang TDT, Vermeulen A, Ragaert P, Devlieghere F (2009) A peculiar stimulatory effect of acetic and lactic acid on growth and fermentative metabolism of *Zygosaccharomyces bailii*. Food Microbiol 26:320–327

Davidson PM, Sofos JN, Branen AL (2005) Antimicrobials in food, 3rd edn. Taylor and Francis, Boca Raton, FL

Daw MA, Falkiner FR (1996) Bacteriocins: nature, function and structure, vol 27. Elsevier, Burlington, pp 467–479

Deegan LH, Cotter PD, Hill C, Ross ÃP (2006) Bacteriocins: biological tools for bio- preservation and shelf-life extension. Int Dairy J 16:1058–1071

Diez L, Bezares BR, Zarazaga M, Rodriguez JM, Torres C, Larrea FR (2012) Antimicrobial activity of pediocin PA-1 against Oenococcus oeni and other wine bacteria. Food Microbiol 31:167–172

Donadio S, Monciardini P, Alduina R, Mazza P, Chiocchini C, Cavaletti L, Sosio M, Puglia AM (2002) Microbial technologies for the discovery of novel bioactive metabolites. J Biotechnol 99:187–198

Elsayed EA, Farid MAF, Enshasy HAE (2013) Improvement in natamycin production by *Streptomyces natalensis* with addition of short-chain carboxylic acids. Process Biochem S1359–5113(13):00542–4

Erginkaya Z, Ünal E, Kalkan S (2011) Importance of microbial antagonisms about food attribution. In: Antonio Méndez-Vilas A (ed) Science against microbial pathogens: communicating current research and technological advances, vol 2. Badajoz, Formatex Research Center-Spain, pp 1342–1348

Eswaranandam S, Hettiarachchy NS, Johnson MG (2004) Antimicrobial Activity of Citric, Lactic, Malic, or Tartaric Acids and Nisin-incorporated Soy Protein film against *Listeria monocytogenes, E. coli* 0157:H7 and *Salmonella gaminara*. J Food Sci 69(3):FMS 79–84

Fagundes PC, Ceotto H, Potter A, Brito MAVP, Brede D, Nes IF, Bastos MCF (2011) Hyicin 3682, a bioactive peptide produced by *Staphylococcus hyicus* 3682 with potential applications for food preservation. Res Microbiol 162:1052–1059

Fajardo P, Martins JT, Fuciños C, Pastrana L, Teixeira JA, Vicente AA (2010) Evaluation of achitosan-based edible film as carrier of natamycin to improve the storability of Saloio cheese. J Food Eng 101:349–356

Galvez A, Abriouel H, Lopez RL, Omar NB (2007) Bacteriocin-based strategies for food biopreservation. Int J Food Microbiol 120:51–70

Galvez A, Abriouel H, Benomar N, Lucas R (2010) Microbial antagonists to food-borne pathogens and biocontrol. Curr Opin Biotechnol 21:142–148

Gao Y, Li D, Sheng Y, Liu X (2011) Mode of action of sakacin C2 against *Escherichia coli*. Food Control 22:657–661

Gao Y, Li D, Liu X (2013) Evaluation of the factors affecting the activity of sakacin C2 against *E. coli* in milk. Food Control 30:453–458

Garcia MT, Omar NB, Lucas R, Pulido RP, Castro A, Grande MJ, Canamero MM, Galvez A (2003) Antimicrobial activity of enterocin EJ97 on *Basillus coagulans* CECT 12. Food Microbiol 20:533–536

Garcia MT, Canamero MM, Lucas R, Omar ND, Pulido RP, Galvez A (2004) Inhibition of *Listeria monocytogenes* by enterocin EJ97 produced by *Enterococcus faecalis* EJ97. Int J Food Microbiol 90:161–170

Garneau S, Martin NI, Vederas JC (2002) Two-peptide bacteriocins produced by lactic acid bacteria. Biochimie 84:577–592

Gomez NC, Abriouel H, Ennahar S, Galvez A (2013) Comparative proteomic analysis of *Listeria monocytogenes* exposed to enterocin AS-48 in planktonic and sessile states. Int J Food Microbiol. doi:10.1016/j.ijfoodmicro.2013.08.019

Gonzalez BJ, Fernandez JF, Tomasini A (2003) Microbial secondary metabolites production and strain ımprovement. Indian J Biotechnol 2:322–333

Grande MJ, Lucas R, Abriouel H, Valdivia E, Omar NB, Maqueda M, Bueno MM, Canamero MM, Ga´lvez A (2006) Inhibition of toxicogenic *Bacillus cereus* in rice-based foods by enterocin AS-48. Int J Food Microbiol 106:185–194

Guerrieri E, Niederhäusern S, Messi P, Sabia C, Iseppi R, Anacarso I, Bondi M (2009) Use of lactic acid bacteria (LAB) biofilms for the control of *Listeria monocytogenes* in a small- scale model. Food Control 20:861–865

Guiotto A, Pozzobon M, Canevari M, Manganelli R, Scarin M, Veronese FM (2003) PEGylation of the antimicrobial peptide nisin A: problems and perspectives. Farmaco 58:45–50

Habimana O, Guillier L, Kulakauskas S, Briandet R (2011) Spatial competition with *Lactococcus lactis* in mixed-species continuous-flow biofilms inhibits *Listeria monocytogenes* growth. Biofouling 27(9):1065–1072

Hawke J (2006) Antimicrobial resistance: ımplications for the food system. Inst Food Technol 5:71–137

He L, Chen W, Liu Y (2006) Production and partial characterization of bacteriocin-like pepitdes by *Bacillus licheniformis* ZJU12. Microbiol Res 161:321–326

Hondrodimou O, Kourkoutas Y, Panagou EZ (2011) Efficacy of natamycin to control fungal growth in natural black olive fermentation. Food Microbiol 28:621–627

Hugas M, Garriga M, Pascual M, Aymerichand MT (2002) Enhancement of sakacin K activity against *Listeria monocytogenes* in fermented sausages with pepper or manganese as ingredients. Food Microbiol 19:519–528

Ibrahim SM, Desouky SG (2009) Effect of antimicrobial metabolites produced by lactic acid bacteria (lab) on quality aspects of frozen tilapia (oreochromis niloticus) fillets. World J Fish Mar Sci 1(1):40–45

Ivey M, Massel M, Phister TG (2013) Microbial ınteractions in food fermentations. Annu Rev Food Sci Technol 2013.4(1):41–62

Jiang T, Feng L, Zheng X, Li J (2013) Physicochemical responses and microbial characteristics of shiitake mushroom (Lentinus edodes) to gum arabic coating enriched with natamycin during storage. Food Chem 138:1992–1997

Jones SE, Versalovic J (2009) Probiotic *Lactobacillus reuteri* biofilms produce antimicrobial and anti-inflammatory factors. BMC Microbiol 9:35. doi:10.1186/1471-2180-9-35

Kaewklom S, Lumlert S, Kraikul W, Aunpad R (2013) Control of Listeria monocytogenes on sliced bologna sausage using a novel bacteriocin, amysin, produced by *Bacillus amyloliquefaciens* isolated from Thai shrimp paste (Kapi). Food Control 32:552–557

Kalmokoff ML, Banerjee SK, Cyr T, Hefford MA, Gleeson T (2001) Identification of a new plasmid-encodedsec-dependent bacteriocin produced by *Listeria innocua*743. Appl Environ Microbiol 67(9):4041–7

Kallinteri LD, Kostoula OK, Savvaidis IN (2013) Efficacy of nisin and/or natamycin to improve the shelf-life of Galotyri cheese. Food Microbiol 36:176–181

Katla T, Moretro T, Aasen IM, Holck A, Axelsson L, Naterstad K (2001) Inhibition of *Listeria monocytogenes* in cold smoked salmon by addition of sakacin P and/or live *Lactobacillus sakei* cultures. Food Microbiol 18:431–439

Kim Y, Oh S, Kim SH (2009) Released exopolysaccharide (r-EPS) produced from probiotic bacteria reduce biofilm formation of enterohemorrhagic *Escherichia coli* O157:H7. Biochem Biophys Res Commun 379:324–329

Klaenhammer TR, Fremaux C, Ahn C, Milton K (1993) Molecular biology of bacteriocins produced by Lactobacillus. In: Hoover DG, Steenson LR (eds) Bacteriocins of lactic acid bacteria. Academic, San Diego, pp 151–180

Krause TA, Liewehr FR, Hahn CL (2007) The antimicrobial effect of MTAD, sodium hypochlorite, doxycycline, and citric acid on *Enterococcus faecalis*. Am Assoc Endod. doi:10.1016/j.joen.2006.08.011

Kumar CG, Anand SK (1998) Significance of microbial biofilms in food industry: a review. Int J Food Microbiol 42:9–27

Kuwano K, Tanaka N, Shimizu T, Nagatoshi K, Nou S, Sonomoto K (2005) Dual antibacterial mechanisms of nisin Z against gram-positive and gram-negative bacteria. Int J Antimicrob Agents 26:396–402

Lacroix C (2011) Protective cultures, antimicrobial metabolites and bacteriophages for food and beverage biopreservation, vol 201, Woodhead publishing series in food science, technology and nutrition. Woodhead Publishing Limited, Oxford

Lancitti R, Patrignani F, Bagnolini F, Guerzoni ME, Gardini F (2002) Evaluation of diacetyl antimicrobial activity against *Escherichia coli, Listeria monocytogenes* and *Staphylococcus aureus*. Food Microbiol 20:537–543

Langa S, Landete JM, Cabrejas IM, Rodriguez E, Arques JL, Medina M (2013) In situ reuterin production by *Lactobacillus reuteri* in dairy products. Food Control 33:200–206

Larsen AG, Knochel S (1997) Antimicrobial activity of food-related Penicillium sp. Against pathogenic bacteria in laboratory media and a cheese model system. J Appl Microbiol 83:111–119

Laws A, Gu Y, Marshall V (2001) Biosynthesis, characterisation, and design of bacterial exopolysaccharides from lactic acid bacteria. Biotechnol Adv 19:597–625

Lebeer S, Verhoeven TLA, Velez MP, Venderleyden J, Keersmaecker SCJD (2007) Impact of environmental and genetic factors on biofilm formation by the probiotic strain *Lactobacillus rhmannosus* GG. Appl Env Microbiol 73(21):6768–6775

Lopez FNA, Gallego JB, Gil VR, Gomez FR, Fernandez AG (2012a) Growth/no growth interfaces of table olive related yeasts for natamycin, citric acid and sodium chloride. Int J Food Microbiol 155:257–262

Lopez EM, Garcia HS, Malo AL (2012b) Organic acids as antimicrobials to control Salmonella in meat and poultry products. Food Res Int 45:713–721

Martinez-Cuesta CM, Bengoechea J, Bustos I, Rodriguez B, Requena T, Pelaez C (2010) Control of late blowing in cheese by adding lacticin 3147- producing *Lactococcus lactis* IFPL 3593 to the starter. Int Dairy J 20:18–24

McIntyre L, Hudson JA, Billington C, Withers H (2007) Biocontrol of foodborne bacteria: past, present and future strategies. Sci Rev Art Food New Zeal 7:25–32

Mollerach ME, Ogueta SB, De Torres RA (1988) Production of Linnocuicina 819, a bacteriocin produced by *Listeria innocua*. Microbiologica 11(3):219–24

Moliz ATM, Luque FMC, Rodriguez EE, Urena LJ, Garcia EM (2008) Bactericidal activity of phosphoric acid, citric acid, and EDTA solutions against *Enterococcus faecalis*. Oral Surg Oral Med Oral Pathol Oral Radiol Endod 106:e84–e89

Moon GS, Pyun YR, Kim WJ (2006) Expression and purification of a fusion-typed pediocin PA-1 in Escherichia coli and recovery of biologically active pediocin PA-1. Int J Food Microbiol 108:136–140

Naidu AS (2000) Natural food antimicrobial systems. CRC, Boca Raton, FL

Nascimento JS, Santos KRN, Gentilini E, Sordelli D, Bastos MCF (2002) Phenotypic and genetic characterization of bacteriocin-producing strains of *Staphylococcus aureus* involved in bovine mastitis. Vet Microbiol 85:133–144

Nascimento JS, Marval MG, Oliveira SS, Ceotto H, Santos KRN, Bastos MCF (2005) Genomic fingerprinting of bacteriocin-producer strains of *Staphylococcus aureus*. Res Microbiol 156:837–842

Netz DJA, Sahl HG, Marcolino R, Nascimento JS, Oliveira SS, Soares MB, Bastos MCF (2001) Molecular characterisation of aureocin A70, a multipeptide bacteriocin isolated from *Staphylococcus aureus*. J Mol Biol 311:939–949

Netz DJA, Pohl R, Sickinger AGB, Selmer T, Pierik AJ, Bastos MCF, Sahl HG (2002) Biochemical characterisation and genetic analysis of aureocin A53, a new, atypical bacteriocin from *Staphylococcus aureus*. J Mol Biol 319:745–756

Nikolic M, Lopez P, Strahinic I, Suarez A, Kojic M, Garcia MF, Topisirovic L, Golic N, Madiedo PR (2012) Characterisation of the exopolysaccharide (EPS)-producing *Lactobacillus paraplantarum* BGCG11 and its non-EPS producing derivative strains as potential probiotics. Int J Food Microbiol 158:155–162

Olivera FC, Caron GR, Brandelli A (2004) Bacteriocin production by *Bacillus licheniformis* strain P40 in cheese whey using response surface methodology. Biochem Eng J 21:53–58

O'Sullivan L, Ross RP, Hill C (2002) Potential of bacteriocin-producing lactic acid bacteria for improvements in food safety and quality. Biochimie 84:593–604

O'Mahony T, Rekhif N, Cavadini C, Fitzgerald GF (2001) The application of a fermented food ingredient containing 'variacin', a novel antimicrobial produced by *Kocuria varians*, to control the growth of *Bacillus cereus* in chilled dairy products. J Appl Microbiol 90:106–114

Otto M (2006) Bacterial evasion of antimicrobial peptides by biofilm formation. Curr Top Microbiol Immunol 306:251–258

Palanivel P, Ashokkumar L, Prabakaran R, Balagurunathan R (2012) Study on antimicrobial metabolite produced by soil bacteria isolated from less explored ecosystem. Life Sciences Leaflets 12:165–173

Papagianni M (2003) Ribosomally synthesized peptides with antimicrobial properties: biosynthesis structure, function and application. Biotechnol Adv 21:465–499

Pattnaik P, Grover S, Batish VK (2005) Effect of environmental factors on production of lichenin, a chromosomally encoded bacteriocinlike compound produced by *Bacillus licheniformis* 26 L-10/3RA. Microbiol Res 160:213–218

Peláez AML, Catano CAS, Yepes EAQ, Villarroel RRG, Antoni LD (2012) Inhibitory activity of lactic and acetic acid on Aspergillus flavus growth for food preservation. Food Control 24:177–183

Pyar H, Liong MT, Peh KK (2013) Characteristics and antibacterial activity of metabolites from *lactobacillus acidophilus* strains produced from novel culture media. Int J Pharmacol 9(1):92–97

Rahman SME, Jin YG, Oh DH (2011) Combination treatment of alkaline electrolyzed water and citric acid with mild heat to ensure microbial safety, shelf-life and sensory quality of shredded carrots. Food Microbiol 28:484–491

Rasch M (2002) The influence of temperature, salt and pH on the inhibitory effect of reuterin on *Escherichia coli*. Int J Food Microbiol 72:225–231

Rattanachaikunsopon P, Phumkhachorn P (2010) Lactic acid bacteria: their antimicrobial compounds and their uses in food production. Ann Biol Res 1(4):218–228

Renye JAJ, Somkuti GA, Garabal JI, Du L (2011) Heterologous production of pediocin for the control of *Listeria monocytogenes* in dairy foods. Food Control 22:1887–1892

Resa CPO, Jagus RJ, Gerschenson LN (2014) Natamycin efficiency for controlling yeast growth in models systems and on cheese surfaces. Food Control 35:101–108

Sabia C, Manicardi G, Messi P, Niederhausern S, Bondi M (2002) Enterocin 416 K1, an antilisterial bacteriocin produced by *Enterococcus casseliflavus* IM 416 K1 isolated from Italian sausages. Int J Food Microbiol 75:163–170

Sabaté DC, Audisio CM (2013) Inhibitory activity of surfactin, produced by different *Bacillus subtilis* subsp. *subtilis* strains, against *Listeria monocytogenes* sensitive and bacteriocin-resistant strains. Microbiol Res 168:125–129

Schillinger U, Becker B, Vignolo G, Holzapfel WH (2001) Efficacy of nisin in combination with protective cultures against *Listeria monocytogenes* Scott A in tofu. Int J Food Microbiol 71:159–168

Settanni L, Corsetti A (2008) Application of bacteriocins in vegetable food biopreservation. Int J Food Microbiol 121:123–138

Silkin L, Hamza S, Kaufman S, Cobb SL, Vederas JC (2008) Spermicidal bacteriocins: lacticin 3147 and subtilosin A. Bioorg Med Chem Lett 18(3103):3106

Speranza B, Sinigaglia M, Corbo MR (2009) Non starter lactic acid bacteria biofilms: a means to control the growth of *Listeria monocytogenes* in soft cheese. Food Control 20:1063–1067

Stevens MJA, Vollenweider S, Mertes P, Lacroix C (2013) Bisulfite as scavenger for enhanced biotechnological production of 3-hydroxypropionaldehyde by *Lactobacillus reuteri*. Biochem Eng J 79:239–245

Stoyanova LG, Ustyugova EA, Netrusov AI (2012) Antibacterial metabolites of lactic acid bacteria: their diversity and properties. Appl Biochem Microbiol 48(3):229–243

Sunazuka T, Hirose T, Omura S (2006) Efficient total synthesis of novel bioactive microbial metabolites. Accounts of Chemical Research 302-314 February 2008 Vol. 41, No.2

Suskovic J, Kos B, Beganovic J, Pavunc AL, Habjanic K, Matosic S (2010) Antimicrobial activity – the most ımportant property of probiotic and starter lactic acid bacteria. Food Technol Biotechnol 48(3):296–307

Tawiah AA, Gbedema SY, Adu F, Boamah VE, Annan K (2012) Antibiotic producing microorganisms from River Wiwi, Lake Bosomtwe and the Gulf of Guinea at Doakor Sea Beach. Ghana BMC Microbiol 12:234

Tiwari BK, Valdramidis VP, Donnell CPO, Muthukumarappan K, Bourke P, Cullen PJ (2009) Application of natural antimicrobials for food preservation. J Agric Food Chem 57(5987–6000):5987

Todorov SD, Dicks LMT (2005) Pediocin ST18, an anti-listerial bacteriocin produced by Pediococcus pentosaceus ST18 isolated from boza, a traditional cereal beverage from Bulgaria. Process Biochem 40:365–370

Trias R, Baneras L, Montesinos E, Badosa E (2008) Lactic acid bacteria from fresh fruit and vegetables as biocontrol agents of phytopathogenic bacteria and fungi. Int Microbiol 11:231–236

Trinetta V, Rollini M, Manzoni M (2008) Development of a low cost culture medium for sakacin A production by *L. sakei*. Process Biochem 43:1275–1280

Tsiraki MI, Savvaidis IN (2014) Citrus extract or natamycin treatments on "Tzatziki" – a traditional Greek salad. Food Chem 142:416–422

Ünal E (2013) Benzelkonium chloride resistance, hydrophobicity and biofilm formation of acid resistant variants of Listeria monocytogenes LO28. Master Thesis, Wageningen University, Holland

Xiraphi N, Georgalaki M, Rantsiou K, Cocolin L, Tsakalidou E, Drosinos EH (2008) Purification and characterization of a bacteriocin produced by *Leuconostoc mesenteroides* E131. Meat Sci 80:194–203

Veen SV, Abee T (2011) Mixed species biofilms of *Listeria monocytogenes* and Lactobacillus plantarum show enhanced resistance to benzalkonium chloride and peracetic acid. Int J Food Microbiol 144:421–431

Wolska KI, Grzes K, Kurek A (2012) Synergy between novel antimicrobials and conventional antibiotics or bacteriocins. Pol J Microbiol 61(2):95–104

Woraprayote W, Kingcha Y, Amonphanpokin P, Kruenate J, Zendo T, Sonomoto K, Benjakul S, Visessanguan W (2013) Anti-listeria activity of poly(lactic acid)/sawdust particle biocomposite film impregnated with pediocin PA-1/AcH and its use in raw sliced pork. Int J Food Microbiol 167:229–235

Wu MH, Pan TM, Wu YJ, Chang SJ, Chang MS, Hu CY (2010) Exopolysaccharide activities from probiotic bifidobacterium: ımmunomodulatory effects (on J774A.1 macrophages) and antimicrobial properties. Int J Food Microbiol 144:104–110

Yang Z (2000) Antimicrobial compounds and extracellular polysaccharides produced by lactic acid bacteria: structures and properties. Auditorium XII of the University Main Building, Aleksanterinkatu 5, Department of Food Technology, University of Helsinki

Yokoyama E, Maruyama S, Katsube Y (1998) Production of bacteriocin-like-substance by *Listeria innocua* against *Listeria monocytogenes*. Int J Food Microbiol 40:133–137

Use of Antimicrobial Edible Films and Coatings as Packaging Materials for Food Safety

Zerrin Erginkaya, Selin Kalkan, and Emel Ünal

1 Introduction

The quality of food products depends on the changes of physical, chemical, and microbiological parameters that occur during storage. Various methods are used such as heating or cooling temperature variation, reduction of water activity, curing, salting, pH control, the addition of antimicrobial agent, controlled atmosphere storage and packaging technologies to extend shelf life and to achieve better quality food products. Packaging technologies are used for protecting food products from some influences such as chemical, air, light, heat, microorganism, and environmental impacts. Packaging also facilitates the transport of food products and offers the consumer that necessary information about the product (Üçüncü 2007; Ayana and Turhan 2010; Mehmetoğlu 2010).

In recent years, due to the increased interest in minimal processed foods depending on consumer demand, use of new technologies and approaches started in the packaging industry. The active packaging technique is the most prominent of these technologies. It can be considered as an emerging technology that could have a significant effect on the shelf life extension and food safety (Perez-Perez et al. 2006). Active packaging takes place engagement of the various active components inside the packaging material. The active components, antimicrobial and become active in various architectures such as the synthetic polymers and the edible films and coatings (Ayana 2007). Organic acids, bacteriocins, antibiotics, fungicides,

Z. Erginkaya (✉) • E. Ünal
Faculty of Agriculture, Department of Food Engineering,
Cukurova University, Adana 01330, Turkey
e-mail: zerriner@cu.edu.tr; unale@cu.edu.tr

S. Kalkan
Bahce Vocational School of Higher Education, Department of Food Technology,
Osmaniye Korkut Ata University, Osmaniye 80500, Turkey
e-mail: selinkalkan@osmaniye.edu.tr

A. Malik et al. (eds.), *Food Processing: Strategies for Quality Assessment*,
Food Engineering Series, DOI 10.1007/978-1-4939-1378-7_10

chelating agents, and parabens may show antimicrobial activity within the food packaging materials (Khairruddin 2005). Use of antimicrobial agents in food packaging can control the microbial population and growth of target specific microorganisms to provide higher safety and quality products (Perez-Perez et al. 2006).

Antimicrobial packaging is a suitable protection method especially for raw meat, processed meat products, poultry, dairy products, and seafood (Suppakul et al. 2003; Karagöz and Candoğan 2007). Controlled release of antimicrobial compounds carried out on these systems can inhibit both of the initial microorganisms and the current microbial growth during storage. These systems form a barrier mechanism for pathogens and/or spoilage microorganism for ensure food safety (Cooksey 2005).

Nowadays, the widespread use of traditional packaging material has synthetic structure. These synthetic materials are safe, convenient, and economical but they aren't biodegradable so they are one of the main factors of environmental pollution. For this reason, researchers have focused on the use of protein, lipid, and polysaccharide polymers as packaging materials which are biologically degradable and also can be consumed with food offering more environmental friendly alternatives. In this context, the use of edible films and coatings containing antimicrobial components is increasingly widespread and adopted (Ayana 2007).

We aimed to review; history of antimicrobial edible films and coatings, evaluation of edible packaging performance, characteristics of biopolymers and antimicrobial substances which are used for preparation of antimicrobial edible film and coatings, functional properties, legal and economic aspects, some food applications, and the place and importance of antimicrobial edible films and coatings in future within this chapter.

2 Edible Films and Coatings

2.1 Definition and Historical Background

Edible films and coatings are defined as continuous matrices prepared from proteins, lipids, and polysaccharides (Çağrı 2002). They become popular in the food industry, due to producing less waste, and being cost effective (Cha and Chinnan 2004). These are thin layers of edible materials and are formed on the surface of a food as coating or between food components. They can control oxygen, moisture, carbon dioxide, flavor, and aroma transfer between food components or the atmosphere surrounding the food (Wen-Xian et al. 2011). They can also carry a wide range of food additives including antioxidants, flavors, preservatives, and antimicrobial agents. If they are prepared properly they could serve all the functional properties as a packaging material (Ko et al. 2001; Skurtys et al. 2010).

Historically, during twelfth and thirteenth century, citrus fruits were dipped in wax in order to slow water loss (Yener 2007). During the fifteenth century, Yuba, the first freestanding edible film, was developed in Japan from soymilk (Çağrı 2002). In order to reduce of the loss of moisture, the surface of the meat coated with oil in sixteenth century in Europe. The introduction of the oil with gelatin coatings had been realized in nineteenth century (Kester and Fennema 1986). In the nineteenth century, almonds, nuts and hazelnuts were coated with sucrose to prevent rancidness and oxidation during storage (Debeaufort et al. 1998). Currently, edible films and coatings are used in variety of applications, including casing for sausages, and chocolate coatings for nuts hazelnuts, almonds, and some fruits (Çağrı 2002).

In order to avoid a negative impact on consumption edible films and coatings, they should be tasteless, odorless, colorless, and transparent as possible. And they should be chosen adaptable with food. Edible films and coatings which are in good property must bear the following conditions: (1) The raw materials which are used to prepare edible films and coating material to be generally recognized as safe (GRAS); (2) Slow, but they should allow controlled ventilation of the product; (3) The structural integrity must provide an appropriate mechanical processing; (4) They reduce of pathogenic and/or saprophyte microorganism without causing deterioration on the surface of product during long-protected storage (Quntavalla and Vicini 2002; Dursun and Erkan 2009).

2.2 *Classification of Edible Films and Coatings*

Edible films can be produced from materials with film forming ability. During manufacturing, film materials must be dispersed and dissolved in a solvent such as water, alcohol, or mixture of water and alcohol, or a mixture of other solvents. Plasticizers such as generally sorbitol or glycerol, antimicrobial agents, colors, or flavors can be added in this process. Adjusting the pH and/or heating the solutions may be done for the specific polymer to facilitate dispersion. Film solution is then casted and dried at a desired temperature and relative humidity to obtain freestanding films. In food applications, film solutions could be applied to food by several methods such as dipping, spraying, brushing, and panning followed by drying (Bourtoom 2008).

According to their components, edible films and coatings can be divided into three categories: hydrocolloids, lipids, and composites. Hydrocolloids include proteins and polysaccharides. Lipids include waxes, acylglycerols, and fatty acids. Composites contain both hydrocolloid components and lipids (Valencia-Chamorro et al. 2011a, b). Several other compounds such as plasticizers and emulsifiers may be added to edible films and coatings to improve their mechanical properties and form stable emulsions when lipids and hydrocolloids are combined. In addition, edible coatings and films can also act as carriers of food additives, including antioxidants, colorants, flavoring agents, and antimicrobial compounds (Cuppet 1994; Baldwin 1999; Franssen and Krochta 2000; Cha and Chinnan 2004; Han and Gennadious 2005).

2.2.1 Polysaccharides

Polysaccharides used for edible films or coatings include cellulose, starch derivatives, pectin derivatives, seaweed extracts, exudate gums, microbial fermentation gums, and chitosan (Bourtoom 2008) Polysaccharides are nontoxic and widely available. They have also selective permeability to carbon dioxide and oxygen, and thus retard the respiration and ripening of many fruits and vegetables by limiting the availability of oxygen. Polysaccharide-based films have a hydrophilic nature. For this reason, they are a poor barrier to water vapor. The poor water vapor barrier property allows for the movement of water vapor across the film, thus, preventing water condensation that can be a potential source of microbial spoilage in horticulture commodities (Yener 2007).

Cellulose and Derivatives

Cellulose is composed of repeating D-glucose units linked through ß-1, 4 glycosidic bonds (Çağrı 2002). It is the most abundant organic compound on earth owing to it is the principal component of plant cell. In esterification process, cellulose is reacted with aqueous caustic, then with methyl chloride, propylene oxide, or sodium monochloroacetate to yield methylcellulose (MC), hydroxypropyl methylcellulose (HPMC), and sodium carboxymethylcellulose (CMC), respectively (Çağrı 2002). These are widely reported as edible films and coatings in the scientific literature and in patents (Debeaufort et al. 1994).

Cellulose derivatives are interesting film-forming compounds, as they are odorless, tasteless, and biodegradable. In addition, their cost of application is low. HPMC is a water-soluble polymer used in the food industry as a gelling and stabilizing agent. It presents excellent film-forming properties with very efficient oxygen, carbon dioxide, and lipid barriers (De Moura et al. 2008; Sanchez-Gonzalez et al. 2009). But HPMC films are highly permeable to water vapor and this condition limits its application. HPMC is approved for food uses by the FDA (21 CFR 172.874) and the EU; its safety in food use has been affirmed by the JECFA (Imran et al. 2010). Another cellulose derivative is methyl cellulose (MC) is of interest to researches because they are able to form a continuous matrix. MC is cellulose that exhibits thermal gelation, forms excellent films, and is used in pharmaceutical and food industries (Turhan and Şahbaz 2004).

However, cellulose derivative films are poor water vapor barriers because of the inherent hydrophilic nature of polysaccharides and they possess poor mechanical properties. A way to improve the moisture barrier would be the incorporation of hydrophobic compounds, such as fatty acids, into the cellulose ether matrix to develop a composite film (Skurtys et al. 2010). In order to improve water barrier properties, lipid compounds such as fatty acids, natural waxes, surfactants, and resins are frequently incorporated in to hydrocolloid-based films (Sanchez-Gonzalez et al. 2009).

Edible coatings made of CMC, MC, HPC, and HPMC have been applied to some fruits and vegetables for providing barriers to oxygen, oil, or moisture transfer (Skurtys et al. 2010).

Chitin and Chitosan

Chitin is the second most abundant naturally occurring biopolymer (after cellulose) and is found in the exoskeleton of crustaceans, in fungal cell walls, and other biological materials. Chitosan is a natural carbohydrate polymer derived by deacetylation of chitin [poly-β-(1$\rightarrow$4)-*N*-acetyl-D-glucose-amine] which is a major component of the shells of crustacean such as crab, shrimp, and crawfish. Chitosan has been reviewed for commercial application in biomedical, food, and chemical industries. It is natural polymer which is nontoxic, biodegradable, and biocompatible. It is insoluble in water, but soluble in various acidic solvent such as dilute hydrochloric, formic, and acetic acids. It is a high molecular weight cationic polysaccharide that exhibits antibacterial and antifungal activity as well as film-forming properties (Beverlya et al. 2008; Ye et al. 2008a, b; Ferreira et al. 2009; Campos et al. 2011). Chitosan antimicrobial activity against bacteria could be due to the polycationic nature of the molecule, which allows interaction and formation of polyelectrolyte complexes with polymers produced at the bacteria cell surface (Moreira et al. 2011). Chitosan is a good choice for antimicrobial films because of its superior film-forming properties, ability to absorb nutrients used by bacteria, and capacity to bind water and inhibit various bacterial enzyme systems (Ye et al. 2008a, b).

Chitosan has been widely used for the production of edible films and coatings (Souza et al. 2010). Chitosan-based films are excellent oxygen barriers; but, due to their hydrophilic nature, they have poor moisture barrier properties (Ojagh et al. 2010). Similarly, the chitosan coatings creates a semi-permeable barrier that controls gas exchange and reduce water loss, thereby maintaining tissue firmness and reducing microbial decay of harvested vegetables for extended periods (Alvarez et al. 2013). Chitosan coatings are usually used on fruit and vegetable products such as strawberries, cucumbers, and bell peppers as antimicrobial coating and on apples, pears, peaches, and plums as gas barrier (Bourtoom 2008).

Starch

Starches are renewable resources widely available and can be obtained from different byproducts of harvesting and industrialization (Sanjurjo et al. 2006). Starches consist of approximately 25 % amylase and 75 % amylopectin (Çağrı 2002; Bourtoom 2008) Amylase is responsible for the film-forming capacity of starch. The largest source of starch is corn (maize) with other commonly used sources being wheat, potato, tapioca, and rice. Starch is used to produce biodegradable films to partially or entirely replace plastic polymer. Starch films are transparent or translucent, flavorless, tasteless, and colorless (Skurtys et al. 2010). Despite starch is the most used agriculture raw material for biodegradable films, few reports are published about starch-based coatings. Amylase is responsible for the film-forming capacity of starches and it is required for film forming and preparation of strong gels. Starches containing 55 or 75 % amylase content are commercially available, with 70 % amylase content; starch is producing stronger, tougher, and more flexible films. In addition, another important component of starch edible films and coatings is the plasticizer, which must be compatible with the film forming polymer. Plasticizers such as glycerol are frequently added to enhance flexibility; they lower glass transition temperature of system and modify the barrier properties of the films (Garcia et al. 1999; Çağrı et al. 2002).

Alginate

Alginate, a polysaccharide derived from marine brown algae (*Phaeophyceae*) and gellan, a microbial polysaccharide secreted by bacterium *Sphingomonas elodea* (formerly referred to as *Pseudomonas elodea*) is finding increasing use in the food industry as texturizing and gelling agents (Rojas-Graü et al. 2007a, b). In molecular terms, alginates are linear water-soluble polysaccharides comprising (1–4)-linked units of R-D-mannuronate (M) and â-L-guluronate (G) at different properties and different distributions in the chain. The chemical composition and sequence of the M and G residues depend on the biological source and the state of maturation of the plant. The hardness of the three blocks decreases in order to GG>MM>MG. The physical properties of alginates depend on the relative proportion of the three types of blocks. For example, formation of gels, by addition of calcium ions, involves the G block so higher the proportion of these, the greater the gel strength; solubility of alginate in acid depends on the proportion of MG blocks present (Cha et al. 2002; Hambleton et al. 2009a, b; Albert et al. 2010; Galus and Lenart 2013).

Alginate films potentially a good option for same fruit such as cut apple, since these films become stronger when cross-linked with Ca, and at the same time, stick to the cut apple surface through this cross-linking (alginate-Ca-pectin). Such biopolymer-based films and coatings can keep good quality and prolong shelf life of foods by increasing water barrier, preventing microbial contamination, maintaining the flavor, reducing the degree of shrinkage distortion, and retarding fat oxidation. Moreover, the coatings may serve as carriers for antimicrobial compounds and antioxidant in order to maintain high concentration of preservatives on the surface of foods (Olivas et al. 2007; Song et al. 2011).

Pectin

Pectin is an ingredient used in food industry with no limitation other than current good manufacturing practice. It is considered as generally recognize as safe (GRAS) by the FDA and it has been used in food mainly as gelling, stabilizing or thickening agent in products such as jams, yoghurt drinks, fruity milk drinks, and ice-cream

(Espitia et al. 2013). Pectin is an anionic polysaccharide, mostly derived from citrus fruits. The main structural features of pectin include a backbone of (1→4)-linked α-D-galacturonic acid units. According to their degree of methylation, pectin is divided into two categories. These are low-methoxyl pectin (LMPs) and high methoxyl pectins (HMPs), with a degree of methylation respectively lower and higher than 50 %. The degree of methylation has a decisive effect on the mechanisms of gelation (Kang et al. 2007; Piazza et al. 2009; Altenhofen Da Silva et al. 2009; Bierhalz et al. 2012). The hydrocolloidal and polyelectrolytic properties of pectin determine its unique abilities, such as strong water retention in colloidal systems together with their stabilization; easy plasticization with glycerol; due to its hydrophobic groups, ability to absorb organic lipoid substance; an expressive cation-exchange ability forming its curative action. The investigation of many authors present the structural and mechanical properties of edible films and coatings containing citric acid and apple pectin in the form of aqueous solutions with concentration up to 1 % (Baeva and Panchev 2005).

Carrageenan

Carrageenans are water-soluble polymers with a linear chain or partially sulfated galactans, which present high potential as film-forming material (Karbowiak et al. 2006). They are extensively used in food, dairy, and pharmaceutical industry as gelling, emulsifying, and stabilizing agents (Seol et al. 2009a, b). The number and position of sulfate groups on the disaccharide repeating unit determine classification in the major types: κ, ι and λ. The κ-, ι- and λ- carrageenans exhibit sulfate contents of 20 %, 33 % and 41 % (w/w), respectively (Fabra et al. 2009). κ-carrageenan has one negative charge per disaccharide with a tendency to form excellent gel and film–forming properties. Films from κ-carrageenan exhibit, therefore, the highest tensile strength when compared with that of λ- and ι-carrageenan films (Seol et al. 2009a, b). Park (1996) reported that κ-carrageenan has excellent film-forming properties with water vapor permeability (WVP) of 1.87×10^{-10} ngm/m^{2}sPa and tensile strength of 22–32 MPa, which has higher than that polyethylene films (Park 1996). And Choi et al. (2005) reported that κ-carrageenan film containing potassium sorbate has great possibility to extend the shelf life or increase the safety of foods when it is used as packaging or coating material (Choi et al. 2005). Iota-carrageenan is a hydrocolloid widely used in the dairy industry as it present significant reactivity with milk proteins (Karbowiak et al. 2006). Iota-carrageenan, a water-soluble polymer with a linear chain mainly composed of alternated (1,3)-D-galactose-4 sulfate and (1,4)-3,6-anhydro-D-galactose-2-sulfate units, is promising as a film-forming material (Hambleton et al. 2009a, b). Edible films made of ι-carrageenans display interesting advantages such as good mechanical properties, stabilization of emulsions, and reduction of oxygen transfer. But, the highly hydrophilic nature of ι-carrageenan films limits their ability to provide a significant moisture barrier. One way to improve barrier properties is to include lipidic materials in their formulation, such as fatty acids or waxes (Seol et al. 2009a, b).

The use of carregeenan as edible films and coating already covers various fields of the food industry such as application on fresh and frozen meat, poultry and fish to prevent superficial dehydration, ham or sausage-casings, granulated-coated powders, dry solids foods, oily foods, etc., but also manufacturing soft capsules, and especially non-gelatin capsules (Karbowiak et al. 2006).

2.2.2 Protein Films

With their availability, different molecular properties and chemical functional proteins are very suitable sources to obtain edible films (Güçbilmez 2005). Several globular proteins, including wheat gluten, corn zein, soy protein, and whey protein, have been investigated for their film properties (Bourtoom 2008). The edible films composed of proteins generally have good gas barrier properties and suitable mechanical and optical properties. However, they are highly sensitive to moisture and show poor water vapor barrier properties than other biopolymers (Güçbilmez 2005). Several protein edible films described and discussed in following section.

Zein

Zein is the most important protein in corn. It is a prolamine protein and for this reason dissolves in 70–80 % ethanol (Guo et al. 2012). Zein is located in small round particles, 1–2 μm in diameter, called protein bodies in maize endosperm. Three distinct fractions, a, b, and g zein, have been identified by differential solubility in aqueous alcohol solutions. Commercial zein is a mixture of proteins with different molecular sizes, solubility, and charge. Commercial preparations usually contain only the fraction of zein (Lai and Padua 1997). Zein is a relatively hydrophobic and thermoplastic material, and has excellent film-forming properties, so it can be used for fabrication of biodegradable films. These film-forming properties have attracted attention in the field of edible film and coating materials. However the characteristic brittleness of zein diminishes its usefulness as a film; some modifications are needed to improve their flexibility. In addition, preparation conditions also affect the properties of zein films (Guo et al. 2012). Zein coatings have been used to coat nuts and candy for increased gloss, and prevention of oxidation and development of off-odors. Zein coatings offer a reasonable alternative to shellac and carnauba wax. The coating can serve as a carrier for antimicrobial compounds and/or antioxidants compounds in order to maintain high concentrations of preservatives on the food surfaces. Performance of zein films as barrier packaging for popcorn, tomatoes, cooked turkey, and shell eggs has been evaluated. In addition, use of zein-based coatings for reducing oil uptake by deep-fried foods, for obtaining controlled release of active ingredients in pharmaceutical tablets, and for masking the taste of bitter orally administered drugs has been discussed in recently awarded patents (Baysal et al. 2009).

Gelatin Films

Collagen is a biopolymeric fiber and the most abundant mammalian and fish protein (Diop 2009). Gelatin versatility and functionality reflects the fact that it is the only food protein that undergoes a thermally reversible helix coil partial transition to resemble its original parent protein structure, collagen (Avena-Bustillos et al. 2006). Gelatin is a protein that is widely used in the pharmaceutical and food industries, and it is produced on a large scale at relatively low prices (De Carvalho and Grosso 2006). Gelatin is an edible biodegradable and biocompatible polymer that is produced by the thermal or physical and chemical degradation of collagen. There are two types of gelatin. These are Type A and Type B. The type of gelatin that is formed from collagen is dependent on whether the collagen is pretreated with an alkaline or an acid. Type A is prepared through acid pretreatment and has an isotonic point between 7 and 9. Gelatin Type B is prepared by alkaline pretreatment of collagen and has an isotonic point between 4.7 and 5.2 (Diop 2009). The use of gelatin in elaboration of edible films or coatings was very well studied until the 1960s, which resulted in many patents, mainly in pharmaceutical area. As a result, these materials characteristics are not easily available in the literature. But gelatin has return to attention of several researches on edible films for food application. Some researches elaborated edible films from bovine hide gelatin and determined their mechanical properties by puncture test, as function of pH and gelatin and sorbitol concentrations in the filmogenic solution, using surface-response methodology. Same researchers studied the drying of films of gelatin plasticized with sorbitol, with special interest in effect of drying conditions on the quality of the formed film in their another investigation. Those recent interest are justified by the ready availability and the low cost of gelatin. However, there is a lack of more detailed works on the plasticizer effect on the thermal and functional properties of gelatin edible films. Therefore, researches focused on to study the WVP and the mechanical and thermal properties of edible films based on bovine hide and pigskin gelatins as functions of sorbitol content (Sobral et al. 2001).

Wheat Gluten Films

Wheat gluten is a general term for water-insoluble proteins of wheat flour which is composed of a mixture of polypeptide molecules, considered to be globular proteins (Bourtoom 2008). Film production from what gluten a mixture of proteins accounting for about 80–85 % of wheat flour proteins, has also been studied (Gennadios et al. 1993). Wheat gluten protein, a renewable resource, is capable of forming a fibrous network which lends strength, elasticity, and plasticity when formed into a film with glycerol. Research on the preparation and properties of wheat gluten films arose owing to the excellent viscoelastic properties of wheat gluten. The two general methods used to prepare wheat gluten films are dry processing or solvent casting. Wheat gluten films have excellent oxygen and carbon dioxide barrier properties compared to plastic films but have low water vapor barrier properties compared to plastic films. A linear relationship between gluten film thickness and film barrier properties for oxygen and carbon dioxide gas was observed by Park and Chinnan (1995). Upwards, Olabarrieta et al. (2006) reported that an increase in oxygen permeability for pH4 wheat gluten films compared to pH11 wheat gluten films, which could be attributed to reduced protein aggregation and a more heterogeneous wheat gluten film structure at pH4 (Olabarrieta et al. 2006). Significant decrease of the WVP for wheat gluten films has proven quite challenging to achieve and represents a major limitation in the application of wheat gluten films for food packaging materials (Cousineau 2012).

Soy Protein Films

The protein content of soybeans (38–44 %) is much higher than the protein content of cereal grain (8–15 %) (Bourtoom 2008). Soy protein is extracted from soybeans used to obtain soy oil. During this process, soy flour is obtained as a secondary product and it can be purified to obtain soy protein concentrate (SPC) and soy protein isolate (SPI), which would add value to agricultural by-products. Soy proteins are composed of a mixture of albumins and globulins. Globulins are protein fractions in which the subunits are associated via hydrophobic and hydrogen bonding (Guerrero et al. 2011). Soy protein is globular in nature and is further classified into 2S, 7S, 11S and 15S fractions; the main components being conglycinin (7S) and glycinin (11S). While both of these fractions are tightly folded, alkaline conditions and heating cause dissociation and subsequent unfolding due to deamination, since soy protein is high in asparagine and glutamine residues (Skudlarek 2012). Soy protein-based edible films have received considerable attention due to their excellent film-forming abilities, low cost, and barrier properties against oxygen, lipid and aroma permeation under low to intermediate humidity conditions. This type of proteins produces smoother, clearer and more flexible films compared to those from other sources. However, due to its inherent hydrophilic nature, this material presents two major disadvantages such as fragility in the wet state and poor properties of moisture barrier. These effects can be minimized using physical, chemical or enzymatic treatments including: blending with hydrophobic additives such as neutral lipids, fatty acids or waxes; changing drying conditions; enzymatic treatment with horseradish peroxidase; heat curing; UV irradiation; and cross-linking (Gonzales et al. 2011).

Collagen Casings

Collagen is a fibrous, structural protein in animal tissue, particularly skin, bones, and tendons and represents about 30 % of the total mass of the body (Yener 2007; Alizadeh and Behfar 2013). Collagen is the most commercially successful edible protein film. Film-forming collagen has been traditionally used in the meat industry, for the production of edible sausage casings. This protein has largely replaced natural gut casings for sausages. Collagen is readily available, non-toxic and provides an excellent basis for biomaterials. Collagen edible films and coatings from animal origin proteins can be dissolved in dilute acid or alkali solutions, and in neutral solutions. Two major components are identified; α (MW 100 000 Da) and β (MW 200 000 Da), and consist of two different types of covalent cross-linked chain pairs α1-α1 and α2-α2. Hydrolysis of collagen results in gelatin. The molecular weight of gelatin covers a broad range, from 3,000 to 200 000 Da, depending on the raw material employed during gelatin production and handling conditions. Edible coatings made with gelatin reduce the migration of moisture, oxygen, and oil. Collagen films are not as strong and tough as cellophane, but have good mechanical properties. Collagen films have an excellent oxygen barrier at 0 % relative humidity; however, the oxygen permeability increases rapidly with increasing relative humidity in a manner similar to cellophane. Different cross-linking chemical agents have been used to improve the mechanical properties, to reduce the solubility, and to improve the thermal stability of these films. Carbodimide, microbial transglutaminase, and glutaraldehyde are usually used as cross-linking agents (Alizadeh and Behfar 2013).

Whey Protein Isolate

Whey proteins are byproducts of the cheese-making industry and have generally been disposed of as animal feed or used in infant formulas and sports food (Zinoviadou et al. 2009). There are several individual proteins within the mixture of whey protein, with 5-lactoglobulin, 3-lactoalbumin, Bovine Serum Albumin (BSA), and immunoglobins being the main proteins. Among them, the most abundant and important protein for film formation is 5-lactoglobulin and the second most abundant whey protein is 3-lactalbumin (Jooyandeh 2011). Whey proteins have exceptional nutritional value and functional properties (Ozdemir and Floros 2008). They have been successfully employed as raw material for biodegradable packaging because they come from a renewable source and are a byproduct of cheese making industry; hence, they are widely available, relatively easy to handle and essentially inexpensive (Ramos et al. 2012a, b). The formation of edible films and coatings from whey proteins can increase the utilization of whey, improve the nutritional value of foods and prolong shelf life (Ozdemir and Floros 2008). Manab et al. (2011) reported that the whey protein-based edible films is usually prepared using whey protein with incorporation of plasticizer, cross linking agent and lipid, before heat denaturation at 90 °C for 30 min, the pH was adjusted to 5,2 and cooled to room temperature before it was template on Teflon plate and semi-vacuum oven for 24 h. The produced edible film had a soft, transparent, and good aroma as well as oxygen-resistant characteristics at low humidity (Manab et al. 2011). Whey protein isolates (WPI) represent the purer form of such whey proteins, and shown promising mechanical features, as well as moderate moisture permeability and good oxygen barrier properties comparable to those exhibited by the best synthetic polymer-based films available, e.g., low-density polyethylene (LDPE), high density polyethylene, ethylene vinyl alcohol, polyvinylidene chloride (PVDC), cellophane, and polyester. Furthermore, those films proved excellent biomaterials for use as carriers of such food additives as antioxidants, antimicrobials, colorants, flavors, fortifying nutrients and spices; the additives improve the functionality of the packaging by bringing about novel (or extra) features (Ramos et al. 2012a, b).

2.2.3 Lipid Films

Lipid compounds utilized as protective coating consist of acetylated monoglycerides, natural wax, and surfactants. The most effective lipid substances are paraffin wax and beeswax. The primarily function of a lipid coating is to block transport of moisture due to their relative low polarity. In contrast, the hydrophobic characteristic of lipid forms thicker and more brittle films. As a result, they must be associated with film forming agents such as proteins or cellulose derivatives. Generally, WVP decrease when the concentration of hydrophobicity phase increases. Lipid-based films are often supported on a polymer structure matrix, usually a polysaccharide, to provide mechanical strength (Bourtoom 2008).

Waxes belong to the non-polar lipid class. Their hydrophobicity is high. They have differences in permeability of wax films. These differences are owing to their chemical composition and crystal type. The waxy skin of fresh fruit and vegetables are applied to reduce dehydration and control the exchange of gases to prolong preservation period. There are some examples of waxes used for coating, including paraffin wax, carnauba wax, beeswax, candelilla wax, polyethylene wax (Yener 2007). Paraffin wax is permitted for use on raw fruit and vegetable and cheese. Carnauba wax is an exudate from palm tree leaves. Mineral oil consists of a mixture of liquid paraffin and naphtheric hydrocarbon. If applied as a thick layer, they must be removed before consumption (certain cheese); when used in thin layers, they are considered edible. Waxes (notably paraffin, carnauba, candellila, and bee wax) are the most efficient edible compounds providing a humidity barrier (Bourtoom 2008).

Monoglycerides are used in edible films as emulsifiers, especially for stabilizing emulsified film and increasing adhesion between two components with different hydrophobicity. Triglycerides are insoluble in bulk water, but will spread at the interface to form a stable monolayer. Water affinity or hydrophobicity of triglyceride depends on its structure. By adding palmitic, stearic, lauric acids, and stearyl alcohols to edible films, the moisture barrier properties are greatly enhanced (Yener 2007).

2.2.4 Composite Films

Edible films and coatings may be heterogeneous in nature, consisting of a blend of polysaccharides, protein, and/or lipids. This approach enables one to utilize the distinct functional characteristics of each class of film former (Bourtoom 2008). Composite films can be designed by combining lipid and hydrocolloid elements. By this way, it can decrease the disadvantages of each film. When a barrier to water vapor is desired, the lipid component can serve this function, while hydrocolloid component provides the necessary durability. Composite films consisting of a conglomerate of casein and acetylated monoglycerides have been studied in many investigations. These films can be used as coatings for processed fruit and vegetables (Yener 2007).

3 Antimicrobial Edible Films and Coatings

Nowadays, studies dealing with edible films with antimicrobial properties are on the increase. These films could prolong the shelf life and safety of foods by preventing growth of pathogenic and spoilage microorganisms as a result of their lag-phage extension and/or their growth rate reduction. Moreover, antimicrobials containing in films can be gradually released on the food surface, therefore, requiring smaller amounts to achieve the target shelf life (Ponce et al. 2008).

3.1 Properties of Antimicrobial Edible Films and Coatings

Microbial contamination in foods occurs due to post-processing to be traded manually on the surface of the food primarily and while reducing the shelf life of foods, may increase the risk of food borne illness. Antimicrobial agents applied to the surface of food directly to prevent or delay the decay of the surface by processes such as spraying or dipping. However, surface application of antimicrobial agents, antimicrobial substances transition to food quickly or neutralize in food. This case is limited the usefulness of what you did in food (Coma et al. 2002). All of these disadvantages and increasing consumer demands have led to the emergence of new food packaging systems for provide longer shelf life of foods and improve food safety. These systems limit the passage of flours, oxygen, and moisture and increase the shelf life of foods by providing antimicrobial activity (Quntavalla and Vicini 2002; Cha and Chinnan 2004). Recently years, antimicrobial edible films which studied extensively to candidate to replace synthetic antimicrobial active packaging system with features such as be consumed with food, being biologically degraded and they are reduce to the use of synthetic materials. Researches began to the mid-1980s work on edible films and coatings containing antimicrobial agent such as sorbic acid and potassium sorbate. In those years, film materials are more commonly methyl cellulose, hydroxypropyl methyl cellulose, polysaccharides, and lipids such as fatty acids but there are few studies on proteins. In the mid 1990s antimicrobial films prepared with organic acid and chitosan. To the end of the 1990s they were used as natural antimicrobial agents in edible protein films (Gernadious et al. 1994).

The antimicrobial food packaging interact with the surface of foods provide food safety by reducing the rate of growth of specific microorganisms in foods. Food composition and the target microorganism to be considered preparation of antimicrobial films and coatings and diffusion kinetics and antimicrobial activity of diffusible substance to food packaging have to be determinate (Appendini and Hotchkiss 2002).

Packaged foods with edible film packaging immediately before or after the process unpacking can be contaminated by microorganisms. These microorganisms settle the surface of food, i.e., the space between food packaging and food. In edible coating applications, on the surface of the food coated with the coating material micro-organisms cannot develop due to the direct interaction of antimicrobial substances and lack of oxygen and microbial growth occurs on the surface of coating. Antimicrobial agent is passed to food layer which not containing antimicrobial originally from the film and coating and consequently decreasing the amount of antimicrobial agent. To reduce the consumption of antimicrobial substances depending on migration on the film and coating, antimicrobial substance migration speed must be controlled by the kinetics of diffusion rate of transition (Appendini and Hotchkiss 2002).

Edible films and coatings have different protection functions. Antimicrobial agent passes slowly to food on edible film systems. Thus, there is no high concentration of antimicrobial agent in the film and food surface and film can effect a longer time against microorganisms (Coma et al. 2002; Çağrı et al. 2002). In edible coating systems, antimicrobial agents must remain on the coating material for the protection of food from microorganisms. Thus, diffusion rate on the coatings should be lower than film for effective antimicrobial activity (Gernadious et al. 1994).

3.1.1 Antimicrobial Substances

Antimicrobial substances inhibit or inactivate of microorganisms. Antimicrobial substances are used in a wide variety such as the organic acid and its salts, fungicides, bacteriocins, antibiotics, enzymes, and alcohols to provide antimicrobial activity in synthetic packaging systems. But, security and edible features are important on edible films and coatings and also type and amount of antimicrobial agent that can be used in these systems are limited (Ayana 2007). Antimicrobial agents can be classified as chemical and natural antimicrobials (Gernadious et al. 1994).

Weak organic acids such as propionic acid, sorbic acid, benzoic acid, tartaric acid, and salts of organic acids such as sodium benzoate, potassium sorbate, and propionate are used as chemical preservatives in foods commonly. Chemical preservatives consumed with the food, thus a number of these chemicals have restriction on using and limited amounts of antimicrobial agent have been used as edible films and coatings (Gernadious et al. 1994; Cha and Chinnan 2004; Ayana 2007).

Enzymes, organic acids, fatty acids, pigments, flavones, and spice oils have antimicrobial effects which are found naturally in foods and natural preservatives. They have effective antimicrobial activity when used critical amount and over. These aspects are not like chemical preservatives. Bacteriocins are antibacterial proteins produced by bacteria to kill or inhibit the growth of other bacteria. Many lactic acid bacteria (LAB) produce a high diversity of different bacteriocins. Nisin, Pediocin, and colicin are other bacteriocins which are used in edible film and coatings. Lysozyme is obtained from various sources and it is an enzyme having the antimicrobial effect by breaking β 1-4 glucozidic bonds in peptidoglican in the gram-positive and gram-negative bacteria cell walls. Lysozyme was used as antimicrobial many studies in the literature for production edible films and coatings (Gernadious et al. 1994; Cleveland et al. 2001; Cha and Chinnan 2004).

Many antimicrobial agents containing edible film and coatings are added to polymer as molten by applying thermal treatment or dissolution in solvent components. Heat-sensitive antimicrobial agents such as enzymes or volatile components can be used by dissolution using the solvent components; this is a convenient method for production antimicrobial edible films and coatings. For instance, lysozyme is used as an antimicrobial prevents denaturation of enzyme when used in cellulose ester film with solvent components. Bacteriocins which are heat resistant antimicrobial agents but they had a higher antimicrobial activity. Antimicrobial agents and polymers are both of must dissolution in the same solvent such cases. Antimicrobial agents can be used with solvents such as water and ethanol in edible films obtained from protein, lipids, and carbohydrates biopolymers (Ayana 2007).

If the non-volatile antimicrobial agents are used in packaging materials, packaging material must be in contact with the surface of food for diffusion of antimicrobials. The multilayer films should be used for diffusion of antimicrobial substances to food slowly. In a multilayer film, the innermost layer while controlling the diffusion rate of active substances, the matrix layer contains active ingredients. Also, barrier layer prevents antimicrobial agents diffusion to out of the packaging material (Appendini and Hotchkiss 2002).

When used as volatile compounds, antimicrobial agent packaging material is not required to direct contact with the food. Porous foods such as hamburger patties and bread or air gap foods such as milk powder prevents the spread of antimicrobial agent due to uneven surface and air gap. In this case only volatile antimicrobial agents are used and they provide effective protection by spreading than heterogeneous foods (Gernadious et al. 1994).

3.2 The Methods Used to Test the Antimicrobial Activity of Antimicrobial Packages

Different methods are used to test the antimicrobial activity of the antimicrobial packaging. These methods are minimum inhibitory concentration test (MIC), agar diffusion test and the rocking flask test. MIC test, which is one of the most widely, compared to antimicrobial activity of polymers and their separate effects. This method based on the principle of incubated different amounts of the antimicrobial agent containing polymers and antimicrobial agents with the target microorganism growth medium and until hold on microbial growth is observed. The lowest concentration of antimicrobial agent is determined for inhibition of microorganisms by MIC test. The agar diffusion method is based on the principle of antimicrobial films after replacing into the solid medium containing the microorganism to be tested and then incubated until the release is observed on the microbial growth medium. The presence of the microorganism to be tested or the presence of antimicrobial any ambient can be determined using this method. During incubation, a zone of inhibition has been observed around the medium by diffusion antimicrobial agent in film to medium. The zone of inhibition that exposes prevented the development of microorganisms. Measuring the diameter of the zone around the film can be expressed the effectiveness of an antimicrobial agent by quantitative terms. Swinging flask test provides detailed information on the kinetics of antimicrobial agents. The target microorganism and containing antimicrobial polymer is placed into the liquid medium such as buffer, growth medium or food and incubated in the stirred medium. Reduction in the rate of growth of microorganisms is measured by this method. Also test provides information about the antimicrobial activities of polymers in buffer (Temiz 2000; Appendini and Hotchkiss 2002).

3.3 *Characteristic Properties of Antimicrobial Edible Films and Coatings*

Edible films and coatings have mechanical properties such as tensile and tear resistance, physical properties such as oxygen, water vapor, flavor permeability, and optical characteristics as well as gloss, haze, transparency. The films and coatings properties can change due to increased heterogeneity in the structure of the film and coating as a result of the addition of antimicrobial substances (Quntavalla and Vicini 2002).

The films optical properties can change due to adding antimicrobial agents to plastic films but the physical strength of these films is not change. The addition of potassium sorbate did not affect the mechanical properties of the films made of polyethylene. However, the addition of water-soluble antimicrobial agent in hydrophobic plastic film caused reduction of film transparency. Film transparency and tensile are not affected when film solution prepared whey protein and antimicrobial substances such as lysozyme because protein and lysozyme compatible with each other and have the high water solubility. But the addition of antimicrobial substances in film solutions prepared such as corn zein protein which soluble in alcohol solutions, the physical resistance of films can reduce. Hence antimicrobial film must be incorporate antimicrobial agents compatible with the film-forming substance. A substance that are used as antimicrobial agents does not affect the film's taste and smell because they used low concentrations in film solutions (Gernadious et al. 1994).

When examine the literature we can see that natural antimicrobial especially essential oils are being used for the preparation of antimicrobial edible films and coatings. The major advantages of using essential oils in film and coatings is that the diffusion rate of the antimicrobial agent can be slowed down, thereby keeping high concentration of active compounds on the product surface for extended periods of time. However, further efforts must be made to control diffusion rate of these active compounds to the product surface during storage. On the other hand, the nature and amount of the essential oils, the essential oils/polymer ratio in the film and the possible interactions between the polymer and the active compounds of essential oils play important role in the film's antimicrobial activity (Sanchez-Gonzalez et al. 2011a, b).

In addition to conferring antimicrobial properties to edible films, the incorporation of essential oils leads to modifications in terms of physical film properties. These modifications are usually similar to those presented when adding more simple lipids to the film matrix. Again, the interactions established between essential oil components and the polymeric matrix become more complex, and it is important to take them into account when optimizing the composition of bioactive coatings (Sanchez-Gonzalez et al. 2011a, b).

Determination of WVP is strongly dependent on measurement conditions, such as temperature and the gradient of water vapor pressure. Coated biomaterials, such as fruits and vegetables, are characterized by irregular shapes and high water contents that make WVP measurements difficult (Garcia et al. 2009). Low WVP values are desirable in order to minimize weight losses in the coated product which, in turn, also directly affects products firmness and appearance. The incorporation of essential oils into polymeric matrices leads to an improvement in the film WVP because of the increment in the hydrophobic compound fraction in the film. Usually, WVP values fall linearly with the increase in essential oils concentration. For example, Sanchez-Gonzalez et al. (2011a, b) reported that pure chitosan films without hydrophobic compounds have poor moisture barrier properties at 20 °C, but the incorporation of bergamot oil (3 %) induces a significant reduction in WVP of nearly 50 % (Sanchez-Gonzalez et al. 2011a, b).

Propionic acid, polylactic acid (PLA), and stearic acids, natural lactoperoxidase system, natamycin, and nisin used in the edible films and coatings production as natural antimicrobials as well as essential oils. Also chemical preservatives such as sodium benzoate and potassium sorbate are widely used in production antimicrobial edible films. Ramos et al. 2012a, b reported that lactic acid, propionic acid, and natamycin incorporated WPI films have lower than those reported elsewhere for edible films manufactured from other materials. In their study WVP values $109 \pm 0{,}75$ for lactic acid, $12{,}8 \pm 0{,}22$ for propionic acid, $11{,}1 \pm 1{,}04$ for natamycin (Ramos et al. 2012a, b). Ko et al. (2001) reported that nisin at the amounts added had no significant effect on the WVP values of WPI. They found that WVP value is $2{,}20 \pm 0{,}24$ when the added nisin in whey protein isolate but if is not added nisin in film solution WVP value is $2{,}41 \pm 0{,}18$ (Ko et al. 2001). Bierhalz et al. (2012) reported that the addition of natamycin caused a significant increase of the permeability coefficient in pectin, pectin and alginate, and only alginate film formulations (WVP values are 2,13–4,11 g mm/d m^2kPa) Researchers reported that this result may be associated to al looser packing of the film macromolecules increasing the free volume of the polymeric structure, which enhances permeability. According to the same study pectin films showed lower WVP values than the respective composite films and alginate films, probably due to observed differences in thickness (Bierhalz et al. 2012). Another study used lactoperoxidase system as antimicrobial Mohamed et al. (2013) low values of WVP ($5{,}61 \pm 0{,}31$–$8{,}42 \pm 0{,}91$) were due to the fact that chitosan films, like many other protein or polysaccharide edible films,

Table 1 Effects of essential oils on water vapor permeability (WVP) of edible films

Essential oil and oil compounds (% w/v)	Film polymers	WVP (g mm/ kPa h m2)	References
Oregano oil (0.1)	Alginate/apple puree	5.25 ± 0.33	Rojos-Graü et al. (2007)
Carvacrol oil (0.1)	Alginate/apple puree	5.02 ± 0.22	Rojos-Graü et al. (2007)
Lemongrass oil (0.5)	Alginate/apple puree	4.91 ± 0.40	Rojos-Graü et al. (2007)
Citral (0.5)	Alginate/apple puree	5.12 ± 0.13	Rojos-Graü et al. (2007)
Cinnamon oil (0.5)	Alginate/apple puree	4.90 ± 0.27	Rojos-Graü et al. (2007)
Cinnamaldehyde (0.5)	Alginate/apple puree	4.37 ± 0.54	Rojos-Graü et al. (2007)
Oregano oil (0.1)	Cassava Starch-Chitosan	0.99 ± 0.04	Pelissari et al. (2009)
Oregano oil (0.5)	Cassava Starch-Chitosan	0.74 ± 0.08	Pelissari et al. (2009)
Oregano oil (1.0)	Cassava Starch-Chitosan	0.62 ± 0.15	Pelissari et al. (2009)
Cinnamon oil (0.4)	Chitosan	2.25 ± 0.074	Ojagh et al. (2010)
Cinnamon oil (0.8)	Chitosan	1.352 ± 0.152	Ojagh et al. (2010)
Cinnamon oil (1.5)	Chitosan	1.014 ± 0.040	Ojagh et al. (2010)
Cinnamon oil (2.0)	Chitosan	1.003 ± 0.067	Ojagh et al. (2010)
Citronella (0.25)	Hake protein	4.12 ± 0.50	Pires et al. (2013)
Tarragon (0.25)	Hake protein	4.2 ± 0.04	Pires et al. (2013)
Thyme (0.25)	Hake protein	3.57 ± 0.11	Pires et al. (2013)
Coriander (0.25)	Hake protein	3.77 ± 0.00	Pires et al. (2013)
Oregano oil (1.0)	Triticale flour protein	0.35 ± 0.04	Aquirre et al. (2013)
Oregano oil (2.0)	Triticale flour protein	0.40 ± 0.05	Aquirre et al. (2013)
Carvacrol oil (3.2)	Chitosan	2.51 ± 0.0003	Rubiler et al. (2013)
Carvacrol oil (20)	Chitosan	1.58 ± 0.0004	Rubiler et al. (2013)
Carvacrol oil (30)	Chitosan	1.69 ± 0.0002	Rubiler et al. (2013)
Satureja hortensis oil (1.0)	κ-carrageenan	1.591 ± 0.112	Shojaee-Aliabadi et al. (2013)
Satureja hortensis oil (2.0)	κ-carrageenan	0.840 ± 0.093	Shojaee-Aliabadi et al. (2013)
Satureja hortensis oil (3.0)	κ-carrageenan	0.556 ± 0.032	Shojaee-Aliabadi et al. (2013)

exhibited relatively low water barrier characteristics due to their high hydrophilic nature (Mohamed et al. 2013). Effects of essential oils on WVP of edible films are summarized in Table 1.

The important study about bio-active edible films carried out by Sanchez-Gonzalez et al. (2013). In this study researches LAB were added directly to the sodium caseinate, pea protein, and methylcellulose and hydroxypropylmethylcellulose film-forming solutions. WVP values were range of 7–34,4 g mm kPa^{-1} h^{-1} m^{-2}. Researchers reported that differences can be attributed to some changes in the experimental conditions; temperature, RH gradient, kind, and amount of plasticizer. It was verified that polymer films are highly permeable to water vapor, which is coherent with the hydrophilic nature of the protein and polysaccharides (Sanchez-Gonzalez et al. 2013).

Rezvani et al. (2013) were prepared edible films using sodium caseniate (6–8 g/100 g) and stearic acid (0–2 g/100 g). WVP values were determined range of 1,368 ± 0,124–1,896 ± 0,047 g mm m^2 h kPa in this study. As shown this values, any increase in ratio of stearic acid to water, decreases the WVP. The effect of stearic acid on decreasing these parameters is mainly due to the addition of hydrophobic groups in formulations (Rezvani et al. 2013). In another study, Bonilla et al. (2013) were prepared films based on PLA and different amounts of chitosan powder (CH). The effects of CH particle size (715 and 180 μm) and the amount of chitosan incorporated in the PLA matrix (5 % or 10 % on PLA basis) were investigated in terms of physicochemical characteristics. In this study, WVP values were determined range of 4–74 (g s^{-1} m^{-1} Pa^{-1}) × 10^{11}. Researchers reported that PLA films showed lower WVP than composite films, which can be due to the greater water affinity of CH could favor the transport of water molecules through the film. According to researchers, water barrier properties of PLA:CH films increased as PLA content was higher as a result of the well-known hydrophobicity of PLA (Bonilla et al. 2013).

Many antimicrobial edible films examples utilized of chemical preservatives in the literature. Potassium sorbate is the one of the chemicals most widely used. Ozdemir and Floros (2008) examined the effect of protein, sorbitol, beewax, and potassium sorbate concentrations in whey protein films on their WVP in a study. According to researches all factor influenced WVP. They reported that mixture proportions of protein 0.53, sorbitol 0.38, beewax 0,08 and potassium sorbate 0.01 would yield an edible film with minimum stickiness and WVP < 9 g mm m^{-2} h^{-1} kPa^{-1}. They stated that the addition of protein and beewax decreased WVP, while the increase in sorbitol increased WVP. Potassium sorbate was the most effective factor that adversely affected WVP (Ozdemir and Floros 2008). Another study potassium sorbate is used as antimicrobial by Arismendi et al. (2013). Xanthan gum and tapioca starch were used as film polymers. In this study WVP values were determined range of 1.7–2.3 × 10^{-9} g/m s Pa. Researchers reported that it was not observed a clear trend concerning the effect of xanthan gum and tapioca starch on WVP (Arismendi et al. 2013).

The packaging material's gases such as carbon dioxide and oxygen, transfer rates and aroma such as volatile compounds are also factors that affect the stability during storage of the food except water vapor. Oxidation of fats in foods due to food rancidity as a result of the presence of oxygen and this situation can cause reduced the acceptability of food. In addition it leads to loss of light-sensitive vitamins. Also, oxygen is important for many different organisms which cause deterioration on foods. Oxygen concentration is effect on microbial growth rate. Out of them volatile compounds that make up the taste and aroma of foods not required out of the transport packaging. The losses will occur in these compounds negatively affect the quality of the food (Ayana 2007). The oxygen barrier is quantified by the oxygen permeability coefficients (OPC) which indicate the amount of oxygen that permeates per unit of area and time in a packaging materials [kg m m^{-2} s^{-1} Pa^{-1}]. So, when a polymer film packaging has a low oxygen permeability coefficient, the oxygen pressure inside the container drops to the point where the oxidation is retarded, extending the shelf life of the product. Generally the biodegradable polymers present a value one or more order of magnitude below the synthetic polymer used in the same field like PET and OPS. Several authors reported in literature the OPC of one of the most commercialized biodegradable polymer like the PLA (Siracusa et al. 2008).

Composite films with essential oil seem to be a better barrier to gases, but little information has been found in the literature. For example, a slight decrease in oxygen permeability of the films based on alginate-apple puree with lemongrass oil (Sanchez-Gonzalez et al. 2011a, b). Altiok et al. 2010 reported that oxygen transmission rate of the films increased with an increasing oil concentration due to the microstructural changes in the film becoming porous with the addition of oil. Oxygen transmission rate of the pure chitosan film was 1.24 cc per m^2 per day. The highest oxygen permeability was obtained as 4.61 cc per m^2 per day for 1.2 % (v/v) thyme oil incorporation which suggests the promising applications of thyme oil loaded chitosan film as a wound dressing (Altiok et al. 2010).

The protection of structural integrity and having a certain impact resistance based on the location to be used are at the beginning of expected properties of a packaging material. Therefore, knowing mechanical properties of packaging materials are important for troubleshooting problems and that may arise from use areas. In general, the mechanical properties of polymeric materials are defined as the elongation, ductility, tensile deformation under the influence of external forces. Commonly used to determine the mechanical properties of polymer films was measured tensile strength and elongation. Tensile strength is the material response which applied on packaging material an external force. Elongation of the material is expressed as geometric change state by the effect of the external forces (Ayana 2007).

The mechanical properties of edible coatings depend on several factors, as the interactions between their components and the polymer matrix are strongly affected by the physical, chemical and temperature conditions, which in turn influence film stability and flexibility. The incorporation of essential oil into a continuous polymeric matrix decreases its mechanical resistance to fracture because of the structural discontinuities caused by the oil-dispersed phage. Elongation at break of pure chitosan films was, for instance, reduced when cinnamon, tea tree, or bergamot oil was incorporated (Sanchez-Gonzalez et al. 2011a, b). Effects of natural antimicrobial compounds on Tensile Strength (TS), Elongation break (E) and Young Module (YM) of antimicrobial edible films are summarized in Table 2.

Optical properties, especially color, are essential in packaging applications since they affect the appearance of the product and make consumers accept or reject the product packaged, so becoming an important quality factor to be taken into account (Ramos et al. 2013; Leceta et al. 2013). Generally, the use of essential oils induces modifications in terms of film transparency, gloss, and color. Addition of essential oils in edible films may change the native color of edible films. The appearances of the coatings are of relevance since their commercial acceptance depends mainly on this attribute. Usually, the incorporation of essential oils into films decreases their gloss and transparency due to the increase in the surface roughness of the composite films as a consequence of the migration of droplets or aggregates to the top the film during film drying, which leads to surface irregularities. Nevertheless, observed differences in terms of color are not significant when low concentrations of essential oils are used

Table 2 Effects of natural antimicrobial compounds on Tensile strength (TS), Elongation break (E), and Young Module (YM) of antimicrobial edible films

Antimcirobial compounds	Film polymers	E (%)	TS (Mpa)	YM (Mpa)	References
Nisin	Whey protein isolate	–	3.50 ± 0.34	–	Ko et al. (2001)
Nisin	Soy protein isolate	–	10.43 ± 2.53	–	Ko et al. (2001)
Nisin	Egg albumin	–	1.53 ± 0.10	–	Ko et al. (2001)
Nisin	Wheat gluten	–	1.96 ± 0.76	–	Ko et al. (2001)
Stearic acid	Hydroxypropyl methylcellulose	6.63 ± 1.28	336 ± 63	191 ± 61	Sebti et al. (2002)
Oregano oil	Alginate/apple puree	56.96 ± 3.86	2.47 ± 0.37	5.75 ± 0.96	Rojos-Graü et al. (2007)
Carvacrol oil	Alginate/apple puree	58.33 ± 4.66	2.58 ± 0.37	5.96 ± 1.12	Rojos-Graü et al. (2007)
Lemongrass oil	Alginate/apple puree	55.95 ± 5.55	2.56 ± 0.46	6.02 ± 1.07	Rojos-Graü et al. (2007)
Citral	Alginate/apple puree	57.38 ± 5.71	2.52 ± 0.44	6.46 ± 1.27	Rojos-Graü et al. (2007)
Cinnamon oil	Alginate/apple puree	57.88 ± 5.37	2.84 ± 0.48	6.86 ± 1.16	Rojos-Graü et al. (2007)
Cinnamaldehyde	Alginate/apple puree	55.50 ± 7.40	2.75 ± 0.42	6.77 ± 0.87	Rojos-Graü et al. (2007)
Grape seed extract	Soy protein isolate	–	10.74 ± 0.88	–	Sivarooban et al. (2008)
Nisin	Soy protein isolate	–	5.11 ± 0.30	–	Sivarooban et al. (2008)
EDTA	Soy protein isolate	–	9.29 ± 0.29	–	Sivarooban et al. (2008)
Grape seed extract + Nisin	Soy protein isolate	–	9.78 ± 0.12	–	Sivarooban et al. (2008)
Grape seed extract + EDTA	Soy protein isolate	–	8.34 ± 0.44	–	Sivarooban et al. (2008)
Nisin + EDTA	Soy protein isolate	–	4.72 ± 0.40	–	Sivarooban et al. (2008)
Grape seed extract + Nisin + EDTA	Soy protein isolate	–	7.78 ± 0.33	–	Sivarooban et al. (2008)
Tea tree essential oil	Hydroxypropyl methylcellulose	0.09	55	1.289	Sanchez-Gonzalez et al. (2009)
Oregano oil	Cassava Starch-Chitosan	27.18 ± 2.85	1.96 ± 0.44	67.72 ± 4.26	Pelissari et al. (2009)
Cinnamon oil	Chitosan	16.57 ± 0.77	13.35 ± 1.23	–	Ojagh et al. (2010)

(continued)

Table 2 (continued)

Antimcirobial compounds	Film polymers	E (%)	TS (Mpa)	YM (Mpa)	References
Natamycin	Alginate	6.84 ± 1.09	106.73 ± 4.99	–	Bierhalz et al. (2012)
Natamycin	Pectin	4.75 ± 0.37	48.46 ± 5.93	–	Bierhalz et al. (2012)
Natamycin	Pectin/Alginate	4.81 ± 0.45	80.87 ± 7.08	–	Bierhalz et al. (2012)
Thyme oil	Hake protein	129.8 ± 51.2	6.67 ± 2.41	–	Pires et al. (2013)
Lactoperoxidase system	Chitosan	24.68 ± 2.21	279.26 ± 46	–	Mohamed et al. (2013)
Lactic acid bacteria (LAB)	Sodium caseinate	5.8 ± 0.6	10.47 ± 0.16	269 ± 77	Sanchez-Gonzalez et al. (2013)
Lactic acid bacteria (LAB)	Pea protein	15 ± 2	5.2 ± 1.2	90 ± 44	Sanchez-Gonzalez et al. (2013)
Lactic acid bacteria (LAB)	Hydroxypropyl methylcellulose	33 ± 2	30.1 ± 1.2	381 ± 115	Sanchez-Gonzalez et al. (2013)
Lactic acid bacteria (LAB)	Methylcellulose	34 ± 2	28.2 ± 1.2	388 ± 92	Sanchez-Gonzalez et al. (2013)
Citronella oil	Hake protein	181.9 ± 31.0	7.06 ± 1.64	–	Pires et al. (2013)
Tarragon oil	Hake protein	161.8 ± 31.0	2.99 ± 1.63	–	Pires et al. (2013)
Thyme oil	Hake protein	111.2 ± 44.8	4.33 ± 0.94	–	Pires et al. (2013)
Coriander oil	Hake protein	178.2 ± 25.5	4.70 ± 1.60	–	Pires et al. (2013)
Oregano oil	Triticale flour protein	12.10 ± 0.26	2.08 ± 0.19	82.53 ± 11.22	Aquirre et al. (2013)
Carvacrol oil	Chitosan	26.79 ± 21.93	37.86 ± 12.07	–	Rubiler et al. (2013)
Resveratrol	Chitosan	6.00 ± 2.00	78 ± 2.00	2.585 ± 151.00	Pastor et al. (2013)
Resveratrol	Methylcellulose	10.00 ± 3.00	65 ± 7.00	1.871 ± 132.00	Pastor et al. (2013)
Satureja hortensis oil	κ-carrageenan	35.82 ± 2.02	19.88 ± 2.37	–	Shojaee-Aliabadi et al. (2013)

in bioactive films (Du et al. 2009; Sanchez-Gonzalez et al. 2011a, b). In reviewing the literature, Rhim et al. (2000) reported that the addition of various compounds that structurally bind with the film-forming solutions changed the native color of the soy protein film (Rhim et al. 2000). Sivarooban et al. (2008) also reported that the incorporation of 1 % grape seed extract into SPI films significantly influenced the L^*, a^* and b^* values (Sivarooban et al. 2008). Du et al. (2009) reported that darker films were produced with the addition of cinnamon, allspice and clove bud oils into apple film-forming solutions, and the $L,^*$ b^* values as well as the whitish index of apple solutions increased as the concentration of the oils increased (Du et al. 2009).

3.4 Applications of Antimicrobial Edible Films and Coatings

In the literature, there are a lot of investigations in vivo and in vitro on edible antimicrobial films and coatings based on natural polymer or polymers mixed in different proportions.

The successes obtained in vitro studies guided the development of edible film and coating applications. Ko et al. (2001) were investigated the effects of hydrophobicity/hydrophilicity of edible films against *Listeria monocytogenes* strain V7 by various nisin concentrations (4.0–160 IU/film disk) and pH values ranging from 2.0 to 8.0. According to these study results, as the nisin concentration increased, the amount of inhibition progressively increased in all tested film. Using nisin, edible films with higher hydrophobicity values of 280–450 units under an acidic environment exerted a greater inhibitory effect against *L. monocytogenes*. Dawson et al. (2005) were examined the antimicrobial activity of nisin-absorbed silica and corn starch powders against *Lactobacillus plantarum* and *Listeria monocytogenes*. And nisin-absorbed powders were highly efficient at both adsorption and release of antimicrobial activity. Zivanovic et al. (2005) were determined the antimicrobial activity of chitosan films enriched with essential oils such as oregano, coriander, basil, and anise against *Listeria monocytogenes* and *Escherichia coli* O157:H7 by an agar diffusion test. The chitosan films and chitosan-oregano essential oil films were applied on inoculated bologna samples and stored 5 days at 10 °C. Pure chitosan films reduced *L. monocytogenes* by 2 logs, whereas the films with 1 % and 2 % oregano essential oil decreased the numbers of *L. monocytogenes* by 3.6 to 4 logs and *E. coli* by 3 logs. These films have the potential to be used as active biodegradable films with strong antimicrobial effects. Seydim and Sarıkuş (2006), in their study, antimicrobial properties of WPI films containing 1.0–4.0 % (wt/vol) ratios of oregano, rosemary, and garlic essential oils were test by in vitro against *Escherichia coli* O157:H7 (ATCC 35218), *Staphylococcus aureus* (ATCC 43300), *Salmonella enteridis* (ATCC 13076), *Listeria monocytogenes* (NCTC 2167), and *Lactobacillus plantarum* (DSM 20174). According to results of these studies, the film containing oregano essential oil was the most effective against these bacteria at 2 % level than those containing garlic and rosemary extracts. The use of rosemary essential oil incorporated into

WPI films did not exhibit any antimicrobial activity, whereas inhibitory effect of WPIs film containing garlic essential oil was observed only 3 % and 4 % level. Sanjurjo et al. (2006) studied the antimicrobial activity of nisin supported in edible films prepared with suspensions of tapioca starch containing glycerol. Studied were performed with *L. innocua*, after equilibration of edible films at a relative humidity (RH) of 57.5 % and at 25 °C. Results obtained showed that nisin supported in starch-based films in active and that the film is a useful barrier to further product contamination. Du et al. (2008)'s study's main objective was to evaluate the antimicrobial activities against *E. coli* O157:H7, storage stabilities of novel edible films made from tomatoes containing carvacrol, the main constituent of oregano oil by two different casting methods. Antimicrobial assays of tomato films indicated that optimum antimicrobial effects occurred with carvacrol levels of approximately 0.75 % added to tomato purees before film preparation.

Applications (Table 3) of antimicrobial edible films and coatings to meat, fish, poultry, fresh fruits, and vegetables, and tree nuts have received increasing interest because films and coatings can serve carriers for various antimicrobials that can maintain fresh quality, extend product shelf life, and reduce the risk of pathogen growth. Edible films with antimicrobial properties could prolong shelf life and safety of foods by preventing growth of pathogenic and spoilage microorganisms as a result of their lag-phage extension and/or their growth rate reduction (Quntavalla and Vicini 2002). Moreover, antimicrobials imbedded in films can be gradually released on the food surface, therefore requiring smaller amounts to achieve the target shelf life (Min and Krochta 2005).

Edible films and coatings have long been known to protect perishable fruits and vegetables from deterioration by retarding dehydration, suppressing respiration, improving textural quality, helping to retain volatile flavor compounds, and reducing microbial growth. The technology for using edible films and coatings as carriers of additives to extend the shelf life has been widely explored. Similarly, edible films and coatings carrying antimicrobials are a promising tool for decreasing the risk of pathogenic bacteria and also for extending minimally processed ready to eat meats shelf life. Minimally processed ready to eat meats and seafood products are a potential source of food-borne pathogenic bacteria such as *Salmonella typhimurim*, *Listeria monocytogenes*, and *E. coli* O157:H7 and studies about antimicrobial edible films and coatings focus on these pathogens mainly. In meats products, application of films and coatings not only is useful as a carrier of the antimicrobial but it can also prevent moisture loss during storage of fresh or frozen meats, reduce the rate of rancidity, and restrict volatile flavors loss. Dairy products, especially cheeses are complex food products that contain casein, fat, and water. In the case of fresh and semi-fresh cheese microbial stability and product safety must be controlled. Antimicrobial edible films and coatings are used mainly controlled microbial growth in the surface and also diminish the risk of pathogens such as *Listeria monocytogenes* and some yeasts and molds (Campos et al. 2011). Table 3 summarizes relevant application of antimicrobial external edible films and coatings to prevent microbial spoilage.

Table 3 Applications of antimicrobial compounds in food system to control pathogens and spoilage micro flora

Antimicrobial compounds	Film polymers	Target Microorganisms	Food Application	References
Propionic acid, Acetic acid, Lauric acid, Cinnamaldehyde	Chitosan	Enterobacteriaceae spp. *Lc. sakei Serratia liquefaciens*	Vacuum-packaged meats processed	Quattara et al. (2000)
Chitosan	Starch/Chitosan	*S. aureus E. coli* Mesophilic aerobes Mold and yeast Psychrotrophic bacteria	Minimally processed carrot	Durango et al. (2006)
Lemongrass oil, Oregano oil, Vanillin	Apple puree-alginate	*L. innocua*	Fuji apples	Rojas-Graü et al. (2007a, b)
Chitosan	Chitosan	Total microorganisms	Sliced mango fruit	Chien et al. (2007)
Oleoresins (olive, rosemary, onion, capsicum, cranberry, garlic, oreganum, and oreganum + carvacrol)	Chitosan Carboxymethyl cellulose Casein	*L. monocytogenes*	Butternut squash	Ponce et al. (2008)
Nisin, Sodium lactate, Sodium diacetate, Potassium sorbate, Sodium benzoate	Chitosan-coated plastic films	*L. monocytogenes*	Ham steaks	Ye et al. (2008a, b)
Acetic acid, Lactic acid	Chitosan	*L. monocytogenes*	Ready-to-eat (RTE) roast beef	Beverlya et al. (2008)
Nisin, Sodium lactate, Sodium diacetate, Potassium sorbate, Sodium benzoate	Chitosan-coated plastic films	*L. monocytogenes*	Smoked-salmon	Ye et al. (2008a, b)
Nisin, Natamycin	Cellulose	Staphylococcus sp. Psychrotrophic bacteria	Sliced mozzarella cheese	Dos Santos Pires et al. (2008)
Ovotransferrin	κ-carrageenan	*Escherichia coli S. typhimurium Staphylococcus aureus Candida albicans*	Chicken breast	Seol et al. (2009a, b)
Oregano oil, Clove oil	Chitosan	*L. monocytogenes*	Kasar cheese	Torlak and Nizamlıoğlu (2009)
Olive leaf extract	Methylcellulose	*S. aureus*	Kasar cheese	Ayana and Turhan (2009)
Oregano oil, Thyme oil	Soy protein	*E. coli* O157:H7	Fresh ground beef patties	Emiroğlu et al. (2010)
		S. aureus		
		P. aeruginosa		
		Lac. plantarum		
Nisin	Sodium caseinate	*L. innocua*	Semi-soft cheese	Cao-Hoang et al. (2010)
Sodium lactate, Sodium diacetate	Alginate	*L. monocytogenes*	Smoked salmon	Neetoo et al. (2010)

(continued)

Table 3 (continued)

Antimicrobial compounds	Film polymers	Target Microorganisms	Food Application	References
Bergamot oil, Lemon oil, Tea tree oil	Chitosan Hydroxypropylmethylcellulose	*L. monocytogenes E. coli S. aureus*	A model solid food system	Sanchez-Gonzalez et al. (2011a, b)
Chitosan	Chitosan Sodium caseinate/ Chitosan	Mezophilic and psychrotrophic bacteria Yeast and Molds	Carrot Cheese Salami	Del Rosario Moreira et al. (2010)
Sodium lactate, Sodium diacetate	Starch Chitosan Alginate Pectin	*L.monocytogenes*	Roasted turkey	Jiang et al. (2011)
Chitosan	Chitosan	Mesophilic and psychrotrophic bacteria Yeast and Moulds Lactic acid bacteria Coliforms *E.coli* O157:H7	Fresh cut broccoli	Del Rosario Moreira et al. (2011)
Sodium benzoate, Sodium propionate, Potassium sorbate	Hydroxypropylmethylcellulose/ lipid	*Penicillium digitatum Penicillium italicum*	Clemenules mandarins	Valencia-Chamorro et al. (2011)
Sodium benzoate, Lactic acid, Chitooligosaccharide	Whey protein isolate	Spoilage microflora	Cheese	Ramos et al. (2012a, b)
Rosmarinic acid extract, Asian essential oil mixture, Italian essential oil mixture	Methylcellulose/ polycaprolactone	*E.coli, S. typhimurium*	Fresh broccoli	Takala et al. (2013a, b)
Tea tree oil, Rosemary oil, Clove oil, Lemon oil, Oreganum oil, Calendula oil, Aloe vera oil, Bee pollen, Ethanolic extract of propolis, Pomegranate extract Resveratrol	Chitosan	*E.coli, L. monocytogenes*	Fresh cut broccoli	Alvarez et al. (2013)
Ethyl-N-dodecanoly-L-arginate	Chitosan	Mesophilic and psychrotrophic bacteria Yeast and Moulds Lactic acid bacteria Coliforms Hydrogen sulfide-producing bacteria Pseudomonas spp.	Chicken breast fillets	Higueras et al. (2013)
Organic acid mixture, Rosemary extract, Asian spice essential oil, Italian spice essential oil	Methylcellulose Polycaprolactone/alginate	*E.coli, S. typhimurium, L. monocytogenes*	Fresh broccoli	Takala et al. (2013a, b)

4 Legal and Economic Aspects of Antimicrobial Edible Films and Coatings

Edible films, including all of its components must be safe to eat or must have the GRAS status. Moreover, edible films when incorporated with antimicrobials are considered as active food packaging (Espitia et al. 2013). Definitions stated in Regulation 1935/2004/EC and in Regulation 450/2009/EC consider that "active materials and articles are intended to extend the shelf life or maintain or improve the condition of packaged food." They are designed to deliberately incorporate components that would release or absorb substance into or from the packaged food or the environmental surrounding the food (Campos et al. 2011).

Edible films can be classified as food products, food ingredients, food additives, food contact substances, or food packaging materials. Thus, their elaboration should follow all required regulations pertinent to food ingredients, since they are an integral part of the edible portion of food products. Moreover, besides the GRAS status of all ingredients used, production of pectin, and other biodegradable polymers, edible films should be done in food processing facilities following with good manufacturing practice (Bierhalz et al. 2012).

According to legislation and labeling in the USA, edible coatings and films are considered part of the food; as a consequence, their ingredients must comply with the Code of Federal Regulations and be declared on the label under the Federal Food, drug, and Cosmetic Act (Frassen and Krochta 2003). The EU considers that an edible film is a special active part of the food and, seen from a legal point of view, it is to be regarded as a foodstuff, along with the food packed in the film, having to fulfill the general requirements for food. Another important topic within regulatory status is the presence of allergens because many edible films and coatings are made with or can contain ingredients that could cause allergic reactions such as wheat protein (gluten) or peanut protein. Therefore, the presence of known allergen on a film or coating on a food must be also stated in the label (Campos et al. 2011).

As is known in each country has clear regulations regarding the addition of preservatives to food, which often include purity requirements, analytical methodology, labeling, and maximum allowed levels. Therefore, at the moment, under such legislation must be ruled the application of edible films containing preservatives. As a consequence, it is important to remark that the edible film formulation proposed must be adapted in order to ensure a content of preservative in the food that is in accordance with maximum values allowed by food legislation of the country of application (Campos et al. 2011).

5 Future Trends of Antimicrobial Edible Films and Coatings

Synthetic polymers are gradually being replaced by biodegradable materials especially those derived from replenishable, natural resources. More than the origin, the chemical structural of biopolymer that determines its biodegradability. Use of such biopackaging will open up potential economic benefits to farmers and agricultural processors.

For example, once considered a waste product in the cheese manufacturing process, whey and whey protein products are "green" alternatives to traditional plastics as edible films. Based on its excellent oxygen barrier properties, whey protein films can be competitive materials replacing EVOH (Ethylene vinyl alcohol), nylon or polyesters, which are typically used as oxygen barriers. Innovative techniques of preserving food safety and structural nutritional integrity as well as complete biodegradability must be adopted. Eventually biopackaging constitutes a niche market and that will be our future. Biodegradable packaging is estimated to grow 20 % over the next few years, taking up a larger share of the packaging market (Tharanathan 2003; Jooyandeh 2011).

The incorporation of essential oils into edible films and coatings allows us to reduce the quantities required to guarantee food safety. However, during the drying stage of the film, significant losses of volatile compounds occur. Micro- and nanoencapsulation of essential oils could be a solution to minimize this problem and improve the effectiveness of active coatings enriched with essential oils (Sanchez-Gonzalez et al. 2011a, b).

Moreover, researches related to the application of nanotechnology to edible films and coatings are scarce. However, when more toxicological studies are published in the coming years, the application of nanotechnology may lead research toward the incorporation of nanoscale materials in edible films and study their potential synergistic effects on films antimicrobial activity as well as their contribution to the improvement of physical-mechanical properties of edible films and coatings (Espitia et al. 2013).

6 Conclusions

The use of edible films and coatings for a variety of foods widespread continue. Edible films and coatings provide benefits as a carrier of antimicrobial agents, flavors, antioxidants, coloring agents, vitamins and probiotics in the field of active packaging. Antimicrobial edible films and coatings can provide effectively inactivation or inhibition of pathogens and spoilage microorganism. These active films and coatings are capable of natural and biodegradable polymers so have competitive offers with synthetic materials both environmentally and economically. But still food packaging industry needs to improve the physical and mechanical properties of edible films and coatings for food applications. A very large part of the studies on the subject laboratory are scale. Many studies have focused on trade are need to provide a more realistic information. Despite limiting factors they use on foods, edible films and coatings particularly containing antimicrobial active substances add value to the food products by extending shelf life of products. Antimicrobial films and coatings applications should be regarded as a hurdle technology such as pulse light, high pressure, and irradiation. But it should be noted that the active ingredients affect to sensory and functional properties of edible films and coatings. Studies about the effects of active edible films and coatings on the sensory properties of the products are very limited. Edible films and coatings formulations providing high-sensory performance and functionality with a long shelf life should develop necessary. Cooperation with regulatory agencies such as government, industry, and research groups will play a key role successfully in application of innovative food packaging technologies such as edible films and coatings.

References

Albert A, Guardeno LM, Salvador A, Fiszman SM (2010) Alginates as Edible Coatings for Microwaveable Food. International Conference on Food Innovation. 25–29 October, Valencia, Spain

Alizadeh A, Behfar S (2013) Properties of collagen based edible films in food packaging: a review. Annal Biol Res 4(2):253–256

Altenhofen Da Silva M, Bierhalz ACK, Kieckbusch TG (2009) Alginate and pectin composite films crosslinked with Ca^{+2} ions: effect of the plasticizer concentration. Carbonhydr Polym 77:736–742

Altiok D, Altiok E, Tihminlioğlu F (2010) Physical antibacterial and antioxidant properties of chitosan films incorporated with thyme oil for potential wound healing applications. J Mater Sci Mater Med 21:2227–2236

Alvarez MV, Ponce AG, Moreire MR (2013) Antimicrobial efficiency of chitosa coating enriched with bioactive compounds to improve the safety of fresh cut broccoli. Food Sci Technol 50:78–87

Appendini P, Hotchkiss JH (2002) Review of antimicrobial food packaging. Innov Food Sci Emerg Technol 3:113–126

Aquirre A, Borneo R, Leon AE (2013) Antimicrobial mechanical and barrier properties of triticale protein films incorporated with oregano essential oil. Food Biosci I:2–9

Arismendi C, Chillo S, Conte A, Del Nobile MA, Flores S, Gerschenson LN (2013) Optimization of physical properties of xanthan gum/tapioca starch edible matrices containing potassium sorbate and exalutaion of its antimicrobial effectiveness. Food Sci Technol 53:290–296

Avena-Bustillos RJ, Olsen CW, Olsan DA, Chiou B, Yee E, Bechtel PJ, McHugh TH (2006) Water vapor permeability of mammalian and fish gelatin films. J Food Sci 71(4):202–207

Ayana B (2007) Antimikrobiyal Yenilebilir Filmlerin Üretimi ve Özelliklerinin Belirlenmesi. Mersin Üniversitesi Fen Bilimleri Enstitüsü Gıda Mühendisliği Anabilim Dalı, Yüksek Lisans Tezi, 60s

Ayana B, Turhan KN (2009) Use of antimicrobila methylcellulose films to control *Staphylococcus aureus* during storage of kasar cheese. Pack Technol Sci 22:461–469

Ayana B, Turhan N (2010) Gida ambalajlamasinda antimikrobiyel madde içeren yenilebilir film-ler/kaplamalar ve uygulamalari. Gıda 2:151–158

Baeva M, Panchev I (2005) Investigation of the retaining effect of a pectin-containing edible film upon the crumb agening of diatetic sucrose-free sponge cake. Food Chem 92:343–348

Baldwin E (1999) Surface treatments and edible coatings in food preservation. In: Rahman MS (ed) Handbook of food preservation. Dekker, New York, pp 557–609

Baysal T, Bilek SE, Apaydın E (2009) The efffect of corn zein edible film coating on intermediate moisture apricot (*Prunus armenica l.*) quality. Gıda 35(4):245–249

Beverlya RL, Janes ME, Prinyawiwatkula W, No HK (2008) Edible chitosan films on ready-to-eat roast beef for the control of *Listeria monocytogenes*. Food Microbiol 25:543–537

Bierhalz ACK, Altenhofen Da Silva M, Kieckbusch TG (2012) Natamycin release from alginate/pectin films for food packaging applications. J Food Eng 110:18–25

Bonilla J, Fortunati E, Vargas M, Chiralt A, Kenny JM (2013) Effects of chitosan on the physico-chemical and antimicrobial properties of pla films. J Food Eng 119:236–243

Bourtoom T (2008) Edible films and coatings: characteristics and properties. Int Food Res J 15(3):1–12
Çağrı A (2002) Antimicrobial whey protein isolate based edible castings. Department of Food Science and Human Nutrition. Doctor of Philosophy Thesis, 241 p
Çağrı A, Ustunol Z, Ryser ET (2002) Inhibition of three pathogens on bologna and summer sausage using antimicrobial edible films. J Food Sci 67(6):2317–2324
Campos CA, Gerschenson LN, Flores SK (2011) Development of edible films and coatings with antimicrobial activity. Food Bioprocess Technol 4:849–875
Cao-Hoang L, Chaine A, Gregorie L, Wache U (2010) Potential of nisin-incorporated sodium caseinate films to control *Listeria* in artificially contaminated cheese. Food Microbiol 27:940–944
Cha DS, Chinnan MS (2004) Biopolymer-based antimicrobial packaging: a review. Crit Rev Food Sci Nutr 44:223–237
Cha DS, Chinnan MS, Choi JH, Parh HJ (2002) Antimicrobial films based on Na-alginate and κ-carrageenan. Lebensm Wissu Technol 35:715–719
Chien PJ, Sheu F, Yang FH (2007) Effects of edible chitosan coating on quality and shelf life of sliced mango fruit. J Food Eng 78:225–229
Choi JH, Choi WY, Cha DS, Chinnan MJ, Park HJ, Lee DS, Park JM (2005) Diffusivity of potassium sorbate in κ-carrageenan-based antimicrobial film. Swiss Soc Food Sci Technol LWT 38:417–423
Cleveland J, Montville TJ, Nes IF, Chikindas ML (2001) Bacteriocins: safe natural antimcirobial for food preservation. Int J Food Microbiol 71:1–20
Coma V, Martial-Gros A, Garreau S, Copnet A, Salin F, Deschamps A (2002) Edible antimicrobial films based on chitosan matrix. J Food Sci 67:1162–1169
Cooksey K (2005) Effectiveness of antimicrobial food packaging materials. Food Add Cont 22(10):980–987
Cousineau JLA (2012) Production and characterization of wheat gluten films. A Thesis Presented to the University of Waterloo in Fulfillment of the Thesis Requirement for the Degree of Master of Applied Science in Chemical Engineering, 118 p
Cuppet S (1994) Edible coatings as carrier of food additives, fungicides and natural antagonists. In: Krochta JM, Baldwin EA, Nisperos-Carriedo MO (eds) Edible coatings and films to improve food quality. Technomic Publishing, Lancaster, PA, pp 121–137
Dawson PL, Harman L, Satthibandhu A, Han IY (2005) Antimicrobial activity of nisin adsorbed silica and corn starch powders. Food Microbiol 22:93–99
De Carvalho RA, Grosso CRF (2006) Properties of chemically modified gelatin films. Brezilian J Chem Eng 23(1):45–53
De Moura MR, Avena-Bustillos RJ, McHung TH, Krochta JM, Mattoso LHC (2008) J Food Sci 73(7):31–37
Debeaufort F, Voilley A, Meares P (1994) Water vapor permeability and diffusivity through methylcellulose edible films. J Membr Sci 91:125–133
Debeaufort F, Quezada-Gallo J, Voilley A (1998) Edible films and coatings: tomorrow's packagings. Crit Rev Food Sci Nutr 38(4):299–313
Del Rosario MM, Pereda M, Marcovich NE, Roura SI (2010) Antimicrobial effectiveness of bioactive packaging materials from edible chitosan and casein polymers: assessment on carrot, cheese and salami. J Food Sci 76(1):54–63
Del Rosario MM, Roura SI, Ponce A (2011) Effectiveness and chitosan edible coatings to improve microbiological and sensory quality of fresh cut broccoli. Food Sci Technol 44:2335–2341
Diop JM (2009) Preparation and characterization of gelatin and chitosan films. An honor Thesis for The Department of Chemical and Biological Engineering. Tufts University, 33 p
Dos Santos Pires AC, Soares NFF, De Andrade NJ, Da Silva LHM, Camilloto GP, Bernardes PC (2008) Development and evaluation of active packaging for sliced mozerella preservation. Pack Technol Sci 21:375–383
Du WX, Olsen CW, Avena-Bustillos RJ, McHugh TH, Levin CE, Friedman M (2008) Antibacterial activity against E.coli o157:h7, physical properties and storage stability of novel carvacrol-containing edible tomato films. Food Microbiol Saf 73(7):378–383

Du WX, Olsen CW, Avena-Bustillos RJ, McHugh TH, Levin CE, Friedman M (2009) Effects of allspice, cinnamon and clove bud essential oils in apple films on antimicrobial activities against *Eschericha coli* o157:h7, *Salmonella enterica* and *Listeria monocytogenes*. J Food Sci 74(7):372–378

Durango AM, Soares NFF, Andrade NJ (2006) Microbiological evaluation of on edible antimicrobial coating on minimal processed carrots. Food Control 17:336–341

Dursun S, Erkan N (2009) Yenilebilir protein filmler ve su ürünlerinde kullanimi. J Fisher Sci 3(4):352–373

Emiroğlu ZK, Yemiş GP, Çoşkun BK, Candoğan K (2010) Antimicrobial activity of soy edible films incorporated with tyhme and oregano essential oils on fresh ground beef patties. Meat Sci 86:283–288

Espitia PJP, Du WX, Avena-Bustillos RJ, Soares NFF, McHung TH (2013) Edible films from pectin: physical-mechanical and antimicrobial properties – a review. Food Hyrodcoll 35:87–296

Fabra MJ, Hambleton A, Talens P, Debeaufort F, Chiralt A, Voilley A (2009) Influence of interactions on water and aroma permeabilities of iota-carrageenan oleic acid-beeswax films used for flavor encapsulation. Carbohydr Polym 76(2):325–332

Ferreira CO, Nunes CA, Delgadillo I, Lopes-Da-Silva JA (2009) Characterization of chitosan-whey protein films at acid pH. Food Res Int 42:807–813

Franssen LR, Krochta JM (2000) Edible coating containing natural antimcirobials for processed foods. In: Roller S (ed) Natural antimicrobial for minimal processing of foods. CRC, Boca Raton, FL, pp 250–262

Frassen LR, Krochta JM (2003) Edible coatings containing natural antimicrobials for processed foods. In: Roller S (ed) Natural antimicrobials for the minimal processing of foods. CRC, Boca Raton, FL

Galus S, Lenart A (2013) Development and characterization of composite edible films based on sodium alginate and pectin. J Food Eng 115:459–465

Garcia MA, Martino MN, Zaritzky NE (1999) Edible starch films and coatings characterization scanning electron microscopy. Water Vapor Gas Permeabil Scann 21:348–353

Garcia MA, Pinotti A, Martino MN, Zaritzky N (2009) Characterization of starch and composite edible films and coatings. In: Embuscado ME, Huber KC (eds) Edible films and coatings for food applications. Springer, Dordrecht. ISBN 978-0-387-92823-4

Gennadios A, Weller C, Testin RF (1993) Modification of physicall and barrier properties of edible wheat gluten-based films. Cereal Chem 70(4):426–429

Gernadious A, McHugh TH, Weller CL, Krochta JM (1994) Edible coatings and films based on proteins. In: Krochta JM, Baldwin EA, Nisperos-Carriedo MO (eds) Edible coatings and films to improve food quality. Technomic Publishing, Lancaster, PA, pp 201–277

Gonzales A, Strumia MC, Igarzabel CIA (2011) Cross-linked soy protein as material for biodegradable films: synthesis, characterization and biodegradation. J Food Eng 106:331–338

Güçbilmez ÇM (2005) Production of functional packaging materials by use of biopreservatives. The Graduate School of Engineering and Science of Izmir Institute of Technology. Master Thesis, 117 p

Guerrero P, Stefani PM, Ruseckaite RA, De la Caba K (2011) Functional properties of films based on soy protein isolate and gelatin processed by compression molding. J Food Eng 105:65–72

Guo X, Lu Y, Cui H, Jia X, Bai H, Ma Y (2012) Factors affecting the physical properties of edible composite film prepared from zein and wheat gluten. Molecules 17:3794–3804

Hambleton A, Debeaufort F, Bonnotte A, Voilley A (2009a) Influence of alginate emulsion-based films structure on its barrier properties and on the protection of microencapsulated aroma compound. Food Hydrocoll 23:2116–2124

Hambleton A, Fabra MJ, Debeaufort F, Dury-Brun C, Voilley A (2009b) Interface and aroma barrier properties of iota-carrageenan emulsion-based films used for encapsulation of active food compouns. J Food Eng 93:80–88

Han JH, Gennadious A (2005) Edible films and coatings: a review. In: Han JH (ed) Innovations in food packaging. Elsevier, Amsterdam, pp 239–262

Higueras L, Lopez-Carbollo G, Hernandez-Munoz P, Gavara R, Rollini M (2013) Development of a novel antimicrobial film based on chitosan with lae (ethyl-nα-dodecanoyl-l-alginate) and its application to fresh chicken. Int J Food Microbiol 165:339–345

Imran M, El-Fahmy S, Revol-Junelles AM, Desorby S (2010) Cellulose derivates based active coatings: effects of nisin and plastizer on physico-chemical and antimicrobial properties of hyhdroxyproply methylcellulose films. Carbonhydr Polym 81:219–225
Jiang Z, Neetoo H, Chen H (2011) Efficacy of freezing, frozen storage and edible antimicrobial coatings used in combination for control *Listeria monocytogenes* on roasted Turkey stored at chiller temperatures. Food Microbiol 28:1394–1401
Jooyandeh H (2011) Whey protein films and coatings: a review. Pak J Nutr 10(3):296–301
Kang HJ, Jo C, Kwon JH, Kim JH, Chung HJ, Byun MW (2007) Effect of a pectin-based edible coating containing green tea powder on the quality of irradiated pork patty. Food Control 18:430–435
Karagöz Z, Candoğan K (2007) Et teknolojisinde antimikrobiyal kaplama. Gıda 32(3):113–122
Karbowiak T, Debeaufort F, Champion D, Voilley A (2006) Wetting properties at the surface of iota-carrageenan-based edible films. J Colloid Interface Sci 294:400–410
Kester JJ, Fennema OR (1986) Edible films and coatings: a review. Food Technol 40:47–59
Khairruddin K (2005) Study on antimicrobial starch-based active packaging system. University of Technology of Chemical Engineering and Natural Resources Engineering, Department of Bioprocess Engineering. A research, 7 p
Ko S, Janes ME, Hettiarachcy NS, Johnson MG (2001) Physical and chemical properties of edible films containing nisin and their action against *Listeria monocytogenes*. J Food Sci 66(7): 1006–1011
Lai HM, Padua GW (1997) Properties and microstructure of plasticized zein films. Cereal Chem 74(6):771–775
Leceta I, Guerrero P, Ibarburu I, Duenas MT, De La Caba K (2013) Characterization and antimicrobial analysis of chitosan-based films. J Food Eng 116:889–899
Manab A, Sawitri ME, Al Awwaly KU, Purnoma H (2011) Antimicrobial activity of whey protein based edible film incorporated with organic acids. Afr J Food Sci 5(1):6–11
Mehmetoğlu ÇA (2010) Yenilebilir filmlerin ve kaplamalarin özelliklerini etkileyen faktörler. Akademik Gıda 8(5):37–43
Min S, Krochta JM (2005) Antimicrobial films and coatings for fresh fruit and vegetables. In: Jongen WNF (ed) Improving the safety of fresh fruits and vegetables. Taylor and Francis, Boca Raton, FL
Mohamed C, Clementine KA, Didier M, Gerard L (2013) Antimicrobial and physical properties of edible chitosan films enhanced by lactoperoxidase system. Food Hyrocoll 30:576–580
Moreira MR, Roura SI, Ponce A (2011) Effectiveness of chitosan edible coatings to improve microbiological and sensory quality of fresh cut broccoli. Food Sci Technol 44:2335–2341
Neetoo H, Ye M, Chen H (2010) Bioactive alginate coatings to control *Listeria monocytogenes* on cold-smoked salmon sliced and fillets. Int J Food Microbiol 136:326–331
Ojagh SM, Rezaei M, Rezavi SH, Hosseini SMH (2010) Development and evaluation of a novel biodegradable film made from chitosan and cinnemon essantial oil with low affinity toward water. Food Chem 122:161–166
Olabarrieta I, Cho S, Gallstedt M, Sarasua J, Johansson E, Hedenguist MS (2006) Aging properties of films of plasticized vital wheat gluten cast from acidic and basic solutions. Biomacromolecules 7(5):1657–1664
Olivas GI, Mattinson DS, Barbosa-Canovas GV (2007) Alginate coatings for preservation of minimally processed "gala" apples. Postharv Biol Technol 45:89–96
Ozdemir M, Floros JD (2008) Optimization of edible whey protein films containing preservatives for water vapor permeability, water solubility and sensory characteristics. J Food Eng 86:215–224
Park HJ (1996) Gas and mechanical barrier porperties of carrageenan-based films. Korean J Food Sci Ind 29:47–53
Park HJ, Chinnan MS (1995) Gas and water barrier properties of edible films form protein and cellulosic meterials. J Food Eng 25(4):497–507
Pastor C, Sanchez-Gonzalez L, Chiralt A, Chafer M, Gonzalez-Martinez C (2013) Physical and antioxidant properties of chitosan methlycellulose based films containing resveratrol. Food Hydrocoll 30:272–280

Pelissari FM, Grossmann MVE, Yamashita F, Pineda EAG (2009) Antimicrobial mechanical and barrier properties of cassava starch-chitosan films incorporated with oregano essential oil. J Agric Food Chem 57:7499–7504

Perez-Perez C, Regolado-Gonzales C, Rodriquez-Rodriquez CA, Barbosa-Roderiquez JR, Villasenor-Ortega F (2006) Incorporation of antimicrobial agents in food packaging films and coatings. In: Guevara-Gonzales RG, Torres-Pacheco I (eds) Advances in agricultural and food biotechnology. Research Signpost, Trivandrum, pp 193–216. ISBN 81-7736-269-0

Piazza L, Dürr-Auster N, Gigli J, Windhab EJ, Fisher P (2009) Interfacial rheology of soy protein-high methoxyl pectin films. Food Hydrocoll 23:2125–2131

Pires C, Ramos C, Teixeira B, Batista I, Nunes ML, Marques A (2013) Hake proteins edible films incorporated with essential oils: physical, mechanical, antioxidant and antibacterial properties. Food Hydrocoll 30:224–231

Ponce AG, Roura SI, Del Valle CE, Moreira MR (2008) Antimicrobial and antioxidant activities of edible coatings enriched with natural plant extracts: in vitro and in vivo studies. Postharv Biol Technol 49:294–300

Quattara B, Simard RE, Piette G, Begin A, Holley RA (2000) Inhibition of surface spoilage bacteria in processed meats by application of antimicrobial films prepared with chitosan. Int J Food Microbiol 62:139–148

Quntavalla S, Vicini L (2002) Antimicrobial food packaging in meat industry. Meat Sci 62:373–380

Ramos OL, Santos AC, Leao MV, Pereira JO, Silva SI, Fernandes JC, Franco MI, Pintado ME, Malcata FX (2012a) Antimicrobial activity of edible coatings prepared from whey protein isolate and formulated with various antimicrobial agents. Int Dairy J 25:132–141

Ramos OL, Silva SI, Soares JC, Fernandes JC, Poças MF, Pintado ME, Malcata FX (2012b) Features and performance of edible films, obtained from whey protein isolate formulated with antimicrobial compounds. Food Res Int 45:351–361

Ramos OL, Reinas I, Silva SI, Fernandes JC, Cerqueira MA, Pereira RN, Vicente AA, Poças MF, Pintado ME, Malcata FX (2013) Effect of whey protein purity and glycerol contant upon properties of edible films manufactured therefrom. Food Hydrocoll 30:110–122

Rezvani E, Schleining G, Sümen G, Taherian AR (2013) Assessment of physical and mechanical properties of sodium caseinate and stearic acid based film forming emulsion and edible films. J Food Eng 116:598–605

Rhim JW, Gennadios A, Handa A, Weller CL, Hanna MA (2000) Solubility, tensile and color properties of modified soy protein isolate films. J Agric Food Chem 48(10):4937–4941

Rojas-Graü MA, Raybaudi-Massilia RM, Soliva-Fortuny RCS, Avena-Bustillos RJ, McHugh TH, Martin-Belloso O (2007a) Apple puree-alginate edible coatings as carrier of antimicrobial agents to prolong shelf-life of fresh-cut apples. Postharv Biol Technol 45:254–264

Rojas-Graü MA, Tapia MS, Rodriquez FJ, Carmona AJ, Martin-Belloso O (2007b) Alginate and gellan-based edible coatings as carriers of antibrowning agents applied on fresh-cut fuji apples. Food Hydrocoll 21:118–127

Rojos-Graü MA, Avena-Bustillos RJ, Olsen C, Friedman M, Henika PR, Martin-Belloso O, Pan Z, McHugh TM (2007) Effects of plant essential oils and oil compounds on mechanical, barrier and antimcirobial properties of alginate-apple puree edible films. J Food Eng 81:634–641

Rubiler JF, Curz RMS, Silva HD, Vincete AA, Khmenlinskii I, Vieira MC (2013) Pyhsico-mechanical properties of chitosan films with carvacrol and grape seed extract. J Food Eng 115:466–474

Sanchez-Gonzalez L, Vargas M, Gonzalez-Martinez C, Chiralt A, Chafer M (2009) Characterization of edible films based on hyhdroxyproply methylcellulose and tea tree essential oil. Food Hydrocoll 23:2102–2109

Sanchez-Gonzalez L, Chafer M, Hernandez M, Chiralt A, Gonzalez-Martinez C (2011a) Antimicrobial activity of polysaccharide films containing essential oils. Food Control 22:1302–1310

Sanchez-Gonzalez L, Vargas M, Gonzalez-Matinez C, Chiralt A, Chafer M (2011b) Use of essential oils in bioactive edible coatings. Food Eng Rev 3:1–16

Sanchez-Gonzalez L, Saavedra JIQ, Chiralt A (2013) Physical properties and antilisterial activity of bioactive edible films containing *Lactobacillus plantarum*. Food Hydrocoll 33:92–98
Sanjurjo K, Flores S, Gerschenson L, Jagus R (2006) Study of the performance of nisin supported in edible films. Food Res Int 39:749–754
Sebti I, Ham-Pichavant F, Coma V (2002) Edible bioactive fatty acid-cellulosic derivative composites used in food packaging applications. J Agric Food Chem 50(15):4290–4294
Seol KH, Lim DG, Jang A, Jo C, Lee M (2009a) Antimicrobial effect of κ-carrageenan-based edible film containing ovotransferrin in fresh chicken breast stored at 5 °c. Meat Sci 83:479–483
Seol KH, Lim DG, Jang A, Jo C, Lee M (2009b) Antimicrobial effect of κ-carrageenan-based film containing ovotransferrin in fresh chicken breast stored at 5 °c. Meat Sci 83:478–483
Seydim AC, Sarıkuş C (2006) Antimicrobial activity of whey protein based edible films incorporated with oregano, rosemary and garlic essential oils. Food Res Int 39:639–644
Shojaee-Aliabadi S, Hosseini H, Mohammadifar NA, Mohammedi A, Ghasemlou M, Ojagh SM, Hosseini SM, Khaksar R (2013) Characterization of antioxidant-antimicrobial κ-carrageenan films containing *Satureja hortensis* essential oil. Int J Biol Macromol 52:116–124
Siracusa V, Pietro R, Ramoni S, Rosa MD (2008) Biodegrable polymers for food packaging: a review. Trends Food Sci Technol 19:634–643
Sivarooban T, Hettiarachchy NS, Jonhson MO (2008) Physical and antimicrobial properties of grape seed extract, Nisin and EDTA incorporated soy protein edible films. Food Res Int 41:781–785
Skudlarek JRG (2012) Antimicrobial efficacy of edible soy protein isolate films and coatings incorporated with hop ethanol extract and the influence on shelf-life and sensory attributes of Bologna. Collage of Agriculture, Department of Animal and Food Sciences at The University of Kentucky. PhD Thesis. Lexington, Kentucky, p. 114
Skurtys O, Acevedo C, Pedreschi F, Osorio F, Agulera JM (2010) Food hydrocolloid edible films and coatings. Nova Science, New York, p 66. ISBN 1616682698
Sobral PJA, Menegalli FC, Hubinger MD, Roques MA (2001) Mechanical, water vapor barrier and thermal properties of gelatin based edible films. Food Hyrocolloids 15:423–432
Song Y, Liu L, Shan H, You J, Yongkang L (2011) Effect of sodium alginate-based edible coatings containing different antioxidants on quality and shelf life of refrigerated bream (*Megalobramo amblycephala*). Food Control 22:608–615
Souza BWS, Cerqueria MA, Martins JT, Casariego A, Teixeira JA, Vincente AA (2010) Influence on electric fields on the structure of chitosan edible coatings. Food Hydrocoll 24:330–335
Suppakul P, Miltz J, Sonneveld K, Bigger SW (2003) Active packaging technologies with an emphasis on antimicrobial packaging and its applications. J Food Sci 68:408–420
Takala PN, Salmieri S, Boumail A, Chan RA, Vu KD, Chauve G, Bouchard J, Lacroix M (2013a) Antimicrobial effect and physicochemical properties of bioactive trilayer polycaprolactone/methylcellulose-based films on the growth of food borne pathogens and total microbiota in fresh broccoli. J Food Eng 116:648–655
Takala PN, Vu KD, Salmieri S, Khan RA, Lacroix M (2013b) Antibacterial effect of biodegradable active packaging on the growth of *Escherichia coli*, *Salmonella typhimurium* and *Listeria monocytogenes* in fresh broccoli stored at 4 °c. Food Sci Technol 53:499–506
Temiz A (2000) Genel mikrobiyoloji uygulama teknikleri. Hatipoğlu Yayınevi, Ankara, 291 p
Tharanathan RN (2003) Biodegradable films and composite coatings: past, present and future. Trends Food Sci Technol 14:71–78
Torlak E, Nizamlıoğlu M (2009) Doğal antimikrobiyal maddeler ile hazirlanan yenilebilir filmlerin *Listeria monocytogenes* üzerine etkileri. Vet Bil Deg 25(1–2):15–21
Turhan KN, Şahbaz F (2004) Water vapor permeability, tensile properties and solubility of methylcellulose-based edible films. J Food Eng 61:459–466
Üçüncü M (2007) Gıda Ambalajlama Teknolojisi. Akademik Yayıncılık, ISBN: 987-483-443-7, p. 896
Valencia-Chamorro SA, Palou L, Angel del Rio M, Perez-Gago MB (2011) Performance of hyhdroxyproply methylcellulose (hpmc)-lipid edible coatings with antifungal food additives during cold storage of clemanules mandarins. LWT Food Sci Technol 44:2342–2348

Wen-Xian D, Roberto J, Avena-Bustillos S, Sheng TH, Tara HM (2011) Antimicrobial volatile essansial oils in edible films for food safety. In: Mendez-Vilas A (ed) Science against microbial pathogens: communicating current research and technological advances, vol 2. Formatex Research Center Publisher, Badajoz. ISBN 978-84-939843-2-8

Ye M, Neetoo H, Chen H (2008a) Control of *Listeria monocytogenes* on ham steaks by antimicrobials incorporated in to chitosan-coated plastic films. Food Microbiol 25:260–268

Ye M, Neettoo H, Chen H (2008b) Effectiveness of chitosan-coated plastic films incorporating antimicrobials in inhibition of *Listeria monocytogenes* on cold-smoked salmon. Int J Food Microbiol 127:235–240

Yener FYG (2007) Development of antimicrobial protective food coating materials from edible alginat films. The Gruduate School of Engineering and Science of Izmir Institute of Technology. Master Thesis, p. 101

Zinoviadou KG, Koutsoumanis KP, Biliadersi CG (2009) Physico-chemical properties of whey protein isolate films containing oregano oil and their antimcirobial action against spoilage flora on fresh beef. Meat Sci 82:338–345

Zivanovic S, Chi S, Draughan AF (2005) Antimicrobial activity of chitosan films enriched with essential oils. Food Microbiol Saf 70(1):45–51

Recent Approaches in Risk Assessment of Foods

Farhana Masood, Samira Umme Aiman, Sana Khan, Showkat Ahmad Lone, and Abdul Malik

Abbreviations

ICMSF	International Commission on Microbiological Specification for Foods
MRM	Microbial Risk Management
SPS	Sanitary and Phytosanitary Measures
CCFH	Codex Committee and Food Hygiene
CAC	Codex Alimentarius Commission
USNACMCF	United States National Advisory Committee on Microbiological Criteria for Foods

1 Introduction

The objective of ensuring safe food for the World's constantly growing population has been a major preoccupation of governments, international organizations (e.g. WHO/FAO CODEX, ILSI, ISO, ICMSF, etc.) and professional and trade bodies over many years. Yet, in deprived areas of the world, there remains a basic need to ensure a reliable and safe food supply. In all countries, especially in developed consumer-oriented countries, the need is to ensure that foods do not present an unacceptable risk to the health and well being of the consumer. Throughout the world, the law imposes a duty of care and responsibility for the safety and quality of foods on those business organizations involved in the procurement, processing, distribution, and retail sale of the products. A risk analysis framework provides a

F. Masood • S.U. Aiman (✉) • S. Khan • S.A. Lone • A. Malik
Faculty of Agricultural Sciences, Department of Agricultural Microbiology,
Aligarh Muslim University, Aligarh 202002, India
e-mail: samiraummeaiman@gmail.com

A. Malik et al. (eds.), *Food Processing: Strategies for Quality Assessment*,
Food Engineering Series, DOI 10.1007/978-1-4939-1378-7_11

process to systematically and transparently collect, analyse, and evaluate relevant scientific and non-scientific information about a chemical, biological, or physical hazard possibly associated with food in order to select the best option to manage that risk based on the various alternatives identified.

1.1 Components of Risk Analysis

As a structured decision-making process, risk analysis includes three distinct but closely connected components: risk management, risk assessment, and risk communication (Fig. 1). Each of these components plays an essential and complementary role in the risk analysis process. Although, risk management and risk communication tended to receive less attention than risk assessment in the past, it is important to stress that risk analysis will only be effective when all three components are successfully integrated.

Risk analysis is a process consisting of three components: risk assessment, risk management, and risk communication. Risk assessment is scientifically based process and focus on estimating the risk that hazardous event or factor will negatively affect a population or subpopulation and consisting of the following steps: (1) hazard identification; (2) hazard characterization; (3) exposure assessment; and (4) risk characterization. Risk management is the process, distinct from risk assessment, of weighing policy alternatives in consultation with all interested parties, considering

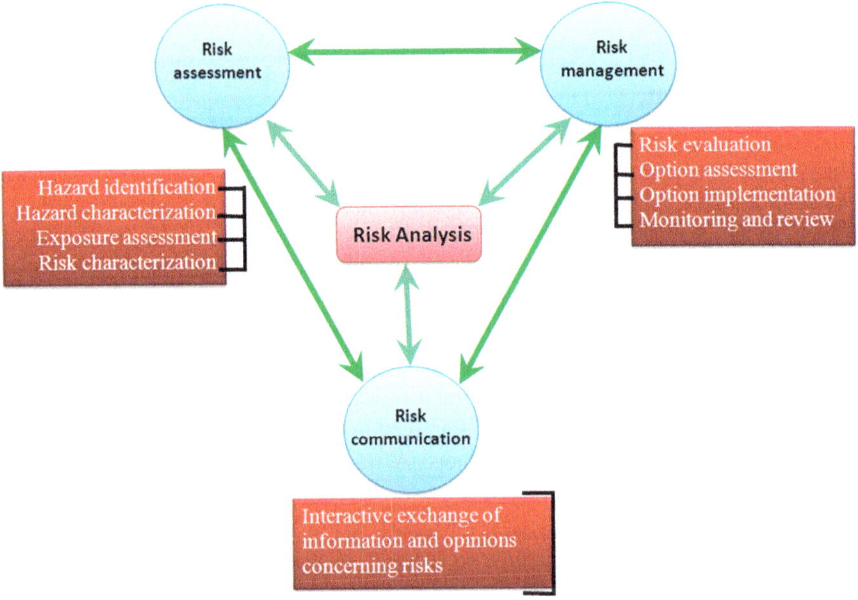

Fig. 1 Components of risk analysis

risk assessment and other factors relevant for the health protection of consumers and for the promotion of fair trade practices, and, if needed, selecting appropriate implementation of prevention and control options. Risk communication is the interactive exchange of information and opinions throughout the risk analysis process concerning risk, risk-related factors and risk perceptions, among risk assessors, risk managers, consumers, industry, the academic community and other interested parties, including the explanation of risk assessment findings and the basis of risk management decisions.

1.2 The Risk Analysis Process

Risk analysis focuses important factors that would enhance protection of human health and minimize the incidence of food-borne diseases through establishing control of food-borne hazards and framing food safety policies and its practical application. Risk analysis provides food safety regulators with the information and evidence they need for effective decision-making. The process normally begins with risk management, which, as a first step, defines the problem, articulates the goals of the risk analysis and defines the questions to be answered by the risk assessment. The science-based tasks of 'measuring' and 'describing' the nature of the risk being analysed (i.e. risk characterization) are performed during the risk assessment. Risk management and assessment are performed within an open and transparent environment based on communication and dialogue. Risk communication encompasses an interactive exchange of information and opinions among risk managers, risk assessors, the risk analysis team, consumers, and other stakeholders. The process often culminates with the implementation and continuous monitoring of a course of action by risk managers.

1.2.1 Essential Characteristics of Risk Analysis

Risk analysis is an iterative and ongoing process in which steps are repeated when needed. The process does not end once a decision is reached. Members of the risk analysis team regularly monitor the success and impact of their decision. Modifications are made as required—on the basis of new data or information or changes in the context of the problem—to achieve further reductions in adverse human health effects. It requires open and effective internal and external communication. Risk managers must interact and communicate frequently with risk assessors and other members of the risk analysis team (internal communication), as well as many different types of stakeholders (external communication) as often as needed.

Risk management has been defined as 'the process, distinct from risk assessment, of weighing policy alternatives, in the light of risk assessment in consultation with all interested parties, considering risk assessment and other factors relevant for the health protection of consumers and for the promotion of fair trade practices, and, if needed, selecting appropriate prevention and control options (Codex). Risk management therefore plays a key role at the beginning of the risk analysis process in identifying food safety problems and considering the best ways to manage them. The consideration of different policy alternatives is a critical part of risk management. This requires a focus on the scientific aspects of the risk (i.e. the detail and the outcome of the risk assessment) as well as any associated economic, legal, ethical, environmental, social, and political factors that are important to people. Risk management is not a linear process. Like the rest of risk analysis, risk management is an iterative process. Therefore, any model for risk management should be flexible enough to enable the various activities to be reviewed, repeated, and adapted as necessary.

2 Food Safety Objectives and Risk Assessment

It confers appropriate level of protection, sanitary, and phytosanitary measures that should be adapted to protect human, animal, and plant life or health. Modern approaches to food safety include the identification of actual, or potential, hazards from microbial contamination, assessing the risk that such contamination may cause disease in the consumer, and then seeking to employ processes that will control and minimize such risks. '*Hazard*' can be defined as something that has the *potential* to cause harm, for instance the contamination of food by pathogenic bacteria. '*Risk*' is defined as the *likelihood* of harm in a defined situation; for instance, consumption of food contaminated with specific pathogenic microorganisms and/or their toxins. *Risk assessment* of foods is therefore concerned with assessing the potential risk that consumption of a food may cause harm to consumers. As is amply demonstrated by ICMSF (2002), risk assessment requires an understanding of microbial contamination per se and also that both food process operations and domestic food handling practices may reduce or increase the risk from a defined hazard for a defined group of consumers (infants, children, the aged, the immuno-compromised, etc.).

The International Commission on Microbiological Specifications for Food (ICMSF) introduced the concept of '*Food Safety Objectives*' (FSOs) that was adopted subsequently by CODEX (CCFH) as part of its MRM document. An FSO provides a means to convert public health goals into parameters that can be controlled by food producers and monitored by government agencies. It is defined (ICMSF 2002) as, '*the maximum frequency and/or concentration of a microbial hazard in a food considered tolerable for consumer protection*'. ICMSF (2002) notes that FSOs are '*typically expressions of concentrations of microorganisms or toxins at the moment of consumption*'. Concentrations at earlier stages of the food chain are considered to be performance criteria. Hence, an FSO seeks to take account of hazards arising both during commercial processing and from unpredictable effects associated with retail and domestic food storage and handling. By contrast, performance criteria relate to the requirement to control hazards at earlier stages of the food chain. So, the FSO is defined as '*the maximum likely level of hazard that is acceptable*' following the integration of several stages in food processing, based on knowledge of microbial associations of foods, processing hurdles

(Leistner and Gould 2001; ICMSF 2005) which may result in death or inhibition of microorganisms and of the likelihood of re-contamination and/or re-growth of organisms during subsequent storage and handling. Thus the FSO concept relates also to the use of the Hazard Analysis and Critical Control Point (HACCP) concept for controlling the effectiveness of food processing operations.

CODEX has now published a Guide for National Food Safety Authorities (Anon 2006) that explains the whole concept of Food Safety Risk Analysis. This report covers all aspects of risk assessment for foods including providing guidance on the four stages of a risk management procedure. Microbiological risk assessment is based on the use of 'quantitative *microbiological metrics*' as a risk management option. 'Quantitative metrics' is defined as the '*quantitative expressions that indicate a level of control at a specific step in a food safety risk management system … the term 'metrics' is used as a collective for the new risk management terms of food safety objective (FSO), performance objective (PO) and performance criteria (PC), but it also refers to existing microbiological criteria*'.

2.1 Understanding Risk Assessment

Codex defines risk assessment as a scientifically based process consisting of four steps: (1) hazard identification; (2) hazard characterization; (3) exposure assessment; and (4) risk characterization. The definition includes quantitative risk assessment, which emphasizes reliance on numerical expressions of risk, and also qualitative expressions of risk, as well as an indication of the attendant uncertainties.

Several aspects of this definition are important to highlight. Firstly, risk assessment is a systematic and science-based process, which involves four major steps. Secondly, risk assessment explicitly addresses uncertainty (i.e. what is not known about the risk) in a logical, transparent, and well-documented manner that is clearly indicated to everyone involved in the risk analysis process. Finally, risk assessment can be descriptive or narrative, qualitative, semi-quantitative, or quantitative. Both qualitative and quantitative risk assessments are important in different circumstances and there is nothing inherently superior or inferior about either.

Qualitative risk assessment is the process of compiling, combining, and presenting evidence to support a statement about risk. While numerical data and analysis may be part of the input into a qualitative risk characterization, the final risk estimate does not necessarily result from attempts to produce a mathematical or computational representation of the risk producing system. Examples of qualitative food safety risk assessments include rating systems used by retail or foodservice establishments.

Quantitative risk assessment is based on numerical data and analysis. It can be deterministic (e.g. food additive safety assessment) or probabilistic (e.g. microbial risk assessment). Quantitative risk assessments should describe uncertainty in numerical terms with uncertainty distributions determined by various statistical methods. A quantitative risk assessment can address risk management questions at a finer level of detail than a qualitative risk assessment. There is no one way to perform a food safety risk assessment. Different models for food safety risk assessment exist and the process will vary according to the type of risk, the model used, and the questions to be answered. Indeed, in some cases (e.g. when the risk management response is obvious and acceptable to all the parties concerned or when there is insufficient data), it may either be unnecessary or impossible to perform a full risk assessment according to Codex guidelines.

2.1.1 Hazard Identification

Various biological, chemical, and physical hazards are at the source of food safety risks. Although the task of identifying a hazard is often considered part of risk management, risk assessors usually also play an important role in hazard identification. In particular, when possible hazards need to be analysed and prioritized on the basis of scientific evidence, risk assessors provide scientific expertise to help risk managers select the hazard of greatest concern. In other cases, where risk managers have already identified the hazard, risk assessors provide supplementary information on the scientific nature of the hazard.

2.1.2 Hazard Characterization

During hazard characterization, risk assessors develop a complete profile of the nature and extent of the adverse health effects associated with the identified hazard. The impact of varying amounts of the hazardous material on human health can be considered quantitatively (in a dose–response relationship) and/or qualitatively in a narrative fashion (Table 1).

2.1.3 Exposure Assessment

The exposure assessment provides scientific insight on the presence of the hazard in the product(s) consumed. It combines information on the prevalence and concentration of the hazardous material in the consumer's food supply and environment and the likelihood that the consumer will be exposed to various quantities of this material in their food. Information on the prevalence and concentration of the hazard could

Table 1 Examples of hazards

Biological	Chemical hazards	Physical hazards
Bacteria	Naturally occurring toxins	Metal, machine filings
Toxin-producing micro-organisms	Direct and indirect food additives	Stones
Moulds	Pesticide residues	Glass
Parasites	Residues of veterinary drugs	Insect parts
Viruses	Chemical contaminants	Jewellery
Other biological hazard		Tools

include estimates of the number of pathogens in a serving of food or the amount of a food additive consumed daily by a representative consumer. Depending on the nature of the problem, exposure assessment takes into account the relevant production, storage, and handling practices along the food chain.

2.1.4 Risk Characterization

During risk characterization, all the evidence from the previous three steps is combined in order to obtain a risk estimate (i.e. an estimate of the likelihood and severity of the adverse health effects that would occur in a given population with associated uncertainties) and respond to the questions posed by the risk managers. In general, the risk characterization includes a summary description of the consequences of exposure to the hazard, as well as an estimate of the likelihood of the adverse consequences of interest in a risk estimate. The outputs of a risk characterization should clearly identify important data gaps, assumptions, and uncertainties in order to help risk managers judge how close the characterization might come to describing reality. Risk characterization rarely gives more than a reasonable estimate or an informed view of the risk in reality.

2.2 *Chemical and Microbial Risk*

Food safety risk assessments are undertaken in response to identified chemical or microbial risks to human health. Chemical risk assessments focus on the presence of chemicals such as food additives, food contaminants, or residues of veterinary drugs. Some chemicals, such as food additives and colourings, are deliberately added to food in small amounts to make food look or taste better, to maintain or improve nutritive value, to help processing or preparation, to maintain freshness, or to help preserve food (direct additives). In addition, indirect additives or 'contaminants' can enter food accidentally during handling, processing (through equipment), or packaging (through migration) or can be generated through chemical processes in the food itself ('chemical reaction'). Technical aids used in primary production (such as pesticides or veterinary drugs) can also remain as residues in food products. As the number of direct and indirect additives to food has increased, so too has public concern about the type and amount of these additives and their potential to cause cancer or other illnesses in people. A microbial risk assessment evaluates the likelihood of adverse human health effects occurring after exposure to a pathogenic microorganism or to the medium in which the organism occurs. The hazard in microbial risk assessment is fundamentally different from the hazard in a chemical risk assessment. In particular, the hazard in a microbial risk assessment is alive, which reorients the focus of the risk assessment significantly. One of the most unique aspects of a living hazard is that the levels of pathogen in a food can change radically over time. Most microbial hazards can grow, decline or die many times before a food is consumed (Table 2).

Table 2 Characteristics of microbial and chemical hazards

Microbial hazard	Chemical hazard
Usually acute and the result of a single exposure	Can be lifetime risk or acute
High degree of variability in both the host and the pathogen	Toxicology does not usually vary greatly from person to person and the toxicity of the chemical itself is invariant
Continuously changing in quantity and characteristics	Tend to be fixed in quantity and hazardous characteristics
Non-homogenous presence in foods (they tend to clump and be distributed non-uniformly throughout a food)	Can be a homogenous presence (e.g. direct food additives), or heterogeneous (chemical contaminants)
Can enter the food chain at many points	Usually enters the food at specific points (e.g. cleaning agent residues during manufacturing, veterinary drugs on the farm)

Given the characteristics of the hazard in microbial risk, there is much more complexity involved in performing a microbial risk assessment than a chemical risk assessment. In addition, because of the potential for a pathogen to enter the food chain at many points, microbial risk assessment often requires a farm-to-table perspective. By comparison most chemical risk assessments focus on a particular part of the food chain. Microbial risk assessments also tend to encounter many more data gaps and greater uncertainties than chemical risk assessments.

3 Techniques Used in Food Safety Risk Assessment

Food safety risk assessment must be based on sound scientific evidence. Food safety regulators must have access to appropriate scientific data, information, and expertise in order to assign a risk assessment. Depending on the nature of the hazard and circumstances in which it occurs, various scientific experts (including biologists, chemists, medical experts, geneticists, epidemiologists, toxicologists, microbiologists, agronomists, botanists, entomologists, zoologists, and others) may be involved. The exact combination of analytical tools and techniques used in qualitative and quantitative risk assessment will vary according to the specific context and type of the risk assessment. In order to apply these techniques and perform risk assessment, certain basic infrastructure (including laboratories, scientific equipment, technology, and research facilities) will be essential.

3.1 Statistical Techniques

Although risk assessment does not usually require expertise in the most advanced and contemporary statistical techniques, a solid understanding of basic statistical techniques is essential for quantitative risk assessment, especially probabilistic risk assessment. Knowledge of the following basic techniques is required for successful risk assessment:

- Descriptive statistical techniques to extract useful information from scientific data and evidence
- Inferential statistical techniques to obtain information about populations from samples
- Different statistical tests to establish the most likely explanation of the observed phenomena

More sophisticated statistical techniques (such as curve fitting, regression analysis, meta-analysis, experimental design, bootstrapping, and the like) can also be used to support risk assessment.

3.2 *Probability*

Probability encompasses variability and uncertainty, both of which are always present in the context of food safety risk assessment. Risk assessors need a good command of basic probability concepts and techniques, including the ability to make basic probability calculations, in order to perform most kinds of quantitative risk assessment. Probabilistic risk assessment also requires a solid understanding of probability distributions and their characteristics since variability and uncertainty are both frequently described using probability distributions.

3.3 *Monte Carlo Process*

The Monte Carlo process has been applied to a large range of complex problems that involve random behaviour. It is a procedure that generates values of a random variable based on one or more probability distributions. It has been used extensively in microbial risk assessment and is increasingly being applied in other types of quantitative risk assessments, e.g. for intake assessment of chemicals in food. The Monte Carlo process encompasses two steps: (1) a random number is generated over the [0,1] interval (2) that number is transformed into a useful value using a probability distribution specified by the individual responsible for the model.

3.4 *Probabilistic Scenario Analysis*

Creating and analysing different scenarios of risk is a useful tool for risk assessment. A scenario can be defined as an outline for any proposed series of events, real or imagined. In other words, a scenario is a series of events that could happen. In risk assessment, a scenario is defined by a set of assumptions about model input values and how those input variables are related. Probabilistic scenario analysis is used to generate different scenarios and undertake a probabilistic analysis of the most likely scenarios and their outcomes. The worst-case scenario is often used in deterministic risk assessment. Scenarios can and have been considered deterministically. However, because of the extent of variability and uncertainty in the world, it is often difficult to identify the full range of possible outcomes of any risk management decision with just a few carefully circumscribed scenarios. It is not unusual for a probabilistic scenario analysis to combine several different tools such as an event tree and the Monte Carlo Process. Probabilistic scenario analysis has been used in most of the quantitative microbial risk assessments completed to date.

3.5 Knowledge Elicitation Techniques

Although risk assessment is based on a scientific and evidence-based approach, it will sometimes be necessary to obtain professional judgements and expert opinions to address data gaps and uncertainty in decision-making processes. Data gaps are encountered frequently during risk assessment. When the missing data are considered important to the decision-making process, risk assessors must try to close the existing data gaps as far as possible. In cases where there is sufficient time and resources, additional research can be undertaken to produce the necessary data. However, in other cases where it is impossible to locate or produce new data, risk assessors can use other techniques—such as knowledge elicitation techniques—to address data gaps.

Knowledge elicitation techniques are used to reveal expert knowledge in these circumstances and help to make expert opinions as evidence-based as possible. A wide variety of techniques can be used to elicit knowledge from experts and improve the quality and transparency of the knowledge gathering process. Traditional methods include the Delphi method, the nominal group approach, scenario analysis, scientific heuristics, rational consensus, indirect elicitation, the direct method, parametric estimation, self-scoring, collective scoring, surveys and questionnaires, interviews, and case studies. Many new knowledge elicitation techniques have been developed in recent years. These include cognitive approaches, contextual approaches and ethnography.

3.6 Ranking Tools

Ranking is a common technique in qualitative risk assessment. Ranking helps risk assessors to prioritize risks. Various kinds of ranking techniques exist. The multi-criteria decision-making literature is rich in methods to rank and sort problems. However, other simpler techniques can also be useful. For instance, criteria and their subjective weights can be used to sort and rank various alternative options. The choices of criteria and weights should be based on as much scientific evidence as possible to make the process as evidence based as possible.

3.7 Sensitivity Analysis

A good risk assessment uses sensitivity analysis to clearly identify and address uncertainty. Sensitivity analysis enables managers to understand how answer(s) to question(s) might change under different conditions or assumptions. It helps risk assessors to systematically investigate and discover which variables have the greatest influence on the outcomes of the risk assessment. A sensitivity analysis can illuminate the option assessment process for risk management by identifying those inputs with the greatest positive and negative effects on outcomes. Complex risk assessments may have dozens of input and output variables that are linked by calculations, systems of equations, assumptions, and so on. Risk assessors and risk managers must understand the relative importance of the various components of a risk assessment and the influence of these variables on the results of the risk assessment. Some outcomes and decisions are sensitive to minor changes in assumptions and input values.

A good sensitivity analysis will aid the risk assessment by revealing the most important variables in the assessment. It will provide insight into the conditions that contribute the most to good and bad outcomes. Once the key inputs are identified, assessors can focus their attention on addressing the uncertainty in these variables or carefully describing their variability. Therefore, sensitivity analysis helps to focus an assessor's attention on the most important inputs. Many different sensitivity analysis techniques exist. One popular approach uses parametric variation of the values of input variables to examine its effects on one or more output variable.

4 Characteristics of a Good Risk Assessment

A good risk assessment helps food safety regulators and other officials to make transparent, science-based decisions about a food safety risk. It improves the quality of the decision-making process and informs the decision for which it was prepared. In general, risk assessments should be as simple as possible whilst meeting the risk manager's needs and should strive to balance greater detail and complexity (e.g. through addressing more questions or alternative scenarios) against having to include the greater set of assumptions that this would entail because more assumptions decrease the reliability of the conclusions. Codex Guidelines (CAC 1999) for microbiological risk assessment contains a list of general principles of microbiological risk assessment, including that:

- Risk assessment be objective and soundly based on the best available science and presented in a transparent manner
- Constraints that affect the risk assessment, such as cost, resources or time, be identified and their possible consequences described
- Microbiological risk assessment should clearly state the purpose of the exercise, including the form of risk estimate that will be the output

- The dynamics of microbiological growth, survival, and death in foods and the complexity of the interaction (including sequelae) between human and agent following consumption as well as the potential for further spread be specifically considered
- Data should be such that uncertainty in the risk estimate can be determined
- Data and data collection systems should, as far as possible, be of sufficient quality and precision that uncertainty in the risk estimate is minimized
- MRA should be conducted according to a structured approach that includes Hazard Identification, Hazard Characterization, Exposure Assessment and Risk Characterization

4.1 Risk Characterization Measures

In assessing food-borne microbiological risks we are principally concerned about the effect of the identified hazard on human health, of which there are a number of possible results from exposure to microbiological pathogens. In any specific individual, there may be no effect, or no measurable effect. However, to be considered a pathogen, there must be possible an adverse health effect in at least a proportion of the exposed population as a result of ingestion of the pathogen or its toxins. Adverse health effects from exposure to pathogens include illnesses of varying severity (morbidity) and duration, ranging from mild self-limiting illness to those requiring hospitalization, or leading to chronic diseases, through to death (mortality). To date, risk assessments have tended to measure risks of microbiological food poisoning or infection as a direct result of exposure to food contaminated with pathogens or their toxins. In population terms, however, the development of asymptomatic carriers of the pathogen may also be classified as an adverse health effect, since this may lead to multiplication, excretion, and spread of the organism, eventually causing illness or death in others (i.e. secondary spread). In addition, there may be adverse health effects of interest specifically at the population level, for example epidemics and pandemics. Risks estimates can be made on an individual risk basis, e.g. risk of illness per serving, or on a population basis, e.g. 'cases per annum'. While the Codex risk assessment framework focuses on severity and probability of disease, measures to compare disease severity are required. The burden of disease can be measured in terms of individual or national economic loss, if required, via probable numbers of days or years of working life lost, cost of treatment, etc. However, the measurement of loss of quality of life is harder to quantify, although various attempts have been made, resulting in the concept of equivalent life years lost through specific types of disability, pain, or other reduced quality of life. This allows the comparison of one health state with another and with mortality itself. Thus it is possible to quantify the adverse health effect of any occurrence in terms of life year equivalents lost and estimate the risk of this from any specified source. Integrated health measures provide information to put diverse risks into context.

There are many potential adverse health effects that a risk manager might be interested in, in addition to those about which the affected individual is directly concerned. This, in turn means that there are many possible ways to measure and express the magnitude of the risk (sometimes called the 'risk metric') that might be selected as the required output from a risk assessment. The selection of the particular measure of risk to be used is therefore not necessarily straightforward and must be discussed between the risk manager, the risk assessor, and other interested stakeholders. In addition, for quantitative modelling, the unit or units required must be defined whilst taking into account the practical aspects of modelling so that the outputs can be produced and reported in those units. Various types of probability models and studies of risk issues have been labelled as 'risk assessments'. FAO/WHO, OIE, and other guidelines advocate decision-making based on a risk assessment. Codex risk assessment guidelines and recommendations have legal significance in terms of what satisfies the food safety risk assessment requirements under the WTO Sanitary and Phytosanitary (SPS) Agreement. Thus, it is of both technical and legal importance to be able to determine whether a particular piece of work can be categorized as a risk assessment.

4.2 Risk Assessment Approaches

This section describes three categories of work that are often labelled 'risk assessment' and discusses when each type of study conforms to the necessary requirements. The three approaches are presented as examples, and other approaches to risk assessment are possible. No 'correct' approach can be recommended or specified: the choice of approach depends on the risk assessment question, the data and resources available, etc. The three categories considered are:

- Estimating an unrestricted or baseline risk
- Comparing risk intervention strategies
- Research-related study or model

4.2.1 Estimating 'Unrestricted Risk' and 'Baseline Risk'

An 'unrestricted risk' estimate is the level of risk that would be present if there were no safeguards; and a 'baseline risk' estimate is the current, standard or reference status, i.e. the point against which the benefits and costs of various intervention strategies can be compared. The concept of unrestricted risk has been most widely used in import-risk analysis, in which it has more obvious utility. A common and practical starting point for a risk assessment is to estimate the existing level of risk, i.e. the level of food safety risk posed without any changes to the current system. This risk estimate is most frequently used as the baseline risk against which intervention strategies can be valued, if desired. This baseline risk may, for example, have utility in determining an Appropriate Level of Protection (ALOP). Using the current risk as a baseline has a number of advantages, among them being that it is the easiest to estimate the effect of changes by estimating the magnitude of the risk after the changed conditions relative to the existing level of risk, i.e. it may obviate the need to explicitly quantify the risk level under either scenario. This approach implicitly accepts the starting point of any risk management actions as being changes to the current system. For some purposes, a baseline other than the existing level of risk might be used as a point of comparison. For example, the baseline risk could be set as that which would exist under some preferred (e.g. least costly) risk management approach and the risk under alternative approaches compared with that.

Estimation of an unrestricted risk, i.e. the level of risk that would be present if no deliberate actions were taken to control the risk, sometimes referred to as inherent risk, may have a role in determining the efficacy of existing microbiological food safety risk management approaches compared with entirely new systems. Over time, as knowledge of the causes of infectious diseases grew, many controls to minimize food-borne illness have been implemented at the level of both consumers and the industry. While it is difficult to imagine being able to realistically assess the risk level in a hypothetical world where all those controls were removed, the principle is valid and takes as its point of departure a 'raw' risk that has been identified, and now quantified, and for which there are many combinations of options to choose from to control the risk. It would, in principle, enable reassessment of what combination of controls (both those in place and new possible interventions) would give the most efficient protection. In practice, one can attempt to estimate a risk where some of the more obvious, and perhaps more costly, interventions currently in place are removed, and then re-evaluate how to address the risk. Using the current risk level as the point of comparison does not encourage one to review the many layers of risk reduction activities that are already present and have evolved over time in the absence of monitoring to evaluate their efficacy and to improve their efficiency. For example, control measures introduced before good information existed about a problem might be expected to be highly conservative. With improved knowledge, better targeted approaches could possibly be devised to deliver the same health protection with fewer disadvantages to consumers or producers.

Estimating a baseline or unrestricted risk may not be for the immediate purpose of managing the risk so much as to measure or bound the severity of a food safety problem. Whilst in theory it may not be necessary to determine a baseline risk in order to evaluate intervention strategies, it is nonetheless almost always carried out in practice.

4.2.2 Comparing Risk Management Strategies

Risk assessment is commonly undertaken to help risk managers understand which, if any, intervention strategies can best serve the needs of food safety, or if current risk management actions are adequate. Ideally, agencies with responsibility for safety of foods would consider all possible risk management interventions along the food chain without regard to who has the authority to enact them, and this objective has led to the creation of integrated food safety authorities in many nations and regions. A farm-to-table model may be most appropriate for this purpose. In practice, however, the scope of the assessment may be limited to those sections of the food chain within the risk manager's area of authority, but a more comprehensive risk assessment might identify relationships outside that area of authority that would motivate the risk manager to seek the new authority required to intervene effectively or to request others with authority to take appropriate actions. For some risk questions, analysis of epidemiological data or a model of part of the food chain may be adequate. In some cases it is possible to estimate the change in risk without producing an estimate of the baseline risk, but caution must be used in these cases. For example, a risk assessment might determine that it is technically feasible to reduce a particular risk 100-fold, but if this risk was negligible at the start, then reducing it 100-fold may not be a worthwhile course of action.

The 'proximity' of a risk is commonly considered in risk analysis applied to management of large construction projects, and in certain circumstances will also be an important factor in food safety risk assessment if unplanned or uncontrolled factors could be expected to change the risk over time, e.g. the increase in average age of populations in many nations is expected to increase overall population susceptibility to many disease, including food-borne diseases, leading to increased incidence. In other situations the risk may be seasonal, or arise only after natural disasters, or be linked to some specific event involving a very large gathering of people, etc. 'Proximity' describes the period or interval of time during which the risk might affect the stakeholders. A natural tendency is to focus on risks that are immediate when we may have a limited ability to manage them: assessing risks that could arise in the future might enable risk management steps to be implemented at a fraction of the cost of that for an emergency response when the risk has been realized.

4.2.3 Research-Related Study or Model

It has already been stated that risk assessment is a decision tool, not a scientific or research tool. Some research-based risk assessments have been produced with the intention of expanding our knowledge and tools for evaluating risks. They may be based on hypothetical or on genuine decisions questions and evaluate the assessment results according to how they respond to those questions. However, they are not always initiated by a 'risk manager'. There are a number of large microbiological food safety models in existence that have been initiated as academic exercises. These models have helped advance the field of microbiological risk assessment by allowing us to see what techniques are necessary, developing new techniques, and stimulating research that can now be seen to have value within a risk assessment context. In some situations, those models have subsequently been used by risk managers to assist in risk management decisions. Such models have also made apparent the changes in collection and reporting methods for microbiological, epidemiological, production, dietary, and other data that would make the data more useful for risk assessment. Early experience with microbiological risk assessments has proven these assessments to be valuable in aiding our understanding of complex systems. The very process of systematically investigating a food chain has contributed to our ability to both appreciate and understand the complexity of the systems that make up the food chain.

The importance of matching the type of risk assessment to its purpose has been emphasized previously. The USA National Advisory Committee on Microbiological Criteria for Foods noted (USNACMCF 2004): 'Risk assessments can be quantitative or qualitative in nature, but should be adequate to facilitate the selection of risk management options. The decision to undertake a quantitative or qualitative risk assessment requires the consideration of multiple factors such as the availability and quality of data, the degree of consensus of scientific opinion, and available resources'.

5 Assessing the Reliability of the Results the Risk Assessment

Every risk assessment has some degree of uncertainty attached to its results. Complying with all the requirements of transparency, of describing model and parameter uncertainties, and all the explicit and implicit assumptions, does not necessarily communicate to risk managers the degree of confidence that the risk assessor has in the results of the risk assessment or limitations in its application. Thus, risk assessors must explain the level of confidence they feel should be attached to the risk assessment results. All assumptions should be acknowledged and made explicit in a manner that is meaningful to a non-mathematician. For example, it would be insufficient to say that 'illnesses were assumed to follow a Poisson process': a better explanation would be 'illnesses were modelled as a Poisson process, which means that each illness is assumed to occur randomly in time, independently of each other, and that the risk of an illness is either constant over time or follows some repeated seasonal pattern'. This type of explanation enables the risk manager to better understand the assumptions, and perhaps pose more informed questions about the effect of any violation of the assumptions. Deciding whether a food is safe or not is a difficult task. Food can never be proven to be entirely safe nor entirely hazardous. It can only be proven to be hazardous to some degree under certain conditions. While demanding completely safe food is unrealistic, it is possible to have food in which potential hazards have been reduced.

6 Conclusions

In summary, the use of a science-based approach will enable governments to develop and implement a range of general improvements and interventions tailored to specific high-risk areas, which will ultimately improve food safety and reduce the burden of food-borne disease. Codex standards are the outcome of multilateral negotiations based upon a risk assessment. It is important to communicate this fact to the public and thus signal that scientific evidence is only one of the determinants of Codex international food safety standards, albeit a very prominent one. The possible trade-offs between economic and political interests on the one hand and public health interests on the other hand, could become more tangible if the outcome of a Codex risk assessment was a ‘menu of policy options’. Existing Codex procedures already allow for this furthermore risk assessors play a more important role in defining the range of policy options to be analysed. Risk analysis will only be effective if it takes place in an environment in which government, industry, academic institutions, and consumers recognize value and participate in the process. Risk analysis must have the support of food safety regulators at the highest level of government. Industry must find value in the results of risk analysis. Academic institutions must produce information that meets the needs of risk analysis. Consumers and businesses must be able to recognize and derive clear benefits from the risk analysis process. Similarly, mechanisms must be in place to enable stakeholders to participate in the development of risk analysis policy, as well as in the various activities performed during risk analysis.

References

Anon (2006) Food safety risk analysis: a guide for national food safety authorities. FAO food and nutrition paper 87. WHO/FAO, Rome

CAC/GL 30 (1999) Principles and guidelines for the conduct of microbiological risk assessment

ICMSF (2002) Microorganisms in foods. 7. Microbiological testing in food safety management. Kluwer Academic/Plenum, New York

ICMSF (2005) Microorganisms in foods. 6. Microbial ecology of food commodities. Kluwer Academic/Plenum, New York

Leistner L, Gould GW (2001) Hurdle technology: combination treatment for food stability, safety and quality. Kluwer Academic/Plenum, New York

USNACMCF (US National Advisory Committee on Microbiological Criteria for Foods) (2004) Response to the questions posed by FSIS regarding performance standards with particular reference to raw ground chicken, Atlanta, GA, USA

Microbiological Quality Systems and Microbial Risk Analysis

Birce Mercanoglu Taban and Sait Aykut Aytac

1 Introduction

Food safety is of increasing global interest to governments and the food industry for both its importance to public health and its potential benefits and impact on domestic and international trades. A joint WHO/FAO expert committee on food safety declares that illnesses due to contaminated foodstuffs are most likely the widespread public health problem in the world and an important cause of reduced economic productivity (Ehiri et al. 1995).

The policies of food industries are increasingly directed towards food safety, ensured by extensive control procedures and accordingly efficient quality management throughout all stages of the agrofood chain, from procurement of raw materials and ingredients to final food product dispatch (Barendsz 1998; Baird-Parker and Tompkin 2000). Therefore, farmers can no longer grow just what they want or use technical aids to farming without considering the effect on the quality of the food produced (Rooney and Wall 2003). In addition to this, foodstuffs from many developing countries cannot contend well in the international food market due to the lack of quality since food consumers also look for certification and reassurance of the foodstuffs' origin and the production methods besides their dietary quality and hygiene (Alemanno 2007; Arvanitoyannis and Kassaveti 2009). As a consequence of this demand for production and distribution of safer foods and the enactment of new agreements through the World Trade Organization, Hazard Analysis and Critical Control Point (HACCP) which is a food safety quality management system has, at first, attracted global support especially in industrialized countries (Barendsz

B.M. Taban (✉)
Department of Dairy Technology, Ankara University, 06110, Diskapi, Ankara, Turkey
e-mail: btaban@ankara.edu.tr

S.A. Aytac
Department of Food Engineering, Hacettepe University, 06532, Beytepe, Ankara, Turkey

A. Malik et al. (eds.), *Food Processing: Strategies for Quality Assessment*, Food Engineering Series, DOI 10.1007/978-1-4939-1378-7_12

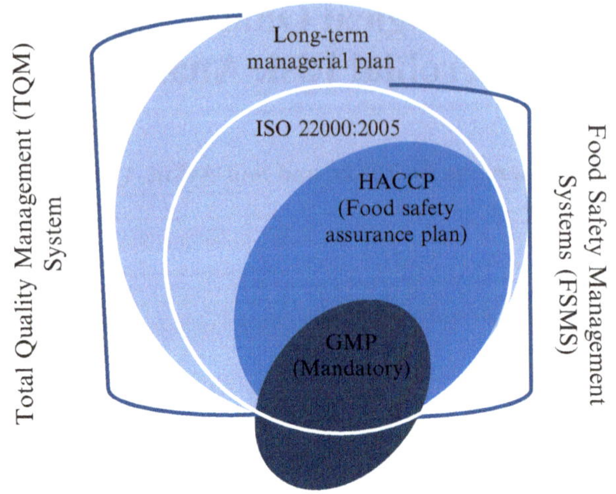

Fig. 1 Food safety within integrated quality management systems

1998; Ropkins and Beck 2000). It aims to prevent product safety problems by identifying biological, chemical, or physical hazards associated with the production, distribution, and retailing of a foodstuff and means for their control to safe levels (Jouve 2000). Starting in the 1990s, various customers in the food chain required their suppliers to have certified HACCP system. From January 1, 2006, the European Union (EU) mandated all food manufacturing and processing plants that produce and process food for the European market to integrate HACCP into their food safety systems (Surak 2003, 2007). However like any other systems, HACCP system has some vulnerable points and cannot manage to solve all food safety and quality-related problems, which will be discussed in detail later in this chapter, and these, may be the major drawback for its non-international application during recent years (Arvanitoyannis and Traikou 2005). This situation enlist the food industries in improving the quality perception which can be achieved by the internationally recognized food safety management systems (FSMS) the International Organization for Standardization (ISO) 22000:2005 that integrates both the quality management system (ISO 9001:2000) and HACCP system (Alemanno 2007) (Fig. 1).

Due to realizing the external benefits that act as an effective entry-to-market tool, opening up new business opportunities around the world as part of their market strategy and also internal benefits as developing massive improvements and efficiency, an increasing number of food industries strengthened their HACCP systems by implementing a certifiable IS0 22000:2005 food safety management system (Surak 2007).

In 2000, the Consumers Food Forum, which is the only independent global food business network, identified the need to enhance food safety, to ensure consumer protection, to strengthen consumer confidence, to set requirements for food safety schemes, and to improve cost efficiency throughout the food supply chain and so published certain guidance documents (Papademas and Bintsis 2010) and there years later, Global Food Safety Initiative (GFSI) was created (GFSI 2010).

In 2008, British Standards Institution (BSI) developed Publicly Available Specification—PAS 220 (2008) which is a new complementary standard to ISO

22000:2005 standard. This specification must be used in conjunction with ISO 22000:2005 standard and applies to management systems designed to help food manufacturing organizations of all sizes meet the ISO 22000:2005 standard. It provides detailed guidelines including construction and layout of buildings and associated utilities; layout of premises, including workspace and employee facilities; supplies of air, water, energy, and other utilities; supporting services, including waste and sewage disposal; suitability of equipment and its accessibility for cleaning, sanitation, and maintenance; how to design and implement relevant food safety management systems. However, it only applies to food manufacturers in the supply chain. We think that bears repeating: PAS 220 2008 supports but does not replace ISO 22000:2005 standard. GFSI agreed that the combination of ISO 22000:2005 standard with PAS 220 2008 contained adequate content for approval, but also highlighted the need for an industry-owned scheme governing the combination of these two standards (Papademas and Bintsis 2010).

Subsequently, FSSC 22000 which incorporates FSMS as well as PAS 220 specification already explained in the previous sections of this chapter was developed (Sansawat and Muliyil 2009).

Nonetheless a Total Quality Management (TQM) system, which complements the above systems, is important for achieving a permanent improvement in food manufacturing and processing plants' performance and so for the long-term success of them with better customer satisfaction (Jouve 2000).

2 Good Manufacturing Practice (GMP) and Hazard Analysis Critical Control Point (HACCP) System

2.1 GMP

For many years many countries have legislated that food producers, processors, and retailers must follow GMP procedures and have created their own GMP guidelines that correspond with their legislation to produce foodstuffs that are microbiologically, chemically, and physically safe and stable. GMP procedures and its requirements for hygienic design, construction, and operation of the food manufacturing and processing plants; hygienic construction and suitable and correct use of food manufacturing and processing equipments and devices; maintenance, cleaning, and disinfection of these equipments and devices; and personnel hygiene training describe the fundamental conditions for the hygienic production of foodstuffs (Jouve 2000).

GMP refers to the GMP Procedures promulgated by the US Food and Drug Administration under the authority of the Federal Food, Drug, and Cosmetic Act. GMP procedures expect a quality approach to food manufacturing and processing plants to minimize or eliminate cases of contamination, mistakes, and faults in order to protect the consumer from purchasing a foodstuff which is not effective or even dangerous by addressing the issues including recordkeeping, personnel qualifications, cleanliness, sanitation, equipment verification, process validation, and complaint handling. Failure of food manufacturing and processing plants to comply with GMP procedures can result in very serious results including recall, seizure, fines, and jail time (http://www.ispe.org/gmp-resources/what-is-gmp).

On the other hand, GMP cannot be used as a quality management system since it is only a part of quality assurance which ensures that foodstuffs are consistently produced and controlled to the quality standards appropriate to their intended use and as required by the marketing authorization and only represents a set of a few standardized guidelines for the safe production of foodstuffs. However, it can be strengthened when control measures are considered by the HACCP team (Barendsz 1998). Besides, HACCP can only be applied to food manufacturing and processing plants applying GMP but cannot replace the need to apply GMP procedures (Jouve 2000).

2.2 *HACCP*

HACCP, which was developed in the late 1960s to ensure the safety of foods for space US National Aeronautics and Space Administration (NASA) flight programs, has both become the generally accepted system for increasing food safety and have been adopted for food assurance in many food manufacturing and processing plants (Buchanan 1990; Griffith 2000). The concepts for examining Critical Control Points (CCPs) and GMP in producing safe foods were introduced during the US National Conference on Food Protection in 1971 (Jouve 2000; Papademas and Bintsis 2010). Then Pillsbury Company organized the first training program to the US Food and Drugs Administration (FDA) under the name of "Food Safety through the Hazard Analysis and Critical Control Point System," with the assistance of NASA in 1973. In the same year, FDA involved HACCP in the regulations applied for low-acid canned foods (Panisello and Quantick 2001; Arvanitoyannis and Kassaveti 2009). In 1987, US National Advisory Committee on Microbiological Criteria for Foods (NACMCF) became responsible for describing HACCP system for its application. After the adoption of the EU Food Hygiene Directive in June 1993, the concept of HACCP was brought into food legislation in EU (Ehiri et al. 1995). Furthermore in 1997, the Codex Alimentarius Commission (CAC) revised and adopted the seven principles of HACCP. On April 29, 2004, Food Hygiene Package, which comprises of Regulations (EC 2004a, b, c) was adopted by the European Parliament and the Council. These new requirements state that everybody, from primary production to retail, including the food business operators, is responsible for the safety of their products and must also apply HACCP principles (Maunsell and Bolton 2004). Meanwhile, British Retail Consortium (BRC) and International Food Standard (IFS), which are also based on HACCP principles, were introduced by the food market (Papademas and Bintsis 2010).

Before application of HACCP to any area of the food chain, that area should have in place prerequisite programmes such as good hygienic practices according to the Recommended International Code of Practice—General Principles of food hygiene, the appropriate Codex Codes of practice, and appropriate food safety requirements. These prerequisite programs to HACCP, including training, should be well established, fully operational, and verified in order to facilitate the successful application of the HACCP system (Codex 2009a) Fig. 2.

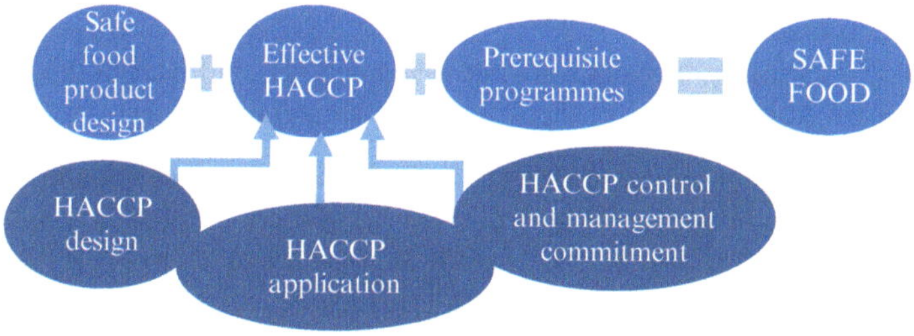

Fig. 2 The purpose of HACCP: safe food framework

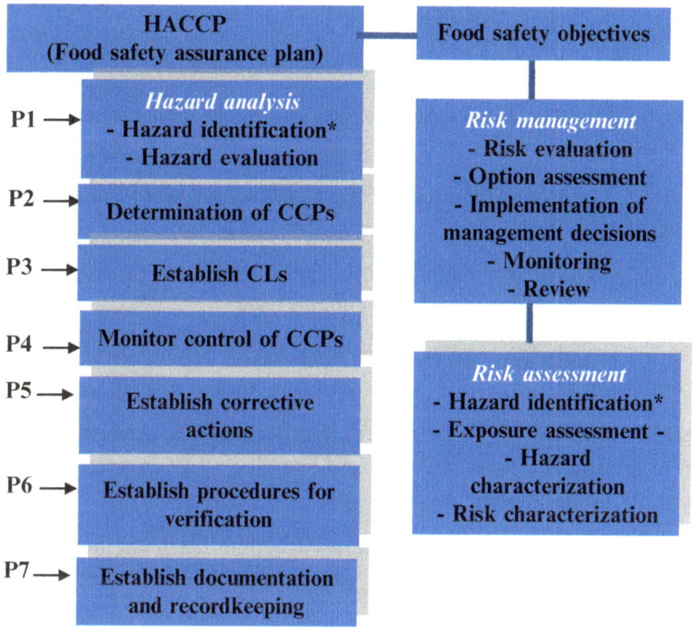

Fig. 3 Hazard identification* step in both HACCP (hazard analysis) and risk assessment

Walker and Jones (2002), van der Spiegel et al. (2004), and (Wallace et al. 2005) stated that HACCP is a good system for the prevention and control of foodborne diseases by identifying hazards and risks at every stage of the food manufacturing and production and determining where controls are needed (Sun and Ockerman 2005) so the task of hazard analysis and risk assessment of foods cannot simply be overemphasized (Barendsz 1998). Therefore hazard analysis is a key element of HACCP that will determine the strength of the resulting HACCP plan with the identification of suitable control measures (Wallace et al. 2014). However, although similarities exist between the inputs in the first principle of HACCP (hazard analysis) and risk assessment (hazard identification) (Coleman and Marks 1999), it is better to state that hazard analysis and risk assessment are profoundly different and independent processes (Sperber 2001; Manning and Soon 2013; Wallace et al. 2014) Fig. 3. Risk assessment, therefore, will be discussed in detail later in this chapter.

Hazards are caused by any of the following (Anonymous 1997):

1. The unacceptable presence of a biological, chemical or physical contaminant in raw food materials or in semifinished or finished food products
2. The unacceptable potential for growth or survival of microorganisms or the unacceptable potential for the generation of chemicals in semifinished or finished food products or in a manufacturing and production line environment
3. The unacceptable contamination or recontamination of semifinished or finished food products with microorganisms, chemicals, or foreign material

The closer examination of the key term definitions from internationally accepted HACCP Guidelines is helpful in understanding this hazard analysis process further (NACMCF 1997; Codex 2009a):

Control (verb): To take all necessary actions to ensure and maintain compliance with criteria established in the HACCP plan.

Control (noun): The state wherein correct procedures are being followed and criteria are being met.

Control measure: Any action and activity that can be used to prevent or eliminate a food safety hazard or reduce it to an acceptable level.

Corrective action: Any action to be taken when results of monitoring at a CCP indicate a loss of control.

Critical control point (CCP): A step at which control can be applied and is essential to prevent or eliminate a food safety hazard or reduce it to an acceptable level that no adverse effect on human health.

Critical limit (CL): A criterion that separates acceptability from unacceptability.

Deviation: Failure to meet a critical limit.

Flow diagram: A systematic representation of the sequence of steps or operations used in the production or manufacture of a particular food item.

HACCP: A system that identifies, evaluates, and controls hazards that are significant for food safety.

HACCP plan: A document prepared in accordance with the principles of HACCP to ensure control of hazards that are significant for food safety in the segment of the food chain under consideration.

Hazard: a biological, chemical, or physical agent in, or condition of, food with the potential to cause an adverse health effect (Codex 2009a).

Hazard: a biological, chemical, or physical agent that is reasonably likely to cause illness or injury in the absence of its control (NACMCF 1997).

Hazard analysis: The process of collecting and evaluating information on hazards and conditions leading to their presence to decide which are significant for food safety and therefore should be addressed in the HACCP plan.

Monitoring: The act of conducting a planned sequence of observations or measurements of control parameters to assess whether a CCP is under control.

Step: A point, procedure, operation, or stage in the food chain, including raw materials, from primary production to final consumption.

Validation: Obtaining evidence that the elements of the HACCP plan are effective.

Verification: The application of methods, procedures, tests, and other evaluations, in addition to monitoring, to determine compliance with the HACCP plan.

Codex (2009a) needs control of hazards that are of such a nature that their elimination or reduction to acceptable levels is essential to the production of a safe food and also says that the process of hazard analysis is intended to identify those hazards that are significant for food safety and therefore should be addressed in the HACCP plan. However, the term significant hazard which is defined by the International Life Sciences Institute (ILSI 1999) is not defined by Codex (Wallace et al. 2014).

Significant Hazard: Hazards that are of such a nature that their elimination or reduction to an acceptable level is essential to the production of safe foods (ILSI 1999).

HACCP certification, which is an international standard defining the requirements for effective control of food safety (Ropkins and Beck 2000), is built around seven principles (http://www.sgs.com/en/Agriculture-Food/Food/Retail-and-Hospitality/Other-Food-Industry-Certification/HACCP-Certification.aspx)

1. Conduct a hazard analysis of biological, chemical or physical food hazards
2. Determine Critical Control Points (CCPs)
3. Establish critical limits (CLs), as an example, minimum cooking/heating temperature and time
4. Establish a system to monitor control of CCPs
5. Establish the corrective actions to be taken when monitoring indicates that a particular CCP is not under control
6. Establish procedures for verification to confirm that the HACCP system is working effectively
7. Establish documentation concerning all procedures and record keeping

These principles consist of the following tasks which provide guidance for the application of them:

1. Assemble an HACCP team
2. Describe the product
3. Identify the intended use
4. Construct a flow diagram
5. Carry out on-site confirmation of the flow diagram

6. List all potential hazards associated with each step, conduct a hazard analysis, and consider any measures to control identified hazards
7. Determine CCPs
8. Establish CLs for each CCP
9. Establish a monitoring system for each CCP
10. Establish corrective actions
11. Establish verification procedures
12. Establish recordkeeping and documentation

An effective HACCP system requires a successful application of the HACCP principles and equally efficient technologies and systems which are up-to-date in order to determine and monitor each of the critical point (Wallace et al. 2014). For all types of food business, management awareness and commitment is necessary for application of an effective HACCP system (Fig. 2). The effectiveness will also rely upon management and employees having enough HACCP knowledge and skills (Codex 2009a). However, what is observed in reality is, due to the poor knowledge of the members of the HACCP team and the managers on how to perform hazard analysis which is the crucial part of the HACCP study, and the usage of unproven structured risk evaluation methods or inappropriate usage of suitable tools by untrained and inexperienced personnel, the HACCP systems in food manufacturing and producing plants are weakened (Wallace et al. 2014). It is therefore food manufacturing and processing plants should recognize the importance of HACCP teams made up of members with the correct blend of technical and HACCP principle application expertise, practical experience, team-working, administration, and leadership skills (Wallace et al. 2012).

In addition to this, Sun and Ockerman (2005) stated that due to the complexity of foods and the preparations involved in food service, it is harder to monitor and control the food safety by HACCP in this branch of the food industry. Also Cerf et al. (2011) showed that the HACCP system is not fully applicable at the primary production level and that food safety is obtained through the careful implementation of good hygiene practice (GHP) at the farm.

3 Risk Analysis

3.1 *Hazard Analysis and Risk Assessment*

Since hazard identification is a common step in the hazard analysis and risk assessment processes (Fig. 3), they have sometimes been confused with each other which could result in inaccurate hazard analysis; with serious consequences for the public health (Voysey and Brown 2000; Sperber 2001). When there is a serious weakness in hazard analysis then assessment of whether the HACCP system is working well or not cannot give an effective measure of food safety since missing hazard and so CCP creates dangerous results to the consumers (Wallace et al. 2014). We think that bears repeating hazard analysis and risk assessment are profoundly different and independent processes. Sperber (2001) gave a very good example of this confusion which triggered a pandemic cholera in South America in 1991. The public health officers who did not want to expose their population to the risk of liver cancer because of the probable formation of trihalomethanes due to the chlorination of drinking water, which was published as a conclusion of a risk assessment study, stopped chlorination of the municipal water supply of their city. Consequently, 1 % of at least 1 million people, who had contracted cholera, died. This could have been prevented by an accurate hazard analysis (Sperber 2001).

Hazard analysis includes a hazard identification step which is an open-ended brainstorming process to determine potential hazards and a hazard evaluation step which is a process to determine which identified hazards are of such significance that a CCP is required to control the hazard (NACMCF 1997). On the other hand, risk assessment is one of the three parts of the greater process of risk analysis, which also includes risk management and risk communication. In this section, we will deal with risk assessment since the structure of risk analysis whose overall objective is to ensure public health protection (FAO/WHO 1995, 1997, 1998) will be discussed in detail in the following section of this chapter.

Although hazard analysis is a qualitative and local process in which the decision is given by an individual HACCP team in each food processing and manufacturing plant, risk assessment is a quantitative and global process in which a numerical degree of risk can be calculated for each hazard by a major consortium [At EU level, the European Food Safety Authority (EFSA) is the body responsible for risk assessment in the field of food and feed safety (Romero-Barrios et al. 2013)], including regulatory, academic, industry, and public health representatives. Besides, hazard analysis lasts a few weeks or months. However, risk assessment lasts several months or years (Baird-Parker and Tompkin 2000; Sperber 2001) which is sometimes unfortunately the reason of usage of traditional hazard assessment through a simple evaluation process rather than this more detailed, structured approaches (Bertolini et al. 2007). We think that bears repeating HACCP teams of the food processing and manufacturing plants conduct their own individual hazard analyses but they do not conduct a risk assessment. They only benefit from the results of global risk assessment for implementation of their hazard analysis. In other words; a HACCP team in a food manufacturing and processing plant may use a published risk assessment as a source of information when conducting a hazard analysis but those conducting a risk assessment would not use the hazard analysis from HACCP plans as a resource material (Baird-Parker and Tompkin 2000; Voysey and Brown 2000). It is useful at this point to define the terms in question:

Dose–response assessment: The determination of the relationship between the magnitude of exposure (dose) to a chemical, biological, or physical agent and the severity and/or frequency of associated adverse health effects (response).

Exposure assessment: The qualitative and/or quantitative evaluation of the likely intake of biological, chemical, and physical agents via food as well as exposures from other sources if relevant.

Hazard characterization (Dose–response assessment): The qualitative and/or quantitative evaluation of the nature of the adverse health effects associated with the hazard. For the purpose of microbiological risk assessment, the concerns relate to microorganisms and/or their toxins.

Hazard identification: The identification of biological, chemical and physical agents capable of causing adverse health effects and which may be present in a particular food or group of foods.

Quantitative risk assessment: A risk assessment that provides numerical expressions of risk and indication of the attendant uncertainties (stated in the 1995 Expert Consultation definition of risk analysis)

Qualitative risk assessment: A risk assessment based on data that, while forming an inadequate basis for numerical risk estimations, nonetheless, when conditioned by prior expert knowledge and identification of attendant uncertainties, permit risk ranking or separation into descriptive categories of risk.

Recall: Removing foods from the market or correcting the problem that may potentially present a significant or serious risk to the consumer or user of the product.

Risk: A function of the probability of an adverse health effect and the severity of that effect, consequential to a hazard(s) in food.

Risk analysis: A process consisting of three components: risk assessment, risk management, and risk communication.

Risk assessment: A scientifically based process consisting of the following steps: (a) hazard identification, (b) hazard characterization, (c) hazard characterization (dose-response assessment), and (d) risk characterization.

Risk characterization: The process of determining the qualitative and/or quantitative estimation, including attendant uncertainties, of the probability of occurrence and severity of known or potential adverse health effects in a given population based on hazard identification, hazard characterization (dose-response assessment), and exposure assessment.

Risk communication: The interactive exchange of information and opinions concerning risk and risk management among risk assessors, risk managers, consumers, and other interested parties.

Risk estimate: Output of risk characterization.

Risk management: The process of weighing policy alternatives in the light of the results of risk assessment and, if required, selecting and implementing appropriate control options, including regulatory measures.

Sensitivity analysis: A method used to examine the behavior of a model by measuring the variation in its outputs resulting from changes to its inputs.

Transparent: Characteristics of a process where the rationale, the logic of development, constraints, assumptions, value judgments, decisions, limitations, and uncertainties of the expressed determination are fully and systematically stated, documented, and accessible for review.

Uncertainty analysis: A method used to estimate the uncertainty associated with model inputs, assumptions, and structure/form.

According to Voysey and Brown (2000), a quantitative risk assessment which produces a mathematical statement that links the probability of exposure to an agent and the probability that the exposure will affect the test individual, should always be carried out to evaluate microbiological hazards. However, if no data are available to do it, then it is not possible to carry out a quantitative risk assessment process (Coleman and Marks 1999). Several quantitative risk assessments for specific microbiological hazards in foodstuffs such as *Escherichia coli* O157:H7 in ground beef hamburgers (Cassin et al. 1998) and *Salmonella* in whole chickens (Oscar 2004) were given in the literature (Manning and Soon 2013).

There are a number of important principles to bear in mind when carrying out a microbiological risk assessment (Voysey and Brown 2000; Codex 2009b):

1. Microbiological risk assessment should be soundly based upon science
2. There should be a functional separation between risk assessment and risk management which ensure that the risk assessment process is unbiased
3. Microbiological risk assessment (MRA) should be conducted according to a structured approach that includes hazard identification, hazard characterization, exposure assessment, and risk characterization
4. A microbiological risk assessment should clearly state the purpose of the exercise, including the form of risk estimate that will be the output. As an example, the purpose may be to estimate the risk of a microbiological hazard in the total food supply or the microbiological hazards associated with a specific commodity and the output may be the estimation of the annual rate of illness in a population or the rate of illness per eating experience, etc. (Baird-Parker and Tompkin 2000)
5. The conduct of a microbiological risk assessment should be transparent which requires systematic documentation, statement of assumptions, value judgments, and a formal record
6. Any constraints that affect the risk assessment, such as cost, resources or time, should be identified and their possible consequences described
7. The risk estimate should contain a description of uncertainty and where the uncertainty arose during the risk assessment process
8. Data should be such that uncertainty in the risk estimate can be determined; data and data collection systems should, as far as possible, be of sufficient quality and precision that uncertainty in the risk estimate is minimized
9. A microbiological risk assessment should explicitly consider the dynamics of microbiological growth, survival, and death in foods and the complexity of the interaction (including sequelae) between human and agent following consumption as well as the potential for further spread
10. Wherever possible, risk estimates should be reassessed over time by comparison with independent human illness data
11. A microbiological risk assessment may need re-evaluation, as new relevant information becomes available

Risk assessment comprises four distinct steps: hazard identification, exposure assessment, hazard characterization (dose-response assessment), and risk characterization which altogether represent a systematic process for identifying adverse consequences and their associated probabilities arising from consumption of foods that may be contaminated with microbial pathogens and/or microbial toxins (Notermans and Jouve 1995; Mayes 1998; Coleman and Marks 1999; Baird-Parker and Tompkin 2000; Jouve 2000; Lammerding and Fazil 2000; Zwietering and van Gerwen 2000; Codex 2009b; Marvin et al. 2009).

3.1.1 Relative Aspects of the Microbial Risk Assessment (MRA) Steps

First Step: Hazard Identification

Hazard identification is mostly a qualitative evaluation of the risk issue and a preliminary examination of information that is analyzed in more detail in the subsequent steps of the risk assessment process. The main focal point of this step in toxicology and environmental health fields is to determine if there is sufficient evidence to consider a substance as the cause of an adverse health effect. On the contrary in MRA, the hazard is already identified as being capable of causing human illness prior to the initiation of the risk assessment (Lammerding and Fazil 2000).

This step identifies the microorganisms or microbial toxins of concern and evaluates whether the microorganism or the toxin is a hazard when present in food. If a pathogen is the focal point in this step, then, available epidemiological and related data need to be used to determine if foodborne transmission is important to the disease and the foods that are implicated. If a food is the focal point in this step, then, the focus will be to use available epidemiological and microbiological data to determine which pathogens could be associated with this food product (Voysey and Brown 2000).

Second Step: Exposure Assessment

Exposure assessment is the estimation of how likely it is that an individual or a population will be exposed to a microbial hazard and what numbers of the microorganism are likely to be ingested (Lammerding and Fazil 2000). An accurate exposure assessment needs the information of the presence of the pathogen in the raw ingredients, food and environment; the effect that food processing, distribution, handling, and preparation steps have on the pathogen; and the consumption patterns (typical serving sizes, weekly or annual consumption rates and circumstances under which the food is prepared and consumed) (Baird-Parker and Tompkin 2000; Voysey and Brown 2000). Thence, to obtain a meaningful microbiological data of numbers of a pathogen present in a food at the point of consumption is seldom possible. Consequently, models [Predictive models such as Food Micromodel, Pathogen Modeling Program, or Ratkowsky-type models were developed to describe the effect of factors in production and distribution on the growth or decline of microorganisms (Zwietering and van Gerwen 2000)] and assumptions are required in order to translate available data into quantitative estimates of the amount of pathogen ingested by an individual at random in the population at risk (Lammerding and Fazil 2000).

Third Step: Hazard Characterization (Dose–response Assessment)

Hazard characterization is the qualitative and/or quantitative evaluation of the nature of the adverse effects associated with biological, chemical, and physical agents that may be present in foods (Lammerding 1997; Voysey and Brown 2000). When considering this step, there is a need to consider the availability of dose–response data, such as the probability of illness and the degree of illness caused by a particular concentration of a microorganism or toxin (Baird-Parker and Tompkin 2000). In other words, a dose–response assessment should be performed if these data are obtainable (Codex 2009b). There are a number of ways in which this information can be obtained:

The effect of an ingested concentration of a microorganism or toxin can be determined by volunteer studies, animal studies, or by outbreak investigations (Zwietering and van Gerwen 2000) but all have limitations (Baird-Parker and Tompkin 2000).

Fourth (Final) Step: Risk Characterization

The integration of hazard identification, exposure assessment, and hazard characterization (dose-response assessment) gives this step which's aim is to give an estimate of the probability of occurrence of illness within a given population, considering the severity of illness possible to be caused, the number of microorganisms or the amount of toxin needed to cause illness, and an assessment of the likelihood of exposure of members of the population (Baird-Parker and Tompkin 2000). In other words, it is for communicating the level of confidence that the risk assessors have in their analysis. Besides, it should summarize the impact that critical assumptions and decisions made in developing the exposure and dose-response assessments have on the interpretation of the overall assessment (Buchanan et al. 2000).

3.2 *Structure of Risk Analysis*

The process of decision-making applied to risk has been termed risk analysis which is used to develop an estimate of risk to human health and safety, to identify and implement appropriate, realistic, and achievable measures to control the risk and to communicate to stakeholders about the risk and measures applied (Schlundt 1999; Toyofuku 2006). The EU food legislation has to be based on risk analysis following

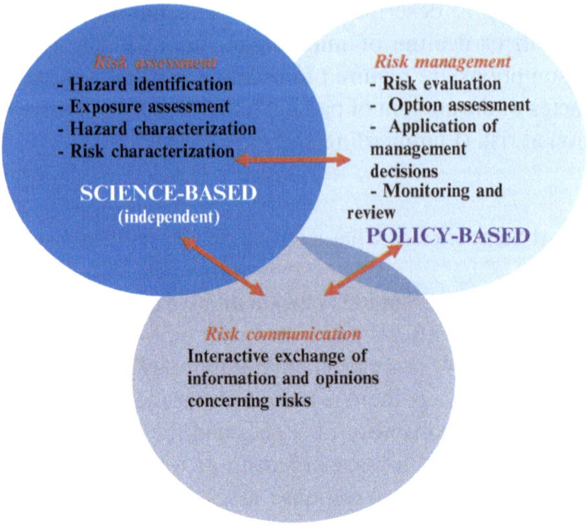

Fig. 4 Risk analysis framework [*Source* Adapted from Baird-Parker and Tompkin (2000)]

Regulation (EC) No 178/2002 (General Food Law), which establishes the general principles governing food and feed safety (Romero-Barrios et al. 2013).

Risk analysis entails three interrelated components, namely risk assessment, risk management, and risk communication (Schlundt 1999; Codex 2003; Marvin et al. 2009; Manning and Soon 2013) Fig. 4. However, the starting point of a risk analysis need not be a risk assessment. Coleman and Marks (1999) stated that risk communication is receiving increasing attention rather than risk assessment as the starting point for risk analyses for transmissible spongiform encephalopathies, *E. coli* O157:H7, and *Salmonella* Enteritidis in the United States of America (the USA) and added that dialogue with industry, academia, government, and the general public is increasingly viewed as essential throughout the entire risk analysis process.

3.2.1 Risk Management

Risk management is a process distinct from risk assessment, of weighing policy alternatives in consultation with interested parties, considering risk assessment and other legitimate factors, and, if need be, selecting appropriate prevention and control options (Marvin et al. 2009). In other words, decisions can be made based on criteria that are not strictly science based and to some extent decouple risk management for the earlier stages of risk assessment which is science based and so independent concern the collection of data that permit the characterization of risk.

According to ISO 17776: 2002, risk management process can be defined as a series of steps (Manning and Soon 2013):

1. Identify the hazards
2. Evaluate the degree of risk and identify risk reduction measures. This implies that attempts to reduce the probability of a risk are carried out in a cost-effective

manner (Notermans and Jouve 1995). As further risk reduction measures are introduced then this will in turn affect the degree of risk with a feedback loop in place. It is at this step that screening criteria are used

3. Set functional requirements

Risk management requires a multidisciplinary approach from the management team and integrated risk management through the supply chain will lead to improved business sustainability (Manning and Soon 2013).

Microbial risk management (MRM) should be governed by the following principles (Schlundt 1999; Codex 2009c):

1. Protection of human health is the primary objective in MRM
2. MRM should take into account the whole food chain
3. MRM should follow a structured approach
4. MRM process should be transparent, consistent, and fully documented
5. Risk managers should ensure effective consultations with relevant interested parties
6. Risk managers should ensure effective interaction with risk assessors
7. Risk managers should take account of risks resulting from regional differences in hazards in the food chain and regional differences in available risk management options
8. MRM decisions should be subject to monitoring and review and, if necessary, revision

3.2.2 Risk Communication

EFSA defines the ultimate goal of risk communication as to assist stakeholders, consumers, and the general public in understanding the rationale behind a risk-based decision, so that they may arrive at a balanced judgment that reflects the factual evidence about the matter at hand in relation to their own interests and values (EFSA 2012). In other words it can be defined as any purposeful exchange of information about risk or perceptions about risk and establishment of an effective dialogue among those responsible for assessing, minimizing, and regulating risks and those who may be affected by the outcomes of those risks. It aims to create communications environment based on trust and credibility and to produce an informed audience that is involved, interested, solution-oriented, and collaborative.

4 ISO 22000:2005 Standard

On September 1, 2005, ISO in cooperation with the European Committee for Standardization (CEN) published a new food safety management system, the ISO 22000:2005 standard—Food safety management systems—Requirements for any organization in the food chain (ISO 2005a, b), providing a framework of internationally harmonized requirements for the global approach to food safety issues. It was developed on the assumption that the most effective food safety systems are designed, operated, and improved within the framework of a structured management system, and incorporated into the overall management activities of the organization (Papademas and Bintsis 2010).

According to the ISO, ISO 22000:2005 standard is intended to confirm that there are no weak links in food supply chains. This standard can be applied to companies ranging from feed producers and suppliers, primary producers through food manufacturers and producers, transport and storage operators and subcontractors to retail and food service outlets—all together with inter-related organizations such as producers of equipments and devices, packaging materials, cleaning and disinfection agents, food ingredients, and additives (ISO 2005a; Cann 2006).

ISO 22000:2005 standard offers the company with competitive efficiencies and lots of benefits (Surak 2003; Arvanitoyannis and Kassaveti 2009; Papademas and Bintsis 2010). However, ISO 22000:2005 standard does not contain the nonexhaustive list of GMP present in the GFSI guidance document (GFSI 2010).

5 Conclusion

Since the results of the end point testing of foodstuffs are obtained by the times that they have been consumed and thence it is hard to trace or recall them, end point testing becomes not enough to ensure food safety in the last two decades (Walker et al. 2003). Advisory Committee on the Microbiological Safety of Food (ACMSF) stated that end product testing is not a suitable instrument for guaranteeing the safety of the food (Manning and Soon 2013).

The design and implementation of FSMS have come a long way since the very early systems of the 1960s, whereupon they are based on generally accepted principles of HACCP and of GMP and then the ISO 22000:2005, the PAS 220 2008, and the FSSC 22000 (Papademas and Bintsis 2010). Instead of tending of previous food safety plans to correct the hazard conditions after they have happened, these systems based on controlling the problems before they happen during manufacturing, processing, and/or serving (McSwane et al. 2003). However, a more dynamic, proactive, science-based, and interactive approach is required, starting with the ability to foresee where food safety and quality-related problems might arise by applying the risk analysis context since food safety legislation requires food manufacturing and processing plants to estimate not only what is an acceptable level of contaminant in a food but also the acceptable level of risk to consumers during distribution, handling and preparing, and consuming that food (Manning and Soon 2013).

The overall objective of safe food is to change the scope of decision making on food safety from single risks to consider foods as the sources of risks, benefits, and costs that are associated with their production and consumption (Marvin et al. 2009). The importance of co-operation between different public health and food safety authorities has already been emphasized in many countries, and the model of a total overview of the problems "from farm to fork" has been obtained.

References

Alemanno A (2007) Trade in food: regulatory and judicial approaches in the EC and the WTO. Cameron May, London, p 540

Anonymous (1997) A simple guide to understanding and applying the hazard analysis critical control point concept, 2nd edn. International Life Sciences Institute (ILSI) Europe, Brussels, p 13

Arvanitoyannis IS, Kassaveti A (2009) HACCP and ISO 22000-a comparison of the two systems. In: Arvanitoyannis IS (ed) HACCP and ISO 22000-application to foods of animal origin. Blackwell, Chichester, pp 3–46

Arvanitoyannis IS, Traikou A (2005) Comprehensive review of the implementation of hazard analysis critical control point (HACCP) to the production of flour and flour-based products. Crit Rev Food Sci Nutr 45:327–370

Baird-Parker TC, Tompkin RB (2000) Risk and microbiological criteria. In: Lund BM, Baird-Parker TC, Gould GW (eds) The microbiological safety and quality of food. Aspen, Maryland, pp 1852–1885

Barendsz AW (1998) Food safety and total quality management. Food Control 9:163–170

Bertolini M, Rizzi A, Bevilacqua M (2007) An alternative approach to HACCP system implementation. J Food Eng 79:1322–1328

Buchanan RL (1990) HACCP: a re-emerging approach to food safety. Trends Food Sci Technol 1:104–106

Buchanan RL, Smith JL, Long W (2000) Microbial risk assessment: dose-response relations and risk characterization. Int J Food Microbiol 58:159–172

Cann O (2006) The food chain. http://www.ifsqn.com/articles_detail.php?newsdesk_id=140&t=the%2bfood%2bchain. Accessed on 1 Aug 2013

Cassin MH, Lammerding AM, Todd ECD, Ross W, McColl RS (1998) Quantitative risk assessment for *Escherichia coli* O157:H7 in ground beef hamburgers. Int J Food Microbiol 41:21–44

Cerf O, Donnat E, The Farm HACCP Working Group 1 (2011) Application of hazard analysis – critical control point (HACCP) principles to primary production: what is feasible and desirable? Food Control 22:1839–1843

Coleman ME, Marks HM (1999) Qualitative and quantitative risk assessment. Food Control 10:289–297

EC (2004a) Regulation (EC) No. 852/2004/EC of the European Parliament and of the Council of 29 April 2004 on the hygiene of foodstuffs (with HACCP principles). Off J Eur Union L 39:1–54

EC (2004b) Regulation (EC) No.853/2004/EC of the European Parliament and of the Council of 29 April 2004 on laying down specific hygiene rules for food of animal origin. Off J Eur Union L 139:55–205

EC (2004c) Regulation (EC) No 854/2004/EC of the European Parliament and of the Council of 29 April 2004 on laying down specific rules for the organisation of official controls on products of animal origin intended for human consumption. Off J Eur Union 139:83–126

Ehiri JE, Morris GP, McEwen J (1995) Implementation of HACCP in food businesses: the way ahead. Food Control 6:341–345

European Food Safety Authority (EFSA) (2012) When food is cooking up a storm: proven recipes for risk communications, Parma. http://www.efsa.europa.eu/en/corporate/pub/riskcommguidelines.htm. Accessed on 11 July 2013

FAO/WHO (1995) Application of risk analysis to food standards issues. Report of the Joint FAO/WHO Expert Consultation. WHO, Geneva, WHO/FNU/FOS/95.3

FAO/WHO (Food and Agricultural Organization/Word Health Organisation) (1997) Risk management and food safety. Report of a joint FAO/WHO expert consultation. FAO, Rome. FAO food and nutrition paper, no.65

Codex (Joint FAO/WHO Food Standards Programme, Codex Alimentarius Commission) (2003) Principles and guidelines for the conduct of microbiological risk assessment. In: Food hygiene basic texts, 3rd edn. Joint FAO/WHO Food Standards Programme, Food and Agriculture Organization of the United Nations, Rome, pp 53–62

Codex (Joint FAO/WHO Food Standards Programme, Codex Alimentarius Commission) (2009a) Hazard analysis and critical control point (HACCP) system and guidelines for its application. In: Food hygiene basic texts, 4th edn. Joint FAO/WHO Food Standards Programme, Food and Agriculture Organization of the United Nations, Rome, pp 23–33, http://www.fao.org/docrep/012/a1552e/a1552e00.htm. Accessed on 15 July 2013

Codex (Joint FAO/WHO Food Standards Programme, Codex Alimentarius Commission) (2009b) Principles and guidelines for the conduct of microbiological risk assessment. In: Food hygiene basic texts, 4th edn. Joint FAO/WHO Food Standards Programme, Food and Agriculture Organization of the United Nations, Rome, pp 44–50, http://www.fao.org/docrep/012/a1552e/a1552e00.htm. Accessed on 15 July 2013

Codex (Joint FAO/WHO Food Standards Programme, Codex Alimentarius Commission) (2009c) Principles and guidelines for the conduct of microbiological risk management (MRM). In: Food hygiene basic texts, 4th edn. Joint FAO/WHO Food Standards Programme, Food and Agriculture Organization of the United Nations, Rome, pp 51–63, http://www.fao.org/docrep/012/a1552e/a1552e00.htm. Accessed on 15 July 2013

Food and Agricultural Organization/Word Health Organisation (FAO/WHO) (1998) The application of risk communication to food standards and safety matters. Report of a joint FAO/WHO expert consultation. FAO, Rome. FAO food and nutrition paper, no.70

Global Food Safety Initiative (GFSI) (2010) Guidance document, 6th edn. http://www.theconsumergoodsforum.com/pdf/workinggroupsaccordion/safety/safety_gfsi/guidance%20documents/gfsi%20guidance%20document%20sixth%20edition%20version%206.2.pdf. Accessed on 22 June 2013

Griffith C (2000) Food safety in catering establishments. In: Farber JM, Todd ECD (eds) Safe handling of foods. Marcel Dekker, New York, pp 235–256

http://www.ispe.org/gmp-resources/what-is-gmp. Accessed on 1 Aug 2013

http://www.sgs.com/en/Agriculture-Food/Food/Retail-and-Hospitality/Other-Food-Industry-Certification/HACCP-Certification.aspx. Accessed on 1 Aug 2013

ILSI (International Life Sciences Institute) (1999) Validation and verification of HACCP. ILSI Europe, Belgium

International Standards Organisation (ISO) (2005a) Food safety management systems – requirements for any organization in the food chain. ISO 22000:2005, Geneva, Switzerland

International Standards Organisation (ISO) (2005b) Food safety management systems – guidance on the application of ISO 22000:2005. Technical specification, ISO/ TS 22004. Geneva, Switzerland

Jouve J-L (2000) Good manufacturing practice, HACCP, and quality systems. In: Lund BM, Baird-Parker TC, Gould GW (eds) The microbiological safety and quality of food. Aspen, Maryland, pp 1852–1885

Lammerding AM (1997) An overview of microbial food safety risk assessment. J Food Protect 60:1420–1425

Lammerding AM, Fazil A (2000) Hazard identification and exposure assessment for microbial food safety risk assessment. Int J Food Microbiol 58:147–157

Manning L, Soon JM (2013) Mechanisms for assessing food safety risk. Br Food J 115:460–484

Marvin HJP, Kleter GA, Frewer LJ, Cope S, Wentholt MTA, Rowe G (2009) A working procedure for identifying emerging food safety issues at an early stage: implications for European and international risk management practices. Food Control 20:345–356

Maunsell B, Bolton DJ (2004) Guidelines for food safety management on farms. The Food Safety Department, Teagasc—The National Food Centre, Dublin

Mayes T (1998) Risk analysis in HACCP: burden bone fit? Food Control 9:171–176

McSwane D, Rue N, Linton R (2003) Essentials of food safety and sanitation, 3rd edn. Pearson Education, New Jersey, pp 169–196

Notermans S, Jouve JL (1995) Quantitative risk analysis and HACCP: some remarks. Food Microbiol 12:425–429

Oscar TP (2004) A quantitative risk assessment model for *Salmonella* and whole chickens. Int J Food Microbiol 93:231–247
Panisello PJ, Quantick PC (2001) Technical barriers to hazard analysis critical control point (HACCP). Food Control 12:165–173
Papademas P, Bintsis T (2010) Food safety management systems (FSMS) in the dairy industry: a review. Int J Dairy Technol 63:489–503
Romero-Barrios P, Hempen M, Messens W, Stella P, Hugas M (2013) Quantitative microbiological risk assessment (QMRA) of food-borne zoonoses at the European level. Food Control 29:343–349
Rooney R, Wall PG (2003) Food safety. In: Caballero B, Trugo L, Fimglas PH (eds) Encyclopedia of food science and nutrition. Academic, Boston, pp 2682–2688
Ropkins K, Beck A (2000) Evaluation of worldwide approaches to the use of HACCP to control food safety. Trends Food Sci Technol 11:10–21
Sansawat S, Muliyil V (2009) Understanding the FSSC 22000 food standard. http://www.sgs.com/foodsafety. Accessed 15 July 2013
Schlundt J (1999) Principles of food safety risk management. Food Control 10:299–302
Sperber WH (2001) Hazard identification: from a quantitative to a qualitative approach. Food Control 12:223–228
Sun Y-M, Ockerman HW (2005) A review of the needs and current applications of hazard analysis and critical control point (HACCP) system in foodservice areas. Food Control 16:325–332
Surak JG (2003) HACCP and ISO development of a food safety management standard. http://www.saferpak.com/iso22000articles/surak_slides.pdf. Accessed on 1 Aug 2013
Surak JG (2007) A recipe for safe food: ISO 22000 and HACCP. Qual Prog 21–27
Toyofuku H (2006) Harmonization of international risk assessment protocol. Mar Pollut Bull 53:579–590
US National Advisory Committee on Microbiological Criteria for Foods (NACMCF) (1997) Hazard analysis and critical control point principles and application guidelines. http://www.fda.gov/Food/FoodSafety/HazardAnalysisCriticalControlPointsHACCP/ucm114868.htm. Accessed on 10 July 2013
van der Spiegel M, Luning PA, Ziggers GW, Jongen WM (2004) Evaluation of performance measurement instruments on their use for food quality systems. Crit Rev Food Sci 44:501–512
Voysey PA, Brown M (2000) Microbiological risk assessment: a new approach to food safety control. Int J Food Microbiol 58:173–179
Walker E, Jones N (2002) An assessment of the value of documenting food safety in small and less developed catering businesses. Food Control 13:307–314
Walker E, Pritchard C, Forsythe S (2003) Hazard analysis critical control point and prerequisite programme implementation in small and medium size food business. Food Control 14:169–174
Wallace C, Powell SC, Holyoak L (2005) Development of methods for standardised HACCP assessment. Br Food J 107:723–742
Wallace CA, Holyoak L, Powell SC, Dykes FC (2012) Re-thinking the HACCP team: an investigation into HACCP team knowledge and decision-making for successful HACCP development. Food Res Int 47:236–245
Wallace CA, Holyoak L, Powell SC, Dykes FC (2014) HACCP - the difficulty with hazard analysis. Food Control 35:233–240
Zwietering MH, van Gerwen SJC (2000) Sensitivity analysis in quantitative microbial risk assessment. Int J Food Microbiol 58:213–221

Value Addition and Preservation by Fermentation Technology

Saghir Ahmad

1 Introduction

History: As the enzymes are the soul of fermentation, it is necessary to give history of enzymes relevant to fermentation. It has been explained that "enzyme" was derived from Greek word (En = in zyme = yeast) (Purohit et al. 1993). Earlier enzymes were supposed to be the source of fermentation because their activities were similar to yeast fermentation. Any substance that is protein wholly or partly regulates the rate of specific biochemical reaction in living organism. It is capable of catalyzing reaction in which substrate is converted to product through process of fermentation of an intermediate enzyme substrate complex. Fermentation in the past was explained as "La Vie sans air" means life with only air. Presently, definition of fermentation has entirely changed with the knowledge of large number of enzymes, discoveries, and numerous new enzymes has made possible to exploit the activity of microorganisms and enzymes to produce the product of commercial value. The research work helped to provide sufficient knowledge of enzyme properties and detection of the potentiality of utilizing enzymes as industrial catalyst. Further development in fermentation was recognized using pure enzyme instead of using enzyme as an integral part of microbial cell. All these factors lead to progress in fermentation. Bacteria, fungi, protozoa are biological sources of many industrially important enzymes, for example, *Bacillus macerans* and *Bacillus polymyxa* are used for production of amylase.

S. Ahmad (✉)
Faculty of Agricultural Sciences, Department of Post-Harvest Engineering and Technology, Aligarh Muslim University, Aligarh, UP 202002, India
e-mail: alsaghir@rediffmail.com

A. Malik et al. (eds.), *Food Processing: Strategies for Quality Assessment*, Food Engineering Series, DOI 10.1007/978-1-4939-1378-7_13

1.1 Definitions

The microbial cells have two types of enzyme, namely, Endozyme and Exozyme.

(a) *Endozyme*: Enzyme which acts within the cell in which they are synthesized.
(b) *Exozyme*: Enzyme which acts outside the cell in which they are synthesized.

Chemical composition and classification of enzymes:

Based on chemical composition, enzymes are classified into two groups

1. *Purely proteinacious enzymes:* They are made up of only proteins, e.g., Protease that splits proteins, amylase that split starch.
2. *Conjugated enzymes:* These are made up of protein molecules to which a non-protein group (prosthetic group) is also attached, e.g., oxidizing enzymes.

Definitions

Substrate: The substance upon which enzyme acts is known as substrate. The enzyme acts on substrate and makes the product. Further enzyme gets released free after product is formed.

Prosthetic group: It indicates the non-protein group of enzyme. The prosthetic group is firmly bound to protein component of the enzyme by chemical bond and therefore not removed by dialysis. Heme biotin and pyridoxine phosphate like flavin, usually function as prosthetic group.

Apoenzyme: If prosthetic group is removed, the remaining protein part of the enzyme is called as apoenzyme and it remains inert.

Coenzyme: It is coworker of prosthetic group and apoenzymes because neither the apoenzyme nor the prosthetic group alone is enzymatically active. The prosthetic group, the coenzyme, as the name indicates, acts as Coworker of apoenzyme. They combine with the enzyme and leave it again in the course of a single catalytic cycle.

Cofactor: When a prosthetic group made up of single atom of some metal such as Mg^{++}, Fe^{++}, Cu^{++}, Mn^{++}, and Zn^{++} then it is known as "cofactor." It can be easily separated from the rest of protein part (Purohit et al. 1993).

1.2 Requirements of Fermentation

Invertase converts sugar into glucose and fructose, zymase converts glucose into alcohol.

$$C_6H_{12}O_6 \xrightarrow{\text{Zymase}} 2C_2H_5OH + 2CO_2\uparrow$$

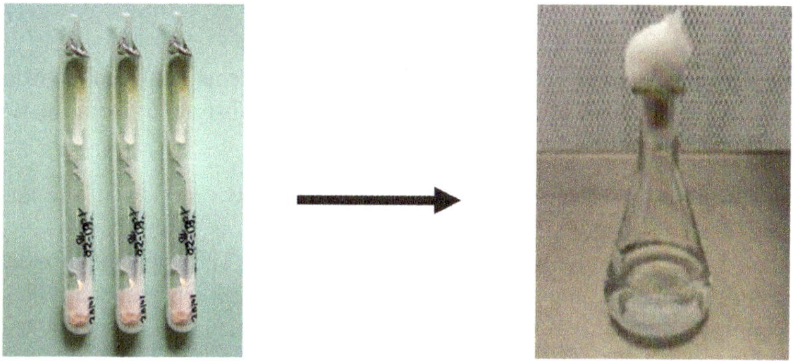

Fig. 1 Preparation of inoculum from the "mother culture" for fermentation process

1.2.1 Inoculums and Culture Propagation

In typical enzyme fermentation, the organism is propagated through several stages of batch culture. Stock cultures from the research laboratory are generally procured in freeze-dried form. And inoculums are usually transferred from stage of ampoule to conical flask generally containing 1 L of media (Fig. 1). Further this media after a growth of 2–3 days are transferred to 50 L tank and propagated till the capacity reaches to commercial level inoculums. The last tank containing inoculums is called seed tank.

1.2.2 Fermentation Methodologies

(a) Surface culture cultivation
(b) Submerged cultivation
(c) Solid-state cultivation

Substrate culture: The microorganisms are cultured on a basic substrate which contains desired nutrients and bulk surface, for example, what bran/cereal meal or rice bran or barley grits. It also contains salts with low water content.

Submerged culture: This type of cultivation is very common. In principle the similar and general methods are used ferments for carrying out fermentation. The substrate is loaded in liquid medium in the fermentors with suitable concentrations of nutrients and culture in inoculated so start fermentation several examples of fermented products like antibiotics, single-cell protein, vinegar fermentation, enzyme production, alcoholic fermentation are carried out by submerged culture fermentation. The capacity of fermenter ranges 10,000 to 100,000 L in batch operation. The time required to complete fermentation varies from 50 to 150 h.

1.2.3 Fermenters

A fermenter/bioreactor is a container designed to provide as optimum environment in which microorganism or enzyme can interact with substrate to form the desired product. The fermenters are of two types.

1. *Open:* It allows continuous processing with substrate entering at one and end product leaving at another.
2. *Closed:* In the types the processing in done in batch. This types of fermenter in used for products of an antibiotic, vinegar, etc. Figure 2 represents a typical fermenter vessel. The microorganisms are grown on nutrient, placed in the vessel at the stage of the fermentation. The vessel is cooled by a water jacket. Air is pumped in to the bottom of the liquid and acid or alkali added as necessary. Stirrer keeps in contents well mixed. Steam lines are provided so that the vessel is sterilized after each fermentation batch.

During fermentation it is as necessary to regulate many factors with predetermined values viz oxygen and carbon dioxide, temperature and media concentration, etc. It's also necessary to mention high degree of sterility within the fermentor. The fermenter should be made of stainless steel or copper so that it can be easily sterilized with steam after each batch.

1.2.4 Fermentation Medium

Fermentation medium includes nutrient essentials so support the growth of microorganism and fermentation of desired end products. The choice of the nutrient should be based on nutritive source and economic aspect. The important sources are

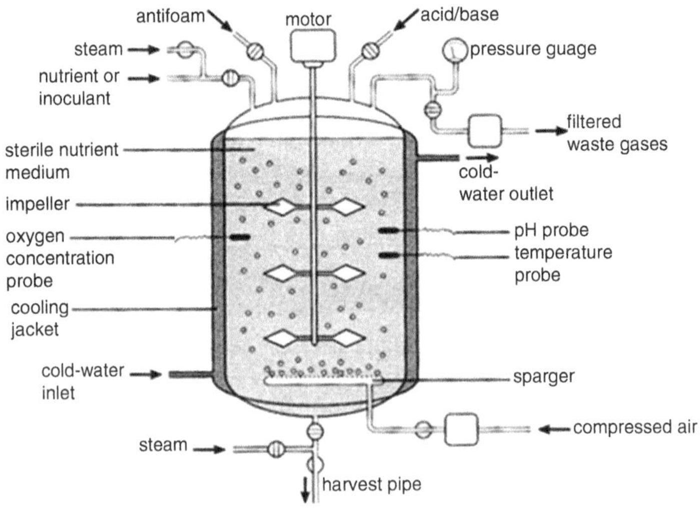

Fig. 2 Schematic and terminological representation of a fermenter vessel

Table 1 Spectrum of substrates used as sources of nutrients for the production of various fermented products

Nutrient	Raw material
Carbon source, glucose maltose, sucrose starch fats and hydrocarbon	Corn, sugar, starch, molasses, vegetable oil, petroleum fractions
Nitrogen source, (protein fraction)	Soy bean meal, corn step, (from corm melting), distiller soluble (alcohol, beverage manufacture)
Ammonia	Ammonia or ammonium salt, nitrate salt
Nitrogen	Air (from nitrogen fixing organisms)
Phosphorus sources	Phosphate salt

Adapted from Purohit et al. (1993)

carbon, nitrogen, and phosphorus. Table 1 presents various nutrient sources used in the industrial fermentation.

There are several factors which should be considered for better fermentation results, like trace element in the medium, quality of water, and nature of pipes used for supply of solution.

2 Value Addition in Fermentation Technology

The fermented food products are not only value-added food but also provide better health benefits. Fermented foods have now become well-known all around the world including developed, developing, and industrialized countries like China, Japan, Thailand, Mexico, and India. Diabetes has wide spread in the civilized world. This disease is tied to other oxidation-linked disease. However, some well-discussed important fermented foods with good taste which control these diseases and also provide health benefits like Yoghurt, Cheese, Sauerkraut, Pickles, Vinegar, etc.

Other fermented products produced commercially in the industries are citric acid, antibiotic, amino acid, enzyme, single-cell protein, baker's yeast, etc.

2.1 Fermentation in Commercial Application

Fermentation as a commercial process can be classified in three groups:

(a) Those commercial application in which enzymatic fermentation remains as essential part of the process such as production of yoghurt, cheese, alcohol, vinegar, beer, etc.
(b) Those commercial application in which enzymatic fermentation remains a part of the process and application of the enzyme only improve the economy of the process such as extraction of fruit juices, after enzymatic treatment conversion of glucose to fructose, production of essential oils, etc.

(c) Those processes in which enzymatic fermentation is used to improve the quality of products such as bread fermentation to improve loaf volume, meat tenderization by papain enzyme, etc.

The fermentation processes are extremely economical and facilitate to produce valuable products from very cheap source of raw material. Food industries uses lot of citric acid (70 % of the total production), although citric acid is constituents of many citrus fruits and can be extracted from these natural sources. If one starts to extract citric acid from fruits, vast quantities (several tonnes) of fruits will be required to produce few kilograms of citric acid. Citric acid has important application in many industries like detergent and pharmaceutical industries (20 % of total production) and 10 % is being used by other industries. The economical production of citric acid is carried out using molasses (by product of sugar industries) both cane and beet molasses are being used in different countries for production of citric acid by fermentation using culture of mold. Further there are three different types of molasses like black strap, refinery, and high test molasses (Prescott and Dunn 1987). These different types of molasses are obtained by different stages of crystallization. Different cultures of molds have been found to accumulate citric acid efficiently. Few of the mold species are *Aspergillus niger*, *A.usami*, *A.awamori*, etc. An example of fermentation process involving mold species is *Aspergillus niger* which is used in the production of citric acid.

Citric acid has varied functions in food products. It acts as antioxidant, improve flavor, eliminate haze, abolish turbidity, and gives a better taste in beverages. In others food products like confectionery, it improves flavors and inverts sugar, stops oxidation, and adjusts pH. The schematic representation of the process of citric acid fermentation is depicted in Fig. 3.

2.2 Fermented Dairy Products

There are several fermented dairy products which have therapeutic value and also good in nutrition. Many people are not able to digest milk, however the milk fermented by using culture of lactic acid bacteria is modified in nature and nutrition value and it can be easily digested. It was reported by Kapoor et al. (1987) that *Lactobacillus bulgaricus*-fermented milk has therapeutic values and suppresses toxins. A good number of fermented dairy products are available in the market, e.g., acidophilus milk, yoghurt (plain, stirred yoghurt, cultured set yoghurt, long life yoghurt), and cultured butter milk. Hence cheese, a fermented milk product is having more than 300 varieties all around the world. The varieties are based on texture, nutrient, flavor, color, and some varieties are location specific like gouda, camembert, and dry cheddar.

Health benefits of fermented products had been verified by several researches, (Speck 1977; Gilliland and Speck 1977). They studied the role of lactic acid bacteria for better health of human being. The ingestion of viable cells of lactic acid bacteria in good number (10^9 per day) helps to make the intestinal microbial balance.

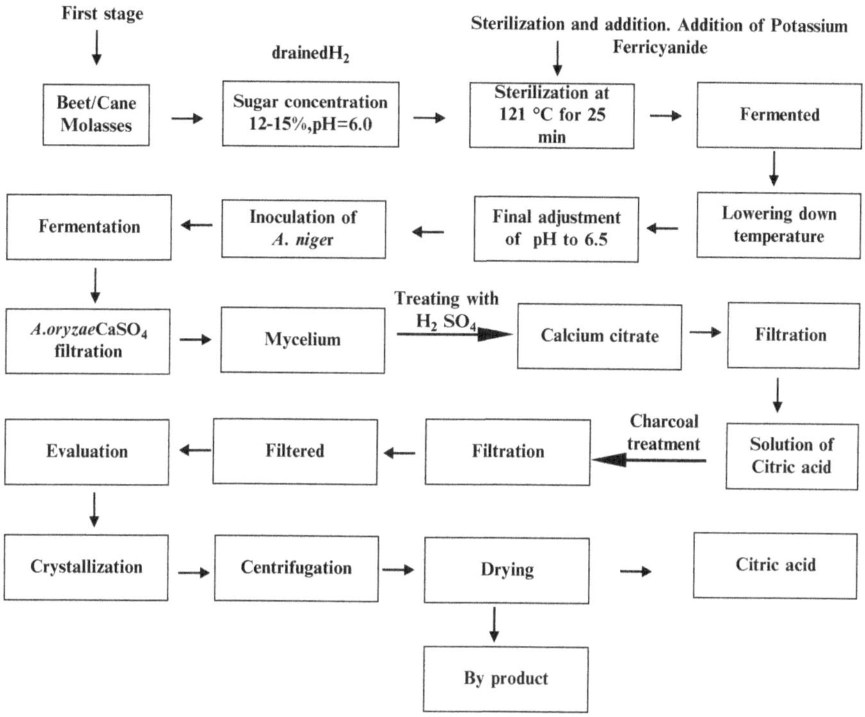

Fig. 3 Process flow diagram of citric acid production using mold *Aspergillus niger*

Qualities of dairy products depend upon the quality of the milk. The milk is subjected to standardization to bring it to a desired level fat and protein contents (assuming the most important constituents to provide solidity). The protein and fat come in the same matrix, when the milk is coagulated by fermentation using rennet enzyme or lactic acid bacteria or both, which help to reduce the pH of the milk. The coagulation of the milk establishes at isoelectric pH, the coagulum contains the major and minor nutrient constituents like protein, fat, vitamins, and minerals. The milk-based fermented products are cheese, yoghurt, and acidophilus milk. Further the quality of product depends upon the process of production. The success of the fermentation is indicated by the rate of change of pH. Several scientists reported about the inhibitory activity of yoghurt particularly on asceles tumor cell proliferation (Reddy et al. 1973). The experiment was carried out on mice and it was found that 28 % inhibition of tumor cell proliferated by ingesting yoghurt to mice (Mitchell and Kenworthy 1976) reported about the antimicrobial activity against the enteric toxicity of yoghurt.

In India curd (dahee) has been recommended for curing loose motion, intestinal disorder, and dysentery. It is also believed that curd improves the appetite and creates antibiotic compounds which stop the growth of pathogenic bacteria and other undesirable microorganism. Dahee produces beneficial effects and helps to enhance the retention of nutrients and improve digestibility.

2.3 *Sauerkraut Fermentation*

Sauerkraut was derived from the Germen word "Sauerkoht" means sour cabbage. Sauerkraut fermentation is well known and the product is produced from surplus cabbage remained after fresh consumption. The cabbage is delicate crop and it will deteriorate within 1–2 days if not properly preserved. It is prepared from the shredded mature cabbage. The white and yellow both varieties are used for making Sauerkraut. The cabbage is shredded in thickness 2–5 mm and 20 cm long. The well-known countries where Sauerkraut is produced and consumed are Germany, United States, Canada, France, Holland, etc. The process flow diagram is given below (Fig. 4).

Out of total production, 90 % of sauerkraut is produced and packed commercially (Steinkraus 1995). It is a very popular food product, cooled and consumed with other food like sausage. The juice of the kraut is also consumed.

However total production of canned sauerkraut in the USA was 133,000 tons in 1974, more than double the production in Germany (61,000 tons) (Almanaz 1978). Temperature at which fermentation is carried out is important for sauerkraut fermentation. Various temperatures have been attempted for fermentation to get the optimum condition for development of acidity. At 7.5 °C fermentation rates is very slow.

Fresh sauerkraut cabbage

↓

rice Trim and clean, remove outer

↓

Core

↓

Shredding (2-5 cm thickness and 20 cm long)

↓

Salt addition (2.25%)

↓

Filing in to vats

↓

Cover and seal with proper weight

↓

Fermented

↓

Pasteurized

↓

in Can or packaged in plastic bags

Fig. 4 Flow diagram of the industrial production of Sauerkraut

L. mesenteroides grow slowly attaining acidity of about 0.4 % in four days and an acidity of 0.8–0.9 in a month. The acidity is important for making preservative effect on kraut. Some lactic acid bacteria do not actively participate at this temperature and kraut may not develop desired acidity in several months. Under optimum concentration of salt 2.25, all lactic acid bacteria grow well in the medium and an acidity of 1.7–2.3 % is obtained within 22 days. The acidity in contributed by lactic acid and acetic acid in the ratio (4:1) on further increasing temperature to 23 °C; the acidity reaches to 1.5 within 10 days. Active growth of *L. brevis* and *L. plantarum* is initiated 3–5 days and kraut is considered to be completely fermented in approximately 30 days. At still high temperature of 30 °C the kraut is rapidly fermented and an acidity of 1.8–2.0% is developed within 8 days. This shows development of acidity will produce better results in respect of quality of product.

2.4 Soy-Fermented Products

Soybean has the highest protein content among the legumes. It is also a leading source of oil. Both constituents are important for food nutrition because of low protein contents of cereal and imbalance of the essential amino acid. They do not supply enough protein for satisfactory growth of babies and children and body maintenance of laborious people. Thus, protein supplementation of cereal is required and soybean has become heavily involved in the part of world food preferences. However soybean has certain antinutritional facts, like more protein components especially phosphorus compounds, the phosphorus in the isolated protein ranges from 0.8 to 1.0 % indicating a reaction of phytin, the principal phosphorus component of the soybean with protein when precipitated at pH 4.5. Another antinutritional factor is trypsin inhibitor activity; trypsin inhibitor activity of soybean can be abolished by autoclaving at desired temperature/pressure of sterilization. As far as sensory characteristics of soybean and soybean products is concerned, it has strong beany flavor, mostly disliked by the people. The fermentation of soybean along with some cereals like wheat and rice has been subjected to making products more nutritional and with better taste and aroma.

2.4.1 Soy Sauce

Soy sauce is a long-term fermented product. In the first step the mold fermentation helps to analyze the protein which is present in complex form in soya bean. It provides good balance of amino acids. Further in a next step, the lactic acid bacteria is inoculated to provide a little acidity, while yeast fermentation provides development of aroma in the soy sauce (Fig. 5). The soy sauce is very popular in Japan and other oriental countries, people sometimes add this product in almost all the dishes made at home.

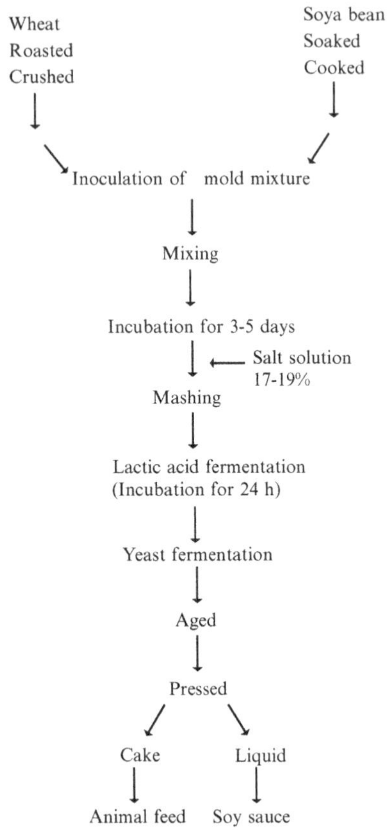

Fig. 5 Process flow chart diagram for soya sauces production

2.4.2 Miso

Miso is a paste-like product popular in Japan and other part of orient. It is prepared by fermenting crushed soya bean and salt. The variety of miso are available in Japan like variation in cheese in the USA, the variation is made by taking rice, soya bean, and salt in different proportion. The cereals used for miso fermentation are rice and barley. The colors of products are white, yellow, and red miso changed by decreasing the proportion of cereals with respect to soya bean. The cereal dilutes the protein content of soya bean and brings light color in the products. Protein is less digestible as compared to carbohydrate and fat. Soya bean contains lot of protein and it is very difficult to digest such high protein food. It is better to combine with cereal in product development. It provides a balance of amino acids and also desired quantity of protein and source of carbohydrate. The process flow diagram for miso production is given below (Fig. 6).

Fig. 6 Process flow chart diagram for Miso production

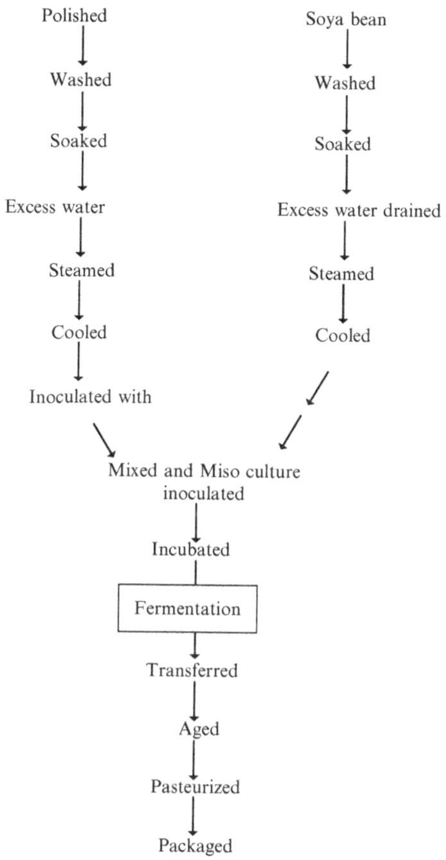

2.4.3 Tempeh

Tempeh is a cake-like product produced from full fat soya bean or grits. It is fermented with culture of mold *Rhizopus oligosporus satio* (NRRL 2710). Fermentation processes completely modify the product and the antinutritional beany flavored characteristic of soya bean is completely changed to very delicious product. Ontzom is a fermented product made from peanut press cake, a by product after extraction of peanut oil. Cassava and residue are produced from soya bean milk may also be used as substrate for tempeh fermentation. Sometimes coconut press cake and residue from soy milk or Tofu is also used a substrate. Materials after soaking are washed properly and steamed for 1 to 1.5 h. Further it is moulded to flat cake and inoculated with *Neurospora* or *Rhizopus* for fermentation. Incubation takes 2 days to bring the product to delicious form and finally packed.

2.4.4 Other Fermented Soya Products

Many other soy-fermented products like Sofu; are highly nutritious and provide health benefits. Sofu is soya milk-fermented product produced from tofu. Tofu is a product made from soy milk as paneer is made from animal milk. There are several health benefits of fermented products. They produce antimicrobial agents from *Lactobacillus bulgaricus* which are antagonistic to the enterotoxic activity of *E. coli*. It has been observed that certain bacteria prevents or ameliorate coliform-associated diarrhea. Fermented dairy products have bacteriostatic and bactericidal effect on *E. coli*, *Streptococcus* sp., *Bacillus sp.*, and other microorganism.

2.5 Meat Fermentation

Fermentation traditionally offers an easy and low-energy preservation method for meats that result in distinctive products that have an important part in the diet of people making them. Such fermented meats contribute both nutritional value and pleasure to meals. However, products are not the same from time to time. Indeed, the product may spoil, cause illness due to pathogenic microorganisms or their toxins, and even become lethal due to botulinum toxin production if the normal beneficial microbial flora does not multiply as usual. To prevent these problems, the use of starter cultures has become common practice in many countries, including developing countries.

The bacteria which play a significant role and commonly found in fermented sausages are lactic acid bacteria (Coppola et al. 1998). These microorganisms are used as starter cultures, promoting meat fermentation (Papamanoli et al. 2003). Lactic acid bacteria improve safety and stability of the product, enhance color stability, prevent rancidity, and release various aromatic substances (Nychas and Arkoudelos 1990; Hammes et al. 1995; Coppola et al. 1998; Papamanoli et al. 2003). Lactobacilli are the predominant lactic acid bacteria and among them the most frequently isolated strains are *Lactobacillus curvatus*, *Lactobacillus sakei*, and *Lactobacillus plantarum* (Hammes 1990). The most promising bacteria for starter cultures are those which are isolated from the indigenous microflora of traditional products. These microorganisms are well adapted in the meat environment and are capable of dominating the microflora of products. The strains selected as starter or protective cultures must have the most important technological properties and/or bacteriocin production capabilities (Hammes 1990).

2.5.1 Fermented Sausages

Sausages consisting of a mixture of small-sized or minced meat, salt, spices, and other ingredients, which was put in a casing, must have been a rather sophisticated invention. The earliest proof of the oldest sausages dates back for more than 5,000 years and was found in the Sumerian culture—the former Babylonia. The first written references about sausages date from approximately 600 BC, both in ancient

China and in Greece. In China a sausage known as "Lachang"—sweetened, seasoned, and smoked—was first mentioned, and in Greece sausages were described in the works of Homer and Epicharmus, the latter having written a comedy entitled "The sausage" (Anon 2006).

Fermented sausages can be either dry or semidry. The most well-known dry sausages, such as Genoa salami, dry salami, and pepperoni, originated from Italy (Ricke and Keeton 1997). In general, days the origin of fermented sausages can be traced back with accuracy to 1730, when salami was first mentioned in Italy (Leistner 1986). The art to produce fermented sausages spread from Italy to other European countries and was established in Germany in 1735 and Hungary in 1835. Today in various parts of the world, a large number of different types of fermented sausage exist. For example, 330 different types are produced in Germany (1975). This very high consumption of fermented meats is an indication that such products have a long tradition of being safe. However, some specific safety aspects deserve consideration. Fermented sausages are stable meat products and normally prepared from a comminuted mixture of meat, fat, salt, and spices which allowed to ferment under controlled temperature and humidity conditions (Ahmad and Srivastava 2000). The typical flavor, texture, and shape of many sausages known today such as frankfurters, braunschweiger, pork sausage, and salami were named due to geographical location of their origin. Dry and semidry sausages represent the largest category of fermented meat products with many of the present day processing practices having their origin in the Mediterranean region.

Sausages have a final pH of 5.0–5.3, lactic acid percent of 0.5–1.0 %, and an moisture protein ratio of <2.3:1. The moisture loss is between 25 and 50 % and the final moisture percent on average is <35 % with water activity (a_w) ranging between <0.85 and 0.91. However, these values may be different due to government and company specifications. German and hard salami are required by FSIS to have an MPR of 1.9:1, with the exception of Genoa salami, which is required to have an MPR of 2.3:1 (FSIS 1986). The pH range for these products is 4.7–4.9, whereas pepperoni has a slightly lower pH range of 4.5–4.8 and a lower MPR of 1.6:1 as required by FSIS (FSIS 1986). The moisture content for these products ranges from 25 to 39 %. All of these products, because of their moisture-to-protein ratios, are considered shelf-stable (FSIS 1986; Ricke and Keeton 1997). Semidry sausages such as summer sausage, cervelat, and Mettwurst typically have a final pH between 4.7 and 5.1, a lactic acid percent of 0.5–1.3 %, and an MPR of >2.3:1 but <3.7:1. The moisture loss ranges from 8 to 15 % and the moisture percent ranges from 45 to 50 %. The water activity range is 0.90–0.94. Again, values will vary depending upon the government and company specifications. For instance, a summer sausage will have a final pH <5.0, a lactic acid percent of 1 %, and an MPR of 3.1:1 with a moisture percent of 41–51 %. Lebanon bologna is unique in that it contains a higher moisture content of 56–62 % (Ricke and Keeton 1997). Due to the higher moisture-to-protein ratios, semidry sausages are required to be refrigerated.

The distinct sensory qualities and remarkable shelf-life characteristics of fermented sausages, as compared to cooked sausages, are largely due to acidification of the meat batter. Traditionally, acidification of the raw meat is the result of a microbial fermentation process. Lactic acid bacteria play a major role in the microbial consortium of fermented and cured meat: they affected both the technological properties and the microbial stability of the final product through the production of lactic and acetic acids (Toldra 2010). Acidification is generally combined with protection from oxygen (stuffing into casings), extensive salting, and curing, and with an ageing stage for product maturation. The latter stage can be absent, short, or long, depending on the type of product, and leads to drying, resulting in a lower water activity, as well as to a complex and desired flavor formation (Campbell Platt and Cook 1995; Lucke 1998). Sometimes, smoking or heating is applied as a last step in the manufacturing process. Heating is common in the USA, where regulations require a core temperature of 58.3 °C before selling the end-product (Lucke 1998). The almost anaerobic environment and the low pH and water activity values that prevail in the sausage are to be considered as the main hurdles that inhibit undesirable microbial growth and lead to a relatively stable end-product.

In the case of spontaneously fermented sausage or sausage prepared through back-slopping, LAB that cause acidification of the meat and, hence, start the fermentation process, originate from the raw material or production environment. However, LAB can also be added deliberately by the sausage manufacturer as a starter culture to the meat batter (Campbell Platt and Cook 1995; Hugas and Monfort 1997). In contrast to spontaneous fermentation, where the manufacturer relies on the presence of a "house microbiota" (counts of 102–103 LAB per gram of fresh batter), the addition of a starter culture leads to high initial LAB counts (106–107 per gram of fresh batter). This enhances acidification, leads to a more standardized and predictable production process, shortens the development of firmness and the overall ripening time, and improves food safety (Lucke 1998).

As an alternative to the use of LAB starter cultures, some manufacturers prefer to apply chemical acidulants, mainly to shorten the production process. Best results have been obtained with glucono-delta-lactone (GdL) (Barbut 2006). However, a disadvantage of chemical acidulates is that they generally induce rapid and poorly controlled acidification, leading to inhibition of flavor development.

2.6 Conclusion

Fermented products have become popular for their nutrition and therapeutic values. They are available almost in every part of the world. All the consumer groups are becoming aware of the importance of the fermented products. Different varieties of dairy-fermented products fermented fruits and vegetable cereal and meat products are consumed all around the world. A large number of fermented foods including cheese, yogurt, acidophilus milk, pickles, sour curd, fermented sausages, etc., offers choice to the consumer.

References

Ahmad S, Srivastava PK (2000) Quality and shelf life evaluation of fermented sausage of buffalo meat with different levels of heart and fat. Meat Sci 75:603–609

Almanaz of the canning, freezing, preserving industries (1978) Edward E. Judge and sons, Inc Westminster

Anon (2006) In Wikipedia the free encyclopedia. Retrieved Mar 2006, at http://cn.wikipedia.org/Wiki/sausage#History

Barbut S (2006) Fermentation and chemical acidification of salami-type products: effect on yield, texture and microstructure. J Muscle Foods 17:34–42

Campbell PG, Cook PE (1995) Fermented meats. Blackie academic and professional, London

Coppola R, Giagnavova B, Iorizzo M, Grazia L (1998) Characterization of lactobacilli involved in the ripening of soppressate molisana a typical southern Italy fermented sausage. Food Microbiol 15:347–353

FSIS (1986) Standards and labeling policy book. Food Safety and Inspection Service, Standard and Labeling Division, Washington, DC

Gilliland SE, Speck ML (1977) Instability of *Lactobacillus acidophilus* in yoghurt. J Dairy Sci 60:1394–1397

Hammes WP (1990) Bacterial starter cultures in food production. Food Biotechnol 4:383–397

Hammes WP, Bosch I, Wolf G (1995) Contribution of *Staphylococcus carnosus* and *Staphloycoccus piscifermentans* to the fermentation of protein foods. J Appl Biol Suppl 79:76–83

Hugas M, Monfort JM (1997) Bacterial starter cultures for meat fermentation. Food Chem 59:547–554

Prescott SC, Dunn CG (1987) Citric acid. In: Prescott SC, Dunn CG (eds) Industrial microbiology. CBS Publisher, New Delhi, p 712d

Leistner L (1986) Allgemeines uber Rohwurst. Fleischwirtsch 66:290–300

Lucke FK (1998) Fermented sausages. In: Wood BJB (ed) Microbiology of fermented foods. Blackie Academic and Professional, London, pp 441–483

Mitchell LDKG, Kenworthy (1976) Investigation on metabolic from lacto *Bacillus bulgaricus* which neutralizes the effect of entero-toxin from *Escherichia coli* pathogenic for pigs. J Bacteriol 41:163–173

Nychas GJE, Arkoudelos JS (1990) *Staphylococci:* their role in fermented sausage. J Appl Bact Suppl 69:167–188

Papamanoli E, Tzanetakis N, Litopoulou- Tzanetaki E, Kotzekidou P (2003) Characterization of lactic acid bacteria isolated from a Greek dry- fermented sausage in respect of their technology and probiotic properties. Meat Sci 65:859–867

Purohit SS, Kothari PR, Mathur SK (1993) Definition for enzyme. In: Basics and agricultural biotechnology. Agro Botanical, Bikaner pp 232–328

Reddy GV, Shahani RM, Banergy MR (1973) Inhibitory effect of yoghurt on enriched ascites tumor-cell proliferation. J Nat Cancer Inst 50:815–817

Ricke SC, Keeton JT (1997) Fermented meat poultry and fish products. In: Doyle MP, Beuchat LR, Montville TJ (eds) Food microbiology: fundamentals and frontiers, 1st edn. ASM Press, Washington, DC, pp 610–628

Speck ML (1977) Acidophilus food products. Soc Ind Microbiol 27:44, abstract

Steinkraus KH (1995) Acid fermented vegetable. In: Handbook of indigenous fermented foods, 2nd edn. Marcel Dekker, New York, p 114

Toldra F (2010) Handbook of meat processing. Blackwell, Ames, pp 1–566

Importance of Yeasts and Lactic Acid Bacteria in Food Processing

Hüseyin Erten, Bilal Ağirman, Cennet Pelin Boyaci Gündüz, Erdem Çarşanba, Selvihan Sert, Sezgi Bircan, and Hasan Tangüler

1 Introduction

Yeasts are non-photosynthetic, relatively sophisticated, living, unicellular fungi. They are substantially beneficial to human culture, in particular for the production of alcoholic beverages and foods. Yeasts also play detrimental role in the spoilage of foods and beverages and some can be pathogenic. Of the yeasts, *Saccharomyces cerevisiae* and related species are widely used in the food and beverage industries. Many species of *Saccharomyces* are safe (GRAS) and the term "yeast" is generally employed as synonymous with *Saccharomyces cerevisiae* (Stewart and Russell 1998). In industry, yeasts are commercially used in the production of alcoholic beverages, industrial alcohols, baker's yeast, enzymes and yeast-derived flavour products (Walker 1999).

Lactic acid bacteria are unicellular prokaryotes of Gram-positive, non-sporing, non-respiring cocci or rods which form lactic acid as the major end-product during the fermentation of sugars. Lactic acid bacteria include the species of *Lactobacillus*, *Carnobacterium*, *Leuconostoc*, *Oenococcus*, *Streptococcus*, *Lactococcus*, *Enterococcus*, *Vagococcus*, *Pediococcus*, *Aerococcus* and *Tetragenococcus*

H. Erten (✉) • B. Ağirman • C.P.B. Gündüz • S. Sert • S. Bircan
Faculty of Agriculture, Department of Food Engineering, Cukurova University, Adana 01330, Turkey
e-mail: herten@cu.edu.tr; bagirman@cu.edu.tr; pgunduz@cu.edu.tr; selvihan_sert@hotmail.com; sezgibircan86@yahoo.com.tr

E. Çarşanba
Altinozu Vocational School of Higher Education, Department of Food Technology, Mustafa Kemal University, Hatay 31750, Turkey
e-mail: carsanba.erdem@gmail.com

H. Tangüler
Faculty of Engineering, Department of Food Engineering, Nigde University, Nigde 51240, Turkey
e-mail: htanguler@nigde.edu.tr

A. Malik et al. (eds.), *Food Processing: Strategies for Quality Assessment*, Food Engineering Series, DOI 10.1007/978-1-4939-1378-7_14

(Bourdichon et al. 2012). Many lactic acid bacteria are safe and they are used extensively in the production and maturation of fermented foods and beverages such as yoghurt, pickles, table olives, sour dough bread (Axelsson 1998; Caplice and Fitzgerald 1999).

This chapter will briefly discuss the beneficial aspects of yeasts and lactic acid bacteria in food processing.

2 Yeasts in Food Processing

2.1 Baker's Yeast and Single-Cell Production

Baker's yeast, *Saccharomyces cerevisiae*, is used for bakery and confectionary processes throughout the world (Akbaria et al. 2012; Romano and Capece 2013).

Baker's yeast can be found in different forms like compressed, granular, cream, dried pellet, instant, encapsulated or frozen (Young and Cauvain 2007).

Some of the basic desired properties of baker's yeast are rapidly utilization of maltose, tolerance to high levels of sucrose, enduring freeze-thawing stress and production of high levels of CO_2. Another desired feature of baker's yeast is using disaccharide melibiose. One of the most desirable characteristic of baker's yeast strains is high fermentation rate. Development of the freeze-tolerance and freeze-thawing survival of yeasts is a property that could be useful for the quality and generation of bakery products. Baker's yeast should be osmotolerant, able to tolerate chemicals (salt, propionates), maintain a high growth capacity, should not aggregate and must have a good storage capability. In addition, during the drying process and after the addition of dry yeast to flour for dough making, the yeast must have a high rate of vitality (Randez-Gil et al. 1999).

High fermentation rate is priority desired for baker's yeast strains since it is completely connected to dough-leavening. Yeasts at the same time encourage gluten network and generate aromatic compounds (Romano and Capece 2013). During the fermentation, *Saccharomyces cerevisiae* metabolises fermentable sugars (glucose, sucrose, maltose and fructose) and results in releasing CO_2 therefore rising dough volume. CO_2 is soluble in water and saturates the aqueous phase. After reaching saturation, all CO_2 produced passes through unsaturated gas phase and permit rising of the volume of bread. Solubilization of CO_2 in water results in decreasing the pH and elevates the acidity of the dough (Boekhout and Robert 2003). In addition, CO_2 affects rheological characteristics of fermented dough (Romano and Capece 2013). The most important two factors for volume of the bread are fermentation activity of yeast (producing CO_2) and ability of the dough to keep gas. The second one is carried out by gluten network. Accordingly these two components have to be in balance to acquire a quality end-product (Boekhout and Robert 2003).

Yeast contributes to the bakery products not only for increasing volume of the dough, but also producing aroma compounds (Birch et al. 2013). Due to the activity of yeasts, aroma compounds are formed widely in the crumb of bread and the most abundant compounds are alcohols, aldehydes as well as 2,3-butanedione (diacetyl), 3-hydroxy-2-butanone (acetoin) and esters (Hazelwood et al. 2008).

Volatile compounds which are responsible for aroma properties of dough and aroma precursors are produced by the thermal reactions during baking. In the course of the dough fermentation performed by yeasts, the chemical phenomena that dominates the fermentation is alcohol fermentation (Pozo et al. 2006).

Single-cell proteins are normally mentioned as source of mixed proteins. They are obtained from microorganisms like algae, fungi, yeasts and bacteria. If microbial cells are composed of significant amount of proteins, these microorganisms are referred as single-cell protein and native protein source. Due to increasing human population and the worldwide shortage of protein microbial biomass as food and feed, single-cell protein has gained importance (Nasseri et al. 2011).

Yeasts are the most suitable microorganisms for the production of single-cell proteins due to their high nutritional value. Some of the yeasts for production of single-cell proteins are *Candida lipolytica, Saccharomyces cerevisiae, Amoco torula, Candida utilis, Candida intermedia* and *Candida tropicalis* (Chandrani and Jayathilake 2000).

2.2 *Production of Alcoholic Beverages*

Yeasts play an important role in the production of fermented alcoholic beverages, especially beer, wine and distilled spirits. *Saccharomyces cerevisiae* is the most important species involved in wine making and brewing (McKay et al. 2011). Fermentation is the core microbial reaction in the production of alcoholic beverages and the microorganisms, which contaminate the raw materials before and after fermentation, can affect the final quality of fermented products. For the quality assurance of alcoholic beverages, controlling of such influences has become an important issue (Fleet 1998).

2.2.1 Beer

Beer is made from starch containing malted cereals, notably barley, hops, water and yeasts. Beer processing is a multistage process and fermentation is one of these stages (Deak 2008). In the brewery fermentation yeasts convert the fermentable sugars (mainly maltose) in the wort to ethanol and carbon dioxide as the major products of metabolism. Also a series of minor metabolites such as esters, higher alcohols, organic acids, aldehydes, ketones, sulphur compounds those contribute to flavour and aroma of beer are produced by yeasts (Briggs et al. 2004; Gibson et al. 2008). Depending on these minor metabolites, yeasts have a fundamental impact on the quality since these compounds play a key role on the organoleptic profile of beer (Pinho et al. 2006; Ferreira et al. 2010).

Brewer's yeast belongs to the group of *Saccharomyces sensu stricto* formed around the species *Saccharomyces cerevisiae* and is mainly divided into two groups, ale brewing yeasts and lager brewing yeasts, according to their use for the production of ales and lagers, respectively (Deak 2008; Lodolo et al. 2008; Nakao et al. 2009). These brewer's yeasts are classified based on flocculation behaviour; top fermenting and bottom fermenting which are called as *Saccharomyces cerevisiae* (ale type) and *Saccharomyces cerevisiae* (lager type) in the literature, respectively (Stewart and Russell 1998; Jentsch 2007). However, in beverage industry, ale yeast is referred as *Saccharomyces cerevisiae* and lager yeast is referred as *Saccharomyces uvarum* (*carlsbergensis*). Currently, yeast taxonomists have assigned all strains used in brewing to the species *Saccharomyces cerevisiae* (Stewart and Russell 1998). Top-fermenting yeasts produce ale beers at fermentation temperatures above 15 °C. At the end of fermentation they are carried to the surface of the wort. The other type is lager beer which is the most widespread beer type throughout the world. Lager beers are produced by bottom fermenting yeasts at fermentation temperatures below 15 °C. Lager yeasts have a good flocculation ability. They form flocculates at the bottom of the vessel by clumping together (Bamforth 2003; Ferreira et al. 2010). Top yeasts and bottom yeasts are collected to be used again from the surface and the fermenter bottom, respectively (Stewart and Russell 1998; Campbell 2003). Moreover, ale and lager yeasts are also different from each other depending on the ability to ferment the disaccharide melibiose (glucose-galactose). Since lager yeasts have the MEL gene, they can produce the extracellular enzyme α-galactosidase (melibiase) and are able to utilise melibiose. On the other hand, ale yeasts do not have the MEL gene, consequently do not produce α-galactosidase, therefore unable to utilise melibiose (Stewart and Russell 1998; Lodolo et al. 2008).

Brewing yeast should produce desirable flavour and aroma metabolites for the final product and fast growth, rapid and efficient fermentation are other essential properties (Campbell 2003). The brewing yeast should tolerate the environmental stresses, such as high ethanol and low oxygen levels (Ferreira et al. 2010). Ethanol is a desirable product of the fermentation process, but accumulation of it can cause significant chemical stress on the yeast cell (Lentini et al. 2003; Hutkins 2006). Suitable flocculation and sedimentation or head formation at the end of the fermentation are other essential requirements depending on the beer type. High final viability for pitching for the next fermentation and high genetic stability are other important properties of suitable brewing yeast (Campbell 2003).

2.2.2 Wine

Yeasts are important microorganisms in wine microbiology. Biotransformation of grape sugars into ethanol, carbon dioxide and several secondary products is a complex process and yeasts are responsible for this conversion (Ciani et al. 2010). Besides the conversion of sugars to ethanol, yeasts also make positive contributions to wine flavour by the synthesis of other minor metabolites that define the flavour and other sensorial properties (Cortes and Blanco 2011; De Benedictis et al. 2011; Navarrete-Bolanos 2012). The main activities of wine yeasts are rapid, complete and efficient conversion of the grape sugars into ethanol and carbondioxide; influencing the quality of the grapes before the harvest by biocontrol of molds; biocatalysis of neutral grape components into flavour active compounds; producing secondary metabolites such as esters, acids, alcohols, aldehydes, ketones, polyols, volatile sulphur compounds which directly impact the wine flavour without development of off-flavours (Lambrechts and Pretorius 2000; Fleet 2003a). Spontaneous fermentation of "must" is initiated by indigenous yeasts and final stages are dominated by the alcohol-tolerant strains of *Saccharomyces cerevisiae*. The non-Saccharomyces yeasts, commonly *Kloeckera* spp. and *Candida* spp., dominate at the beginning of the fermentation and affect the sensorial characteristics of wine (Ganga and Martinez 2004). The other isolated yeasts are *Metschnikowia, Dekkera, Pichia, Kluyveromyces, Issatchenkia, Saccharomycodes, Zygosaccharomyces, Torulaspora, Debaryomyces,* and *Schizosaccharomyces* (Fleet 2003a). It was reported that *Kloeckera apiculata* (its telemorph *Hanseniaspora uvarum*) is the predominant non-*Saccharomyces* yeast present in grape must (Fleet and Heard 1993). When the ethanol content starts to increase during the fermentation, *Saccharomyces cerevisiae,* the main wine yeast, degrades the sugar and can tolerate high ethyl alcohol. It takes over the fermentation and completes the process (Fleet and Heard 1993; Blanco et al. 2012).

Fast fermentation of grape juice sugars to high ethanol concentrations is essential for wine yeasts. Wine yeasts should exhibit uniform dispersion and produce minimal foam. At the end of the fermentation, sediment should be quickly taken from the wine. It is also important that the yeast should not give slow, sluggish or stuck fermentations (Bisson 1999). Desirable characteristics of wine yeasts include some properties such as rapid initiation of fermentation, high fermentation efficiency, high ethanol tolerance, high osmotolerance, moderate biomass, high genetic stability, high sulphite tolerance, low sulphite binding activity, low foam formation, compacts sediment, resistance to desiccation, killer activity, genetic marking, proteolytic activity and low nitrogen demand. Other properties which are related to the flavour characteristics are also reported as low volatile acidity production, less higher alcohol production, liberation of glycosylated flavour precursors, high glycerol production, hydrolytic activity, enhanced autolysis and modified esterase activity. Moreover, the yeast must give a good flavour, free of sensory faults, and allow the grape character to be perceived by the consumer (Lambrechts and Pretorius 2000; Swiegers et al. 2005; Ciani et al. 2010). Low sulphite and biogenic amine formation and low ethyl carbamate (urea) potential are metabolic properties related to health implications (Pretorius 2000; Curtin et al. 2011).

2.3 Yeasts in Table Olives

Table olive is an important fermented product in the food industry. Depending on some of its unique characteristics such as the bitter component called oleuropein, low sugar concentration and high oil content, it cannot be consumed directly and needs to be processed (Arroyo-López et al. 2008). During the processing of table olives, microorganisms, both lactic acid bacteria and yeasts, play important roles.

Yeasts act as both desirable and spoilage microorganisms in table olives (Garrido-Fernández et al. 1997; Arroyo-López et al. 2012). It was reported that the species of yeast mainly present in the table olive fermentation are in the genera of *Saccharomyces, Pichia, Debaryomyces, Candida* and *Kluyveromyces* (Arroyo-López et al. 2008; Tofalo et al. 2013). Yeasts produce compounds such as alcohols, ethyl acetate, acetaldehyde and organic acids which enhance the organoleptic characteristics (Garrido-Fernandez et al. 1995; Arroyo-López et al. 2008; Alves et al. 2012). Moreover, yeasts in olive fermentation could improve the aromatic profile of fermented olives by increasing their free fatty acid content, which could be the precursors to the formation of diverse volatile compounds, such as propanol or 2-butanol (Hernández et al. 2007; Rodriguez-Gómez et al. 2010; Rodríguez-Gómez et al. 2012). Another positive effect of yeasts on the olive fermentation is acting as biocontrol agents which are achieved by the toxic proteins also called as killer toxins those enable the inhibition of the growth of fungi (Viljoen 2006; Arroyo-López et al. 2012). It was reported that a considerable number of killer strains of *Debaryomyces, Pichia* and *Candida* species were isolated from table olives (Marquina et al. 1997). Especially, inhibitory activities of species *Wickerhamomyces anomalus* and *Pichia membranaefaciens* were proven against the fungi (Santos et al. 2000; Arroyo-López et al. 2012). Moreover, yeasts can improve growth of lactic acid bacteria which are the essential organisms of fermentation by the synthesis of nutritive compounds (Viljoen 2006; Alves et al. 2012; Tofalo et al. 2013).

Nevertheless, yeasts sometimes may be associated with different kinds of olive spoilage such as gas pocket formation depending on the production of CO_2, softening of the fruits due to pectinolitic activity, clouding of brines, biofilm production and sometimes production of off-flavours (Arroyo-López et al. 2008; Alves et al. 2012).

2.4 Yeasts in Meat-Based Fermented Foods

Yeasts are found on meat and processed meat products since meat is a suitable media for growth of them. They have also positive effects on fermented meat products and many yeast species such as *Candida, Debaryomyces, Pichia, Trichosporon, Cryptococcus, Rhodotorula* and *Yarrowia* have been isolated from fermented meat products, especially sausages (Dillon and Board 1991; Deak 2008). Lipolytic and proteolytic activities, which contribute to flavour due to the production of volatiles, of those yeasts were described (Olesen and Stahnke 2000; Selgas and Garcia 2007). Most frequently isolated yeasts are *Yarrowia lipolytica* and *Debaryomyces hansenii* (its anamorph *Candida famata*) (Gardini et al. 2001; Deak 2004). *Debaryomyces hansenii* is also used as commercial starter culture in fermented meat products due to its positive contributions on final product (Hammes and Hertel 1998; Toldrá 2002; Selgas and Garcia 2007).

2.5 *Yeasts in Dairy-Based Fermented Foods*

Various types of yeasts are naturally found in milk and fermented dairy products such as kefir, koumis, viili, longfil, yoghurt and all types of cheese (Wouters et al. 2002; Frölich-Wyder 2003; Cantor et al. 2004). Yeasts play essential role due to their important functions in dairy based products such as contributing to the ripening of cheese, speeding up the maturation, improving texture and aroma characteristics of certain milk products, increasing pH of cheese, manufacturing of some metabolites like ethanol, acetaldehyde, CO_2, amino acids and vitamins, removing toxic end-products of metabolism, taking part in some interactions and contributing to the fermentation by supporting the starter cultures, preventing some undesired microorganisms those cause product quality default, inducing the growing of starter cultures by means of the utilization of organic acids, contributing to the flavour characteristic of dairy products due to the strong proteolytic and lipolytic activity, fermenting lactose and utilizing citric acid (Fleet 1990; Spinnler et al. 2001; Jakobsen et al. 2002; Ferreira and Viljoen 2003).

Most common yeast species found in dairy products are as follows: *Candida lusitaniae, Candida krusei, Kluveromyces lactis, Debaryomyces hansenii, Yarrowia lipolytica, Kluyveromyces marxianus, Saccharomyces cerevisiae, Galactomyces geotrichum, Candida zeylanoides* and various *Pichia* species. These yeast species play a key role in the processing of dairy-based fermented products by the contribution to flavour and colour (Jakobsen and Narwnjs 1996; Viljoen 2001; Samelis and Sofos 2003; Jacques and Caserogola 2008).

There are numerous studies to identify yeast species in various cheese types. *Debaryomyces hansenii, Candida lipolytica, Candida kefyr, Candida intermedia, Saccharomyces cerevisiae, Cryptococcus albidus* and *Kluyveromyces marxianus* are the most prevelant species in Camembert and Blue-veined cheese (Roostita and Fleet 1996), *Debaryomyces hansenii, Galactomyces candidum, Issatchenkia orientalis, Kluyveromyces lactis, Kluyveromyces marxianus, Saccharomyces cerevisiae, Yarrowia lipolytica* and *Candida catenulata* were identified as predominant yeast species in various cheeses from Austria, Denmark, France, Germany and Italy (Prillinger et al. 1999). Additionally, *Saccharomyces unisporus* and *Kluyveromyces marxianus* are utilised for the manufacture of kefir (Abdelgadir et al. 2001; Gadaga et al. 2001; Strehaiano et al. 2006).

Especially, *Debaryomyces hansenii* and *Yarrowia lipolytica* are suitable for generation of starter cultures because of their proteolytic and lipolytic activity and also make favorable contributions on cheese ripening (Van den Tempel and Jacobsen 2000; Guerzoni et al. 2001; Ferreira and Viljoen 2003). In a study on feta cheeses, various aroma compounds were investigated and it was shown that yeast species are effective in the formation of aromatic substances such as 2,3-bütandiol (McSweeney and Sousa 2000), 1-bütanol, 1-heptanol, hexanal and nonanal (Bintsis and Robinson 2004; Kesenkaş and Akbulut 2006).

2.6 Yeasts in Cereal-Based Fermented Foods

Cereal and cereal crops are accepted as significant nutrients all over the world. Cereal grains and legumes are utilised as raw material for many foods and beverages in different countries and cultures (Blandino et al. 2003; Heredia et al. 2009). The basic cereals utilised for nourishment are corn (maize), wheat, barley, rice, oats, rye, millet, sorghum and soybeans. Cereals are substrates for some fermentation products such as; beer, sake, spirits, boza, fura, malt vinegar, tarhana, idli and baked goods made from doughs leavened by yeasts or sourdough. Cereal based foods are usually performed by natural fermentations including mixed cultures of yeasts, bacteria and fungi (Gotcheva et al. 2000; Hammes et al. 2005; Settani et al. 2011).

Sourdough is a significant product for bakeries and it is characterized as combined activity of yeasts and lactic acid bacteria (Giannou et al. 2003; Chavan and Jana 2008). The most important function of yeasts in sourdough fermentation is metabolizing fermentable sugars for generating CO_2, increasing gas formation capacity, improving flavour and aroma, contributing to the texture of the crumb and the nutritional value (De Vuyst and Vancanneyt 2007; Vogelmann et al. 2009; Vrancken et al. 2010; Chavan and Chavan 2011). The generated CO_2 plays an important role in the formation of dough volume and is used for leavening agent and affects bread texture, density and volume (Decock and Cappelle 2005).

2.7 Bioethanol Production (Industrial Ethyl Alcohol)

The importance of biofuels continues to increase worldwide. Alternative energy sources are necessary for all over the world because of political instability of oil producer countries, global environmental concerns, volatile oil price and negative effects of fosil fuels (Wyman 2007; Bai et al. 2008; Mussatto et al. 2010). Bioethanol is commonly thought as the most promising biofuels among renewable sources (Sanchez and Cardona 2008; Moona et al. 2012).

Bioethanol can be produced from various raw materials. Wide variety of renewable feedstock can be classified in three main groups: (1) simple sugars those containing significant amounts of easily fermentable sugar (sugar cane, sugar beets, sweet sorghum), (2) starches and fructosans (corn, potatoes, rice, wheat, agave, inulin) and (3) cellulosics (stover, grasses, corn cobs, wood, sugar cane bagasse) (Demirbas 2007; Sanchez and Cardona 2008; Amorim et al. 2009; Chi et al. 2011).

Yeasts for bioethanol fermentation can be defined in terms of their performance parameters such as temperature range, pH range, alcohol tolerance, growth rate, productivity, osmotic tolerance, specificity, yield, genetic stability and inhibitor tolerance (Dien et al. 2003). Conventially, *Saccharomyces cerevisiae* is used for bioethanol fermentation. *Saccharomyces cerevisiae* can ferment glucose into bioethanol, but unable to ferment xylose (Keshwani and Cheng 2009). Xylose-fermenting yeasts, such as *Pichia stipitis, Candida shehatae* and *Candida parapsilosis*, can metabolise xylose via the action of xylose reductase and xylitol dehydrogenase. Hereby, using recombinant *Saccharomyces cerevisiae*, carrying heterologous xylitol reductase and xylitol dehydrogenase from *Pichia stipitis* and xylulokinase from *Saccharomyces cerevisiae*, bioethanol production from xylose can be successfully done (Katahira et al. 2006).

2.8 Yeast-Derived Products

A reliable source of ingredients and additives for food processing is obtained from yeasts (Demain et al. 1998). Preparations of baker's and brewer's yeasts have been available for many years as dieatary supplements due to their high content of proteins, peptides, amino acids, B vitamins and trace minerals. At the present time, numerous products are derived from yeasts which are antioxidants, autolysates, enzymes, minerals, vitamins, colour and flavour compounds (Stam et al. 1998).

Yeasts and yeast extracts are known as source of antioxidant compounds for years (Abbas 2006). It is believed that yeast peroxisomes play nearly the same role to plant peroxisomes. Consequently, the response in yeasts to oxygen-derived radicals would include various enzymes, involving catalases, superoxide dismutases and glutathione, besides several NADP-dependent dehydrogenases (Del Rio et al. 2003).

Interest to the biotechnological generation of natural aroma compounds is rapidly increasing. Yeasts contribute significantly to the aroma of fermented foods. During fermentation yeasts synthesize a vast number of aroma compounds (Suomalainen and Lehtonen 1979; Berry 1995). The numerically and quantitatively largest groups of aroma compounds synthesized by yeasts consist of fusel alcohols, fatty acids, acids, esters, carbonyl compounds, acetals, phenols, hydrocarbons, nitrogen compounds, sulphur compounds, lactones, sugars, and a diversity of other unclassified compounds (Suomalainen and Lehtonen 1978, 1979; Berry 1989, 1995; Garafolo 1992; Dickinson 2003).

Yeast-formed flavours can be generally classified into three categories as yeast metabolic products which include products synthesized or derived through yeast biocatalysis, yeast cell mass-derived products which include products prepared through yeast autolysis and complex products resulting from the interaction of yeast-derived products with other food matrix ingredients (Kollar et al. 1992; Stam et al. 1998).

Yeast extracts can be produced using autolysis, plasmolysis and hydrolysis but the most frequently production practise is autolysis (Tanguler and Erten 2008). They are concentrates of the soluble fraction of yeast cells and mainly produced from baker's or spent brewer's yeasts, both *Saccharomyces cerevisiae*. In Europe, the main raw material for yeast extract is baker's yeast which is primary grown high protein yeast. In the UK and the USA, debittered brewer's yeast is used. Other yeasts, in particular, *Candida utilis* and *Kluyveromyces marxianus* are also sometimes used (Sommer 1998). Recently, temperature-sensitive autolysing strain of *Saccharomyces cerevisiae* which showed increased autolysis at 37 °C has been used for yeast extract by Asahi Breweries in Japan (Stam et al. 1998).

To produce a product by autolysis, viable yeast slurry containing 15–20 % yeast solids is maintained at an elevated temperature in the range of 45–60 °C for up to 36 h or more at about pH 5.5 (Nagodawithana 1992; Joseph 1999). At elevated temperatures, the yeast cells die but their native enzymes are still remain active (Sommer 1998). During autolysis, intracellular yeast enzymes located in the general matrix of the cell cause breakdown of mainly proteins, peptides, carbohydrates, nucleic acids (mainly RNA) and cell wall materials into free amino acids (mainly glutamate), peptides, sugars and nucleotides (Stam et al. 1998). The most important flavour enhancing components are glutamate and 5′-nucleotides, especially 5′-guanylate (5′-GMP). Glutamate has 100–300 mg/l of taste threshold, for 5′-inosinate (5′-IMP) and 5′-guanylate 120 and 35 mg/l, respectively. Glutamate and 5′-nucleotides are chemically stable substances but enzimatically active foods such as raw meats, raw fish and vegetable tissues can degrade these compounds (Sommer 1998).

Plasmolysis is generally used for rapid initiation of the cell degradation process for the production of yeast extract especially in Europe but less popular in the USA. During plasmolysis, yeast cells start to lose water to equilibrate their osmotic pressure with the surrounding medium in the presence of high amounts of promoters (Nagodawithana 1992, 1994).

Commonly used plasmolysing agents are common salt (sodium chloride), sucrose, ethanol, ethyl acetate, amyl acetate, chloroform, toluene and combinations of salts such as potassium chloride (Peppler 1982). The most frequently utilised agent is common salt (Reed and Nagodawithana 1991) at the ratio of 1–3 %. However, using salt as a plasmolysing agent leads to an extract with high salt concentration (Reed and Nagodawithana 1991; Nagodawithana 1994).

Yeast cells are also rich source of vitamins such as thiamine, pantothenic acid, riboflavin, vitamin B6, and vitamin B12 (Harrison 1970; Peppler 1970; Reed 1981; Halasz and Laszity 1991). Yeast cells are also good sources of biotin, folic acid and ergosterol.

Intracellular yeast enzymes can be prepared from whole yeast cell mass by mechanical disruption and other means. These enzymes have found several food uses (Peppler 1979; Reed 1981; Halasz and Laszity 1991). Invertase obtained from *Saccharomyces cerevisiae* and other sucrolytic food yeasts are used in the confectionary industry to break down sucrose for manufacturing liquid-centre candies (Reed 1981; Halasz and Laszity 1991). Lactase from *Kluyveromyces* spp. is also important for several food uses, especially to hydrolyse lactose. Ribonuclease obtained from baker's yeast is used for RNA denaturation during the manufacture of yeast nucleotides (Sanchez et al. 2003).

A number of yeasts can produce carotenoids which are used as food colorants including species of *Rhodotorula* (*Rhodotorula glutinis, Rhodotorula lactis, Rhodotorula gracilis, and Rhodotorula rubra*), *Rhodosporidium, Phaffia rhodozyma* and *Sporobolomyces pararoseus* (Cang et al. 2002; Squina et al. 2002; Simova et al. 2003; Frengova et al. 2003, 2004; Cheng et al. 2004).

2.9 Yeasts as Biocontrol Agents

There is a good relationship between the terms of sustainable agriculture and biocontrol, because biological control concept benefits from natural biological cycles with the minimal environmental effect in the field of crop production (Spadaro and Gullino 2004). All foods and beverages have a wide variety of microbial species and interactive responses which affect the product quality. The major underlying response in biocontrol concept is antagonistic interactions (Fleet 2006). Antogonism effect is the inhibition of undesired or pathogenic microorganisms by competition for space or nutrients via the manufacture of toxic materials and by providing environmental exchange and by production of antimicrobial metabolites thus survival of desirable species (Huber 1997; Fleet 2006; Pometto et al. 2006; Satyanarayana and Kunze 2009).

Biocontrol activity of antagonism could be increased using many applications such as combining organic and inorganic additives with antogonistic yeasts (Mecteau et al. 2002). The most interesting microorganisms in biological control programmes are yeasts. Because they have some important properties which make them reasonable to be used as biocontrol agents (Satyanarayana and Kunze 2009).

In the last 20 years, some yeasts which can be utilised as potential biocontrol agent were identified. Some fungi such as *Botrytis*, *Penicillium*, *Aspergillus*, *Rhizopus* spp. give rise to pre and postharvest spoilage of fruits and vegetables. Certain yeasts can be used as biocontrol agents to these spoilage microorganisms and therefore chemicals can be less frequently used (Punja and Utkhede 2003; Fleet 2003b; Spadaro and Gullino 2004). Some of prominent potential biocontrol yeasts have been commercialised; e.g. *Candida oleophila* and *Pseudozyma flocculosa.* Some of the other species which are used as potential biocontrol agents are determined as *Metschnikowia pulcherrima*, *Pichia guilliermondii*, *Candida sake*, *Sporobolomyces roseus*, *Aureobasidium pullulans* and various *Cryptococcus* species (Fleet 2003b).

2.10 Probiotic Yeasts

Although lactic acid bacteria are well-known probiotic organisms, some yeast species also recognised as probiotics due to their health benefits. *Saccharomyces cerevisiae* var. *boulardii* and *Saccharomyces cerevisiae,* have been reported as the major probiotic yeasts. Moreover, there is an interest to some other non-*Saccharomyces* species such as *Debaryomyces hansenii, Yarrowia lipolytica, Issatchenkia orientalis, Kluyveromyces marxianus* and *Kluyveromyces lactis* to be used as probiotics (Fleet 2006; Fleet and Balia 2006).

3 Lactic Acid Bacteria in Food Processing

3.1 *Plant-Based Fermented Products*

Plant-based fermented foods are popular all over the world and consumer demand for these fermented products is increasing. Olives, cucumber and sauerkraut are commercially important plant-based fermented vegetables eventhough most vegetables are fermented at small-scale level.

3.1.1 Table Olives

Table olives are one of the most important fermented foods obtained by mainly the action of lactic acid bacteria. The main processing types are lye-treated green olives in brine, untreated black olives in brine and ripe olives. First two methods include the lactic acid fermentation (Harris 1998; Hurtado et al. 2012).

Lactobacillus plantarum and *Lactobacillus pentosus* are the predominant species in most of the fermentations. However, the other lactobacilli or genera can take partial responsibility for this essential role or even can be the major actor of the fermentations depending on the olive cultivar, processing method and the geographical origin (Hurtado et al. 2012). Both of them are suitable for fermenting various table olive cultivars (Sánchez et al. 2001; Panagou et al. 2008) but cultivar and processing method are the major actors of a successful inoculation (Panagou and Tassou 2006; Hurtado et al. 2010).

In order to achieve an enhanced and more predictable fermentation process, brine inoculation with an appropriate starter culture of lactic acid bacteria can be used. Lactic acid bacteria convert carbohydrates into lactic acid, CO_2 and other organic acids without the need for oxygen in the medium. However, higher concentrations of phenolic compounds in olive fruit, mainly oleuropein, could inhibit lactic acid bacteria (Amiot et al. 1990; Sánchez-Gómez et al. 2006; Landete et al. 2008; Hurtado et al. 2009; Rodriguez et al. 2009; Ghabbour et al. 2011). Moreover these phenolic compounds, especially oleuropein, give bitterness, therefore they are removed from fruit to become edible by the treatment with sodium hydroxide for lye-treated Spanish style green olives production. Unlike alkali treatment, to hydrolyse the bitter-tasting oleuropein, *Lactobacillus pentosus* can be used as starter due to its glycosidases and esterase activities (Servili et al. 2006).

Starters can be chosen based on a large variety of criteria like homo- and hetero-fermentation, acid production, salt tolerance, flavour development, temperature range, oleuropein-splitting ability and bacteriocin-production (Ruiz-Barba and Jimenez-Diaz 1995; Durán Quintana et al. 1999; Delgado et al. 2005).

Olives' fermentation is done by the natural biota of olives consisting of a variety of bacteria, yeasts, and molds. The lactic acid bacteria become prominent during the intermediate stage of fermentation. Initially *Leuconostoc mesenteroides* and

Pediococcus cerevisiae (now called *Pediococcus pentosaceus*) become prominent, and then lactobacilli, with *Lactobacillus plantarum* and *Lactobacillus brevis* become the most important.

Growth of *Lactobacillus plantarum* in the fermentation provides the necessary lactic acid formation for preservation and also for its characteristic flavour (Rodriguez De Le Borbolla et al. 1979, 1981). Also using suitable *Lactobacillus plantarum* starter cultures potentially improve the microbiological control of the process, increase the lactic acid yield and highly qualified fermented green olives are produced (Fernandez Diez 1983; Roig and Hernandez 1991; Ruiz-Barba et al. 1991; Garrido-Fernandez et al. 1995; Ruiz-Barba and Jimenez-Diaz 1995). Lactic acid bacteria can form the flavour during the fermentation. Numerous volatile compounds make a significant contribution to the final flavour of table olives (Sabatini et al. 2008).

3.1.2 Pickled Vegetables

Cucumber

Cucumbers are typically fermented in brine solutions in large tanks. Lactic acid bacteria may be involved during the primary fermentation of cucumbers. *Lactobacillus plantarum* and *Pediococcus pentosaceus* have been chosen as the desired species of lactic acid bacteria for commercial cucumber fermentations. These homofermantative species are preferred for the fermentation to minimize purging requirements to remove CO_2. Cucumber sugars are converted into lactic acid by the fermentation which is carried out by primarily *Lactobacillus plantarum* and pH is decreased (Ic and Ozcelik 1995, 1999).

During brine fermentation, keeping the structure integrity of whole cucumbers is very important. As a result of respiration and malolactic fermentation by *Lactobacillus plantarum*, CO_2 may be formed. In order to prevent the serious economic losses due to gaseous spoilage (bloater damage), cucumbers may be purged with air. It should be taken into consideration that such kind of practice may increase the risk of growth of molds and yeasts (Tamang et al. 2005).

The potential involvement of *Lactobacillus buchneri* is indicated in the study of spoilage of fermented cucumber (Fleming et al. 1989, 2002; Kim and Breidt 2007). *Lactobacillus buchneri* has been isolated from fermented cucumbers those had undergone spoilage, characterized by decreased concentrations of lactic acid, increased pH, and increased concentrations of acetic and propionic acids (Franco and Pérez-Díaz 2012; Johanningsmeier et al. 2012). Only *Lactobacillus buchneri* was found to initiate lactic acid utilization in fermented cucumber media after several lactic acid bacteria have been isolated from spoiled fermented cucumber (Johanningsmeier et al. 2012). Therefore, *Lactobacillus buchneri* plays the major role in the initiation of secondary fermentations which lead to spoilage of fermented cucumber.

Sauerkraut

Sauerkraut is a commonly consumed vegetable in some European countries. It is a fermentation product of fresh cabbage. The starter for sauerkraut production is generally the normal flora of cabbage. *Leuconostoc mesenteroides* and *Lactobacillus plantarum* are the two most preferred lactic acid bacteria in sauerkraut fermentation. As well as *Lactobacillus plantarum*, *Lactobacillus brevis* also affects the final stages of sauerkraut production (Kalac et al. 1999).

Sauerkraut production is generally based on a sequential microbial process that involves heterofermentative and homofermentative lactic acid bacteria. In this process *Leuconostoc* species and *Lactobacillus*, *Pediococcus* species involve as the first and second group, respectively (Font De Valdez et al. 1990). *Leuconostoc* primarily uses glucose and fructose for its growth and produce lactic and acetic acids, ethanol, mannitol and CO_2 (Aukrust et al. 1994). These bacteria slow down and begin to die off, when the acidity reaches to 0.25–0.3 % as lactic acid. The activity started by *Leuconostoc mesenteroides* is continued by *Lactobacillus plantarum* and *Lactobacillus cucumeris* until an acidity level of 1.5–2 % as lactic acid is reached. Finally, *Lactobacillus pentoaceticus* continues the fermentation and the acidity reaches to 2–2.5 %, so completes the fermentation.

The end-products of a normal sauerkraut fermentation are mainly lactic acid, smaller amounts of acetic and propionic acids, a mixture of gases of principally carbon dioxide, small amounts of alcohol and a mixture of aromatic esters. However the acidity helps to control the growth of spoilage and undesired microorganisms.

3.2 Sour-Dough Breads

The dough properties (Collar 1996), organoleptic characteristics (Hammes et al. 1996), nutritional value (Lopez et al. 2001) and the shelf life of bread (Lavermicocca et al. 2000) are improved by lactic acid bacteria in bread making.

Sourdough is prepared with flour and water, containing a wide variety of lactic acid bacteria (Gobbetti 1998; Hammes and Gaenzle 1998). High numbers of lactic acid bacteria found in cereal sourdoughs, including mainly *Lactobacillus, Leuconostoc* and *Lactococcus* species (Hounhouigan et al. 1993; Johansson et al. 1995). It is reported that the dominant *Lactobacillus* species in wheat sourdoughs are *Lactobacillus sanfranciscensis* (which is reported as identical to *Lactobacillus brevis* var. *lindneri*) (Hutkins 2006), *Lactobacillus brevis, Lactobacillus fermentum* and *Lactobacillus fructivorans* (Gobbetti et al. 1994; Corsetti et al. 2001, 2003). Some described species such as *Lactobacillus plantarum, Lactobacillus alimentarius, Lactobacillus acidophilus, Lactobacillus delbrueckii* subsp. *delbrueckii* (Gobbetti et al. 1994; Corsetti et al. 2001, 2003), *Lactobacillus spicheri* (Meroth et al. 2003), *Lactobacillus mindensis* (Ehrmann et al. 2003), *Lactobacillus frumenti* (Müller et al. 2000) and *Lactobacillus paralimentarius* (Cai et al. 1999) were also isolated from sourdough. In this complex system, the synthesis of bacteriocins and other antimicrobial compounds could regulate the interactions between the starter and the contaminant microflora of the sourdough (Corsetti et al. 2004).

The performance of lactic acid bacteria strains in the food matrix has been studied by characterization of the acidification parameters and lactic and acetic acid production during sourdough fermentation (Hammes and Gaenzle 1998). The organic acids would also contribute to the production of aroma compounds (Meignen et al. 2001).

In sourdoughs, the sugar usage by lactic acid bacteria strains is related to the microorganism, the type of sugar, the presence of yeasts, and the manufacturing conditions (Hammes and Gaenzle 1998; Martinez-Anaya 2003). On the whole, *Lactobacillus reuteri* strains, which were isolated from homemade doughs, ferment different sugars, e.g., sucrose, melibiose, rafinose, fructose, while most *Lactobacillus sanfranciscensis* strains only ferment maltose (Corsetti et al. 2001).

3.3 Lactic Acid Bacteria and Wine

Lactic acid bacteria are responsible for malolactic fermentation (MLF) in wines which can be beneficial in some cases and undesirable in the others (Krieger 1993).

Primary importance of lactic acid bacteria in wine making is malolactic fermentation. The main malolactic fermentation reaction is the decarboxylation of L-malic acid to L-lactic acid. In this reaction the acidity decreases and pH raises by 0.3–0.5 units. The malolactic fermentation is done both by Lactobacilli and Leuconostoc.

Malolactic bacteria growing in wine must be capable of tolerate low nutrient concentration, low pH and high concentrations of ethanol and SO_2. When wines involve residual glucose and fructose, there is undesirable acidification.

Malolactic fermentation could be conducted by the species of *Lactobacillus* or *Pediococcus,* but this reaction usually results in non-acceptable wines when pH is higher than 3.5. These genera usually can not tolerate low pH and produce undesirable flavours along with high levels of acetic acid (Murphy et al. 1985; Krieger et al. 1990).

To conduct malolactic fermentation, *Oenococcus oeni* (formerly *Leuconostoc oenos*) is primarily preferred bacterial species, rather than yeast or other lactic acid bacteria. *Oenococcus oeni* is especially adapted to the harsh environment of wine and is capable of converting malic acid to lactic acid quickly. Hence different strains of *Oenococcus oeni* can have particularly different effects on the final product, and some strains are more beneficial to the properties of wine than others (Henick-Kling 2002).

3.4 Lactic Acid Bacteria in Fermented Dairy Products

Especially mesophilic *Lactococcus lactis* and thermophilic *Lactobacillus delbrueckii* subsp. *bulgaricus* and *Streptococcus thermophilus* are important lactic acid bacteria in the process of dairy products; for example, yogurt, cheese and buttermilk. By lactic acid fermentation, acetaldehyde and diacetyl production increase and this causes flavour formation in yogurt and buttermilk. Lactic acid bacteria also produce exopolysaccharides which help to increase structural properties of fermented dairy products. In addition, one of the health benefits of dairy products is the conversion of lactose and galactose to the sweetener such as L-alanine (Hugenholtz et al. 2000).

Lactococcus lactis, especially sub-species of *lactis* and *cremoris* are the commonly used starter bacteria for some cheese types (Gouda and Cheddar cheese), butter and buttermilk. This bacterium can grow in milk easily and converts lactose to lactic acid. With a nitrogen source in medium, it hydrolyses casein. This biochemical process occurs during ripening of cheese and it helps to form flavour by the result of amino acid release. Other important characteristic of *Lactococcus lactis* in some strains is that they can convert citric acid to diacetyl (flavour of butter) and CO_2 (Hugenholtz 1993).

Streptococcus thermophilus is another lactic acid bacteria used in dairy fermentation as starter culture especially in yogurt. It is also used for some cheeses such as Swiss (Emmenthaler) and Italian (Parmesan) types. Its growing temperature is 40–45 °C. Special characteristic of this bacterium is that, it metabolises only the glucose-moiety from lactose and residual galactose is exerted from cell. *Streptococcus thermophilus* forms only L (+) lactate. During milk fermentation of *Streptococcus thermophilus*, a flavour compound, acetaldehyde is also produced (Caplice and Fitzgerald 1999; Hugenholtz et al. 2000).

Lactobacillus delbrueckii subsp. *bulgaricus* is used together with *Streptococcus thermophilus* for the production of yoghurt. It is homofermantative yoghurt bacterium with growing temperature of 40–45 °C. *Lactobacillus delbrueckii* subsp. *bulgaricus* produce D (-) lactic acid. Although lactic acid is the main end-product of yoghurt fermentation, flavour compounds such as acetaldehyde, acetone, diacetyl (2,3-bütanedione) and acetoin can also be formed in very low amounts (Caplice and Fitzgerald 1999; Chaves et al. 1999).

Lactobacillus acidophilus as obligate homofermantative, is a probiotic lactic acid bacterium. Hexose sugars are metabolised primarily to lactic acid. It is used for the production of acidophilus in milk mixed culture with *Lactobacillus delbrueckii* subsp. *bulgaricus.*

Lactic acid bacteria contribute to the structural characteristic of the fermented dairy products by the production of exopolysaccharides. Especially in yoghurt and in some Scandinavian dairy products, for example viili and longfil, both *Streptococcus thermophilus* and *Lactobacillus delbrueckii* subsp. *bulgaricus* produce these sugar polymers (Van Kranenburg et al. 1999).

3.5 Lactic Acid Bacteria in Meat Products

Lactic acid bacteria help meat products to protect them from pathogenic microorganism and improve their sensory quality during fermentation. Increasing acidification leads to prevent development of spoilage and pathogenic activities. It also contributes to stabilization of colour and texture. Lactic acid bacteria can produce bacteriocins which help to preserve fresh and processed meat (Kröckel 2013). The predominant lactic acid bacteria during lactic acid fermentation of sausages are *Lactobacillus sakei* and *Lactobacillus curvatusare*. However, some species of *Lactobacillus* spp. and *Pediococcus* spp. are used as starter cultures for most European fermented sausages formulated with nitrite (Caplice and Fitzgerald 1999).

3.6 Lactic Acid Bacteria in Traditional Turkish Fermented Foods and Beverages

3.6.1 Tarhana

Tarhana, a traditional cereal-based lactic acid fermented food product, is widely consumed in Turkey and Middle East. Tarhana is obtained primarily by mixing yoghurt, wheat flour, yeast, salt, depending on the region raw or cooked vegetables (tomato, onion, pepper, etc.) and spices (mint, basil, dill, tarhana herb, etc.). Fermentation is usually carried out by yoghurt bacteria and fermentation lasts for 1–7 days (Ibanoglu and Ibanoglu 1998). *Lactobacillus delbrueckii* subsp. *bulgaricus, Streptococcus thermophilus, Lactobacillus fermentum, Pediococcus acidilactici, Lactobacillus plantarum, Lactobacillus brevis, Enterococcus faecium, Pediococcus pentosaceus, Leuconostoc pseudomesenteroides* and *Weissella cibaria* are identified in tarhana fermentations (Sengun et al. 2009; Settani et al. 2011).

The resulting product is listed among the acidic fermented foods characterised by sour taste and strong yeast flavour (Ibanoglu and Ibanoglu 1999; Sagdic et al. 2002; Dağlioğlu et al. 2002; Sengun et al. 2009).

The dominant microbiota of tarhana is mainly lactic acid bacteria and yeasts. Lactic acid bacteria are the most important microbial group for tarhana fermentation. They play the main role in the production of aromatic compounds those are typical for the final product (Settani et al. 2011). Also, they increase acidity and control the mechanism to enhance the safety (Settanni and Corsetti 2008).

3.6.2 Boza

Boza is a mildly alcoholic beverage produced from the fermentation of barley, oats, millet, maize, wheat or rice. It is a traditional Turkish-fermented beverage. *Lactobacillus acidophilus, Lactobacillus brevis, Lactobacillus coprophilus, Lactobacillus coryniformis, Lactobacillus fermentum, Lactobacillus paracasei, Lactobacillus pentosus, Lactobacillus plantarum, Lactobacillus rhamnosus, Lactobacillus sanfrancisco, Lactococcus lactis* subsp. *lactis, Leuconostoc mesenteroides, Leuconostoc mesenteroides* subsp. *dextranicum, Leuconostoc raffinolactis, Pediococcus pentosaceus, Oenococcus oeni, Weissella confusa* and *Weissella paramesenteriodes* were found in boza samples (Arici and Daglioglu 2002; Todorov and Dicks 2006). Only a few papers reported the isolation of yeasts and moulds from boza. It was reported that boza is a good source of bacteriocin-producing lactic acid bacteria (Todorov and Dicks 2006).

3.6.3 Kefir

Kefir is an acidic and low-alcoholic fermented dairy product with its functional properties (Farnworth 1999, 2006; Farnworth and Mainville 2003). Traditionally, the people of the Caucasus prepared the kefir by fermenting milk in tulum made from fur of animals (Yaygin 1995).

Kefir differs from other fermented dairy products, because "kefir grains" have a complex microflora include heterofermentative and homofermentative lactic acid bacteria, acetic acid bacteria and yeasts (Marshall et al. 1984; Toba et al. 1987; Piodux et al. 1990). Manufacturing a quality kefir beverage with stable starter culture is difficult because type of microorganisms and their ratio differ from the origin of "kefir grains" and there are different opinions for the type of microorganisms in "kefir grains" (Yaygin 1995).

The main producer of kefiran polymer in kefir grains is *Lactobacillus kefiranofaciens* and other species of lactobacilli (Frengova et al. 2002; Irigoyen et al. 2005). Lactic acid bacteria present in kefir grains or kefir products were isolated and identified, including *Lactobacillus acidophilus* (Angulo et al. 1993), *Lactobacillus brevis* (Simova et al. 2002), *Lactobacillus paracasei* subsp. *paracasei* (Simova et al. 2002), *Lactobacillus delbrueckii* (Simova et al. 2002; Witthuhn et al. 2004), *Lactobacillus helveticus* (Angulo et al. 1993; Lin et al. 1999; Simova et al. 2002), *Lactobacillus kefiri* (Angulo et al. 1993; Takizawa et al. 1998; Garrote et al. 2001), *Lactobacillus kefiranofaciens* (Takizawa et al. 1998), *Lactobacillus plantarum* (Garrote et al. 2001), *Leuconostoc mesenteroides* (Lin et al. 1999; Garrote et al. 2001; Witthuhn et al. 2004), *Lactococcus lactis* (Garrote et al. 2001; Simova et al. 2002; Witthuhn et al. 2004), *Streptococcus thermophilus* (Simova et al. 2002).

3.6.4 Shalgam

Shalgam, a traditional Turkish lactic acid fermented beverage, is mainly produced in some provinces of Southern Turkey (Tangüler and Erten 2012a). The raw materials used for shalgam production are black carrot, turnip, rock-salt, sourdough, bulgur flour and drinkable water. It is a red coloured, cloudy and sour non-alcoholic drink (Erten et al. 2008).

Lactic acid bacteria are the main fermentation agents of shalgam and they are responsible for the acidification process by converting sugars into mainly lactic acid and other end compounds which give the typical taste and flavour to the shalgam (Erten et al. 2008).

There is limited information about the microflora of shalgam and its microbiology is complex. *Lactobacillus plantarum*, the predominant lactic acid bacteria, *Lactobacillus brevis*, *Lactobacillus delbrueckii* subsp. *delbrueckii* and *Lactobacillus fermentum* were found in shalgam samples (Erginkaya and Hammes 1992; Arici 2004; Tangüler and Erten 2012a, b).

References

Abbas CA (2006) Production of antioxidants, aromas, colours, flavours, and vitamins by yeast. In: Querol A, Fleet GH (eds) Yeasts in food and beverages, vol 2, The yeast handbook. Springer, Heidelberg, pp 285–334

Abdelgadir WS, Hamad SH, Møller PL, Jakobsen M (2001) Characterisation of the dominant microbiota of Sudanese fermented milk Rob. Int Dairy J 11:63–70

Akbaria H, Karimia K, Lundin M, Taherzadeh M (2012) Optimization of baker's yeast drying in industrial continuous fluidized bed dryer. Food Bioprod Process 90:52–57

Alves M, Gonçalves T, Quintas C (2012) Microbial quality and yeast population dynamics in cracked green table olives' fermentations. Food Control 23:363–368

Amiot MJ, Tacchini M, Fleuriet A, Macheix JJ (1990) Le processus technologique de désamérisation des olives: caractérisation des fruits avant et pendant le traitement alcalin. Sci Aliment 10:619–631

Amorim HV, Basso LC, Lopes ML (2009) Sugar cane juice and molasses, beet molasses and sweet sorghum: composition and usage. In: Ingledew WM, Kelsall DR, Austin GD, Kluhspies C (eds) The alcohol text book, 5th edn. Nottingham University Press, Nottingham, pp 39–46

Angulo L, Lopez E, Lema C (1993) Microflora present in kefir grains of the Galician region (North-West of Spain). J Dairy Res 60:263–267

Arici M (2004) Microbiological and chemical properties of a drink called shalgam. Ernahrungs-Umschau 51(1):10–11

Arici M, Daglioglu O (2002) Boza: a lactic acid fermented cereal beverage as a traditional Turkish food. Food Res Int 18:39–48

Arroyo-López FN, Querol A, Bautista-Gallego J, Garrido-Fernández A (2008) Role of yeasts in table olive production. Int J Food Microbiol 128:189–196

Arroyo-López FN, Romero-Gil V, Bautista-Gallego J, Rodríguez-Gómez F, Jiménez-Díaz R, García-García P, Querol A, Garrido-Fernández A (2012) Yeasts in table olive processing: desirable or spoilage microorganisms. Int J Food Microbiol 160:42–49

Aukrust TW, Blom A, Sandtorv BF, Slinde E (1994) Interactions between starter culture and raw material in lactic acid fermentation of sliced carrot. LWT Food Sci Technol 27:337–341

Axelsson L (1998) Lactic acid bacteria: Classification and physiology. In: Salminen S, Von Wright A, (eds) lactic acid bacteria- Microbiology and functional aspects. Markel Dekker, New York, pp 1–72

Bai FW, Anderson WA, Moo-Yong M (2008) Ethanol fermentation technologies from sugar and starch feedstocks. Biotechnol Adv 26:89–105

Bamforth C (2003) Yeast and fermentation. In: Bamforth C (ed) Beer tap into the art and science of brewing. Oxford University Press, New York, pp 144–159

Berry DR (1989) Manipulation of flavor production by yeast: physiological and genetic approaches. In: Piggott JR, Paterson A (eds) Distilled beverage flavour. Harwood, Chichester, UK, pp 299–307

Berry DR (1995) Alcoholic beverage fermentations. In: Lea AGH, Piggott JR (eds) Fermented beverage production. Blackie, Glasgow, UK, pp 32–44

Bintsis T, Robinson RK (2004) A study of the adjunct cultures on the aroma compounds of feta-type cheese. Food Chem 88(3):435–441

Birch AN, Petersen MA, Hansen AS (2013) The aroma profile of wheat bread crumb influenced by yeast concentration and fermentation temperature. LWT Food Sci Technol 50:480–488

Bisson LF (1999) Stuck and sluggish fermentations. Am J Enol Viticult 50:107–119

Blanco P, Mirás-Avalos JM, Orriols I (2012) Effect of must characteristics on the diversity of *Saccharomyces* strains and their prevalence in spontaneous fermentations. J Appl Microbiol 112(5):936–944

Blandino A, Al-Aseeri ME, Pandiella SS, Cantero D, Webb C (2003) Cereal-based fermented foods and beverages. Food Res Int 36:527–543

Boekhout T, Robert V (2003) Yeasts in foods. Beneficial and detrimental aspects. CRC Press, Boca Raton, p 488
Bourdichon F, Casaregola S, Farrokh C, Frisvad JC, Gerds ML, Hammes WP, Harnett J, Huys G, Laulund S, Ouwehand A, Powell IB, Prajapati JB, Seto Y, Ter Schure E, Van Boven A, Vankerckhoven V, Zgoda A, Tuijtelaars S, Hansen EB (2012) Food fermentations: microorganisms with technological beneficial use. Int J Food Microbiol 154:87–97
Briggs DE, Boulton CA, Brookes PA, Stevens R (2004) Fermentation technologies. In: Briggs DE, Boulton CA, Brookes PA, Stevens R (eds) Brewing science and practice. Woodhead; CRC Press, England, pp 509–542
Cai Y, Okada H, Mori H, Benno Y, Nakase T (1999) *Lactobacillus paralimentarius* sp. nov., isolated from sourdough. Int J Syst Bacteriol 49:1451–1455
Campbell I (2003) Microbiological aspects of brewing. In: Priest FG, Campbell I (eds) Brewing microbiology. Springer, New York, pp 1–17
Cang Y, Lu L, Wang G (2002) Antioxidant role of carotenoids in *Rhodotorula* sp. and screening of carotenoid hyperproducing yeasts. Junwu Xitong 21(2):84–91
Cantor MD, Van Den Temple T, Hansen TK, Ardö Y (2004) Blue cheese. In: Fox PF, McSweeney P, Cogan T, Guinee T (eds) Cheese: chemistry, physics and microbiology, major cheese groups, vol 2, 3rd edn. Chapman and Hall, London, pp 175–198
Caplice E, Fitzgerald GF (1999) Food fermentations: role of microorganisms in food production and preservation. Int J Food Microbiol 50:131–149
Chandrani W, Jayathilake AN (2000) Characteristics of two yeast strain (*Candida tropicalis*) isolated from Caryora urens (Khitul) toddy for single cell protein production. Natl J Sci Found (Sri Lanka) 28:79–86
Chavan RS, Chavan SR (2011) Sourdough technology: a traditional way for wholesome foods: a review. Compr Rev Food Sci Food Saf 10:169–182
Chavan RS, Jana A (2008) Frozen dough for bread making: a review. Int J Food Sci Technol Nutr 2:9–27
Chaves ACSD, Kleerebezem M, Lerayer ALS, Hugenholtz J (1999) Improved yoghurt flavour by metabolic engineering of *Streptococcus thermophilus*. Sixth symposium on lactic acid bacteria, book of abstracts, G2
Cheng Q, Rouviere PE, Tao L (2004) Mutant carotenoid-producing microorganisms and their use in carotenoid fermentation. PCT Int Appl WO 2004056974
Chi ZM, Zhang T, Cao TS, Liu XY, Cui W, Zhao CH (2011) Biotechnological potential of inulin for bioprocesses. Bioresour Technol 102:4295–4303
Ciani M, Comitini F, Mannazzu I, Domizio P (2010) Controlled mixed culture fermentation: a new perspective on the use of non-*Saccharomyces* yeasts in winemaking. FEMS Yeast Res 10:123–133
Collar C (1996) Biochemical and technological assessment of the metabolism of pure and mixed cultures of yeast and lactic acid bacteria in breadmaking applications. Food Sci Technol Int 2:349–367
Corsetti A, Lavermicocca P, Morea M, Baruzzi F, Tosti N, Gobbetti M (2001) Phenotypic and molecular identification and clustering of lactic acid bacteria and yeasts from wheat (species *Triticum durum* and *Triticum aestivum*) sourdoughs of Southern Italy. Int J Food Microbiol 64:95–104
Corsetti A, De Angelis M, Dellaglio F, Paparella A, Fox PF, Settanni L, Gobbetti M (2003) Characterization of sourdough lactic acid bacteria based on genotypic and cell-wall protein analysis. J Appl Microbiol 94:641–654
Corsetti A, Settanni L, Van Sinderen D (2004) Characterization of bacteriocin-like inhibitory substances (BLIS) from sourdough lactic acid bacteria and evaluation of their in vitro and in situ activity. J Appl Microbiol 96:521–534
Cortes S, Blanco P (2011) Yeast strain effect on the concentration of major volatile compounds and sensory profile of wines from *Vitis vinifera* var. *treixadura*. World J Microbiol Biotechnol 27:925–932

Curtin C, Chambers P, Pretorius S (2011) Wine fermentation. In: Wheeler MB, Hoover DG, Heldman DR (eds) Encyclopedia of biotechnology in agriculture and food. Taylor & Francis, Boca Raton, pp 689–694
Dağlioğlu O, Arici M, Konyali M, Gumus T (2002) Effects of tarhana fermentation and drying methods on the fate of *Escherichia coli* O157:H7 and *Staphylacoccus aureus*. Eur Food Res Technol 215:515–519
De Benedictis M, Bleve G, Grieco F, Tristezza M, Tufariello M, Grieco F (2011) An optimized procedure for the enological selection of non-*Saccharomyces* starter cultures. A Van Leeuw J Microb 99(2):189–200
De Vuyst L, Vancanneyt M (2007) Biodiversity and identification of sourdough lactic acid bacteria. Food Microbiol 24:120–127
Deak T (2004) Spoilage yeasts. In: Steele R (ed) Understanding and measuring the shelf-life of food. Woodhead, Boca Raton, pp 91–110
Deak T (2008) Handbook of food spoilage yeasts. CRC Press, Boca Raton, 325 pp
Decock P, Cappelle S (2005) Bread technology and sourdough technology. Trends Food Sci Technol 16:113–120
Del Rio L, Corpas J, Sandalio L, Palma J, Barroso J (2003) Plant peroxisomes, reactive oxygen metabolism and nitric oxide. IUBMB Life 55(2):71–81
Delgado A, Brito D, Peres C, Noé-Arroyo F, Garrido-Fernández A (2005) Bacteriocin production by *Lactobacillus pentosus* B96 can be expressed as a function of temperature and NaCl concentration. Food Microbiol 22:521–528
Demain AL, Phaff HJ, Kurtzman CP (1998) The industrial and agricultural significance of yeasts. In: Kurtzman CP, Fell JW (eds) The yeasts – a taxonomic study, 4th edn. Elsevier, Amsterdam, pp 13–19
Demirbas A (2007) Producing and using bioethanol as an automotive fuel. Energ Sour B 2:391–401
Dickinson JR (2003) The formation of higher alcohols. In: Smart KA (ed) Brewing yeast fermentation performance, 2nd edn. Blackwell, Oxford, UK, pp 196–205
Dien BS, Cotta MA, Jeffries TW (2003) Bacteria engineered for fuel ethanol production: current status. Appl Microbiol Biotechnol 63:258–266
Dillon VM, Board RG (1991) Yeast associated with red meats. J Appl Microbiol 71:93–108
Durán Quintana MC, García García P, Garrido-Fernández A (1999) Establishment of conditions for green olive fermentation at low temperature. Int J Food Microbiol 51:133–143
Ehrmann MA, Muller MR, Vogel RF (2003) Molecular analysis of sourdough reveals *Lactobacillus mindensis* sp. *nov*. Int J Syst Evol Microbiol 53:7–13
Erginkaya Z, Hammes WP (1992) Şalgam suyu fermantasyonu sırasında mikroorganizmalarin gelişimi ve izole edilen laktik asit bakterilerinin tanımlanmaları üzerine bir araştırma. Gida 17(5):311–314
Erten H, Tangüler H, Canbaş A (2008) A traditional turkish lactic acid fermented beverage: Shalgam (Salgam). Food Rev Int 24:352–359
Farnworth ER (1999) Kefir: from folklore to regulatory approval. J Nutraceut Funct Med Foods 57–68
Farnworth ER (2006) Kefir a complex probiotic. Food Sci Technol Bull Funct Foods 2:1–17. http://www.foodsciencecentral.com/fsc/bulletin-ff-freeS
Farnworth ER, Mainville I (2003) Kefir: a fermented milk product. In: Farnworth ER (ed) Handbook of fermented functional foods. CRC Press, Boca Raton, pp 77–112
Fernandez Diez MJ (1983) Olives. In: Rehm HJ, Reed G (eds) Biotechnology: food and feed production with microorganisms. Verlag Chemie, Florida, pp 379–397
Ferreira AS, Viljoen BC (2003) Yeasts as adjunct starters in maturated Chedar cheese. Int J Food Microbiol 86:131–140
Ferreira IMPLVO, Pinho O, Vieira E, Tavarela JG (2010) Brewer's Saccharomyces yeast biomass: characteristics and potential applications. Trends Food Sci Technol 21:77–84
Fleet GH (1990) Yeasts in dairy products. J Appl Bacteriol 68:199–211
Fleet GH (1998) The microbiology of alcoholic beverages. In: Fleet GH, Wood BJB (eds) Microbiology of fermented foods. Thomson Science, London, pp 217–255

Fleet GH (2003a) Yeast interactions and wine flavour. Int J Food Microbiol 86:11–22
Fleet GH (2003b) Yeasts in fruit and fruit products. In: Boekhout T, Robert V (eds) Yeasts in foods: beneficial and detrimental aspects. Behr's Verlag, Hamburg, pp 267–287
Fleet GH (2006) The commercial and community significance of yeasts in food and beverage production. In: Queol A, Fleet GH (eds) Yeasts in food and beverages. Springer, Heidelberg, pp 1–13
Fleet GH, Balia R (2006) The public health and probiotic significance of yeasts in foods and beverages. In: Queol A, Fleet GH (eds) Yeasts in food and beverages. Springer, Heidelberg, pp 381–397
Fleet GH, Heard GM (1993) Yeasts – growth during fermentation. In: Fleet GM (ed) Wine microbiology and biotechnology. Harwood Academic, London, pp 27–54
Fleming HP, Da Eschel MA, Mcfeeters RF, Pierson MD (1989) Butyric acid spoilage of fermented cucumbers. J Food Sci 54(3):636–639
Fleming HP, Humpries EG, Fasina OO, McFeeters RF, Thompson RL, Breidt J (2002) Bag in box technology: pilot system for process-ready, fermented cucumbers. Pickle Pak Sci 8:1–8
Font De Valdez G, De Giori GS, Garro M, Mozzi F, Oliver G (1990) Lactic acid bacteria from naturally fermented vegetables. Microbiol Alim Nutr 8:175–179
Franco W, Pérez-Díaz IM (2012) Development of a model system for the study of spoilage associated secondary cucumber fermentation during long term storage. J Food Sci 78(2):264–269
Frengova GI, Simova ED, Beshkova DM, Simov ZI (2002) Exopolysaccharides produced by lactic acid bacteria of kefir grains. Z Naturforsch 57(9–10):805
Frengova G, Simova E, Beshkova D (2003) Carotenoid production by lactoso-negative yeasts co-cultivated with lactic acid bacteria in whey ultrafiltrate. J Biosci 58(7/8):562–567
Frengova G, Simova E, Beshkova D (2004) Use of whey ultrafiltrate as a substrate for production of carotenoids by the yeast *Rhodotorula rubra*. Appl Biochem Biotechnol 112(3):133–141
Frölich-Wyder M-T (2003) Yeasts in dairy products. In: Boekhout T, Robert V (eds) Yeasts in food beneficial and detrimental aspects. Behr's Verlag, Hamburg, pp 209–237
Gadaga TH, Mutukumira AN, Narvhus JA (2001) Int J Food Microbiol 70:11–19
Ganga MA, Martinez C (2004) Effect of wine yeast monoculture practice of biodiversity of non-Saccharomyces yeasts. J Appl Microbiol 96:76–83
Garafolo A (1992) The ester synthesizing activity of yeast under different nutrient conditions. Riv Vitic Enol 45(3):41–58
Gardini F, Suzzi G, Lombardi A, Galgano F, Crudeleb MA, Andrighetto C, Schirone M, Tofalo R (2001) A survey of yeasts in traditional sausages of Southern Italy. FEMS Yeast Res 1:161–167
Garrido-Fernandez A, Garcia PG, Brenes MB (1995) Olive fermentations. In: Rehm HJ, Reed G, Puhler A, Stadler P (eds) Biotechnology: enzymes, biomass, food and feed. VCH Verlag, Weinheim, pp 593–627
Garrido-Fernández A, Fernández Díaz MJ, Adams RM (1997) Table olives: production and processing. Chapman & Hall, London
Garrote GL, Abraham AG, De Antoni GL (2001) Chemical and microbiological characterization of kefir grains. J Dairy Res 68:639–652
Ghabbour N, Lamzira Z, Thonart P, Cidalia P, Markaouid M, Asehraou A (2011) Selection of oleuropein-degrading lactic acid bacteria strains isolated from fermenting Moroccan green olives. Grasas Aceites 62:84–89
Giannou V, Kessoglou V, Tzia C (2003) Quality and safety characteristics of bread made from frozen dough. Trends Food Sci Technol 14:99–108
Gibson BR, Boulton CA, Box WG, Graham NS, Lawrence SJ, Linforth RST, Smart KA (2008) Carbohydrate utilization and the lager yeast transcriptome during brewery fermentation. Yeast 25:549–562
Gobbetti M (1998) Interactions between lactic acid bacteria and yeasts in sourdoughs. Trends Food Sci Technol 9:267–274
Gobbetti M, Corsetti A, Rossi J, La Rosa F, De Vincenzi S (1994) Identification and clustering of lactic acid bacteria and yeasts from wheat sourdoughs of central Italy. Ital J Food Sci 1:85–94
Gotcheva V, Pandiella SS, Angelov A, Roshkova Z, Webb C (2000) Microflora identification of the Bulgarian cereal-based fermented beverage boza. Process Biochem 36:127–130

Guerzoni ME, Lanciotti R, Vannini L, Galgano F, Favati F, Gardini F, Suzzi G (2001) Variability of the lipolytic activity in *Yarrowia lipolytica* and its dependence on environmental conditions. Int J Food Microbiol 69:79–89

Halasz A, Laszity R (1991) Use of yeast biomass in food production. CRC Press, Boca Raton, p 319

Hammes WP, Gaenzle MG (1998) Sourdough breads and related products. In: Wood BJB (ed) Microbiology of fermented foods, 2nd edn. Blackie Academic and Professional, London, pp 199–216

Hammes WP, Hertel C (1998) New developments in meat starter cultures. Meat Sci 49(1): 125–138

Hammes WP, Stolz P, Gaenzle M (1996) Metabolism of lactobacilli in traditional sourdoughs. Adv Food Sci 18(5/6):176–184

Hammes WP, Brandt MJ, Francis KL, Rosenheim J, Seitter MFH, Vogelmann SA (2005) Microbial ecology of cereal fermentations. Trends Food Sci Technol 16:4–11

Harris LJ (1998) The microbiology of vegetable fermentations. In: Wood BJB (ed) Microbiology of ferment food. Academic Professional, London, pp 45–72

Harrison JS (1970) Miscellaneous products from yeast. In: Rose AH, Harrison JS (eds) The yeasts, vol III. Academic, London, pp 529–545

Hazelwood LA, Daran JM, van Maris AJ, Pronk JT, Dickinson JR (2008) The Ehrlich pathway for fusel alcohol production: a century of research on *Saccharomyces cerevisiae* metabolism. Appl Environ Microbiol 74:2259–2266

Henick-Kling T (2002) Malolactic fermantation. In: Fleet GH (ed) Wine microbiology and biotechnology. Taylor and Francis, London, pp 289–326

Heredia N, Wesley I, Garcia S (2009) Microbiologically safe foods. Wiley, Canada, p 667

Hernández A, Martín A, Aranda E, Pérez-Nevado F, Córdoba MG (2007) Identification and characterization of yeast isolated from the elaboration of seasoned green table olives. Food Microbiol 24(4):346–351

Hounhouigan DJ, Nout MJR, Nago CM, Houben JH, Rombouts FM (1993) Composition and microbiological and physical attributes of mawè, a fermented maize dough from Benin. Int J Food Sci Tech 28:513–517

Huber JT (1997) Probiotics in cattle. In: Fuller R (ed) Probiotics 2: applications and practical aspects. Chapman and Hall, London, pp 162–186

Hugenholtz J (1993) Citrate metabolism in lactic acid bacteria. FEMS Microbiol Rev 12:165–178

Hugenholtz J, Starrenburg M, Boels I, Sybesma W, Chaves AC, Mertens A, Kleerebezem M (2000) Metabolic engineering of lactic acid bacteria for the improvement of fermented dairy products. In: Proceedings of the international meeting of biothermokinetics. Stellenbosch University Press, pp 285–290

Hurtado A, Reguan TC, Bordons A, Rozès N (2009) Influence of fruit ripeness and salt concentration on the microbial processing of Arbequina table olives. Food Microbiol 26:827–833

Hurtado A, Reguant C, Bordons A, Rozès N (2010) Evaluation of a single and combined inoculation of a *Lactobacillus pentosus* starter for processing cv. *Arbequina* natural green olives. Food Microbiol 27:731–740

Hurtado A, Reguant C, Bordons A, Rozes N (2012) Lactic acid bacteria from fermented table olives. Food Microbiol 31:1–8

Hutkins RW (2006) Microbiology and technology of fermented foods. Blackwell, Ames, p 473

Ibanoglu S, Ibanoglu E (1998) Rheological characterization of some traditional Turkish soups. J Food Eng 35(2):251–256

Ibanoglu S, Ibanoglu E (1999) Rheological properties of cooked Tarhana, a cereal based soup. Food Res Int 32:29–33

Ic E, Ozcelik F (1995) Hiyar turşusu fermentasyonunda görülen mikroorganizmalar. Gida (Food) 20(3):173–178

Ic E, Ozcelik F (1999) Hiyar turşularinin düşük tuzlu salamurada fermentasyonu üzerine bir araştirma. Gida (Food) 24(2):77–87

Irigoyen A, Arana I, Castiella M, Torre P, Ibanez FC (2005) Microbiological, physicochemical and sensory characteristics of Kefir during storage. Food Chem 90:613–620
Jacques N, Caserogola S (2008) Safety assessment of dairy microorganisms: the hemiascomycetous yeasts. Int J Food Microbiol 126:321–326
Jakobsen M, Narwnjs J (1996) Yeasts and their possible beneficial and negative effects on the quality of dairy products. Int Dairy J 6:755–768
Jakobsen M, Larsen MD, Jespersen L (2002) Production of bread, cheese and meat. In: Osiewacz HD (ed) A comprehensive freatise on fungi as experimental systems and applied research, The Mycota, industrial applications. Springer, Berlin, pp 3–22
Jentsch M (2007) Top-fermented beer specialities in focus. Brauwelt Int 5:332–334
Johanningsmeier SD, Franco W, Pérez-díaz IM, Mcfeeters RF (2012) Influence of sodium chloride, pH and lactic acid bacteria on anaerobic lactic acid utilization during fermented cucumber spoilage. J Food Sci 77(7):397–404
Johansson ML, Quednau M, Molin G, Ahrné S (1995) Randomly amplified polymorphic DNA (RAPD) for rapid typing of *Lactobacillus plantarum* strains. Lett Appl Microbiol 21:155–159
Joseph R (1999) Yeasts: production and commercial uses. In: Robinson RK, Patel P, Batt CA (eds) Enc food microbiol. Academic, London, pp 2335–2341
Kalac P, Špička J, Křížek M, Steidlová S, Pelikánová T (1999) Concentration of seven biogenic amines in sauerkraut. Food Chem 67:275–280
Katahira S, Mizuike A, Fukuda H, Kondo A (2006) Ethanol fermentation from lignocellulosic hydrolysate by a recombinant xylose- and cellooligosaccharide-assimilating yeast strain. Appl Microbiol Biotechnol 72:1136–1143
Kesenkaş H, Akbulut N (2006) Destek Kültür Olarak Kullanılan Bazı Mayaların Beyaz Peynir Aroması Üzerine Etkileri. Ege Üniv Ziraat Fak Derg 43(2):73–84
Keshwani DR, Cheng JJ (2009) Switchgrass for bioethanol and other value-added applications: a review. Bioresour Technol 100:1515–1523
Kim JH, Breidt F (2007) Development of preservation prediction chart for long term storage of fermented cucumber. J Life Sci 17(2):1616–1621
Kollar RE, Sturdik E, Sajbidor J (1992) Complete fractionation of *Saccharomyces cerevisiae* biomass. Food Biotechnol 6:225–237
Krieger S (1993). The use of active dry malolactic starter cultures. In: Proceedings of ASVO seminar, 30 Jul 1992, McLaren Vale, pp 56–62
Krieger SA, Hammes WP, Henick-Kling T (1990) Management of malolactic fermentation using starter cultures. Vineyard Winery Manage (Nov/Dec):45–50
Kröckel L (2013) The role of lactic acid bacteria in safety and flavour development of meat and meat products. In: Kongo M (ed) Lactic acid bacteria- R&D for food, health and livestock purposes. InTech, Croatia, p 658
Lambrechts MG, Pretorius IS (2000) Yeast and its importance to wine aroma—a review. S Afr J Enol Vitic 21:97–129
Landete JM, Curiel JA, Rodríguez H, De Las RB, Muñoz R (2008) Study of the inhibitory activity of phenolic compounds found in olive products and their degradation by *Lactobacillus plantarum* strains. Food Chem 107:320–326
Lavermicocca P, Valerio F, Evidente A, Lazzaroni S, Corsetti A, Gobbetti M (2000) Purification and characterization of novel antifungal compounds from the sourdough *Lactobacillus plantarum* strain 21B. Appl Environ Microbiol 66:4084–4090
Lentini A, Rogers P, Higgins V, Chandler DI, Stanley M, Chambers P (2003) The impact of ethanol stress on yeast physiology. In: Smart K (ed) Brewing yeast fermentation performance. Blackwell, Oxford, pp 25–38
Lin CW, Chen HL, Liu JR (1999) Identification and characterisation of lactic acid bacteria and yeasts isolated from kefir grains in Taiwan. Aust J Dairy Technol 54:14–18
Lodolo EJ, Kock JLK, Axcell BC, Brooks M (2008) The yeast *Saccharomyces cerevisiae*- the main character in beer brewing. FEMS Yeast Res 8:1018–1036

Lopez HW, Krespine V, Guy C, Messager A, Demigne C, Remesy C (2001) Prolonged fermentation of whole wheat sourdough reduces phytate level and increases soluble magnesium. J Agric Food Chem 49:2657–2662
Marquina D, Toufani S, Llorente P, Santos A, Peinado JM (1997) Killer activity in yeast isolated from olive brines. Adv Food Sci 19:41–46
Marshall VM, Cole WN, Brooker BE (1984) Observations on the structure of kefir grains and the distribution of the microflora. J Appl Bacteriol 57(3):491–497
Martinez-Anaya MA (2003) Associations and interactions of microorganisms in dough fermentations: effects on dough and bread characteristics. In: Lorenz K, Kulp K (eds) Handbook of dough fermentations. Marcel Dekker, New York, pp 63–95
McKay M, Buglass AJ, Lee CG (2011) Alcoholic fermentation. In: Buglass AJ (ed) Handbook of alcoholic beverages: technical, analytical and nutritional aspects. Wiley, Chichester, pp 72–95
McSweeney PLH, Sousa MJ (2000) Biochemical pathways for the production of flavour compounds in cheeses during ripening: a review. Lait 80:293–324
Mecteau MR, Arul J, Tweddell RJ (2002) Effect of organic and inorganic salts on the growth and development of *Fusarium sambucinum*, a causal agent of potato dry rot. Mycol Res 106:688–696
Meignen B, Onno B, Gelinas P, Infantes M, Guilois S, Cahagnier B (2001) Optimization of sourdough fermentation with *Lactobacillus brevis* and baker's yeast. Food Microbiol 18:239–245
Meroth CB, Walter J, Hertel C, Brandt MJ, Hammes WP (2003) Monitoring the bacterial population dynamics in sourdough fermentation processes by using PCR-denaturing gradient gel electrophoresis. Appl Environ Microbiol 69:475–482
Moona SK, Kimb SW, Choi GW (2012) Simultaneous saccharification and continuous fermentation of sludge-containing mash for bioethanol production by *Saccharomyces cerevisiae* CHFY0321. J Biotechnol 157:584–589
Müller MRA, Ehrmann MA, Vogel RF (2000) *Lactobacillus frumenti* sp. *nov.*, a new lactic acid bacterium isolated from rye-bran fermentations with a long fermentation period. Int J Syst Evol Microbiol 50:2127–2133
Murphy MG, O'Connor L, Walsh D, Condon S (1985) Oxygen dependent lactate utilization by *Lactobacillus plantarum*. Arch Microbiol 121:75–79
Mussatto SI, Dragone G, Guimaraes PMR, Silva JPA, Carneiro LM, Roberto IC, Vicente A, Domingues L, Teixeira JA (2010) Technological trends, global market, and challenges of bio-ethanol production. Biotechnol Adv 28:817–830
Nagodawithana T (1992) Yeast derived flavors and flavor enhancers and their probable mode of action. Food Technol 46:138–144
Nagodawithana T (1994) Savory flavours. In: Gabelman A (ed) Bioprocess production of flavour fragrance and color ingredients. Wiley, New York, pp 135–168
Nakao Y, Kanamori T, Itoh T, Kodama Y, Rainieri S, Nakamura N, Shimonaga T, Hattori M, Ashikari T (2009) Genome sequence of the lager brewing yeast, an interspecies hybrid. DNA Res 16:115–129
Nasseri AT, Rasaul S, Morowvat MH, Ghasemi Y (2011) Single cell protein: production and process. Am J Food Technol 6(2):103–116
Navarrete-Bolanos JL (2012) Improving traditional fermented beverages: how to evolve from spontaneous to directed fermentation. Eng Life Sci 12(4):410–418
Olesen PT, Stahnke LH (2000) The influence of *Debaryomyces hansenii* and *Candida utilis* on the aroma formation in garlic spiced fermented sausages and model minces. Meat Sci 56:357–368
Panagou EZ, Tassou CC (2006) Changes in volatile compounds and related biochemical profile during controlled fermentation of cv. *Conservolea* green olives. Food Microbiol 23:738–746
Panagou EZ, Schillinger U, Franz C, Nychas GJE (2008) Microbiological and biochemical profile of cv. *Conservolea* naturally black olives during controlled fermentation with selected strains of lactic acid bacteria. Food Microbiol 25:348–358
Peppler HJ (1970) Food yeasts. In: Rose AH, Harrison JS (eds) The yeasts, vol 3, Yeast technology. Academic, London, pp 421–462
Peppler HJ (1979) Production of yeasts and yeast products. In: Peppler HJ, Perlman D (eds) Microbial technology, vol I, 2nd edn. Academic, New York, pp 157–185

Peppler HJ (1982) Yeast extracts. Econ Microbiol 7:293–312

Pinho O, Ferreira IMPLVO, Santos LHMLM (2006) Method optimization by solid-phase microextraction in combination with gas chromatography with mass spectrometry for analysis of beer volatile fraction. J Chromatogr A 1121:145–153

Piodux M, Marshall VM, Zanoni P, Brooker B (1990) Lactobacilli isolated from sugary kefir grains capable of polysaccharide production and minicell formation. J Appl Bacteriol 69:311–320

Pometto A, Shetty K, Paliyath G, Robert E (2006) Food biotechnology, 2nd edn, Food science and technology. CRC Press, Boca Raton, p 1903

Pozo MA, Guichard E, Cayot N (2006) Flavor control in baked cereal products. Food Rev Int 22(4):335–379

Pretorius IS (2000) Tailoring wine yeast for the new millennium: novel approaches to the ancient art of winemaking. Yeast 16:675–729

Prillinger H, Molnar O, Eliskases-Lechner F, Lopandic K (1999) Phenotypic and genotypic identification of yeasts from cheese. A Van Leeuw J Microb 75:267–283

Punja ZK, Utkhede RS (2003) Using fungi and yeasts to manage vegetable crop diseases. Trends Biotechnol 21:400–407

Randez-Gil F, Sanz P, Prieto JA (1999) Engineering baker's yeast: room for improvement. Trends Biotechnol 17:237–244

Reed G (1981) Use of microbial cultures: yeast products. Food Technol 35(1):89–94

Reed G, Nagodawithana TW (1991) Yeast technology, 2nd edn. Avi Publishing, New York, p 765

Rodriguez De Le Borbolla Y, Alcala JM, Rejano Navarro L (1979) Sobre la preparacion de las aceitunas estilo sevillano la fermentacion I. Grasas Aceites 30:175–185

Rodriguez De Le Borbolla Y, Alcala JM, Rejano Navarro L (1981) Sobre la preparacion de las aceitunas estilo sevillano la fermentacion II. Grasas Aceites 32:103–113

Rodriguez H, Curiel JA, Landet EJM, Rivas DLB, López DF, Gómez- Cordovés C, Mancheño JM, Muñoz R (2009) Food phenolics and lactic acid bacteria. Int J Food Microbiol 132:79–90

Rodriguez-Gómez F, Arroyo-López FN, López-López A, Bautista-Gallego J, Garrido-Fernández A (2010) Lipolytic activity of the yeast species associated with the fermentation/storage phase of ripe olive processing. Food Microbiol 27:604–612

Rodríguez-Gómez F, Romero-Gil V, Bautista-Gallego J, Garrido-Fernández A, Arroyo-López FN (2012) Multivariate analysis to discriminate yeast strains with technological applications in table olive processing. World J Microbiol Biotechnol 28:1761–1770

Roig JM, Hernandez JM (1991) El uso de microorganismos iniciadores (starters) en la fermentacion de aceitunas. Olivae 37:20–28

Romano P, Capece A (2013) *Saccharomyces cerevisiae* as bakers' yeast. In: Heldman DR, Wheeler MB, Hoover DG (eds) Encyclopedia of biotechnology in agriculture and food. Taylor and Francis, New York, pp 1–4

Roostita R, Fleet GH (1996) The occurrence and growth of yeast in camembert and blueveined cheeses. Int J Food Microbiol 28:393–404

Ruiz-Barba JL, Jimenez-Diaz R (1995) Availability of essential B group vitamins to *Lactobacillus plantarum* in green olive fermentation brines. Appl Environ Microbiol 61:1294–1297

Ruiz-Barba JL, Jimenez-Diaz R, Piard JC (1991) Plasmid profiles and curing of plasmids in *Lactobacillus plantarum* strains isolated from green olive fermentations. J Appl Bacteriol 71(5):417–421

Sabatini N, Mucciarella MR, Marsilio V (2008) Volatile compounds in uninoculated and inoculated table olives with *Lactobacillus plantarum* (*Olea europaea* L., cv. Moresca and Kalamata). LWT- Food Sci Tech 41:2017–2022

Sagdic O, Kuscu A, Ozcan M, Ozcelik S (2002) Effects of Turkish spice extracts at variousconcentrations on the growth of *Escherichia coli* O157:H7. Food Microbiol 19:473–480

Samelis J, Sofos JN (2003) Yeasts in meat and meat products. In: Boekhout T, Robert V (eds) Yeasts in food: beneficial and detrimental aspects. Behr V, Hamburg, pp 239–266

Sanchez OJ, Cardona CA (2008) Trends in biotechnological production of fuel ethanol from different feedstocks. Bioresour Technol 99:5270–5295

Sánchez AH, Rejano L, Montaño A, De Castro A (2001) Utilization at high pH of starter cultures of lactobacilli for Spanish-style green olive fermentation. Int J Food Microbiol 67:115–122
Sanchez B, Reverol L, Galindo-Castro I, Bravo A, Rangel-Aldao R, Ramirez JL (2003) Brewer's yeast oxidoreductase with activity on Maillard reaction intermediates of beer. Tech Quart MBAA 40(3):204–212
Sánchez-Gómez AH, García P, Rejano L (2006) Trends in table olives production, elaboration of table olives. Grasas Aceites 57:86–94
Santos A, Marquina D, Leal JA, Peinado JM (2000) (1→6)-β-D-Glucan as cell wall receptor for *Pichia membranifaciens* killer toxin. Appl Environ Microb 66:1809–1813
Satyanarayana T, Kunze G (2009) Yeast biotechnology: diversity and applications. Springer, New Delhi, p 766
Selgas MD, Garcia ML (2007) Starter cultures: yeasts. In: Toldrá F (ed) Handbook of fermented meat and poultry. Blackwell, Ames, pp 159–170
Sengun IY, Nielsen DS, Karapinar M, Jakobsen M (2009) Identification of lactic acid bacteria isolated from Tarhana, a traditional Turkish fermented food. Int J Food Microbiol 135:105–111
Servili M, Settani L, Veneziani G, Esposto S, Massitti O, Taticchi A, Urbani S, Montedoro GF, Corsetti A (2006) The use of *Lactobacillus pentosus* 1MO to shorten the debittering process time of black table olives (Cv.Itrana and Leccino): a pilot-scale application. J Agri Food Chem 54:3869–3875
Settani L, Tangüler H, Moschetti G, Reale S, Gargano V, Erten H (2011) Evolution of fermenting microbiata in tarhana produced under controlled technological conditions. Food Microbiol 28:1367–1373
Settanni L, Corsetti A (2008) Application of bacteriocins in vegetable food biopreservation. Int J Food Microbiol 121:123–138
Simova E, Beshkova D, Angelov A, Hristozova TS, Frengova G, Spasov Z (2002) Lactic acid bacteria and yeast in kefir grains and kefir made from them. J Ind Microbiol Biotechnol 28:1–6
Simova ED, Frengova G, Beshkova DM (2003) Effect of aeration on the production of carotenoid pigments by *Rhodotorula rubra-Lactobacillus casei* subsp. Casei co-cultures in whey ultrafiltrate. J Biosci 58(3/4):225–229
Sommer R (1998) Yeast extract: production, properties and components. Food Aust 50:181–183
Spadaro D, Gullino ML (2004) State of the art and future prospects of the biological control of postharvest fruit diseases. Int J Food Microbiol 91:185–194
Spinnler HE, Berger C, Lapadatescu C, Bonnarme P (2001) Production of sulphur compounds by several yeasts of technological interest for cheese ripening. Int Dairy J 11:245–252
Squina FM, Yamashita F, Pereira JL, Mercadante AZ (2002) Production of carotenoids by *Rhodotorula rubra* and *R. glutinis* in culture medium supplemented with sugar cane juice. Food Biotechnol 16(3):227–235
Stam H, Hoogland M, Laane C (1998) Food flavours from yeast. In: Wood BJ (ed) Microbiology of fermented foods, vol 2, 2nd edn. Blackie, London, pp 505–542
Stewart GG, Russell I (1998) An introduction to brewing science and technology series III Brewer's yeast. Institute of Brewing, London, p 108
Strehaiano P, Ramon-Portugal F, Taillandier P (2006) Yeasts as biocatalysts. In: Queol A, Fleet GH (eds) Yeasts in food and beverages. Springer, Heidelberg, pp 243–285
Suomalainen H, Lehtonen M (1978) Yeast as a producer of aroma compounds. Dechema Monogr 82:207–220
Suomalainen H, Lehtonen M (1979) The production of aroma compounds by yeast. J Inst Brew 85(3):149–156
Swiegers JH, Bartowsky EJ, Henschke PA, Pretorius IS (2005) Yeast and bacterial modulation of wine aroma and flavour. Aust J Grape Wine R 11:139–173
Takizawa S, Kojima S, Tamura S, Fujinaga S, Benno Y, Nakase T (1998) The composition of the *Lactobacillus* flora in kefir grains. Syst Appl Microbiol 21:121–127

Tamang JP, Tamang B, Schillinger U, Franz CMAP, Gores M, Holzapfel WH (2005) Identification of predominant lactic acid bacteria isolated from traditionally fermented vegetable products of the Eastern Himalayas. Int J Food Microbiol 105:347–356

Tanguler H, Erten H (2008) Utilisation of spent brewer's yeast for yeast extract production by autolysis: the effect of temperature. Food Bioprod Process 86:317–321

Tangüler H, Erten H (2012a) Occurence and growth of lactic acid bacteria species during the fermentation of Shalgam (Şalgam), a traditional Turkish fermented beverage. LWT Food Sci Technol 46:36–41

Tangüler H, Erten H (2012b) Chemical and microbiological characteristics of Shalgam (Şalgam); a traditional Turkish fermented beverage. J Food Qual 35:298–306

Toba T, Arihara K, Adachi S (1987) Comparative study of polysaccharides from kefir grains, an encaosuied homofermentative *Lactobacillus* species and *Lactobacillus kefir*. Milchwissenschaft 42(9):565–568

Todorov SD, Dicks LMT (2006) Screening for bacteriocin-producing lactic acid bacteria from boza, a traditional cereal beverage from Bulgaria. Process Biochem 41:11–19

Tofalo R, Perpetuini G, Schirone M, Suzzi G, Corsetti A (2013) Yeast biota associated to naturally fermented table olives from different Italian cultivars. Int J Food Microbiol 161:203–208

Toldrá F (2002) Fermentation and starter cultures. In: Toldrá F (ed) Dry-cured meat products. Food & Nutrition Press, Trumbull, pp 89–112

Van den Tempel T, Jacobsen M (2000) *Debaryomyces hansenii* and *Yarrowia lipolytica* as potential starter cultures for production of blue cheese. Int Dairy J 10:263–270

Van Kranenburg R, Van Swam II, Marugg JD, Kleerebezem M, de Vos WM (1999) Exopolysaccharide biosynthesis in *Lactococcus lactis* NIZO B40: functional analysis of the glycosyltransferase genes involved in synthesis of the polysaccharide backbone. J Bacteriol 181:338–340

Viljoen BC (2001) The interaction between yeasts and bacteria in dairy environments. Int J Food Microbiol 69:37–44

Viljoen BC (2006) Yeast ecological interactions: yeast–yeast, yeast–bacteria, yeast fungi interactions and yeasts as biocontrol agents. In: Querol A, Fleet GH (eds) Yeasts in food and beverages. Springer, Berlin, pp 83–110

Vogelmann SA, Seitter M, Singer U, Brandt MJ, Hertel C (2009) Adaptability of lactic acid bacteria and yeasts to sourdough prepared from cereals, pseudocereals and cassava and use of competitive strains as starters. Int J Food Microbiol 130:205–212

Vrancken G, Vuyst LD, Van der Meulen R, Huys G, Vandamme P, Daniel HM (2010) Yeast species composition differs between artisan bakery and spontaneous laboratory sourdoughs. FEMS Yeast Res 10:471–481

Walker GM (1999) Yeast physiology and biotechnology. Wiley, London, p 350

Witthuhn RC, Schoeman T, Britz TJ (2004) Isolation and characterization of the microbial population of different South African kefir grains. Int J Dairy Technol 57:33–37

Wouters JTM, Ayad EH, Hugenholtz J, Smit G (2002) Microbes from raw milk for fermented dairy products. Int Dairy J 12:91–109

Wyman CE (2007) What is (and is not) vital to advancing cellulosic ethanol. Trends Biotechnol 25(4):153–157

Yaygin H (1995) Yoğurt, III. Süt ve Süt Ürünleri Sempozyumu 2-3 Haziran 1994 Istanbul, Milli Prodüktivite Merkezi Yayınları, No:548

Young LS, Cauvain SP (2007) Technology of breadmaking, 2nd edn. Springer, New York, p 398

Application of Membrane Technology in Food Processing

Saghir Ahmad and SK. Mushir Ahmed

1 Introduction

Membrane separation processes use semipermeable membrane of definite physical and chemical nature to separate molecules primarily on the basis of size, shape, and chemical composition. A membrane separation system separates an influent stream into two effluent streams known as permeate and the concentrate. The permeate is the portion of the fluid that has passed through the semipermeable membrane but the concentrate stream contains the constituents that have been rejected by the membrane.

Membrane separation processes have many advantages over conventional separation processes:

1. Environmentally safe.
2. Produces high quality products.
3. Greater flexibility in designing system.
4. Easy to operate.
5. Clean technology.

Membrane separation systems have been used extensively in the chemical process industry but their use in the food industry is growing day by day. In food industry, membrane separation process is mainly used for clarification of fruit juices by using microfiltration and ultrafiltration. Membrane separation also used for concentrating fruit juice and dairy products and also for purification of water.

S. Ahmad (✉) • SK. Mushir Ahmad
Faculty of Agricultural Sciences, Department of Post Harvest Engineering and Technology, Aligarh Muslim University, Aligarh, UP 202002, India
e-mail: alsaghir@rediffmail.com

A. Malik et al. (eds.), *Food Processing: Strategies for Quality Assessment*, Food Engineering Series, DOI 10.1007/978-1-4939-1378-7_15

2 Principle in Membrane Separation

A membrane process is a separation process that covers a broad range of problem from particles to molecules and a wide variety of membranes are available to design a process. Although the membranes may vary in material (organic vs. inorganic) and structure (porous vs. nonporous) the basic principle of membrane separation is the same. The separation process and mass transport through the membranes is a function of the membrane being used and the constituents being separated, and subsequently the theory used to describe the process and mechanisms. However, common for all systems is the principle illustrated in Fig. 1 where a membrane is considered a perm selective barrier, or interface between two phases, and the separation process takes place due to a specific driving force transporting a compound through the membrane from the one phase to the other.

The membrane separation process is defined by which compound is more readily transported (selectivity) through the membrane and by the flow of the specific compound (flux). Though this may occur by various mechanisms the performance and efficiency of the process is described by these parameters. The selectivity of a membrane is generally expressed as a retention factor (R) or by a separation factor (α). The definition of retention factor is given by Eq. (1):

$$R = \frac{Cfeed - Cpermeate}{Cfeed} = 1 - \frac{Cpermeate}{Cfeed} \tag{1}$$

Retention is often used for dilute aqueous solutions where the solvent is the water. As the retention is a dimensionless unit it expresses a percent that varies between 100 and 0 %, where $R = 100$ % means a complete retention of the solute, while $R = 0$ % means both the solute and the solvent pass equally through the membrane. The separation efficiency of gas mixtures and organic mixtures is generally expressed by the separation factor. For a binary mixture of compounds A and B, the

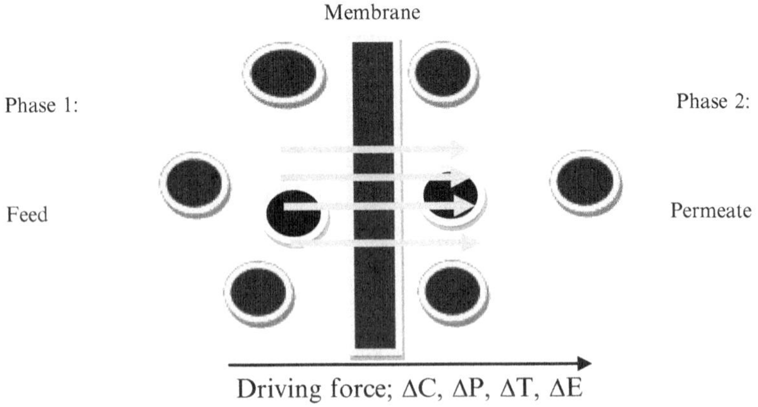

Fig. 1 Schematic of membrane separation process with different driving forces that are present

separation factor is expressed by the respective concentrations in the feed (xA, xB) and permeate (yA, yB) as expressed by Eq. (2);

$$\alpha A/B = \frac{yA/\,yB}{xA/\,xB} \quad (2)$$

The selectivity is chosen such that its value is greater than unity and it is expressed by which component passes through the membrane, i.e., for $\alpha A/B$ the permeation rate of compound A is greater than compound B. If $\alpha A/B = \alpha B/A = 1$, no separation is achieved (Leiknes, Ph.D Thesis).

3 Some Important Membrane Processes

There are several membrane separation processes which are of industrial importance as on today. They are briefly discussed below with respect to their principal characteristics.

3.1 Reverse Osmosis (RO)

The Reverse Osmosis (RO) is a process wherein a relatively pure solvent is separated from a salt solution by using a semipermeable membrane by the application of hydrostatic pressure. The hydrostatic pressure can vary from 2 MPa to around 6 MPa depending upon the salt content of the feed mixture. The solvent permeates preferentially through the membranes whereas the solutes, particularly electrolytes and low molecular weight nonelectrolytes are retained by the membranes. For effectively retaining microsolutes having molecular weight less than 300 or effective size less than 10 Å, reverse osmosis process is used. The process is used to produce relatively pure water or a concentrated solution of microsolutes from a given salt solution. A simple schematic of the process is given in Fig. 2. The most notable example of reverse osmosis process is the production of drinking water from naturally occurring saline waters.

3.2 Nanofiltration (NF)

The process of nanofiltration is slightly different from the reverse osmosis process in the sense that the permeating species in this case is solvent as well as low molecular solutes or low valency solutes. This process also operates with hydrostatic pressure difference across a semipermeable membrane having pore sizes which are slightly larger than that of reverse osmosis membranes. The pore sizes of NF membranes are

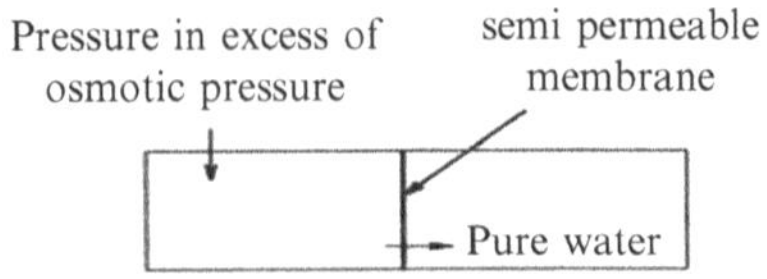

Fig. 2 The schematic representation of a reverse osmosis process

in the range of 10–30 Å. The hydrostatic pressure used in this process can vary from 1.5 to 2 MPa. This process is essentially used to fractionate solutes based on valency of either cation or anion and also to separate various organic solutes of low molecular weights. Various applications of NF process are briefly outlined in latter part of this unit.

3.3 Ultrafiltration (UF)

The ultrafiltration (UF) is a process wherein the solvent along with micro-solutes permeates through the membrane and macrosolutes are retained by the membranes. This process is similar to sieving and the driving force is the hydrostatic pressure across the membrane. The process is used to fractionate the solutes in a solution based on their size or molecular weight difference. Size or the molecular weight difference of the macrosolute retained by the membrane depends upon the pore size of the membranes. Microsolutes whose effective sizes are smaller than the pore size of the membranes permeate along with the solvent, whereas macrosolutes whose effective sizes are larger than the pore size of the membranes are retained. The driving force used in ultrafiltration processes is of the order of 500 kPa or so. The membranes used in ultrafiltration processes have pore sizes ranging from around 30 to 200 Å. Typical applications of the ultrafiltration process are the concentration of protein in milk for cheese making and separation of colloidal particles, oils, and macromolecules from effluent waters as well as from surface waters.

3.4 Microfiltration (MF)

The microfiltration (MF) is a process mainly used for the separation of submicron size (<0.1 m) particulate matter from solution. This process also requires hydrostatic pressure gradient across the membrane and the pressure used is of the order of 100 kPa or so. The pore size of the membranes decides the size of the particulate matter retained. The process is similar to ultrafiltration and separation takes place by sieving. The essential difference between ultrafiltration and microfiltration is the size of the macrosolute retained, pore size of the membranes, and the hydrostatic pressure needed across the membranes. Typical applications of microfiltration process are in the removal of bacteria from water samples and removal of submicron size suspended dust and particulate matters from gas streams. Removal of chemical oxygen demand from effluent waters is another important application of the MF process (Table 1; Fig. 3).

Table 1 Membrane filtration range

Filtration spectrum	Diameter of pores in membrane	Molecular weight (of solute)	Filtrate
Microfiltration	0.05–5.0	>1,000,000	Latex, blood, paint pigment, indigo dye, yeast, bacteria, plant gums, amylopectin
Ultrafiltration	0.005–0.1	4,000–10,000	Colloidal silica, virus, enzymes, protein, gelatin, amylose
Nanofiltration	0.0005–0.01	100–500	Synthetic dye, antibiotics, colorant, amino acids, sugars
Reverse osmosis	0.0001–0.001	<800	Atoms, metal ions, fragrance, flavors, salts

Source Adapted from Mannapperuma (1997)

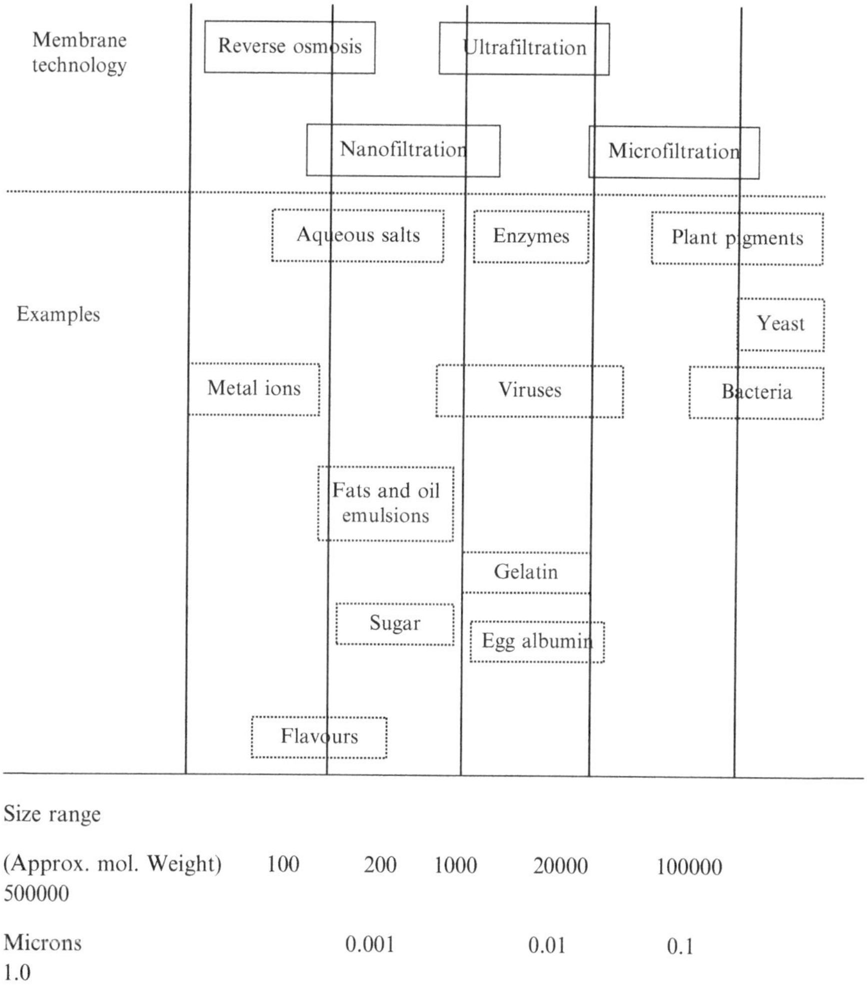

Fig. 3 Size separation capabilities of different membrane systems (from Anon 1997)

4 Membrane Performances

The flow of water through a membrane is described by

$$m_w = \frac{K_w A(\Delta P - \Delta II)}{t}, \tag{3}$$

where m_w is water flow rate (kg/s), ΔP is the hydraulic pressure differential across the membrane (kPa), ΔII is the osmotic pressure difference across the membrane (kPa), t is time (s), A is area (m^2), and kw is the coefficient of water permeability through the membrane (kg/[m^{2}kPa]).

The flow of a solute through a membrane is given by

$$m_s = \frac{K_s A \Delta c}{t}, \tag{4}$$

where m_s is the solute flow rate, Δc is the differential of solute concentration across the membrane (kg/m^3), and K_s is the coefficient of solute permeability through the membrane (L/m).

It is evident that water flow rate through the membrane is increased by increasing the hydraulic pressure gradient across the membrane. The hydraulic pressure gradient has no effect on the solute flow rate. The solute flow is influenced by the concentration gradient across the membrane.

The performance of a membrane system is often described by the retention factor, R_f

$$R_f = \frac{(c_f - c_p)}{c_f}, \tag{5}$$

where c_f is the concentration of a solute in the feed stream (kg/m^3) and c_p is the concentration of a solute in the permeate stream (kg/m^3).

Another factor used to describe the performance of a membrane system is the "rejection factor," R_j

$$R_j = \frac{(c_f - c_p)}{c_p} \tag{6}$$

Membrane performance may be expressed as molecular weight cutoff or the maximum molecular weight for the solute to pass through the membrane.

Another term used to denote membrane performance is conversion percentage, Z.

$$Z = \frac{m_p \times 100}{m_f}, \tag{7}$$

where m_p is product flow rate and m_f is the feed flow rate. Thus, operating a membrane at a conversion percentage of 70 % means that a feed of 100 kg/h will yield 70 kg/h of product (permeate) and 30 kg/h of retentate (Sing and Heldman 1992).

5 Membrane Modules

A wide variety of materials are used for manufacturing of membranes, including sintered metals, ceramics, and polymers (Table 2). The membrane structure varies in its chemical nature, microcrystalline structure, pore size and pore size distribution, and degree of asymmetry. Two simple parameters—membrane permeates flux and solute rejections—are used to describe the characteristics of membranes. Since the properties of membrane material can be influenced by environmental conditions and time, secondary properties such as resistance to compaction, temperature and chemical stability, and resistance to microbial attack are also important. Additional requirements for food processing are good tolerance of cleaning and disinfecting solutions and lack of toxicity of the contact materials (Cheryan 1992). Membranes are assembled as modules that are easily integrated into systems containing hydraulic components. The modules are designed to contain large membrane area in a small volume, withstand the pressures required during separation, and crossflow velocities required to maintain a clean membrane surface (Mannapperuma 1997). The most common module configurations are flat plate, tubular, hollow fiber, and spiral wound (Fig. 4).

Table 2 Polymer, ceramic, and metallic base membranes and their filtration range and modules

Membrane	Filtration material	Range module
Polymers		
Cellulose acetate (CA)	NF, RO	FP, TU, HFF, SW
Polyamide (PA)	NF, RO	FP, TU, HFF, SW
Sulfonated polysulfone	NF, RO, UF	FP, TU, HFF, SW
Polysulfone	MF, UF	FP, TU, SW
Polyethersulfone	MF, UF	FP, SW
Polyviniylidenefluoride	MF, UF	FP, TU, HF, SW
Polytetrafluoroethylene	MF	FP, TU, SW
Polypropylene	MF	FP, HF
Polyacrylonitrile	MF, UF	FP, HF, SW
Polycarbonate	MF	
Polyester	MF	
Ceramics/metallic		
Zirconia/alumina	MF, UF	TU
Alumina	MF	TU
Silicon carbide	MF	TU
Zirconia/metal	MF	TU
Titanium oxide/metal	MF	TU
Sintered steel	MF	TU, FP
Zirconia/carbon	MF	TU
Sintered alloys	MF	TU, FP
Silica	MF	TU

Note: *MF* microfiltration, *UF* ultrafiltration, *NF* nanofiltration, *RO* reverse osmosis, *FP* flat plate, *TU* tubular, *HF* hollow fiber, *HFF* hollow fine fiber, *SW* spiral wound
Source Adapted from Mannapperuma (1997)

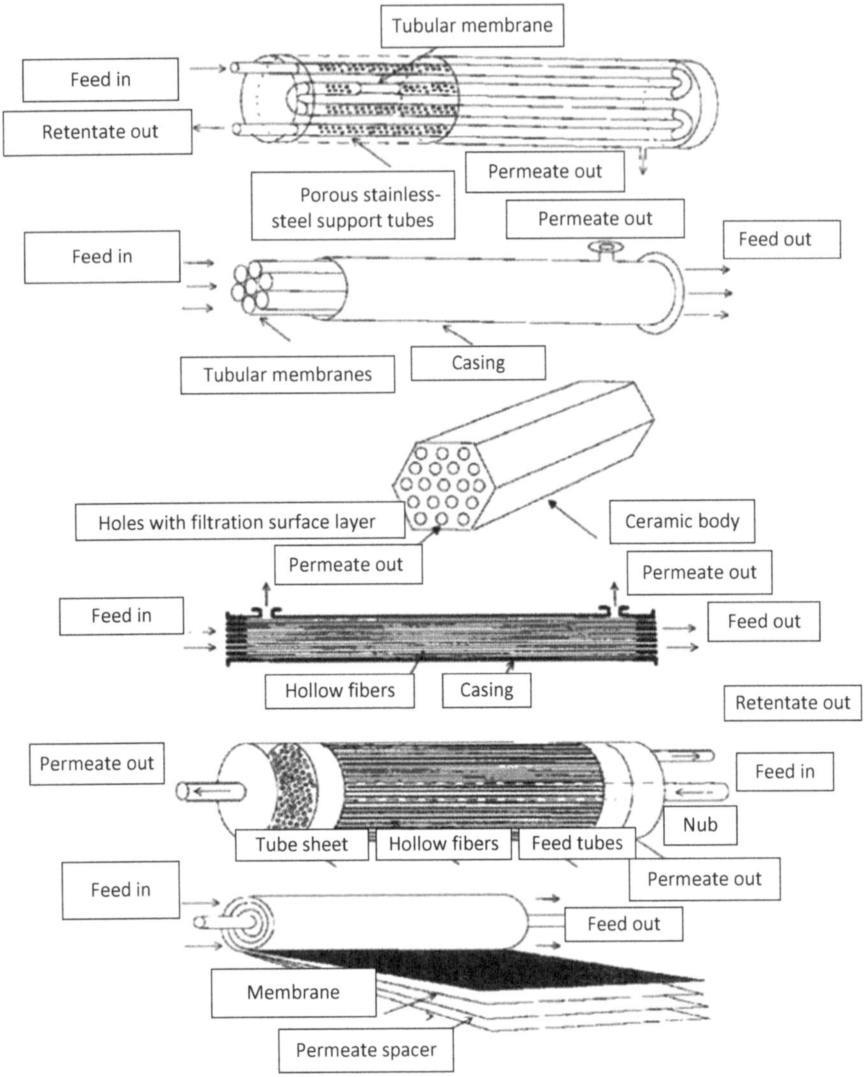

Fig. 4 Membrane modules (Mannapperuma 1997)

Each design has its own advantages and disadvantages. In flat plate modules, two flat sheets are separated by a support plate that also contains the permeate channels. These membrane sandwiches are separated by a spacer that also has feed-flow channels. Alternate layers of membrane sandwiches and spacers are assembled and held together by bolts. Advantages of such system are (1) fairly low holdup and moderately high packing density, (2) easy membrane replacement, (3) flexibility with regard to membrane usage, (4) simplicity in the capacity, and (5) ability of module to withstand high pressures. The main disadvantages are susceptibility to fouling by suspended particles and high initial capital cost. Tubular modules consist of membrane casted inside a porous support tube, typically 6–25 mm in diameter. Several of such tubes are housed within one pressure vessel in a shell and tube arrangement. Advantages of tubular modules include (1) high turbulence, (2) ability to handle suspended particles of 1–1.5 mm, and (3) easy cleaning. The major disadvantages are low surface area to volume and high energy costs for pressures. Hollow fine fiber modules are made with a strand of fine fiber about 50–100 mm in diameter. A bundle of fibers are formed into U shape, the ends are formed into a single header, and the U bundle is placed in a tube. The feed liquid is outside the fibers, while the permeate flows into the fibers. Some advantages of this configuration are (1) compactness, very high packing density; (2) relatively low holdup; and (3) high resistance to compression and hence can withstand high pressures. Some disadvantages are (1) extremely susceptible to fouling by suspended particles, (2) difficulty in operating in sanitary mode and in cleaning, and (3) individual membrane elements (i.e., fibers) cannot be replaced when damaged. Spiral wound modules utilize flat sheet membranes. Two membrane sheets are sandwiched with a permeate spacer between them and three edges are sealed. The fourth is connected to a central perforated tube. A feed channel spacer is placed on top of one layer, and the membrane-screen composite is rolled into a spiral configuration around the central collection tube. The module is placed inside the tubular pressure vessel. Feed flows longitudinally in the feed channel, while permeate flows between the membrane sandwich and spirally around to the permeate collection tube (Cheryan 1992). Advantages of such systems are (1) relatively high packing density, (2) low cost per unit membrane area, (3) easy replacement of modules from the pressure vessels, and (4) low energy consumption. The disadvantage is difficulty in cleaning when fouled with large amount of suspended matter and hence pre filtration is needed. The selection of module configuration and membrane material depends on the feed type and economics.

6 Applications in the Food Industry

6.1 Removal of Bacteria and Spores from Milk, Whey, and Cheese Brine

The removal of bacteria and spores from milk to extend its shelf-life by MF is an alternative way to ultra-pasteurization. In this approach, the organoleptic and chemical properties of the milk are unaltered. The first commercial system of this so-called Bactocatch was developed by Alfa Laval (Hansen 1988) and marketed by Tetra Pak under the name Tetra Al cross_ Bactocatch. In this process, the raw milk is separated into skim milk and cream, see Fig. 5. The resulting skim milk is microfiltered using ceramic membranes with a pore size of 1.4 mm at constant transmembrane pressure (TMP). Thus, the retentate contains nearly all the bacteria and spores,

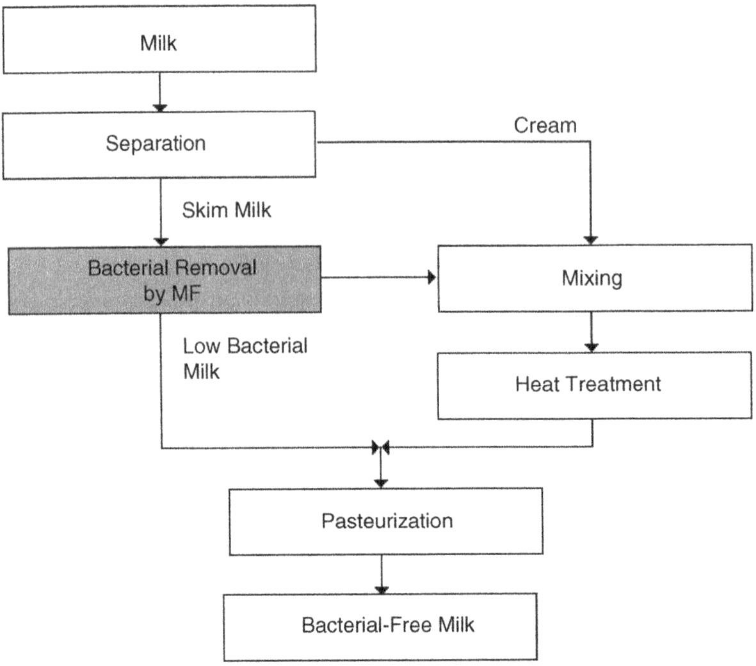

Fig. 5 Bacterial removal from milk by MF

while the bacterial concentration in the permeate is less than 0.5 % of the original value in milk. The retentate is then mixed with a standardized quantity of cream. Subsequently, this mix is subjected to a conventional high heat treatment at 130 °C for 4 s and reintroduced into the permeate, and the mixture is then pasteurized. Since less than 10 % of the milk is heat treated at the high temperature, the sensory quality of the milk is significantly improved. MF for the removal of bacteria and spores can be further applied in the production of other dairy products. In the production of cheese, the use of low bacterial milk improves also the keeping quality of cheese due to the removal of spores, thus eliminating the need of additives (e.g., nitrate). While in the production of whey protein concentrates (WPC) and isolates (WPI), this MF concept is used to remove bacteria and spores giving a high quality product. Hence, by applying MF the heat treatment of the WPC/WPI is kept to a minimum, which preserves the functional properties of the whey proteins. Finally, in the manufacture of cheese the concentrated curd is submerged in a salt solution to improve the cheese preservation and to develop the flavor and other cheese properties. This process is called brining. Efficient sanitation of cheese brine has become a major concern to the dairy industry in recent years. This results from the possibility of postcontamination of cheeses in the brine, especially by pathogenic bacteria. The application of MF for sanitation of cheese brine, using ceramic or spiral wound membranes, results in a superior cheese quality compared to the traditional processes of heat treatment and kieselguhr filtration. MF has the advantages of being simple to perform, of maintaining the chemical balance of the brine and of eliminating filter aids. In the brine treatment by MF it is normally necessary to make a pre filtration of the brine solution, which is easily done by dead-end filter bag or cartridge with a pore size of 100 mm (Ottosen and Kønigsfeldt 1999).

6.2 Cheese Manufacturing

Another early application of membrane technology in the dairy industry was in cheese manufacturing for production of Feta cheese and brine treatment by UF. Nowadays, membrane-processed milk is also successfully used in the manufacturing of quark and cream cheeses. Together with WPC production, the use of UF milk for the production of cheese is the most widespread application of membranes in the dairy industry. The advantages of UF concentrated milk in cheese making compared to traditional methods are the following: Increases the total solid, which increase the cheese yield and therefore decreases the production costs in terms of energy and equipment; reduces the rennet and starter culture requirements since UF-milk has a good ability of enzymatic coagulation; reduces the wastewater processing costs of the cheese plant; improves the quality and composition control; increases the nutritional value due to the incorporation of the whey protein in the cheese. UF in cheese processing can be used in three ways (Rosenberg 1995):

1. Pre-concentration—The standardized cheese milk is concentrated by a factor of 1.2–2 and it can be used for most cheese types. This allows the capacity of the cheese vats and whey draining equipment to be doubled. However, the cheese yield will not be significantly improved since only 4.5–5 % of the protein content is increased. It is used to produce Cheddar, Cottage Cheese, and Mozzarella, and it can be used to standardize cheese milk and manipulate its mineral composition, resulting in a more consistent quality in the final product.
2. Partial concentration—The standardized cheese milk is concentrated by a factor 2–6. It is used in the manufacture of Cheddar cheese by using for example, the APV-Siro Curd process, in which the milk is concentrated five times with DF in order to standardize the salt balance (Tamime 1993). It is also used to produce other cheese types like Queso Fresco, structure Feta, Camembert, and Brie.
3. Total concentration—The standardized cheese milk is concentrated to the total solids content in the final cheese. This provides the maximum yield increase and since there is no whey drainage, the cheese can be manufactured without the need for a cheese vat. It is used to produce cast Feta, quark, cream cheese, Ricotta, and Mascarpone. The UF permeate, which contains mainly lactose, can be concentrated by RO. The permeate from the RO process can be polished by another RO unit. After pasteurization or UV light treatment, the permeate from the polisher can be used at the plant as process water, thus reducing the water costs of the plant. Although UF has advantages in cheese production, the increase

of whey content in the cheese due to the concentration of all milk proteins can have a negative effect on the ripening of semi hard and hard cheeses (Qvist et al. 1986) Therefore, UF should be viewed as a complementary process to cheese manufacturing and not as an alternative process.

6.3 Beer

The conventional brewing process starts in the brew house with the stepping of the malt with hot water to produce wort, a thick sweet liquid. The wort is then passed to the wort boiler in which it is brewed/boiled for up to 2 h followed by clarification and cooling. The clarified and cooled wort is combined with yeast and passed on to the fermentation tanks in which the yeast converts the grain sugar to alcohol and as such produces beer. Before being transferred to the bright beer tanks, the beer is commonly clarified. The finished beer might then be fine-filtered and pasteurized before bottling. In the case of beer de-alcoholisation, the alcohol removal takes place before the beer clarification. The overall brewing process with potential applications of crossflow membrane filtration is shown in Fig. 6.

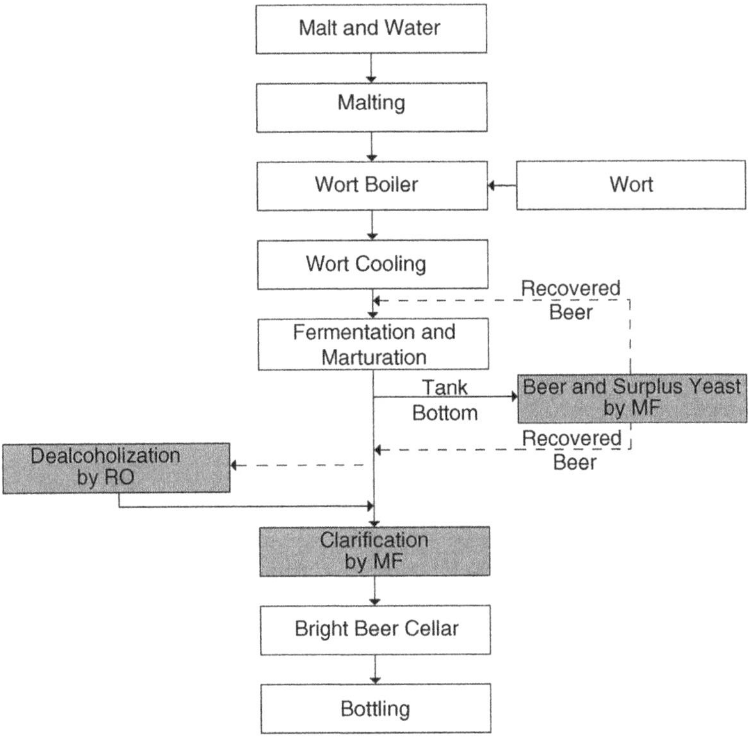

Fig. 6 Beer production with membrane technology

Fig. 7 Membrane processes in fruit juice production

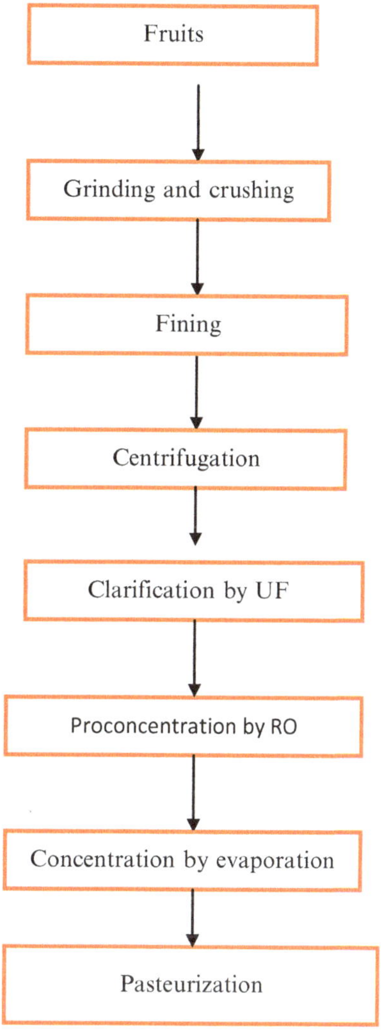

6.4 *Fruit Juices*

The general production flow in the fruit juice industry starts with grinding or crushing of the fruits into an optimal and uniform size of particles and then pressing out the fruit mash. The traditional fining process consists of long retention time in tanks followed by kieselguhr filtration and requires large amounts of enzymes, gelatin and other chemicals. After clarification/fining, the fruit juice is concentrated to reduce costs for transportation and storage. The common approach to concentrate fruit juice is by using an evaporator combined with an aroma-recovery unit concentrating the apple juice from originally 11–12 Brix to over 70 Brix. The concentrated fruit juice can then be optionally pasteurized before transportation. The general fruit juice production process including membrane processes is shown in Fig. 7

6.5 *Fruit Juice Clarification*

The clarification of fruit juice, mainly apple but also grape, pineapple, and orange juice by UF has proven to be an attractive substitute for the traditional fining and filtering process from an economic and qualitative point of view since the 1970s. The UF process removes the suspended solids and other high molecular solids and the filtered juice obtains a clarity and excellent quality, which has not previously been obtainable. Thus, the UF process substitutes the fining step in the traditional process. In order to achieve high yield, high capacity and excellent quality, an enzyme treatment and proper pre filtration must be carried out before the UF system is utilized. Until now, the industrial standard is to use polymeric and ceramic tubular modules for the clarification of the juice. However, this module type is associated with low packing density and high membrane replacement costs. Furthermore, this process is commonly run in batch mode and difiltration water has to be added in the final stage of the clarification to maximize the process yield. More recently, a new concept has been developed, which combines a high-speed separator with spiral-wound UF modules to overcome these limitations (Gebhardt 2001), see Fig. 7.

6.6 *Fruit Juice Concentration*

For the concentration of apple juice, the combination of RO and evaporation can provide an interesting process combination. RO as initial step can remove more than 50 % of the water content prior to evaporation, while maintaining 98–99 % of sugar and acid as well as 80–90 % of volatile flavors in the concentrate see Fig. 1.5.2. By applying RO, concentration levels of 20–25 Brix can be achieved, while the subsequent evaporation can boost these levels to above 75 Brix. By applying this concept, only 7–9 kWhperm^3 fruit juice are required, which represents an energy saving of 60–75 % compared to direct evaporation. Furthermore, the permeate from the RO unit can be recycled as process water as process water.

6.7 *Other Membrane Applications in the Food Industry*

The development of new applications of the established membrane processes MF, UF, NF, and RO will be driven by economical and environmental targets. An additional driver for membrane processes is the high growth rate of the market for functional foods, a segment in which membranes has a high potential. In Table 3, some of the most recent research trends on membrane applications for MF, UF, NF, and RO in the food industry are summarized.

Table 3 Applications of MF, UF, NF, and RO in the food industry

Application	Membrane processes
Concentration of whole and skim milk	RO
Partly demineralized WPC (baby food, special WPC products)	NF/UF
Production of whey protein concentrates and isolates	UF
Concentration of chicken blood plasma	MF/UF
Filtration of extra-virgin olive oil	MF/UF
Dry degumming of vegetable oil	UF/NF

6.8 *Future Scope of Membrane Processing*

Countries are implementing legislation for sustainable waste management practices. In the European Union, (Smith 2002) organic waste disposal must be reduced by 80 % in 2010. Food companies are investigating the use of membrane separations in new products or reduction of waste streams. New material Such as ceramics and synergies with biotechnology demonstrate the potential of membranes is far from exhausted (Pap et al. 2004).

7 Conclusion

Membrane processes allow separation of solutes from one another or from solvent with no phase change are interface mass transfers. There are many kinds of membrane processes as discussed earlier. The classification of these methods is based on the driving forces that cause transfer of solutes through the membrane. The membrane technology is versatile and it can be used for purification of fruit juices and concentration of milk.

References

ANON (1997) Cost-effective membrane technologies for minimizing wastes and effluents, Guide GG54, Environmental Technology Best Practice Programme, UK Government

Cheryan M (1992) Concentration of liquid foods by reverse osmosis. In: Heldman DR, Lund DB (eds) Handbook of food engineering. Marcel Dekker, New York, pp 393–436

Gebhardt W (2001) Weltforum derweinund saftbereitung. F&S Filt Sep 15:234

Hansen R (1988) Better market milk, better cheese milk, better low heat milk powder with Bactochatch treated milk. North Eur Food Dairy J 54:39–41

Leiknes T Theory of transport in membrane. Ph.D Thesis (ISBN: 82-5591-2, Chapter 2)

Mannapperuma JD (1997) Handbook of food engineering practice. CRC Press, Boca Raton, FL

Ottosen N, Kønigsfeldt P (1999) Microfiltration of cheese brine. Preliminary APV Nordic, Membrane Filtration, Silkeborg

Pap NE, Pongracz L, Myiiykoski KR (2004) Waste minimization and utilization in the food industry. Oulu University Press, Oulu, pp 159–168

Qvist KB, Thomsen D, Kjærgard JG (1986) Fremstilling af havarti ost ud fra mælk, der er koncentreret ca. 5 gange ved ultrafiltrering. 268, Beretning/Statens Mejeriforsøg, Hillerød, Denmark

Rosenberg M (1995) Current and future applications of membrane processes in the dairy industry. Trends Food Sci Technol 6:12–19

Sing RP, Heldman DR (1992) Membrane separation. In: Introduction to food engineering. Academic, London, pp 636–637

Smith C (2002) Applications of reverse osmosis in wine making, www.vinovation.com

Tamime AY (1993) Modern cheese making: hard chesses. In: Robinson RK (ed) Modern dairy technology. Elsevier, New York

Technology for Value Addition and Preservation of Horticultural Produce

R.K. Gupta

1 Introduction

Horticultural crops not only provide human beings with nutritional and healthy foods but also generate a considerable cash income for growers in many countries. India ranks second in fruits with annual production of 48.57 million tons, accounting for about 10.3 % of total fruit production (India is the Second largest producer of vegetables next to China, with an estimated production of about 93.92 million tons, accounting for about 13.5 % of the world production). The percentage of world share in banana is 29 %, mango 44 %, and papaya 30 %. In vegetable production, India is one of the largest producers of cauliflower, sharing about 34 % of the world and pea 38 % of the world share. It is estimated that fruit production will touch 98 million tons by the year 2020–2021 with vegetable production of 220 million tons. In fact horticultural crops covering only 6.1 % of the area under cultivation, contribute as much as 18.8 % of the country's gross domestic product.

Food security, both in terms of availability and access to food, poses a challenge to rapidly growing populations, in environments of dwindling land and water resources. The horticultural sector has established its credibility for improving land use and generating employment and nutritional security. Horticulture, which includes the production of fruits, vegetables, flowers, spices, medicinal and aromatic plants, and plantation crops has emerged as a major economic activity in India.

R.K. Gupta (✉)
Director, Central Institute of Post Harvest Engineering and Technology, Ludhiana, India
e-mail: rkguptaciphet@gmail.com

A. Malik et al. (eds.), *Food Processing: Strategies for Quality Assessment*,
Food Engineering Series, DOI 10.1007/978-1-4939-1378-7_16

2 Importance of Fruits, Nuts, and Vegetables in Human Nutrition and Health

Fruits and vegetables are of immense significance to man. Fruits and vegetables contribute approximately 91 % of vitamin C, 48 % of vitamin A, 27 % of vitamin B_6, 17 % of niacin, 16 % of magnesium, 19 % of iron, and 9 % of calories to the human diet. Fruit and vegetable consumption has increased in response to growing health consciousness. Their consumption has been strongly linked to reduced risk of some forms of cancer, heart disease, stroke, and other chronic diseases. Fruits and vegetables are sources of antioxidants which modify the metabolic activation and detoxification/disposal of carcinogens, or even influence processes that alter the course of tumor cell growth. Although antioxidant capacity varies greatly among fruits and vegetables, consumption of a variety of fruits and vegetables is preferred, over limiting fruit and vegetable consumption to those having the highest antioxidant capacity.

Apart from their nutritive value, other constituents of fruits and vegetables which deserve attention include antioxidants, bioflavonoids, flavor compounds, and dietary fiber. Often, the leafy portion of some important vegetables is discarded, while the fleshy portion is consumed with little recognition for the fact that rich sources of nutrients such as calcium, iron, vitamin-C, and carotene go to the waste. Apart from the common and generally costly fruits, a large number of indigenous fruits, examples of which include "Aonla" and "woodapple" are rich sources of ascorbic acid (Vitamin-C) and riboflavin (Vitamin-B_2). Some tropical fruits and vegetables are known to have therapeutic properties and are popularly used in traditional medicine in several countries of the region.

3 Postharvest Losses

It should be noted that the production of fruits and vegetables is of significance only when they reach the consumer in good condition and at a reasonable price. The concept of placing exclusive emphasis on increased production of fruits and vegetables is self-defeating. It is important to see how much of the produce goes through marketing channels and finally reaches the consumer. Efforts should be made to integrate production with postharvest management since postharvest loss reduction and utilization have considerable bearing on food availability. It is known that food loss reduction is normally less costly than equivalent increases in food production. Reduction of postharvest losses is essential in increasing food availability from existing production. The success of production lies in the proper distribution of produce and its subsequent utilization by the consumer with zero waste in the process, i.e., 100 % utilization of production in one form or another should be the motto. Opportunities exist in both domestic and international markets for fresh and processed fruits and vegetables (Kader 1992).

4 Causes of Postharvest Losses

Factors affecting postharvest food losses of perishables vary widely from place to place and become more and more complex as systems of marketing become more complex. The farmer must give careful attention to:

- Market demand for the products he will grow; he must know the market and his buyers
- Cultivation
- Harvesting and field handling
- Packing or packaging
- Transport
- Market handling; possibly storage or refrigeration
- Sales to consumers, wholesalers, or agents
- Perishability of the produce

All fruits, vegetables, and root crops are living plant parts containing 65–95 % water, and they continue their living processes after harvest. Their postharvest life depends on the rate at which they use up their stored food reserves and their rate of water loss. When food and water reserves are exhausted, the produce dies and decays. Anything that increases the rate of this process may make the produce inedible before it can be used. The principal causes of loss are mentioned above, but in the marketing of fresh produce they all interact, and the effects of all are influenced by external conditions such as temperature and relative humidity (Sudher and Indira 2007).

Physical damage to fresh produce can come from a variety of causes, the most common being:

5 Mechanical Injury

The high moisture content and soft texture of fruits, vegetables and root crops make them susceptible to mechanical injury, which can occur at any stage from production to retail marketing because of:

- Poor harvesting practices
- Unsuitable field or marketing containers and crates, which may have splintered wood, sharp edges, poor nailing, or stapling
- Overpacking or underpacking of field or marketing containers
- Careless handling, such as dropping or throwing or walking on produce and packed containers during the process of grading, transport, or marketing

Injuries caused can take many forms:

- Splitting of fruits or roots and tubers from the impact when they are dropped
- Internal bruising, not visible externally, caused by impact
- Superficial grazing or scratches affecting the skins and outer layer of cells
- Crushing of leafy vegetables and other soft produce

Injuries cutting through or scraping away the outer skin of produce will:

- Provide entry points for molds and bacteria causing decay
- Increase water loss from the damaged area
- Cause an increase in respiration rate and thus heat production

Bruising injuries, which leave the skin intact and may not be visible externally, cause:

- Increased respiration rate and heat production
- Internal discoloration because of damaged tissues
- Off-flavors because of abnormal physiological reactions in damaged parts

6 Injuries from Temperature Effects

All fresh produce is subject to damage when exposed to extremes of temperature. Commodities vary considerably in their temperature tolerance. Their levels of tolerance to low temperatures are of great importance where cool storage is concerned:

- Freezing injury—All produce is subject to freezing at temperatures between 0 and −2 °C. Frozen produce has a water-soaked or glassy appearance. Although a few commodities are tolerant of slight freezing, it is advisable to avoid such temperatures because subsequent storage life is short. Produce which has recovered from freezing is highly susceptible to decay.
- Chilling injury—Some types of fresh produce are susceptible to injury at low but nonfreezing temperatures. Such crops are mostly of tropical or subtropical origin, but a few temperate crops may be affected.

Sensitivity varies with the commodity, but with each there is a temperature below which injury occurs: the lowest safe temperature (LST). Within a single commodity type, the LST may vary between varieties. Fruit is generally less sensitive when ripe.

Symptoms of chilling injury may not develop until the produce is removed from cold storage to normal market (i.e., ambient) temperature. When susceptible produce has to be held for some time in storage, it must be kept at a temperature just above its LST. This means that such crops will have a shorter marketing life than nonsensitive crops because respiration has continued at a relatively fast rate during storage at higher than normal cold-storage temperatures.

- High temperature injury—If fresh produce is exposed to high temperatures caused by solar radiation, it will deteriorate rapidly. Produce left in the sun after harvest may reach temperatures as high as 50 °C. It will achieve a high rate of respiration and, if packed and transported without cooling or adequate ventilation, will become unusable. Long exposure to tropical sun will cause severe water loss from thin-skinned root crops such as carrots and turnips and from leafy vegetables.

7 Diseases and Pests

Diseases caused by fungi and bacteria commonly result in losses of fresh produce. Viral diseases, which can cause severe losses in growing crops, are not a serious postharvest problem.

Insect pests, that are mainly responsible for wastage in cereals and grain legumes, are rarely a cause of postharvest loss in fresh produce. Where they do appear, they are often locally serious, e.g., the potato tuber moth.

Diseases: Losses from postharvest disease in fresh produce fall into two main categories.

Loss in quantity, the more serious, occurs where deep penetration of decay makes the infected produce unusable. This is often the result of infection of the produce in the field before harvest.

Loss in quality occurs when the disease affects only the surface of produce. It may cause skin blemishes that can lower the value of a commercial crop. In crops grown for local consumption, the result is less serious since the affected skin can often be removed and the undamaged interior can be used.

Fungal and bacterial diseases are spread for the most part by microscopic spores, which are widely distributed in the air, soil and on dead and decaying plant materials.

Produce can become infected:

- Through injuries caused by careless handling, by insect or other animal damage, or through growth cracks
- Through natural pores in the above- and below-ground parts of plants, which allow the movement of air, carbon dioxide and water vapor into and out of the plant
- By direct penetration of the intact skin of the plant. The time of infection varies with the crop and with different diseases. It can occur in the field before harvest or at any time afterwards

Field infections before harvest may not become visible until after harvest. For example, decay of root crops caused by soil molds will develop during storage. Similarly, tropical fruits, infected at any time during their development, may show decay only during ripening (Paull and Armstrong 1994).

Infection after harvest can occur at any time between the field and the final consumer. It is for the most part the result of invasion of harvesting or handling injuries by molds or bacteria.

Postharvest diseases may be spread in the field before harvest by the use of infected seed or other planting material. Many diseases can survive by using weed plants or other crops as alternate or alternative hosts. They are also spread by means of infected soil carried on farm implements, vehicles, boots, etc., and from crop residues or rejected produce left decaying in the field.

Postharvest diseases can also be spread by:

- Field boxes contaminated by soil or decaying produce or both
- Contaminated water used to wash produce before packing
- Decaying rejected produce left lying around packing houses
- Contaminating healthy produce in packages

Pests: Although relatively few postharvest losses of fresh produce are caused by attacks of insects or other animals, localized attacks by these pests may be serious.

- Insect damage is usually caused by insect larvae burrowing through produce, e.g., fruit fly, sweet potato weevil, potato tuber moth. Infestation usually occurs before harvest. Postharvest spread is a problem where produce is held in store or is exposed to lengthy periods of transport.
- Rats, mice, and other animal pests again are sometimes a problem when produce is stored on the farm.

An increase in the rate of loss because of normal physiological changes is caused by conditions that increase the rate of natural deterioration, such as high temperature, low atmospheric humidity, and physical injury. Abnormal physiological deterioration occurs when fresh produce is subjected to extremes of temperature, of atmospheric modification, or of contamination. This may cause unpalatable flavors, failure to ripen or other changes in the living processes of the produce, making it unfit for use.

Careless handling of fresh produce causes internal bruising, which results in abnormal physiological damage or splitting and skin breaks, thus rapidly increasing water loss and the rate of normal physiological breakdown. Skin breaks also provide sites for infection by disease organisms causing decay.

8 Remedies/Technologies for Improving the Quality and Safety of Horticultural Commodities

8.1 Raw Material

No matter how perfect postharvest operations are, good returns cannot be obtained from poor quality raw materials. The selection of suitable varieties is, therefore, essential. Linking production to postharvest operations is essential to optimize results. Preharvest parameters such as selection of proper planting material, crop management, and disease and pest control must be geared toward producing high quality produce. Once the crop is ready for harvest, attention must be paid to the harvesting technique/procedure. Poor harvesting practices can lead to irreparable damage to horticultural produce. It is therefore necessary to standardize maturity indices and harvesting techniques for each and every fruit and vegetable in order to minimize damage at the time of harvest.

8.2 Packing House

There is an absolute lack of the concept of packing house establishments in the country. Fruits and vegetables are generally packed in the field without any pretreatment. Some are even transported without any packaging. In developed countries, on the other hand, fruits and vegetables are generally selected, cut, placed in bulk containers and transported to packing stations where they are trimmed, sorted, graded, packed in cartons or crates, and cooled. They are temporarily placed in cool storage for subsequent loading or are loaded directly onto refrigerated vehicles and transported to the market. A number of important operations are also carried out at packing stations. These include SO_2 fumigation, fungicidal dipping, surface coating with wax, degreening of citrus, ripening and conditioning, vapor heat treatment, etc. Due to the lack of proper packaging systems in India, large volumes of the inedible portions of vegetables such as cauliflower, peas, etc., are transported to wholesale markets from the field. They are discarded to various degrees and large quantities of biomass which could be used as value added products are wasted. Removal of these inedible vegetable portions prior to marketing would reduce both transportation costs and environmental pollution. These inedible vegetable parts ultimately undergo decomposition, cause sanitation problems and produce gases which are detrimental to the environment. Farmer's cooperatives and other agencies should, therefore, be encouraged to establish packing stations at nodal points to augment the marketing of fresh horticultural produce.

8.3 *Primary Processing*

Unlike durable crops such as cereals, pulses and oilseeds, fresh fruits, and vegetables are highly perishable and must be marketed immediately after harvesting without primary processing. Fruits and vegetables generate large quantities of valuable waste that ends up as garbage. However, if they are gainfully utilized at the proper time they can become value-added products. Vegetables such as cauliflower, peas, leafy vegetables, etc., can be minimally processed at packing stations immediately after harvesting, through the removal of inedible parts, following which they can be marketed in metro city markets in unit packs. Between 10 and 60 % of the fresh fruits and vegetables, marketed and purchased by consumers in India, are rejected as inedible. In villages or small towns the inedible portions of fruits and vegetables are either fed to animals or are discarded as garbage by consumers in metro cities. Primary processing of food crops other than horticultural crops has its origin from the dawn of civilization. It was a necessary step to the consumption of foods such as rice, wheat, oilseeds, etc. Processing not only renders these commodities edible but also adds value to them. Value addition to horticultural crops was never considered essential, owing to the fact that many of these fruits and vegetables, e.g., tomato, melon, cucumber, carrot, etc., could be directly consumed after harvesting. Today, there is considerable interest in processing to add value, as well as to reduce losses in fruits and vegetables.

8.4 *Packaging*

Packaging is an integral element in the marketing of fresh horticultural produce. It provides an essential link between the producer and the consumer. Fresh fruits and vegetables are generally packed in bamboo baskets, plastic crates, plastic bags, or nylon sacks for transportation, in many developing countries. Often, they are transported in an unpackaged form. After harvest, fresh fruits and vegetables are generally transported from the farm to either a packing house or distribution center. Farmers sell their produce either in fresh markets or in wholesale markets. At the retail level, fresh produce is sold in an unpackaged form, or is tied in bundles. This type of market handling of fresh produce greatly reduces its shelf-life if it is not sold quickly. The application of proper postharvest technologies, would, however, extend postharvest shelf-life, retain fresh quality, and reduce losses.

Packaging plays a very important role in protecting fresh produce:

- It provides protection from dust
- It reduces microbial contamination from the surrounding environment and from consumer contact
- It helps to maintain the freshness of produce
- It extends the postharvest shelf-life
- It increases the sale of fresh produce

The development of packaging which is suited to the handling of fresh produce necessitates an understanding of the physiological characteristics of the produce. Fruits and vegetables may be characterized as being either climacteric or non-climacteric, depending on their respiratory pattern:

- *Non-climacteric fruits* ripen only while still attached to the parent plant. Their eating quality suffers if they are harvested before they are fully ripe because their sugar and acid contents do not increase further. Their respiration rate gradually declines during growth and after harvesting. Maturation and ripening are a gradual process. Examples of climacteric fruit include cherries, cucumbers, grapes, lemons, and pineapples.
- *Climacteric fruit* can be harvested when mature but before the onset of ripening. These fruits may undergo either natural or artificial ripening. The onset of ripening is accompanied by a rapid rise in respiration rate, generally referred to as the respiratory climacteric. After the climacteric, the respiration rate slows down as the fruit ripens and develops good eating quality. Examples of climacteric fruit include apples, bananas, melons, papaya, and tomatoes.

Ethylene gas is a plant hormone which initiates senescence or ripening. It is active in very small amounts at 0.1 ppm. Ethylene action and binding to the receptor site is impeded by oxygen levels of less than 8 %. Fresh produce is generally packed in plastic crates, plastic bags, or corrugated paper boxes in Thailand. Corrugated paper boxes are used for the packaging of fixed counts of uniformly sized produce such as avocadoes, mangoes, and oranges. Boxes protect the commodity by cushioning and immobilizing it. They are easily handled throughout distribution and marketing and can minimize the impact of rough handling. Boxes should serve as a moisture barrier and should be designed with appropriate ventilation capacity. Corrugated paper boxes cannot, however, significantly extend shelf-life even when used for the low temperature storage of fresh produce. Much of the plastic material used in the bagging of fresh produce is unsuitable owing to poor moisture and gas permeability. This often leads to condensation, high CO_2 and low O_2 levels in bagged produce, and results in flavor deterioration and fermentation, or failure of the fruit to ripen. Thus use of plastic packaging designed for the marketing of fresh produce should incorporate consideration for factors such as O_2 uptake, CO_2 production, and the production of heat and ethylene by the produce. Low density polyethylene film is generally used for the packaging of fresh fruits and vegetables, owing to its high permeability and softness when compared to high density film. Polyethylene can be easily sealed, has good O_2 and CO_2 permeabilities, low temp durability, and good tear resistance and is of a good appearance. This film is therefore used for the production of modified atmosphere packaging (MAP) which can be manipulated to match the characteristic respiration of produce by reducing O_2 levels in order to slow down the respiration rate, metabolic rate, and senescence of the produce. Effective O_2 levels must be maintained at between 2 and 10 % in MAP systems, if fermentation of the produce is to be prevented. Elevated CO_2 levels reduce the sensitivity of fresh produce to ethylene, and slow down the loss of chlorophyll. At CO_2 levels ranging between 1 and 5 %, however, fruits fail to ripen, internal breakdown occurs and off-flavor development ensues. Oxygen and carbon dioxide transmission rates for MAP films should, therefore, match the respiration rate of the produce to be stored.

Owing to its favorable properties, wood has remained the main packaging material for fruits and vegetables. Timber conservation is, however, critical in order to maintain an ecological balance, and there is an urgency to identify substitutes for the use of timber in an effort to protect forest resources in many developing countries. Packaging has been identified as one of the most important areas where substitution of wood is not only possible but also obviously desirable. Considerable work has been done by different agencies in introducing alternative types of packaging. Corrugated fiber board (CFB) containers consume one third of the wood required for producing timber boxes of the same size. CFB boxes can also be fabricated from kraft paper produced from bamboo, long grasses, and many other types of agricultural residues like bagasse, paddy, cotton stalk, jute stick, wheat straw, and recycled paper and cardboard. Packaging produced from timber is often used as a source of firewood, owing to the severe shortage of fuel wood in India. CFB cartons, on the other hand, are recycled as pulp or paper. Thus switching over from wood to CFB boxes for the packaging of horticultural produce is a very practicable and environmentally friendly option. Increased use of corrugated cartons for local distribution of produce could be accomplished with improvement in the quality of boxes produced in India. The ventilated CFB box, which contains ventilated partitions; and which was developed at the IARI, was found to be ideal for the packaging and transportation of fruits, owing to the comparably minimal level of bruising observed in these boxes. Cushioning materials used in the packaging of fruits and vegetables in wooden boxes include dry grass, paddy straw, leaves, sawdust, paper shreds, etc., all of which end up as garbage and add to environmental pollution in cities. Molded trays or cardboard partitions used in CFB boxes are, however, easily recycled.

8.5 Storage

The lowest temperature that does not cause chilling injury is the ideal storage temperature for fresh fruits and vegetables. Mechanical refrigeration is generally used for the storage of fruits and vegetables. This refrigeration is, however, energy intensive and expensive, involves considerable initial capital investment, and requires uninterrupted supplies of electricity which are not always readily available, and cannot be quickly and easily installed (Liu 1991a, b). Available cold storage in India is used primarily for the storage of potatoes.

8.6 On Farm Storage

On farm storage is required in remote and inaccessible areas of India, to reduce losses in highly perishable fresh horticultural produce. The high cost and high energy requirements of refrigeration, and the difficulty of installing and running refrigerated facilities in remote areas of India, precludes the use of refrigerated storage in many parts of India. Low-cost, low-energy, environmentally friendly cool chambers made from locally available materials, and which utilize the principles of evaporative cooling, were therefore developed in response to this problem. These cool chambers are able to maintain temperatures at 10–15 °C below ambient, as well as at a relative humidity of 90 %, depending on the season. Fruits and vegetables are stored in plastic crates within the chamber. The shelf-life of the fruits and vegetables maintained in the cool chamber was reported to be increased from 3 days at room temperature to 90 days.

8.7 Control Atmosphere/Modified Atmosphere Storage

Controlled atmosphere (CA) or modified atmosphere (MA) storage involves adjustment of the atmospheric composition surrounding commodities by removal (mainly O_2) or addition (mainly CO_2) of gases from the environment surrounding the fruits and vegetables. MA does not differ in principle from CA storage except for the fact that the concentrations of the gases are less precisely controlled. Basic requirements for CA storage include a gas tight chamber and control of the concentrations of CO_2 and O_2. When combined with refrigeration, CA markedly enhances the shelf-life of fruits and vegetables.

Controlled atmospheric storage: CA storage consists of reduction of oxygen by elevating nitrogen and/or carbon dioxide concentrations in storage units, so as to enable preservation of the quality of food products during storage. The temperature is controlled by mechanical refrigeration and the composition of atmosphere is constantly analyzed for CO_2 and O_2 levels. When the oxygen has reached the level required for the particular crop being stored, it is maintained at that level by frequently introducing fresh air from outside. The same applies to carbon dioxide. When a predetermined level is reached the atmosphere is passed through a chemical, which removes carbon dioxide, and then back into the store. This is called as active scrubbing. Alternatively the carbon dioxide absorbing chemical may be placed inside the store, where it can keep the level low (usually about 1 %), this is called passive scrubbing.

It has been shown that low oxygen levels in store can affect the crop in several ways: (1) it can reduce respiration rate as well as substrate oxidation, (2) delays ripening of climacteric fruits and prolongs storage life, (3) delays chlorophyll breakdown, (4) reduces rate of production of ethylene, (5) reduces degradation rate of soluble pectins, (6) formation of undesirable flavurs and odors, (7) altered texture, and (8) development of physiological disorders. The effect of carbon dioxide in extending the storage life of crops appears to be by reducing respiration. It has been shown that carbon dioxide could inhibit the enzyme succinate dehydrogenase in the tricarboxylic acid (TCA) cycle, which is part of the crop's respiratory pathway. Increased carbon dioxide level (1) delays the initiation of ripening, (2) inhibits some enzymatic reactions, (3) decreases production of some organic volatiles, (4) reduces the rate of breakdown of pectic substances, (5) inhibits chlorophyll breakdown, (6) induces physiological disorders, (7) retards fungal growth on the crop, (8) inhibits the effect of ethylene, and (9) changes in sugar content. Carbon monoxide if added to controlled atmosphere stores, with levels of oxygen between 2 and 5 %, can inhibit discoloration of lettuce on the cut butts or where leaves have been mechanically damaged. The respiration rate of lettuce was reduced during a 10-day storage period at 2.5 °C when carbon mono-oxide was added to the store. A combination of 4 % oxygen, 2 % carbon dioxide, and 5 % carbon mono-oxide was shown to be optimum for delaying ripening and maintaining good quality in mature green tomatoes stored at 12.8 °C. In peppers and tomatoes, the level of chilling injury symptoms could be reduced but not eliminated when carbon monoxide was added to the store. The technology has made possible extension of the product shelf-life to meet the demands of long distance transport, distribution and marketing as well as maintenance of product quality through reduction of respiration rate, delaying ripening, and controlling other metabolic reactions.

Modified Atmosphere Storage: The MA technique consists of enclosing respiring produce in polymeric films in which the gaseous environment is actively or passively altered to slow down respiration, reduce moisture loss, and extend the shelf-life of the produce. MA techniques provide low O_2 and high CO_2 regimes similar to those achieved by using CA storage and are theoretically able to maintain desired atmospheres throughout the marketing chain. The degree to which atmospheric modification takes place in packages is dependent upon several variables such as film permeability to O_2 and CO_2, produce respiration and the influence of storage temperature.

Individual shrink wrapping (ISW), a form of modified atmosphere packaging, is also used to enhance the storage life and maintain the harvest freshness of fruits and vegetables. This technique involves sealing of individual fruit in a flexible film followed by heat shrinking to confine the shape of the fruit making it as another layer of protective cuticle. Shrink wrapping of individual fruit helps to greatly reduce weight loss, maintain firmness, delay ripening and senescence, and alleviate chilling injury.

8.8 Containerization

The use of containers for the transportation of goods was recently introduced into India. Relatively little attention has, however, been given to the use of containers for the transport of fresh horticultural produce. Containerization provides an excellent system for the shipment of goods from one place to another. Refrigerated containers are used in the transportation of fruits, vegetables, and flowers in many developing countries. The design and fabrication of ventilated containers which incorporate evaporative cooling systems should be considered for the Indian context. One of the greatest advantages of the container is that it can be placed on truck or rail, without interfering with the movement of the vehicle. Palletization and containerization will go a long way in developing local and international trade.

8.9 Rapid Transportation Systems

Railways and roads are two important transportation systems for the movement of goods in India. The use of railways for the transportation of fruits and vegetables in India could be greatly enhanced by: making provisions for cooling and ventilation, providing improved handling facilities at platforms and providing storage space to accommodate the goods upon arrival at their destinations. Similarly road services could be considerably improved by widening roads, upgrading surfaces and through the introduction of one-way traffic. Long waits at level crossings should be avoided by introducing flyovers at intersections in order to increase the speed and movement of goods by road. Facilities such as sheds, for temporary cool storage should be available on highways and major roads.

8.10 Cold/Cool Chain

The adoption of cold chain systems has been pivotal to trade in fruits and vegetables in developed countries. In India, there has been a spectacular development in horticulture during the last decade. Production and productivity have increased manifold and availability of fruits and vegetables are on the increase. India occupies second position in production of fruits and vegetables, with an annual production of 94 million tons of vegetables and 45 million tons of fruits. However, losses in these commodities accounting for 25–30 % are of great concern.

Fruits and vegetables are living commodities even after harvest and low temperature storage is the only widely used method to extend the storage life, maintenance of quality, and control of spoilage. Cold storage is a system with thermal insulation and refrigeration in which perishables commodities can be stored for a set period of time under controlled conditions of temperature and humidity.

Why Cold Chain Is Necessary?

- For preservation
- For maintaining nutritional quality
- To increase shelf-life
- To ensure availability of the produce throughout the year for direct consumption as well as processing
- To reduce wastage losses
- To preserve the seasonal produce in cold storage and selling during off season to fetch higher returns

9 Factors Involved for Effective Cold Storage of the Produce

Product quality: Fresh fruits and vegetables intended for storage should be free from physical damage, of optimum maturity and free from infections.

Temperature: Low temperature storage is recommended for perishables as it retards respiration and metabolic activity, aging due to ripening, softening and textural and color changes, moisture loss, spoilage due to diseases and undesirable growth (sprouting/cooling).

Relative humidity (RH): The relative humidity of the air in storage rooms directly affects the keeping quality of the products held in them. If it is too low, wilting or shriveling is likely to occur and if it is too high, it may favor the development of decay. An RH of 85–90 % is recommended for most perishables.

Air circulation and package spacing: Air must be circulated to keep a cold storage room at an even temperature throughout the storage. This is required to remove respiratory heat. Entry of outside air and proper spacing of containers on pellets are also important.

Respiration rates, heat evolution, and refrigeration: When the storage of fresh fruits and vegetables is considered, it should be remembered that these commodities are alive and carry on all activities of living tissues, the most important being respiration. During this process, energy is released in the form of heat which varies with the commodity and the temperature.

Weight loss in storage: Loss of water from harvested horticultural crops is a major cause of deterioration in storage. Some loss can be tolerated but losses great enough to cause wilting or shriveling must be avoided. Under good handling conditions with recommended humidity and temperature, moisture loss can be held under control.

Sanitation and Air purification: Good air circulation alonc is of considerable value in minimizing surface molds. Accumulation of odors and volatiles may contribute to off flavors and hasten deterioration.

Precooling is the process of removing field heat surroundings during harvesting. For most fresh horticulture commodities, 1 h time loss at the field temperature of 35 °C between harvest and precooling can reduce quality as much as 20 h in storage under proper conditions.

Benefits from precooling

- Helps in slowing down the respiration
- Reduces ethylene production and delays ripening of fruits
- Reduces moisture loss and shriveling of the produce
- Reduces/inhibits growth of decay causing microorganisms
- Reduces the load on the cooling system (refrigeration) of transport or storage chamber

To achieve maximum storage for a crop or to reduce losses during its marketable life it is essential to keep it at the most appropriate temperature; this is usually just above the temperature which will cause chilling or freezing injury. To maximize the effect, the crop should be brought to that temperature as quickly as possible after harvest. This is known as crop precooling. The rate of precooling depends on:

1. The difference in temperatures between the crop and cooling medium.
2. Accessibility of the cooling medium to the crop.
3. The nature of the cooling medium.
4. The velocity of the cooling medium.
5. Rate of transfer of heat from the crop to the cooling medium.

Heat can be transferred from the crop in three ways: conduction, convection, or radiation. In precooling, this is usually achieved by conduction, which is the direct movement of heat from one object or substance to another by direct contact. If two objects or substances are touching and they are of different temperatures, heat will always move from the warmer to the cooler. If a crop has a relative large mass and a small surface area, the crop will be cooled more slowly than if it had a smaller mass and a larger surface area. This is because in the former the heat from the inside of the crop has to move to the surface before it is transferred. The heat within the crop comes either from convection from the air surrounding it—radiation, mainly from the sun—or from metabolic heat from chemical reactions within the crop. If a crop is harvested in the early morning it will be cooler, since the sun has not been able to warm the air or the crop. It will also have a lower level of metabolic heat since, generally, the warmer the crop the higher its metabolic rate. Another factor, which affects cooling, is the packaging material used. If the cooling medium is air, the produce boxes should have ventilation holes which ensure good circulation through the crop. These holes may be placed to coincide with an air stream, as in forced-air cooling. For hydrocooling and icing, the box used must withstand direct contact with water without disintegrating. Plastic boxes are suitable for this purpose or fiberboard cartons which have been waxed or treated with some other substance which will render them waterproof. Where heat is removed by evaporative cooling, the crop must not be sealed in moisture-proof film such as polyethylene bags.

The bags must have adequate perforations to allow the water vapor to flow out. The methods commonly used to achieve precooling are dealt with below.

Precooling crops promptly after harvest may have a large benefit on their subsequent maximum storage life. In many cases, such as curing root crops, drying onions, and quailing oranges, there are definite storage advantages in delaying precooling (Liu 1991a, b). In a study of *Asparagus*, where spears were held at 18 °C and 50 % RH for different periods until they lost 0, 2, 4, 6, or 8 % in weight and then stored at V_2°C, it was found that spears which had lost 8 % in weight were unsaleable after 14 days storage, while those that had lost 0 % in weight were still saleable after 28 days.

Several methods of precooling are used commercially and which method is most suitable depends on the crop, the marketable or storage life required, and the economics. Different methods include placing ice over the produce, passing cold air around the produce, immersing it in water, or reducing the air pressure around it. The different methods have been shown to give widely different cooling times.

The different methods used for precooling are as follows:

- Icing
- Room cooling
- Forced air cooling
- Hydro cooling
- Vacuum cooling

9.1 Icing

This has been used for many decades and is also called contact or top icing. It is commonly applied to boxes of produce by placing a layer of crushed ice directly on top of the crop. The ice melts and the cold water runs down through the crop, cooling it. It can also be applied as ice slurry which is hosed onto the top of the crop from a tank. A typical slurry is made from 60 % finely crushed ice, 40 % water, and usually 0.1 % sodium chloride to lower the melting point of the ice, although water–ice ratios may vary from 1:1 to 1:4. Ice slurries give greater contact between produce and ice than top icing and should therefore result in quicker cooling. The main use of top icing is for road transport and it can be applied shortly after harvest to, for example, field-packed lettuce, to begin precooling as soon as possible.

9.2 Room Cooling

This method simply involves placing the crop into a cold store. This may be the same cold store where the crop is to be stored for long periods or it may be to hold the crop, briefly, before it enters the marketing chain, to facilitate accumulation of sufficient produce to send to market. The type of room used may vary, but generally they consist of a refrigeration unit across which cold air is passed from a fan. The air circulation may be such that the air is blown across the top of the room and falls through the crop by convection. It may also be blown in at the bottom of the store, in which case some type of plenum chamber is usually used. The drawback to these methods is the length of time it can take for the temperature of the crop to be reduced. The main advantage is its cost, because no special facilities or equipment are required.

9.3 Forced-Air Cooling

The principle of this type of precooling is to place the crop into the cold room but to arrange the airflow pattern so that it is directed through the crop. The heat given out from the surface of the crop is then carried away in the stream of cold air, thus setting up a temperature gradient and cooling the crop more quickly. In a comparison between two airflow rates on the half-cooling time of single apple fruits, it is mentioned in literature that it to be 1.25 h in an air velocity of 40 m min^{-1} and 0.5 h at 400 m min^{-1}. To achieve this rapid cooling, the cooled air may be forced into a plenum chamber and the boxes of crops placed at exit to the chamber. A common type of precooler, which uses this method, is the letterbox system. The plenum chamber takes up one wall of the cold store, with the wall of the chamber fitted with slots which are closed with removable covers. These slots coincide with pallet bases so that the cooled air is directed into the base. Horizontal exits from the base are blocked off and the base of the pallet is slatted, so that the air is forced up through it, up through the produce, and recycled back through the cooling unit. A variation of this method is to have the plenum chamber set into the floor of the cold room so that when the produce is stacked the air is directed up through it. It is essential to ensure that as much air as possible is blown directly through the produce, so all possible escape channels must be blocked off with inflatable bags. It is also important to achieve an even distribution of air through the crop to ensure uniform cooling.

An efficient forced-air cooling system can pass air of a very high velocity and thus lead to desiccation of the crop. To reduce this effect various methods of humidifying the cooling air have been devised. One of these was the "ice bank cooler." In cold stores air is blown over cooling coils, which are metal pipes of various designs through which a cooled liquid is passed. Where rapid cooling is necessary the surface area of the pipes must be large and the temperature of the cooling liquid must be considerably lower than the air temperature to achieve good heat exchange. This can result in moisture condensing or freezing on the surface of the coils, which reduces their efficiency but, perhaps even more important, it can also increase the speed with which the crop is being desiccated. The ice bank cooler has the cooling coils immersed in water so that the water freezes, building up ice around them. Water is then pumped over the ice and sprayed in a fine mist in a countercurrent of air entering the plenum chamber. The air passes through a filter to take out liquid water particles from it and is then passed through the crop. The effect is that the air is both cooled and humidified. Humidity is close to 100 % RH in a well-constructed system and will result in minimum desiccation of the crop during cooling. Portable ice bank coolers have been constructed. These are small (about 1 ton capacity) insulated trailers which can be towed into the field and driven from the power take-off of a tractor. The ice bank is built up overnight and the water pump and air circulation fan are driven from the tractor engine in the field. It is useful for crops such as strawberries, where it may be important to begin cooling directly after harvest.

9.4 Hydrocooling

The transmission of heat from a solid to a liquid is faster than the transmission of heat from a solid to a gas. Therefore cooling crops with cooled water can be very quick and result in no loss of weight. If the crop is simply immersed in iced water, the water in direct contact with the crop will heat up and the rate of cooling will be slow. To achieve maximum effect the cooled water must constantly be passed over the crop. This can be done by submersing the crop in cold water which is constantly being circulated through a heat exchanger. Where crops are being transported around a pack house in water, the transport can be the hydrocooler. A common design is to transport the crop on a perforated conveyer belt over a tank of water. Water is pumped from the tank over cooling coils or blocks of ice and allowed to fall on the crop and then through to the tank below. This system has the advantage that the speed of the conveyer can be adjusted to the time required to cool the crop. Hydrocooling has the advantage over other precooling methods in that it can help clean the produce. However, the water can become contaminated with microorganisms, which can result in increased levels of spoilage during subsequent storage or marketing. Chlorine may be added to the water if this is a problem. Hydrocooling is unsuitable for many crops. Capsicums which were hydrocooled had a higher incidence of rots during subsequent storage than those which were not, even when chlorine was added to the water. This was due to water being trapped between the fruit and the calyx. Cooling time depends on the size of the produce. Asparagus spears, being long and narrow can be hydrocooled in about 2 min but capsicums, which are large and globular, took about 10 min to cool. Tomatoes, melons, and leafy vegetables can also be hydrocooled.

9.5 Vacuum Cooling

Cooling is achieved by the latent heat of vaporization rather than conduction. At normal air pressure of 760 mmHg, water will boil at 100 °C. As air pressure is reduced so the boiling point of water is reduced, and at 4.6 mmHg water boils at 0 °C. For every 5 or 6 °C reduction in temperature, under these conditions, the crop will lose about 1 % in weight. The speed and effectiveness of cooling is related to the ratio between the mass of the crop and its surface area, so it is particularly suitable for leaf crops such as lettuce. In forced-air cooling the air passes over the surface of the crop, cooling the outside, while the inside is cooled by heat transfer from inside to the outside of the crop. In contrast, with vacuum cooling of leaf vegetables the reduced pressure is exactly the same around the leaves in the center as it is around the leaves on the outside. This means the cooling is very even and quick throughout the crop. Where there is a low ratio between mass and surface area, and an effective barrier to water loss from the crop surface, vacuum cooling can be slow. Crops like tomatoes, which have a relatively thick wax cuticle, are not suitable for vacuum cooling. Vacuum coolers have to be strongly constructed to withstand the vacuum and therefore tend to be expensive to construct. They are made from heavy-duty steel and are usually cylindrical in shape to withstand the low pressure. The vacuum is usually achieved by a vacuum pump attached to the cylinder and the moisture from the crop is condensed on cooling coils situated on the outlet from the cylinder to the pump. In spite of the high cost, they are probably used for most lettuce marketed in Europe and the USA because the method is so quick. Loss of water during marketing of crops not only affects their value where they are sold by weight, but can also affect their value because of loss of quality. In order to reduce this problem the crop may be sprayed with water before loading it into the vacuum cooler. Special vacuum coolers have been developed called "hydrovac" coolers, which have a built-in water spray which is activated toward the end of the cooling operation.

9.6 Indigenous Fruit Processing

India produces a range of indigenous tropical fruits, of excellent flavor and color. Many of these fruits are, however, underutilized. A large number are known for their therapeutic/medicinal and nutritive value. Consumers today are becoming increasingly conscious of the health and nutritional benefits of the food they consume, and there is an increasing tendency to avoid the consumption of chemically treated foods. Indigenous fruits can play an important role to in satisfying the demand for nutritious, delicately flavored and attractive natural foods of high therapeutic value. Some of the fruits produced in India are, however, unappealing in the fresh form, but offer considerable potential for processing and marketing. Wood apple fruit, for example, which has a hard shell, mucilaginous texture, and numerous, is not popularly consumed as a dessert fruit in India. Kokum is yet another fruit which is not acceptable in the fresh form owing to its high level of acidity, while fresh aonla has a strong astringent taste. Though some value-added fruit products are currently being manufactured on a small scale, no systematic approach has been made to utilize the potential of indigenous fruit on a large scale, owing primarily to the lack of adequate quantities of raw materials. The development of orchards and the systematic collection of raw materials are of utmost importance in developing fruit processing industries. Rather than competing in markets where products are already established, India must break new ground and create markets for value-added products from its indigenous fruit, which could offer a competitive advantage.

10 Traditional Food Processing Technologies

Traditional processing technologies, applied in the conservation of horticultural produce employ a gradient of technologies, ranging from artisanal to intermediate to high technologies. Major categories of processed products produced with the use of these technologies include fruit preserves, fruit and vegetable juices, fermented products (wines and vinegars), candied products, and frozen and dried products. While these processing technologies are generally effective in rendering fruits and vegetables microbiologically stable, they alter the flavor, color, and texture characteristics of the product.

11 Freezing

Freezing is a fairly widespread technology which when properly carried out causes minimal changes in the shape, texture, flavor, and color of foods. Freezing is a quick, convenient, and popular way to preserve fruits and vegetables, and it reserves more nutrients in the food if properly done. Frozen foods are easy to serve on short notice. Freezing retards the growth of bacteria, molds, and yeasts. Once the food is thawed, microorganisms may continue to grow. Natural enzymes in foods cause changes in flavor, color, texture, and nutritive value. Freezing slows this activity but does not stop it. To prevent further enzyme activity, vegetables need to be blanched in boiling water or steamed before freezing (Khetarpaul 2005). Some nutrient loss occurs when vegetables are blanched before freezing. By comparison, however, the nutrient losses from enzymatic activity are greater if vegetables are not blanched. Enzymatic browning in light colored fruits can be prevented by using ascorbic acid mixtures. If fruits and vegetables are not properly packaged, air can cause changes that affect flavor. Moisture from the food can evaporate, causing the food to become dry and tough. Off-flavors will develop. To prevent this "freezer burn," use moisture vapor-proof or resistant packaging, such as "can or freeze" glass jars, plastic freezing containers, heavy weight aluminum foil, plastic coated freezer paper, and plastic wraps. The water in fruits and vegetables expands during freezing and breaks the cell walls. That's why thawed foods sometimes have a soft, mushy texture. To maintain as much of the crisp texture as possible, fruits and vegetables should be frozen at the lowest possible setting on the freezer. Store frozen foods at 0 °F or lower. Constant storage temperature helps in retention of quality. Fluctuating temperature will damage the texture of frozen fruits and vegetables as the ice melts and then refreezes in the cells. Nutrient loss is lower when stored at 0 °F or below than at higher temperatures. Many fruits and vegetables will retain good quality for up to 12 months, when stored at temperatures of −180 °C. The keeping quality of frozen products is, however, dependent on their storage conditions. Energy requirements for freezing operations are high and thus the cost of this technology and the storage of its products are high.

12 Thermal Processing

Thermal processing is still the major technology applied in the shelf-stable preservation of fruits and vegetables. Thermal processing can be carried out at a range of technical levels, from the cottage through to the industrial level, and is widely applied in the production of jams, jellies, and canned and bottled fruits and vegetables. Recent developments in thermal processing technologies include the use of aseptic processing systems which make use of sterile laminated packaging. While these developments have improved product quality, they have increased product price.

13 Various Degrees of Preservation

There are various degrees of preservation by heating; a few terms have to be identified and understood.

(a) Sterilization: By sterilization we mean complete destruction of micro-organisms. Because of the resistance of certain bacterial spores to heat, this frequently means a treatment of at least 121 °C (250 °F) of wet heat for 15 min or its equivalent. It also means that every particle of the food must receive this heat treatment. If a can of food is to be sterilized, then immersing it into a 121 °C pressure cooker or retort for the 15 min will not be sufficient because of relatively slow rate of heat transfer through the food in the can to the most distant point.
(b) "Commercially sterile": Term describes the condition that exists in most of canned or bottled products manufactured under Good Manufacturing Practices procedures and methods; these products generally have a shelf-life of 2 years or more.
(c) Pasteurization: It means a comparatively low order of heat treatment, generally at a temperature below the boiling point of water. The more general objective of pasteurization is to extend product shelf-life from a microbial and enzymatic point of view. Pasteurization is frequently combined with another means of preservation—concentration, chemical, acidification, etc.
(d) Blanching: It is a type of pasteurization usually applied to vegetables mainly to inactivate natural food enzymes. Depending on its severity, blanching will also destroy some microorganisms.

14 Drying and Dehydration Techniques

Drying or dehydration is the removal of the majority of water contained in the fruit or vegetable and is the primary stage in the production of dehydrated fruits and vegetables. Several drying methods are commercially available and the selection of the optimal method is determined by quality requirements, raw material characteristics, and economic factors. There are three types of drying processes: sun and solar drying; atmospheric dehydration including stationary or batch processes (kiln, tower, and cabinet driers) and continuous processes (tunnel, continuous belt,

belt-trough, fluidized-bed, explosion puffing, foam-mat, spray, drum, and microwave-heated driers); and sub atmospheric dehydration (vacuum shelf, vacuum belt, vacuum drum, and freeze driers).

Sun drying (used almost exclusively for fruit) and solar drying (used for fruit and vegetables) of foods use the power of the sun to remove the moisture from the product. Sun drying of fruit crops is limited to climates with hot sun and dry atmosphere, and to certain fruits, such as prunes, grapes, dates, figs, apricots, and pears. These crops are processed in substantial quantities without much technical aid by simply spreading the fruit on the ground, racks, trays, or roofs and exposing them to the sun until dry. Advantages of this process are its simplicity and its small capital investment. Disadvantages include complete dependence on the elements. Solar drying utilizes black-painted trays, solar trays, collectors, and mirrors to increase solar energy and accelerate drying. Atmospheric forced-air driers artificially dry fruits and vegetables by passing heated air with controlled relative humidity over the food to be dried, or by passing the food to be dried through the heated air, and are the most widely used method of fruit and vegetable dehydration. Various devices are used to control air circulation and recirculation. Stationary or batch processes include kiln, tower (or stack), and cabinet driers. Continuous processes are used mainly for vegetable dehydration and include tunnel, continuous belt, belt-trough, fluidized-bed, explosion puffing, foam-mat, spray, drum, and microwave-heated driers. Tunnel driers are the most flexible, efficient, and widely used dehydration system available commercially. Sub atmospheric (or vacuum) dehydration occurs at low air pressures and includes vacuum shelf, vacuum drum, vacuum belt, and freeze driers. The main purpose of vacuum drying is to enable the removal of moisture at less than the boiling point under ambient conditions. Because of the high installation and operating costs of vacuum driers, this process is used for drying raw material that may deteriorate as a result of oxidation or may be modified chemically as a result of exposure to air at elevated temperatures. There are two categories of vacuum driers. In the first category, moisture in the food is evaporated from the liquid to the vapor stage and includes vacuum shelf, vacuum drum, and vacuum belt driers. In the second category of vacuum driers, the moisture of the food is removed from the product by sublimation, which is converting ice directly into water vapor. The advantages of freeze drying are high flavor retention, maximum retention of nutritional value, minimal damage to the product texture and structure, little change in product shape and color, and a finished product with an open structure that allows fast and complete rehydration. Disadvantages include high capital investment, high processing costs, and the need for special packing to avoid oxidation and moisture gain in the finished product.

15 Postdehydration Treatments

Treatments of the dehydrated product vary according to the type of fruit or vegetable and the intended use of the product. These treatments may include sweating, screening, inspection, instantization treatments, and packaging. Sweating involves holding the dehydrated product in bins or boxes to equalize the moisture content.

Screening removes dehydrated pieces of unwanted size, usually called "fines". The dried product is inspected to remove foreign materials, discolored pieces, or other imperfections such as skin, carpel, or stem particles. Instantization treatments are used to improve the rehydration rate of the low-moisture product. Packaging is common to most of all dehydrated products and has a great deal of influence on the shelf-life of the dried product. Packaging of dehydrated fruits and vegetables must protect the product against moisture, light, air, dust, microflora, foreign odor, insects, and rodents; provide strength and stability to maintain original product size, shape, and appearance throughout storage, handling, and marketing; and consist of materials that are approved for contact with food. Cost is also an important factor in packaging. Package types include cans, plastic bags, drums, bins, and cartons, and depend on the end-use of the product.

16 Fermentation

Fermentation is the slow bioconservation process of foods induced by microorganisms, or by enzymes of microbial, plant, or animal origin. It is one of the oldest forms of food preservation/processing technologies in the world. Fermentation technologies in developing countries have evolved through years of experience rather than through scientific breakthroughs. A variety of fruits and vegetables are preserved by fermentation.

Although fermentation of foods has been in use for thousands of year, it is likely that the microbial and enzymatic processes responsible for the transformations were largely unknown. It is only recently that there has been a development in the understanding of these processes and their adaptation for commercialization. There is tremendous scope and potential for the use of micro-organisms toward meeting the growing world demand for food, through efficient utilization of available natural food and feed stocks and the transformation of waste materials.

17 Benefits of Fermentation

17.1 Food Preservation

Fermentation is a cheap and energy efficient means of preserving perishable raw materials. When harvested, fruit and vegetables undergo rapid deterioration, especially in the humid tropics where the prevailing environmental conditions accelerate the process of decomposition. Fermentation requires very little sophisticated equipment, either to carry out the fermentation or for subsequent storage of the fermented product. It is a technique that has been employed for generations to preserve food for consumption at a later date and to improve food security. There are examples from around the world of the role; fermented foods have played in preserving food to enhance food security.

17.2 Salvaging Waste Foods

Fermentation can salvage waste food which otherwise would not be usable as food by changing the consistency of the product and making it digestible. This increases the range of raw materials available as food.

17.3 Removal of Antinutritional Factors

Many fruits and vegetables contain naturally occurring toxins and antinutritional compounds. These can be removed or detoxified by the action of microorganisms during fermentation. For instance the fermentation process that produces the Sudanese product *Kawal* removes the toxins from the leaves of *Cassia obtusifolia* and fermentation is an important step in ensuring that cassava is safe to eat.

17.4 Minimal Processing

Minimal processing employs an integrated approach wherein the handling, processing, packaging, and distribution of raw fruits and vegetables is properly managed with the application of appropriate food safety principles of Good Manufacturing Practices (GMP) and Hazard Analysis and Critical Control Point (HACCP).

The consumer demand for high quality foods requiring only minimum amount of effort and time for preparation has led to the introduction of ready to use, convenience foods preserved by mild methods (so called minimal processing methods) only. There is now a wide choice of prepacked fresh produce available, meeting the twin consumer demands for unadulterated freshness and maximum convenience.

Fruits and vegetables are living, respiring, edible tissues which continue to be metabolically active when harvested. Losses due to respiration, breakdown of organic substrates, and transpiration water are not replaced once the fruit or vegetable is detached from the parent plant and so deterioration occurs. This physiological deterioration can be affected by intrinsic (i.e., climacteric vs. non-climacteric commodities) and extrinsic [i.e., temperature, ethylene (C_2H_4), oxygen (O_2), and carbon dioxide (CO_2) concentrations] factors. Other types of deterioration which occur include chemical and enzymatic changes, microbial deterioration, and attack by macro-organisms. Physical damage, such as improper harvesting, handling, processing, and packaging can also affect the shelf-life of fruits and vegetables.

Minimal processing of fruits and vegetables may involve cooling, washing, trimming, peeling, shredding, or slicing before packaging. Such products include prepared salads, pre-cut vegetables, peeled fruits, and whole vegetables, sold within 7 8 days of preparation, after storage at a low temperature. A minimally processed product must have a consistent, fresh appearance, be of acceptable color, and be reasonably free of defects.

17.5 Processing of Fruits and Vegetables for Value Addition

The fruit and vegetable sector has grown substantially both in volume and in variety of outputs traded globally. Rising incomes, falling transportation costs, improved technologies, and evolving international agreements, have all contributed to this level of growth. This increased level of fruit and vegetable production has, unfortunately, not been matched by developments in supply chain management, or by vertical integration of production with processing in India. Processing activities are of critical importance to expansion and diversification within the fruit and vegetable sector in that they increase market opportunities for fresh fruits and vegetables and add value while minimizing postharvest losses. Furthermore, processing improves the viability, profitability, and sustainability of fruit and vegetable production systems by increasing farm incomes and generating rural employment and foreign exchange. Traditional processing technologies such as thermal processing (bottling and canning), freezing, dehydration (salting, brining and candying), drying, and fermentation are widely applied in the processing of fruits and vegetables at various levels (artisanal, intermediate and high) and scales (cottage, small, medium and large). Tropical juices and fruit pulps, canned pineapples, tomato paste, and canned and dried mushrooms are examples of fruit and vegetable products produced using traditional processing technologies and which are increasingly entering in international trade. Dried and canned mushrooms produced in China, currently account for 52 % of world trade in processed mushrooms, while canned pineapples produced in Thailand accounts for approximately 45 % of that product in world trade. Minimal processing technologies, specialized packaging, and natural preservation systems are increasingly being applied in the preservation of fruits and vegetables for both developed and developing country markets, in response to growing consumer demand for convenience and for "fresh-like" fruits of high quality which are nutritious, flavorful, and stable. These processing technologies focus on adding value with comparatively little product transformation while increasing product diversity. While minimal and traditional processing technologies present considerable opportunities for innovation and vertical diversification in the fruit and vegetable sector, relatively few small and medium enterprises (SMEs) are able to tap into and benefit from these opportunities. Many SMEs lack the capacity to operate competitively in the current globalized market environment owing to problems of scale, the poor quality of input supplies, poor access to technology, limited technical expertise and research capacity, low production efficiency, high marketing cost, lack of knowledge, and consequently inability to comply with international standards for processed products.

18 Modern Fruit and Vegetable Processing Technologies

Growing consumer demand for convenience and for safe foods of high quality with "fresh like attributes" has led to considerable innovation and diversification in fruit and vegetable processing. New product lines which include high care products such as trimmed and packed beans, ready prepared salads, pre-prepared stir fry mixes, and prepared fruits are increasingly entering in supermarkets in developing countries and in export trade. While value addition of this type generally requires relatively little product transformation, it necessitates investment in technology, equipment, management systems, and stringent adherence to food safety principles and practices if product quality is to be ensured. Water of an appropriate quality and cold storage during the processing, packaging, distribution, and retailing are key requirements of minimal processing operations. Minimal processing applications which employ the hurdle concept are generally inexpensive, energy efficient, simple, and satisfactory for the in situ preservation of fruits and vegetables.

19 Irradiation

Ionizing radiation can be applied to fresh fruits and vegetables to control microorganisms and inhibit or prevent cell reproduction and some chemical changes (Patterson 1993). It can be applied by exposing the crop to radioisotopes (normally in the form of gamma-rays but X-rays can also be used) and from machines which produce a high-energy electron beam. Radiation doses are measured in Grays (Gy). In older literature they were quoted in rads, where 1 Gray = 100 rads.

Radioisotopes cannot be switched on and off so they are immersed in a pool of water to allow operators to enter the processing area. When food is to be irradiated, the radioisotope is raised out of the water and the material to be irradiated is usually passed through the radiation field on conveyor belts. The whole processing area is surrounded by thick concrete to contain the radiation field.

Another method of irradiation is electron gun, in which radiation is produced from a machine source which accelerates electrons produced in a heated cathode by a high-voltage electrostatic field. These electrons are guided to form a beam, which may be used directly on the crop. Alternatively they may be used to strike a heavy metal such as tungsten to produce X-rays or gamma-rays, which in turn can be used to irradiate the crop. The main advantage of this form of irradiation is that it can be switched off when not in use, but it tends to be expensive and have limited penetration.

20 Uses of Irradiation

Irradiation can be used to suppress sprouting in potatoes. It is described in literature that irradiated tubers did not sprout during 15 months of storage at 4.4 °C. A dosage rate of 100 Gy was found to be adequate, but lower rates of about 35 Gy can be used to delay sprouting without the side effects which were observed on tubers exposed to higher doses. Another study demonstrated interactions between temperature and dosage of irradiation. They also showed that a dose of 100 Gy was effective in preventing spouting of potatoes in storage for 6 months at 10 °C and 21.1 °C, but 50 Gy was equally effective when the tubers were stored at 4.4 °C. Yam tubers did not sprout during 3 months of storage after being exposed to 50 Gy radiation compared with untreated tubers all sprouting during the same storage period. In onions, doses of 20–30 Gy could completely prevent sprouting if applied as soon as possible after harvest. If irradiation was delayed, it was much less effective. A combination of hot water treatment (55 °C for 5 min) followed by 30 Gy irradiation was found to be the best treatment in terms of shelf-life extension and quality of mangoes. After this treatment mangoes had a storage life of 38 days (at 15 °C), 28 % rotting and no irradiation injury.

Irradiation can be used to control postharvest diseases of fresh fruit and vegetables (Diehl 1992). It can also be used as a postharvest treatment to disinfest fruit of insects (Lalaguna 1998). 300 Gy was shown to control mango seed weevil but more work is needed; 150 Gy was shown to control 11 species of Tephritid fruit fly and 75 Gy prevents the adults emerging from the fruit. Doses of irradiation in excess of 600 Gy caused lenticels spotting, surface discoloration and retardation of ripening of Kensington Pride mangoes, but irradiation at this level contributed only minor improvements in disease control. However, irradiation followed immediately by hot benomyl treatment controlled anthracnose and stem-end rot during storage at 20 °C for 15 days.

21 High Pressure Processing

High pressure processing (HPP), also known as high hydrostatic pressure, is a non-thermal food preservation technique that has the potential to meet these demands. It is an opportunity to preserve food, by applying intensive pressure in the range of 300–900 MPa, without adversely affecting organoleptic, textural, and nutritional qualities as thermal processing like pasteurization and sterilization may do. In a typical high pressure batch cycle, the food prepackaged in a high-barrier flexible pouch or a plastic container is loaded into a perforated basket that goes into the pressure vessel; the pressure is then increased to the processing target pressure (come-up time); the product is held at the desired pressure for 3–10 min (pressure holding time); after which the pressure is released in usually few seconds (decompression time) and the product can be unloaded at this point. The pressure is applied uniformly in all directions simultaneously and this is known as isostatic pressure (Cheftel 1995). Pressurization is usually accompanied by a moderate and uniform temperature increase called adiabatic heating. However, the food product usually rapidly returns to its initial temperature at decompression. With the recent shift in consumer lifestyle toward healthy living and healthier food, the consumption of raw fruits and vegetables has increased in popularity. However, as per the Centers of Disease Control and Prevention, fruits and vegetables have recently been associated with multiple food-borne disease outbreaks; the effect of high pressure processing on microbial safety, quality, and sensory characteristics of fruits and vegetables has therefore been widely investigated as an alternative to traditional food processing and preservation methods. HPP inactivates microorganisms and quality-deteriorating enzymes and has limited effects on covalent bonds resulting in minimal modifications of food quality attributes such as color, flavor, and nutritional values (Knorr 1993). However, depending on the fruit or vegetable, high pressure could induce chemical or biochemical reactions that can affect their quality attributes.

22 Ohmic Heating

Ohmic heating is an advanced thermal processing method wherein the food material, which serves as an electrical resistor, is heated by passing electricity through it. Electrical energy is dissipated into heat, which results in rapid and uniform heating (Sarkis et al. 2013). Ohmic heating is also called electrical resistance heating, Joule heating, or electro-heating, and may be used for a variety of applications in the food industry.

Ohmic heating can be used for heating liquid foods containing large particulates, such as soups, stews, and fruit slices in syrups and sauces, and heat-sensitive liquids. The technology is useful for the treatment of proteinaceous foods, which tend to denature and coagulate when thermally processed. For example, liquid egg can be ohmically heated in a fraction of a second without coagulating it. Juices can be treated to inactivate enzymes without affecting the flavor. Other potential applications of ohmic heating include blanching, thawing, on-line detection of starch gelatinization, fermentation, peeling, dehydration, and extraction.

Like thermal processing, ohmic heating inactivates microorganisms by heat. Additional nonthermal electroporation-type effects have been reported at low-frequency (50–60 Hz), when electrical charges can build up and form pores across microbial cells, however, it is not necessary to claim such effects since heating is the main mechanism.

The shelf-life of ohmically processed foods is comparable to that of canned and sterile, aseptically processed products.

Pulsed Electric Field (PEF) processing is a method for processing cells by means of brief pulses of a strong electric field. PEF holds potential as a type of low temperature alternative pasteurization process for sterilizing food products (McDonald et al. 2000). In PEF processing a substance is placed between two electrodes and then the pulsed electric field is applied. The electric field enlarges the pores of the cell membranes which kills the cells and releases their contents. When exposed to high electrical field pulses, cell membranes develop pores either by enlargement of existing pores or by creation of new ones. These pores may be permanent or temporary, depending on the condition of treatment. The pores increase membrane permeability, allowing loss of cell contents or intrusion of surrounding media, either of which can cause cell death. PEF has limited effects on spores and only appears to affect a few enzymes (McNamee et al. 2010). Enzymes are important in juice processing because surviving enzymes can reduce pectin, which then can be less effective in keeping fruit particles suspended. Some sedimentation is common in fruit juices, but too much is unattractive. Some surviving enzymes may also enable discoloration and production of off-flavors. PEF offers a 5-log reduction of most pathogens and is considered a pasteurization process, so products must be refrigerated. An important process consideration is prevention of postprocess contamination.

23 Microwave Processing

Microwave heating has vast applications in the field of food processing such as cooking, drying, pasteurization, and preservation of food materials (Coronel et al. 2005). Microwave pasteurization has the ability to achieve destruction of microorganisms at temperatures less than that of conventional pasteurization due to significant enhancement or magnification of thermal effects. Applications of microwave drying include microwave assisted hot air drying, microwave vacuum drying, and microwave freeze drying. Microwave drying combined with other conventional methods of drying enhances the drying characteristics of the sole effect of microwave drying. Microwave modeling can be used to predict the temperature and moisture distributions during microwave heating of food materials (Steed et al. 2008). It is required to obtain better end product qualities of food materials by conducting more research at pilot scale levels. It is also necessary to eliminate hot spots or nonuniform temperature distribution during microwave heating of food materials.

24 Ultrasound Technology

Ultrasound is one of the nonthermal methods that are used for foods in the last decades. It can be applied to solid, liquid and gas systems for different purposes. Its instrumentation can be fully automated and make rapid and precise measurements. The principle aim of this technology is to reduce the processing time, save energy and improve the shelf-life and quality of food products. The advantages of ultrasound over the heat treatment include; minimization of flavor loss, greater homogeneity, and significant energy savings. Other advantages include high productivity, enhanced quality, reduced chemical and physical hazards, and are environmentally friendly. When it is applied with pressure and/or temperature its efficiency increases but cautions needed to determine and control nutritional loss. Also, process parameters and applied material change the results. Consequently, ultrasound is a good alternative method for the food preservation and processing and also no adverse effect on human health has been proven. Although there are many studies relating ultrasonic application in laboratory scale, its application in the food industry is not sufficiently common. Future studies should be focused on scale-up and standardization of treatment processes.

25 Recommendations for Improvement of Existing Techniques/Technologies in India

The suggested recommendations to improve postharvest management are described as follows:

25.1 Postharvest "Specific" Problems

Although postharvest is not a new discipline, there is still a total lack of postharvest activity in India, weak activity in some and reasonably active research and application in others. However, in India, specific postharvest problems are not fully identified and characterized and therefore solutions will not be easily found or implemented.

25.2 Research

Few specialized postharvest institutions exist in India, and postharvest research, education, and extension are not very active here. Improvement in the postharvest sector is pivotal to stabilization of the food supply. There is an urgent need for the enhancement of agricultural and rural development through fostering agricultural research and technology development/transfer, and by strengthening inter- and intraregional collaboration. Research needs to be justified and made more efficient. Researchers need to be more multidisciplinary and collaborate with researchers of other disciplines (engineering, biology, ecology, chemistry, physics, etc.). Collaboration within the same region should be increased. Federal and private sector investment in research needs to be increased very significantly.

25.3 Postharvest Education/Extension

The availability of appropriate postharvest technologies, and the difficulties in India to easily establish sound research programs, requires that more attention be given to postharvest education and extension, which is very weak here. Growers and distributors should be "made aware" and should be trained adequately to exploit the existing knowledge and technologies.

25.4 Technology Transfer

Technology transfer programs in India are either totally lacking, or are not operating adequately, and therefore should be improved significantly.

25.5 *Increase Private Sector Participation/Reduce Government Involvement*

It is clear that the successful stories in agriculture in general and in postharvest application in particular here have been due to private sector leadership. In contrast, many of the failures have been due to government controls.

25.6 *Adequate Collaboration Between Researchers Especially at the Regional Level*

There is more collaboration between researchers in developing countries and developed countries than between researchers within the developing countries. Although collaboration between researchers in all countries is important, Quality standards and quality control measures are not commonly used, especially for products intended for national markets. Adequate transportation facilities are reasonably good in some countries, but are either not available or not functioning adequately in several other countries. In general, there is a scarcity of trained service technicians.

There are major weaknesses and difficulties in communication, which hampers several aspects of development, including food availability. There is a lack of sufficiently qualified personnel for information collection, and still very weak access to the internet and to international networks on information, and lack of necessary information and communication facilities.

Technical assistance to producers/handlers is adequate, but very scarce and deficient in several others. Marketing systems and infrastructure are improving and in fact are reasonably strong, but still face major difficulties.

Although postharvest research is reasonably active, it is scarce or even totally lacking in some regions; it is without defined objectives. Research infrastructure is slightly improving, but still poor in many of these areas. Education on postharvest has improved, but is still poor or nonexistent in several regions. There is clearly a very weak collaboration between researchers, and between the research and industrial sectors. Postharvest extension is still very weak. There is no clear linkage between research and extension in postharvest. There is still a lack of available, locally written publications on postharvest, and most are still written in foreign languages (especially English) which make it difficult for many people to understand and utilize the information.

References

Cheftel JC (1995) High pressure, microbial inactivation and food preservation. Comptes Rendus de l_Academie d_Agriculture de France 81:13–38

Coronel P, Truong VD, Simunovic J, Sandeep KP, Cartwright GD (2005) Aseptic processing of sweetpotato purees using a continuous flow microwave system. J Food Sci 70:531–536

Diehl JF (1992) Food irradiation: is it an alternative to chemical preservation? Food Addit Contam 9:409–416

Kader AA (ed) (1992) Postharvest technology of horticultural crops. Division of Agriculture and Natural Resources, University of California, Oakland, CA, 296 pp

Khetarpaul N (2005) Freezing of Foods. In: Food processing and preservation. Daya Publishing House, Delhi, India, 338 pp

Knorr D (1993) Effects of high-hydrostatic-pressure processes on food safety and quality. Food Technol 47(156):158–161

Lalaguna R (1998) Response of "galia" muskmelons to irradiation as a quarantine treatment. Hort Sci 33:118–120

Liu FW (1991a) Precooling of horticultural crops. In: Yahia EM, Higuera CI (eds) Memorias Simposio Nacional Fisiologia Y Tecnologia Postcosecha deProductos Horticolas en Mexico. Noriega Editores, Mexico, pp 235–240

Liu FW (1991b) Storage systems for horticultural crops. In: Yahia EM, Higuera CI (eds) Memorias Simposio Nacional Fisiologia y Tecnologia Postcosecha de Productos Horticolas en Mexico. Noriega Editores, Mexico, pp 241–247

McDonald CJ, Lloyd SW, Vitale MA, Petersson K, Innings F (2000) Effects of pulsed electric fields on microorganisms in orange juice using electric field strengths of 30 and 50 kV/cm. J Food Sci 65:984–989

McNamee C, Noci F, Cronin DA, Lyng JG, Morgan DJ, Scannell AGM (2010) PEF based hurdle strategy to control *Pichia fermentans*, *Listeria innocua* and *Escherichia coli* k12 in orange juice. Int J Food Microbiol 138:13–18

Patterson MF (1993) Food irradiation and food safety. Rev Med Microbiol 4:151–158

Paull RE, Armstrong JW (1994) Introduction: insect pests and fresh horticultural products: treatment and responses. CAB International, Wallingford, pp 1–33

Sarkis JR, Jaeschke DP, Tessaro IC, Marczak LDF (2013) Effect of ohmic and conventional heating on anthocyanin degradation during the processing of blueberry pulp. LWT Food Sci Technol 51:79–85

Steed LE, Truong VD, Simunovic J, Sandeep KP, Kumar P, Cartwright GD, Swartzel KR (2008) Continuous flow microwave-assisted processing and aseptic packaging of purple-fleshed sweetpotato purees. J Food Sci 73:E455–E462

Sudher KP, Indira V (2007) Post harvest technology of horticultural crops. New India Publishing Agency, New Delhi, pp 49–60

Advancements in Post-harvest Management of Fruits and Vegetables

P.K. Srivastava and Saghir Ahmad

1 Introduction

Fruits and vegetables have a significant role in human nutrition, especially as source of vitamins, minerals, dietary fibres and antioxidants. Increased consumption of variety of fruits and vegetables on regular basis is recommended because of associated health benefits which include reduced forms of cancer, heart disease and strokes besides some other chronic diseases.

2 Nutritional Importance of Fruits and Vegetables

Among horticultural crops, fruits and vegetables play a significant role in human nutrition, especially as sources of vitamin C (ascorbic acid), vitamin A, vitamin B_1 (thiamine) and B_2 (niacin), pyridoxine (vitamin B_6), folacin (vitamin B_9), vitamin E, minerals and dietary fibres. Their contribution as a group is estimated as 91 % of vitamin C, 48 % of vitamin A, 30 % of vitamin B9, 27 % of vitamin B6, 17 % of thiamine and 15 % of vitamin B_2 in the diets. Fruits and vegetables remain important source of nutrients in many parts of the world and offer advantages over dietary supplements because of low cost and wide availability. There is increasing evidence that consumption of whole food is better than isolated food components such as

P.K. Srivastava (✉)
College of Agricultural Engineering and Post Harvest Technology,
Central Agricultural University, Ranipool 737135, Gangtok, Sikkim, India
e-mail: Dean.caepht@gmail.com

S. Ahmad
Faculty of Agricultural Sciences, Department of Post Harvest Engineering and Technology,
Aligarh Muslim University, Aligarh, UP 202002, India

A. Malik et al. (eds.), *Food Processing: Strategies for Quality Assessment*,
Food Engineering Series, DOI 10.1007/978-1-4939-1378-7_17

dietary supplements and neutraceuticals. For example threshold consumption of carotenoids rich fruits and vegetables is more effective than carotenoid dietary supplements in increasing LDL oxidation. Thus demand for fresh fruits and vegetables, inspite of emphasis on processing and value addition, needs to be met through advancement in post-harvest management techniques.

3 Post-harvest Losses Due to Injuries and Diseases and Pests

All horticultural crops including and fruits vegetables are living plant materials containing 65–95 % water. They continue their living processes even after harvest. Their post-harvest life depends on the rate at which they use their stored food reserves and the rate of their water loss. When food and water reserves are exhausted, these crops/produces die and decay. Anything that increases the rate of this process may make the produce inedible before it can be consumed.

Physical damage to fresh produce can come from a variety of causes, the most common being mechanical injury. The high moisture content and soft texture make fruits/vegetables susceptible to mechanical injury.

(a) ***Mechanical injury***

Fruits and vegetables as well as root crops are susceptible to mechanical injuries which can occur at any stage from production to retail marketing. Owing to their high moisture content and soft texture, the mechanical injuries could be because of:

- the product being dropped on hard surface during transportation, loading or unloading or during movement of transport system. Such injury is known as impact injury.
- vibration or abrasion injuries result when produce is able to move within a container because of vehicles with small wheels and poor/bad shock-absorbers, weak crates, bad roads, transmission vibrations, etc.
- compression injuries are caused by improper packaging or inadequate package performance resulting from over packaging of boxes/crates, too high stacking of crates, weak packaging, etc.
- puncturing injuries resulting from nails or splinters from the boxes/crates, fingers or nails of a person, other crates, fork lift, hard and sharp stalks of fruits, etc.

Injuries caused can take many forms—for example, splitting due to impact when they are dropped, internal bruising not visible externally, caused by impact, superficial grazing or scratches affecting the skins and outer layers of the cells and crushing of leafy vegetable and soft produces.

Injuries cutting through or scraping away the outer skin of produce:

(a) provide entry points for moulds and bacteria causing decay,
(b) increase water loss from the damaged area, and
(c) cause an increase in respiration rate and thus heat production

Brushing injuries which leave the skin intact and may not be visible externally will cause:

- increased respiration rate and heat production
- internal discoloration because of damaged tissues, and
- off-flavours because of abnormal physiological reaction in damaged parts.

In addition to mechanical injuries, there may be injuries from the temperature effects also as all fresh produce is subjected to damage when exposed to extremes of temperature, though commodities vary considerably in their temperature tolerance. Their tolerance levels to low temperature are of great importance when cold storage is concerned. These injuries are chilling injuries, freezing injuries and high temperature injuries as described below:

Chilling Injuries

Some types of fresh produces are susceptible to injury at low but non-freezing temperatures. Such horticultural crops/produces are mostly of tropical origin though a few temperate crops are also affected.

Sensitivity varies with the commodity but with each there is a specific temperature, the lowest safe temperature (LST), below which injury occurs. With a single commodity type, the LST may vary between varieties. Fruits are generally less sensitive when ripe. Symptoms of chilling injury (CI) may not develop until the produce is removed from the cold storage to normal ambient temperatures. When susceptible produce is to be held for sometimes in storage, it must be kept at a temperature just above its LST. This means that such crops will have a shorter marketing life than non-sensitive crops because respiration may continue at relatively faster rate during storage at higher than normal cold storage temperatures.

Freezing Injury

All produces are subjected to freezing at temperatures between 0 and −2 °C. Frozen produce has a water soaked or glassy appearance. Although a few commodities are tolerant of slight freezing, it is advisable to avoid such temperatures because subsequent storage life is short. Produce which has been recovered from freezing is highly susceptible to decay.

High Temperature Injury

If fresh produce is exposed to high temperature caused by solar radiation, it will deteriorate rapidly. Produce left in the sun after harvest may reach temperature as high as 50 °C. It will achieve a high rate of respiration and if packed and transported without cooling or adequate ventilation, it will become unstable. Long exposure to sun will cause severe water loss from thin-skinned vegetables, viz. carrots, turnips, leafy vegetables, etc.

In addition to mechanical injuries, there are post-harvest losses due to diseases and pests as described below:

(b) **Losses due to diseases and pests**

Diseases caused by fungi and bacteria commonly result in loss of fresh produce. Virus diseases, which can cause severe losses in growing crops, are not a serious post-harvest problem. Insects-pests that are mainly responsible for wastage in cereals and grain legumes are rarely a cause of post-harvest loss in fresh horticulture produces. Where they do appear, they are often locally serious, viz. the potato tuber moth.

Losses from post-harvest disease in fresh produce fall in to two main categories:

- **Losses in quality**: This type of more serious losses occurs where deep penetration of decay makes the infected produce unstable. This is often the result of infection of produce in field before harvest.
- **Losses in quality**: Such losses occur when the disease affects only the surface of the produce. It may cause skin blemishes that can lower the value of a commercial crop. In crops grown for local consumption, the result is less serious since the affected skin can often be removed and the undamaged interior can be used. Fungal and bacterial diseases are spread for most part by microscopic spores which are widely distributed in the air and the soil and on dead and decaying plant material. Produce can become infected:
  - through injuries caused by careless handling, insect or other animal damage or through growth cracks.
  - through natural pores in the above and below-ground parts of the plants which allow the movement of the air, carbon-di-oxide and water vapour in and out of the plant.
  - by direct penetration of the intact skin of the plant. The time of infection varies with the crop and with different diseases it can occur in the field before harvest or at any time afterwards.

Although relatively few post-harvest losses of fresh produce are caused by attacks of insects or other animals, localized attacks by these pests may be serious. Insect damage is usually caused by insect larvae burrowing through the produce, e.g. fruit fly, sweet potato weevil, potato tuber moth, etc. Infestation is usually before harvest. Post-harvest spread is a problem where produce is held in store or is exposed to lengthy period of transport.

Rats, mice and other animal pests again are sometimes a problem when the produce is stored on the farms.

4 Post-harvest Management for Supply of Wholesome Fruits and Vegetables

Because of poor and faulty operations during harvesting and post-harvest management, processing and value addition, post-harvest losses in horticultural crops are very high, which range between 20 and 40 % or more in certain developing countries. These losses vary from fruits to fruits and vegetables to vegetables and their varieties and depend upon various other factors. They can be minimized through adoption of scientific post-harvest management practices. These practices have been standardized and can be presented with the following flow diagram:

The benefits of scientific post-harvest management and value addition can be realized in terms of minimal harvest and post-harvest losses, better shelf-life, improved nutritional and sensory qualities and higher marketability with enhanced profit.

Horticultural crops

↓

Harvesting at optimum maturity

↓

Sorting and grading

↓

Washing

↓

Pre-cooling

↓

Drying/dehydration for removal of moisture

↓

Surface treatments viz. coating/waxing

↓

Further drying if required

↓

Packaging (in CFB boxes/plastic trays)

↓

Transportation (preferably by refrigerated vehicles)

↓

Whole sale marketing

↓

Retail marketing

↓

Consumers

Quality in further processing can be achieved through primary, secondary and tertiary processing operations for production of a variety of value added diversified products, while the by-products/co-products and residues obtained during post-harvest processing could be utilized with higher economic returns. Even the remaining biomass could be finally converted to useful products like manure/vermin compost or as fuel and fibre.

5 Injury Free Harvesting–Handling–Packaging of Fruits and Vegetables: Strategies

The strategy of decreasing losses is more economical because it requires smaller inputs per unit of the final product than a strategy of increasing production extensively, especially in the short term. The inputs that would be used for losses reduction are usually accessible and in most cases do not require expensive inputs. There are several simple and advanced technologies to minimize or avoid post-harvest losses. In this reference it may be noted that post-harvest losses increase proportionately as the time between harvest and consumption increases; hence the knowledge of the post-harvest management and on-farm value addition technologies become pertinent. Some important technological strategies are described below:

5.1 Harvesting and Harvest Handling

Quality of produce cannot be improved after harvest but only maintained. Therefore it is important to harvest fruits and vegetables at the proper stage/at peak quality. Immature or over mature produce may not last as long in storage as that picked at proper maturity. Harvesting should be completed during the coolest time of the day, which is usually in the early morning and produce should be kept shaded in the field. Appropriate maturity indices should be used and produce handled gently. Produce destined for the storage should be as free as possible from skin breaks, bruises, spots, rots, decay and other deteriorations. Bruises and other mechanical damage not only affect appearance, but provide entrance to decay organism as well.

Post-harvest rots are more prevalent in fruits and vegetables that are bruised or otherwise damaged. Mechanical damage also increases moisture loss. The rate of moisture loss may be increased by as much as 400 % by a single bad bruise on an apple and skinned potatoes may lose three to four times as much weight as non-skinned potatoes. Damage can be prevented by harvesting at proper maturity, harvesting dry whenever possible, handling each fruit/vegetable no more than necessary, installing padding inside bulk bins and avoiding over or packaging of containers.

5.2 *Pre-cooling*

Pre-cooling is the first step in good temperature management. The field heat of a freshly harvested crop, the heat the produce holds from the sun and ambient temperature, is usually high and should be removed as quickly as possible, before shipping, processing or storage. Refrigerated trucks are not designed to cool fresh commodities but only maintain the temperature of pre-cooled produce. Likewise most refrigerated rooms have neither the refrigeration capacity nor the air movement needed for rapid cooling. Therefore, pre-cooling is generally a separate operation requiring special equipment and/or rooms. Rapid pre-cooling to the products lowest safe temperature is most critical for crops with inherently high respiration rates. These include artichokes, Brussels sprouts, cut flowers, green onions, snap beans, asparagus, broccoli, mushroom, peas and sweet corn. Crops with low respiration rates include nuts, apples, grapes, garlic, onion, potatoes and sweet potatoes.

- **Different types of pre-cooling devices are described below:Hydro-cooling**: Hydro-cooling removes the heat and cleans the product at the same time. In addition it also reduces water loss and wilting. It is done by dipping/drenching/rinsing/immersing/spraying cold water over the fruits and vegetables.
- **Forced air pre-cooling**: Forced air pre-cooling has been reported to be most efficient and suitable method for certain fruits like Nagpur Mandarin as compared to hydro-cooling and room cooling. Temperature range of 6–7 °C with 90–95 % relative humidity during pre-cooling and storage is reported to be optimum. For export, Nagpur Mandarins are usually wax treated, packed in CFB boxes and then forced air pre-cooled before shipment.
- **Vacuum cleaning:** This system works best for leafy vegetables/crops having high surface to volume ratio.
- **Top or Liquid Icing:** These methods work well with high respiration commodities, viz. asparagus, beets, broccoli, carrots, cauliflower, radish, spinach, etc. Commodities sensitive to icing are strawberries, raspberries, tomatoes, green beans, cucumber, garlic, onions, okra, etc.

The choice of cooling methods is influenced by nature of the products, viz. sensitivity to chilling and moisture, product packaging requirement, etc. It may be noted that chemical treatment of the fruits and vegetables in combination with pre-cooling, viz. hydro-cooling, is very effective. For example calcium compounds in general and calcium nitrate in particular have been reported to extend the shelf-life of many fruits by maintaining their freshness and minimizing their respiration rate, proteolysis and tissue breakdown (Gupta et al. 1980). It also acts as anti-senescent agent by preventing cellular disorganization by maintaining protein and nucleic acid synthesis. Here hydro-cooling is used to remove field heat. A study on the effect of calcium nitrate and hydro-cooling on cold storage life of peach Cv. Shan-e-Punjab may be considered (Brar et al. 1998) as reference. Calcium nitrate spray (2.5 %) thrice at weekly intervals before anticipated date of harvest on peach and post-harvest hydro-cooling to 15–20 °C by dipping the fruits in ice cold water followed by packaging in corrugated fibre board (CFB) boxes and storage at 3.3 °C was examined. Pre-harvest spray of calcium nitrate improved the colour of fruits during storage. Hydro-cooling reduced the spoilage and physiological loss in weight. It was reported that 2 % calcium nitrate spray combined with hydro-cooling at 15 °C retained in highest fruit firmness. Evaporative cooling is an inexpensive and simple way to help reduce temperature. The simplest way is to wet a porous material that will hold water and move air through the wilted material (working a desert cooler). Air may be moved by fans or by using prevailing winds. Since cool air is heavier than hot air, it is also effective to place the wetted matter over the top of the container.

5.3 *Post-harvest Hauling*

Farm roads should be kept in good condition because great damage can be inflicted on produce carried over rough roads in unsuitable vehicles. Containers must be loaded on vehicles carefully and stacked in such a way that they should shift or collapse damaging their contents. Vehicles need good shock absorbers and low-pressure tyres and must move with care. Jolting of laden containers can aggravate damage to produce on rough roads, even at low vehicle speed.

Produce carried by trucks may be in palletized field containers, in bulk bins or in hand-loaded sacks or wooden or plastic boxes. Where vehicles wait in sun or rain for long periods before unloading, only top part of the load should be protected by a covering. Grasses or leaves are not recommended for this purpose because they restrict ventilation and may be a source of disease. Complete enclosing of the load with tarpaulin is disastrous because it restricts ventilation and the temperature of the produce rises rapidly.

During on-farm handling and transport of the fruits and vegetables, suitable packaging and handling techniques can reduce the amount of the damage to which fresh produce is exposed during transport and marketing. Following issues are to be taken care:

- to avoid the package itself damaging produce during handling and transport, wooden boxes or cardboard cartons must be properly assembled. Nails, staples and splinters are always a danger in wooden boxes.
- individual items of the produce should be packaged to avoid rubbing against each other during handling and transport. Loose fill packs are particularly susceptible to vibration damage.
- bruising results from overfilling containers or from the collapse of boxes. Collapse may be caused by weak walls of the boxes, by softening of the cardboard walls because of moisture or by the failure to stack boxes in such a manner that the sides and walls support those above. Stacks of the boxes should never exceed height that has been recommended by the vehicle maker. Produce in woven jute sacks or nets is especially susceptible to shock damage. Sacks of 25

Table 1 Journey hazards during transportation and handling of horticultural crops

Sl. No.	Hazards	Samples
1.	Biological	Insects, rodents, moulds and microorganism
2.	Compression, static	Stacks in warehouse or vehicles
3.	Human	Pilferage, inspection
4.	Impact, vertical	Package dropped when unloading trucks/wagons
5.	Impact, horizontal vibration	Shunting, rail track
6.	Puncturing, snagging	Projection on vehicles
7.	Racking or deformation	Uneven support due to poor floor or uneven lifting
8.	Temperature, high	Sun, ships, boilers
9.	Temperature, cold	Cold store
10.	Water, liquid	Rain, spray, condensation
11.	Water vapour	Humidity of atmosphere, natural or artificial

or 50 kg capacity are normally used for relatively low value produce, viz. root or tuber crops and are often roughly handled on account of their weight, so wherever possible, handling of bagged produce should be minimized by stacking sacks in unit loads on pallets or in pallet boxes.

Fruits and vegetables and their products may be scientifically transported for marketing and distribution by using different modes of transport, viz. road, rail, sea and air. However, care needs to be taken to minimize physical and mechanical losses during various journey hazards. These hazards are listed in Table 1. In transportation proper attention needs to be given to palletization and containerization as desired below:

(a) ***Palletization***: Loading and unloading of fruits and vegetables are very important steps in post-harvest handling which are quite often neglected. The individual handling of packaged produce in many cases leads to mishandling and to high post-harvest losses. With the introduction of CFB boxes, serious consideration should be given to palletization and mechanical handling of the produce with the use of fork lift trucks in order to minimize produce mishandling.

(b) ***Containerization***: The use of containers for the transport of horticultural produce is relatively new in countries like India. Containerization provides an excellent system for the shipment of the goods from one place to another. Refrigerated containers are used in transportation of fruits/vegetables. Very large numbers of containers are shipped through cargo under refrigerated condition. Containers are transported by refrigerated trucks without interfering the movement of the vehicle. Handling of the produce package is generally done by break bulk, unit loading or through load as described below:

- ***Break bulk:*** It is the traditional system in which packages are handled individually; handling is essentially manual and no special requirements are imposed on vehicle.

- ***Unit loading*****:** In this system, packages are assembled into larger unit and it requires mechanical handling, viz. fork lift truck as 0.5–1.0 tonnes load is placed on a pallet and transported throughout the journey as one unit.
- ***Through load*****:** In this system consignment is usually over 5.0 tonnes and move together to ultimate destination in a container which is loaded at point of origin. In practice a combination of system is adapted that consists of break bulk used first followed by the unit load on pellet and ultimately shipped containers.

6 Improved/Advanced Post-harvest Technologies

6.1 Harvesting at Optimum Maturity and Controlled Ripening

Maturity indices help judging the appropriate stage for harvesting of fruits and vegetables from the point of view of the market value, shelf-life, nutritional and sensory qualities. The harvesting maturity is judged by:

- Visual colour change in the fruit
- Total soluble solids
- Size of the fruits

Maturity indices of the certain fruits produced in India are given in Table 2.

Controlled ripening is practised in climacteric fruits and vegetables, viz. tomatoes, banana, mango, guava, etc. These fruits are capable of generating ethylene, the hormone required for ripening even when detached from the mother plants.

Climacteric fruits and vegetables are harvested at slight immature stage and treated with ripening agents like ethylene gas or ethepton (2-chloro ethyl phosphoric acid) before shipment so that the product reaches the consumers at the right stage of maturity. Ripening agents are generally used to speed up the ripening process with uniform ripening; examples may be given of mangoes as well as papaya.

In case of mangoes, both premature and mature fruits are uniformly ripened with the help of ethrel (Ethepton). The period of ripening is also reduced by this treatment. Papaya ripens satisfactorily between 20 and 25 °C, but temperature above 32.2 °C causes uneven colour development, rubbery pulp texture and browning of fruit surface. Exposure of mature green to 1/4th yellow colour papayas to 100 ppm ethylene at 20–25 °C and 90–95 relative humidity for 24–48 h results in faster and uniform ripening.

6.2 Fumigation

In case of certain fruits, fumigation is practised during storage which helps in extension of their shelf-life. For example, Litchi (*Litchi chinensis sonn*) fruits are fumigated with sulphur dioxide following a dip in dilute hydrochloric acid (HCL)

Table 2 Maturity indices of selected fruits and vegetables produced in India

Sl. No.	Crop wise indices	Values
1.	**Khasi mandarin**	
	Colour of the rind	Yellow–orange, orange
	Days from flowering to harvesting	230–250
	Juice yield, %	>49.0
	TSS, oBrix	9.5–10
	Titrable acidity, % of citric acid	0.75–0.81
	TSS: acid ratio	12.30–12.97
2.	**Pineapple (Kew variety)**	
	Average fruit weight	>1.0 kg
	Specific gravity	>0.93
	Days from flowering to harvesting	141–150 days after flowering when the colour changes from green to yellow in ¼ to ½ portion of the fruits from its base
	TSS (°Brix)	>12
	TSS: acid ration	>20
Note In case of such pineapples it is recommended that the stem end of the fruits should be trimmed to 2.0 cm along with 1/3rd of the crown.		
3.	**Jack fruit**	
	Appearance	Dull
	Sound on tapping	Dull
	Colour	Changes from green to yellow and brown
4.	**Papaya**	
	Latex colour	Changes from white to watery
	Skin colour	Changes from dark green to light green with some yellowness at the blossom end (colour break)
Note Papayas are usually harvested at colour break to 1/4th yellowness for export or ½ to 3/4th yellowness for local markets		
5.	**Passion fruits**	
	Fruit surface colour	Amount of yellow or purple colour
Note Development of complete yellow/orange or purple colour is a better indicative of full maturity		

solution. This practice controls browning which is mainly due to the degradation of red anthocyanin pigments as well as oxidation of phenolics caused by poly-phenol oxidase. As an alternative, use of red phosphorous which has characteristics similar to sulphur and can be burnt easily without toxicity has been recommended. Litchi fruit can be fumigated with red phosphorous at 25 g/m^2 for 60 min and stored under ambient air temperature of 28–33 °C and 95–100 % relative humidity conditions. This treatment delays the increase in polyphenol oxidase (PPO) activity and pH value and decrease of colour and eating quality and maintains the level of anthocyanin and total phenolics. The treatment particularly inhibits the decay of the fruits during storage. Red phosphorous fumigation together with other post-harvest technologies viz. proper packaging and air-conditioning has shown promising results in extending the shelf-life of Litchi fruits.

6.3 Disinfectant Treatment

Following methods are used for disinfectant treatment of the fruits and vegetables:

(a) ***Sanitization***: Sensitization not only protects the produce against post-harvest disease but also protects consumer from food-borne disease. Use of a disinfectant in wash water helps to prevent both post-harvest diseases and food-borne illness. For most of the vegetables, chlorine (75–150 ppm) or hydrogen peroxide (0.5 %) in wash water can be used effectively.

(b) ***Ozonisation***: Ozone not only kills food-borne pathogens, but it also destroys microbes responsible for food spoilage. In fact, ozone is the most effective natural bactericide of all disinfecting agents. Ozone is now being used in food processing and storage of perishables as an antimicrobial agent and as a food processing aid. It is safe and natural purifier and disinfecting agent which is strong and ideal sanitizer, germicide, sterilizer and vermicide, antimicrobial, bactericide–fungicide, deodorizer, and detoxifying agent. Ozone oxidizes metabolic products and neutralizes the odour generated during the ripening stage in storage of fruits. This helps to preserve and almost double the shelf-life of fresh produce. It also enhances the taste by retaining the original flavour of the products. Ozone enhances the taste of most of the perishables by oxidizing pesticides and neutralizing ammonia and ethylene gases produced by ripening or decay. The reduction of ethylene gas increases the shelf-life and reduces shrinkages. It changes the chemical's complex molecular structure back to its safe and original basic elements. Its use does not leave any toxic by-products or residues, does not affect healthy cells or alter its chemistry and is non-carcinogenic. Factors like age, crispness, quality, humidity, temperature, the condition on receiving and the reduction of the pathogens during ozonisation will determine the shelf-life of each different type. Positive effects will show at low and constant levels between 0.05 and 0.1 ppm and it allows workers to enter storage area and carry out their works comfortably (Perez et al. 1993).

Ozone should be constantly consumed and absorbed during the oxidation process. The effectiveness is influenced (lowered) due to the presence of the steam or 100 % humidity levels. The microorganisms have to be in a certain condition of swelling in order to be affected when the humidity level is below 50 %. The efficiency of it slows down as a bacterial medium (Castillo et al. 2006).

Growers and processors can use ozonized water to wash the fruits replacing chlorine. Application of it can be safely done as process water in hydro-cooling system, bin, dump and dip tanks, flumes, spray wash, wastewater, processing and storage at an affordable cost.

6.4 Hot Water Treatment/Blanching

This process is also known as scalding, parboiling or pre-cooking. It is usually done in case of vegetables by exposing them to boiling water or steam for 2–5 min, followed by cooling. Fruits on the other hand are generally not blanched. Blanching inactivates most of the plant enzymes which cause toughness, discoloration (polyphenoloxidase), mustiness, off-flavour (peroxidase), softening and loss of nutritive value.

The other advantages of blanching are given below:

- it reduces the area of leafy vegetables such as spinach by shrinkage or wilting, making their packaging easier.
- it removes tissue gases which reduce sulphides.
- it also reduces the number of microorganism by as much as 90 %.
- it enhances the green colour of vegetables such as peas, broccoli, spinach, etc.
- it removes saponins in peas.
- it removes undesirable acids and astringent taste of the peel and thus improves the flavour.
- it also removes the skin of the vegetables such as beetroot and tomatoes which help in their peeling.

6.5 Pretreatment with Chemicals

Certain examples of pretreatment of fruits and vegetables are given below:

- Minimum spoilage and loss in weight of peach (Cv. Shan-e-Punjab) is reported for fruits which are pretreated with 2.5 % calcium nitrate. 2 % calcium nitrate spray combined with hydro-cooling at 15 °C retains significantly high fruit freshness (Brar et al. 1998).
- Treatment of cauliflower curds with 1 % Potassium meta-bisulphite, onion rings with 0.5 % $NaHCO_3$ and potato slices with 0.5 % Sodium benzoate and 0.5 % Potassium meta-bisulphite solution, respectively, followed by solar dehydration and packaging in laminated foils enhance the shelf-life of these products up to 6 months. At the time of use, the rehydration values are reported to be high, viz. 6.5–7.0 in cauliflower, and 8.6 in onion.
- The combination of bavistin (0.1 %) dip for 5 min followed by packaging in polyethylene pouches/bags reduces the physiological weight loss of citrus fruits and improves the quality under cold and evaporative cooled structures (Kaushal and Thakur 1996).
- Kinnow fruits given a post-harvest dip in 0.1 % bavistin solution for 5 min and packed in 150 gauge LDPE bags could be stored under ambient conditions as well as in evaporative cooled chambers or cold storage conditions. Fruits stored under ambient conditions remain in marketable condition up to 8 weeks whereas in other conditions up to 10 weeks. Bavistin treatment followed by packaging of fruits in

LDPE bags is effective in retaining fruit quality during storage. Best results are obtained when four fruits are packed in a single unit (Thakur et al. 2002).

- Papaya fruits (Cv. CO-2) may be treated with gibberellic acid. Gibberellic acid at 150 pm is most effective in reducing physiological loss in weight, TSS and total sugar content by maintaining fruit firmness. Ripening parameters like colour and total carotenoid content are delayed, thereby increasing the shelf-life of papaya fruits.

6.6 *Surface Coating and Treatment with Wax*

Surface coating using wax is one of the traditional methods of the fruits and vegetables' preservation practised since the twelfth/thirteenth century, when it was applied to citrus fruits in China. Surface coating is considered to be an easy to use and one of the most economical methods for prolonging shelf-life of variety of fruits and vegetables. Application of wax as surface coating material slows down the permeability of water vapour and other gases, retards the ripening and checks the microbial interaction. Nowadays carnauba wax, sugar cane wax, bees wax, shellac wax, resins and certain thermoplastics are most commonly used coating/waxing substances.

As people have become more health conscious and concerned with safety, there is now greater demand of food grade coating materials which should not be synthetic or chemical based.

Surface coating of fruits and vegetables has been found suitable for preservation of fruits and vegetables due to the following advantages:

- **Improved appearance**: Coating provides fruits and vegetables a gloss/shine which improves their appearance.
- **Lesser moisture loss**: Coating blocks the pores in the cuticle which results in significant reduction of water loss from fruits and vegetables.
- **Reduced post-harvest decay**: Surface coating establishes a barrier against the entrance of fungal and bacterial pathogens into the product. Post-harvest pathogens typically require a film of free moisture on the products skin to grow. A hydrophobic (non-water compatible) surface on the fruits/vegetables surface is not conducive to pathogen growth and development.
- **Longer shelf-life**: Fruits and vegetables are living organisms that continue to respire even after harvest. Coating helps in delaying post-harvest decay in those farm products which remain physiologically active after harvest by creating a modified atmosphere around it. This results in reduction in the product's respiration rate and an increase in the post-harvest shelf-life. A prolonged shelf-life allows for an extension in the marketing period for the crop.
- **Lesser susceptibility to chilling injury**: Surface coating reduces the severity of the chilling injury (CI) and allows for the storage of CI-sensitive commodities at slightly lower temperatures without incurring damage. It may, however, be noted that waxing does not eliminate C I on the susceptible commodities.

- **Reduced economic losses**: Water is the principal component of all fresh fruits and vegetables. Growers/producers often sell their horticultural produces based on their weight. The less is the weight, lower is the price and lower is the economic return. Coating helps to limit this loss of water and thus reduces the loss of profit margin. The mechanism, by which the surface coating helps in preservation of fruits and vegetables, is by producing a modified atmosphere surrounding the product. This modified atmosphere serves several purposes including reduced oxygen availability and increased internal carbon-di-oxide concentration in produces. Modified atmosphere created by surface coating is produced by physical trapping of carbon-di-oxide gas within the horticultural products tissues during respiration. The edible surface coating materials which have been investigated include Methyl cellulose, Hydroxy propyl methyl cellulose (HPMC), Stearic acid, Palmatic acid/wax, high amylose starch, collagen, gellatin, Zein, gluten, soy protein isolate, casein, casein bees wax, whey protein isolate, bees wax, shellac, etc. (Raj 2007). Different commercial coatings recommended for fruits are given in Tables 3 and 4.

Carnauba and sugarcane wax, thermo-plastic terpene, resins, chitosan-based coatings and shellac are used commercially. However, waxes of plant origin are now being preferred. Development in wax emulsion for surface coating is not

Table 3 List of some commercially available edible coatings

Sl. No.	Coating materials	Fruits for which used
1.	Semper fresh	Banana, apple, guava
2.	Brilloshine	Apple, avocado, citrus fruits
3.	Citrashine, Nu-coatflo	Citrus fruits
4.	Tal prolong, apple wax	Apple
5.	Nutrisave	Golden delicious Apple
6.	Vapour guard	Mango
7.	Banseel	Banana, plantains
8.	Chitosan	Strawberry, raspberry

Table 4 List of specific applications of coating of different fruits

Sl. No.	Coating materials	Application on fruits
1.	Prolong	Banana
2.	Semperfresh	Banana, Granny, Smith apple, Guava
3.	Semperfresh with organic acid	Banana
4.	Ban seel	Banana, and Plantains
5.	Tal prolong semperfresh and apple wax	Apple
6.	Nutrisave	Golden delicious apples
7.	Brilloshine and citrashine, Nu.coatflo	Citrus fruits
8.	Palm oil	Guava
9.	Vapour guard	Mango
10.	Chitosan	Strawberry and raspberry
11.	N_2O—Carboxy methyl chitosan	Fruits

limited to the formulation of new types of wax, but also includes the addition of chemical fungicides. For example, benzalconic chloride in association with surface coating of wax successfully controls rot caused by *Colletotrichum gloeosporides* in mango fruits. The effect of hydro dispersion of malto-dextrin, carboxyl methyl cellulose, propylene, glycol and sorbitan esters as coating for mango on reduction or prevention of spoilage by cogloeo-sporides which produce anthracnose, has also been established. Polysaccharides-based coatings have also been investigated.

6.6.1 Problems Associated with Edible Coating

Even though some edible coatings have been successfully applied to fresh produce, other applications adversely affect quality. Modification of the internal atmosphere by the use of edible coatings can increase disorders associated with high carbon-dioxide or low oxygen concentration. Smock (1940) had earlier indicated that waxing apples and pears inhibit normal ripening rate; if sufficient wax is applied, respiration is greatly inhibited and alcoholic flavours are developed by anaerobic fermentation. Smith and Stow (1984) reported that apples (cv. Cox's orange Pippin) coated with sucrose fatty acid ester have fewer detrimental changes in terms of firmness, yellowing and weight loss with increased incidence of core flush. Smith et al. (1987) summarized the effects of physiological disorders associated with modification of internal atmosphere by use of coatings, as score flush, flesh breakdown and accumulation of ethanol and alcoholic off-flavours.

It may be noted that wax and some other commercially available mixtures are not equally effective for all produce. Another problem is that consumers tend to be wary of the waxy coatings. Therefore, development of alternative edible coatings which do not impart a waxy taste is desirable. The effect of edible coating on internal gas composition and their interaction with quality parameters must be determined for coated fresh produce. For example, colour change and firmness are very important quality parameters in fruits. In this reference it has been suggested that colour change, loss of firmness, ethanol fermentation, decay ratio and weight loss of edible-film coated fruits are all important quality attributes for various products.

6.6.2 Edible Coating Materials

Following are some are edible coatings materials which have been studied by various researchers.

1. Sucrose polyester (SPE)
2. Cornz-ein
3. Methyl cellulose (MC)

4. Hydroxy propyl cellulose (HPC)
5. Chitosan
6. Wheat gluten

6.7 Treatment with Essential Oils as Coating Materials

Application of essential oils as coating materials to control post-harvest pathogens and maintenance of fruit quality as an alternate to chemical fungicides has been suggested. This process eliminates needs of synthetic fungicides and organic requirements and reduces environmental pollution. Essential oils are volatile, natural complex compounds characterized by a strong odour and are formed by aromatic plants as secondary metabolites. These are registered food grade materials and have the potential to be applied as an alternative treatment to control post-harvest decay of fruits. An important characteristic of essential oil and their components is their hydrophobicity, which enables them to partition the lipids of the bacterial cell membrane and mitochondria, disturbing the structures and rendering them more permeable. As a result, damage of the membrane proteins and depletion of proton motive forces takes place. Then leakage of ions and other cell contents can occur which after attaining extensive loss of cell contents or critical molecules and ions lead to death of the bacterial cells. A number of essential oil compounds have been identified as effective antibacterial, viz. eugenol, carvacrol, thymol, menthol, etc., in several fruits like apple, peach, sweet cherry, strawberry, grapes, citrus, mango, etc. Table 5 describes use of some essential oils for control of post-harvest diseases of fruits.

Table 5 Essential oils used for control of diseases of fruits

Sl. No.	Essential oils	Major components	Bacterial spp.	Fruit crop
1.	Clove oil	Eugenol	*P. expansum* *M. fructigena* *B. cinera* *P. vagabunda*	Apple, grape
2.	Mint oil	Menthol	*P. italicum* *B. cinerea* *R. stolonifer*	Orange, strawberry
3.	Thyme oil	Thymol, Carvacerol	*B. cinerea* *R. stolonifer* *M. fructicola*	Strawberry, grape, sweet cherry
4.	Cinnamon oil	Cinnamaldehyde	*Natural flora*	Kiwi fruit
5.	Lemon grass oil	Citral	*C. gloesporiodes* *P. expansum* *B. cinerea* *R. stolonifer*	Mango, peach

6.8 Treatment with Salicylic Acid

Salicylic acid (SA) or ortho-hydroxy benzoic acid is a ubiquitous simple phenolic compound involved in the regulation of many processes in plant growth and development. It is considered as plant hormone because of its role in regulating some aspect by disease resistance in plants. Recently the involvement of SA as a single molecule in systemic acquired resistance associated with the production of pathogenesis-related proteins has been extensively shown. Moreover, dietary salicylates from fruits and vegetables are described as bioactive compounds with health care potential and considered as generally safe (GRAS).

There are several reports on beneficial effects of SA treatment in fruits. For example during ripening of Kiwi fruits, the pattern of decrease in endogenous SA levels is to related to accelerated softening, while the application of acetylsalicylic acid (ASA), a derivative of salicylic acid, slows down the softening rate of Kiwi fruit by inhibiting ethylene production and maintaining higher endogenous SA levels. On the other hand, SA application, either pre-harvest or post-harvest reduces fungal decay in sweet cherry through induction of the defence resistance system and simulation of antioxidant enzymes (Xu and Tian 2008). In addition, in chilling injury-sensitive fruits, pretreatment with SA reduces chilling injury symptoms in peaches and pomegranate.

6.9 Treatment with Methyl Jasmonate

Jasmonates are a class of endogenous plant growth regulators that have unique and potential useful properties which affect plant growth and development in response to environmental stresses. Methyl Jasmonate (MeJA) is a sweet smelling compound in Jasminium gradiflorum flower extract. Its main effects in post-harvest management of fruits are to control post-harvest diseases and decay of fruit, alleviate chilling injury, regulate fruit ripening and senescence, maintain fruit quality, develop colour and aroma volatiles, etc. MeJA can be effectively used to control grey mould rot in strawberry caused by Botrytis cinerea (Zhang et al. 2006) and to enhance disease resistance in peaches. Loquat (Cao et al. 2000) and raspberries (Chanjirakul et al. 2006). MeJA has also been implicated in delaying the onset of fruit ripening on the tree. Jasmonate has been found to have close interaction with plant hormone ethylene in this regard. Jasmonate induced ripening delay is associated with upregulation of polyamine levels in peach fruit. The impact of Jasmonate application on volatile compounds production is dependent on fruit ripening stage. It increases the volatiles in preclimacteric fruits but decreases the volatiles in climacteric fruits (Kando 2009).

6.10 Treatment with Bio-control Agents

Bio-control strategy has picked up in last few decades for post-harvest management of fruits and vegetables. As a result several bio-control treatments have been approved for commercial applications. Now there is a focus on improving bio-efficiency of the antagonists. One of the approaches has been the selection of combination of antagonists, which may work more effectively. It is a very challenging work as microorganisms have differential growth habits, requirement for nutrition and cultural conditions.

Naturally occurring microorganisms, which are found to be adhered on the fruits and vegetables surfaces, have been shown to posses potential to protect the fresh produce against the postharvest disease causing pathogens. Of recent, several products viz. Serenade (*Bacillus subtilis* based), Messenger (*Erwinia amylovora* based). Biosave (*Pseudomonas syringae* strain 10 LP), Aspire (*Candida oleophila* strain I-18) AQ-10 Biofungicide (*Ampelomyces quisqualis*), etc., have been isolated and reported in the USA and Germany. Use of some safe bioactive compounds has been proved beneficial in bringing down the physiological activity of fruits during transportation, storage and minimizing the overall qualitative and quantitative losses (Asrey and Berman 2011).

6.11 Nitric Oxide Treatment

Nitric oxide is a gaseous free radical with relatively long shelf-life, lasting in biological systems up to 3–5 s. It is very reactive and forms other oxides, viz. NO_2, N_2O_3 and N_2O_4 in presence of atmospheric oxygen. Post-harvest treatment of fruits with a low concentration of NO can extend postharvest shelf-life, but the application of NO by release from a gas cylinder is practically difficult.

In the above reference, the solid NO donor compound diethylenetriamine nitric oxide may be used in sachets for liberating nitric oxide in storage chambers. The use of nitric oxide can be made for delaying fruit ripening and improvement in retention of texture during storage (Zhu et al. 2010). Better retention of cellular components such as pigments, titrable acidity and antioxidants are retained well due to reduction in the degree of disintegration of cellular membranes with lesser electrolyte leakage. The chilling injury is reduced in the fruits kept in cold storage and symptoms like fresh browning and translucency are reduced. It is highly potent molecule for induction of resistance in produce against systemic infection. It also protects from microbial infections such as *Penicillium italicum, Rhizopus nigricans, Aspergillus niger* and *Monilinia fructicola*. This results in extension of shelf-life of the treated fruits. Due to maintenance of cellular compartmentation, the enzyme does not come into contact with the substrate, thus preventing the rate of reaction of enzymes such as polyphenol oxidase.

This new direction emanated from research on NO would allow more recent formulations for the use in enhancing shelf-life of the fruits in an eco-friendly manner (Asrey and Berman 2011).

6.12 Application of Ethanol

Ethanol is also known as ethyl alcohol, pure alcohol, and grain alcohol or drinking alcohol. It is a volatile, flammable and colourless liquid. It is a small molecule produced either by chemical synthesis or by microbial fermentation. The production of two anaerobic metabolites, acetaldehyde and ethanol in fruit while still attached on tree or during post-harvest storage, leads to dramatic changes in fruit ripening. Ethanol is a volatile compound naturally produced by plant tissues under anaerobic conditions. It is also accumulated in a short period of anaerobically stored fruits, without adversely affecting fruit quality.

Ethanol can be applied simply by dipping the fruits in an ethanol solution or as vapour. In some fruits, viz. grapes, ethanol can be applied as spray. It can be used in controlling decay of the fruits.

Ethanol also acts as a precursor of natural aroma compounds. It is converted to acetaldehyde by enzyme alcohol dehydrogenase. Acetaldehyde is the precursor for the acetate esters. Storage of red delicious apples for 24 h in an atmosphere containing ethanol vapours results in more than threefold increase in the ethyl ester formation.

The effect of ethylene on a range of climacteric fruits has been shown to enhance or inhibit ripening depending on the type of the fruit. Another application is to enhance anthocyanin content in fruit tissues as reported in baybery fruits when treated with 1,000 ml/lit ethanol (Zhang et al. 2007). The exogenous application of ethanol can be beneficially applied to many fruits for improving their aroma, controlling decay, delaying ripening and ethylene production and reduction of CI symptoms (Asrey and Berman 2011).

6.13 Treatment with Polyamines

Polyamines are low molecular weight small aliphatic amines that are ubiquitous in living organisms and have been implicated in a wide range of biological processes, including plant growth, development and response to stress. The polyamine treatment of the fruits inhibits biosynthesis of ethylene, increases fruit firmness, reduces chilling injury and respiration, induces resistance to mechanical damage and retards colour changes. Polyamines are applied by dipping the fruit in aqueous solution and vacuum/infiltration. Polyamines can be isolated commercially from plant as well as microbial sources which include leaves and stems of corn/maize, cucumber, oat, radish, etc.

6.14 Vapour Heat Treatment

Vapour heat treatment (VHT) is an effective non-chemical method of treatment to get rid of the oriental fruit-fly problem in fruits and vegetables. VHT is suitable for disinfestation of tropical and subtropical fruits, viz. mango, papaya, guava, citrus, etc. VHT treated mangoes are exported from India to Japan.

6.15 Degreening/Sweating/Curing

Degreening constitutes removal of chlorophyll from the peel when the night temperatures are not low enough (in case of citrus fruit) for the fruit peel to develop its characteristics colour. This process also called grassing, sweating or curing is accomplished by post-harvest treatment of fruits with dip in Ethephon (2,000–4,000 ppm) at 20–25 °C temperature and 92–95 % relative humidity. Pre-harvest spray with Ethephon (150–250 ppm) at colour break stage also improves the colour of mandarins, but calcium acetate 1 % should be added in the spray solution to check the leaf drop in this treatment. The fruits after degreening need to be washed by dipping fruits in disinfectant for 1 min in a tank filled with sanitized water at room temperature. Care should be taken that fruits are not washed before degreening. Depending upon the amount of the green colour, size of the fruits and some cultural practices, degreening time varies; e.g. excessive nitrogen fertilization promotes vigorous growth and intense green colour. For oranges maximum degreening time ranges between 48 and 50 h.

6.16 Enzymatic Hydrolysis

Enzymatic hydrolysis is a well-known pretreatment process which increases the juice yield from fruit and vegetables. In this process, enzymes of microbial origin are used for disintegration of fruit and vegetable pulp for more juice yield and clarification of juices (Birch et al. 1981).

Enzymatic degradation of biomaterials depends upon the type of enzyme, application temperature, incubation time, agitation, concentration, pH, and use of different enzymes combination (Baumenn 1981).

These parameters require optimization for maximum bio-conversion to get a quality product. In this reference the pure enzyme preparations that are highly specific and uncontaminated are very costly due to their extraordinarily and expansive downstream processing used for extraction and purification of enzymes from crude mass. Different enzymes with less purity and activity are derived from microbial sources and with minimum downstream processing may be a step towards cutting losts of enzymes.

The above approach may be understood in reference to enzymatic hydrolysis of carrot for increased juice recovery (Chadha et al. 2003). The effects of temperature, incubation time, concentration pectolytic and cellulolytic enzymes and their ratio on enzyme hydrolysis of carrot mash were studied using response surface methodology. The parameter ranges were 35–55 °C, 50–90 min, 0.2 = 0.4 mg enzyme protein/kg of carrot mash and the pectolytic and cellulolytic enzymes were mixed in the ratio 3.7–7.3. Results showed that all the parameters affected the juice yield, pectin content and crude fibre content significantly and enzyme concentrations were more pronounced than those of the incubation time and enzyme ratio. The juice yield increased by 5–10 %. Viscosity of juice decreased, while pH, colour index and total solids increased due to enzymatic hydrolysis.

7 Storage

The storage of fruit and vegetables has assumed great significance all over the world because of their continued and increased demand at all times of the year as well as need to spread peaks of the production over long periods to maximize profit and reduce wastage. Certain crops like apples and potatoes can be stored for long period whereas in case of certain crops like tomato and peach, the storage period is very short. There are few important factors which need to be taken into account before storage of the produce. These are:

- Suitability of crop for storage
- Knowledge of appropriate storage conditions, and
- Compatibility with other crop for storage

Different methods of fruits and vegetables storage are described below:

7.1 Storage at Low Temperature

Low temperature is achieved by taking away heat from the storage area. Heat is transferred from the storage area by the principles of latent heat of vaporization. The low temperature methods can be classified in the following ways:

Refrigeration: In the refrigeration technique, the fruits and vegetables are stored at low temperature (4–5 °C) by means of refrigerants. Refrigerants commonly used are ammonia, freon, methyl chloride, ethyl chloride, sulphur dioxide and carbon-di-oxide. Ammonia is used in large commercial applications because of its economy and the amount of heat carried away per kg of ammonia evaporated is quite high. However, for small household system and installations for food, Freon is used because it is inert and operates at low pressure difference. Storage of polyethylene packed citrus fruits under refrigerated conditions is reported to enhance their storage life considerably (Ben-Yehoshua 1985; Cohen et al. 1990). Papaya fruits have a maximum storage life of up to 7 days under tropical ambient conditions. However, the mature green to 1/4th yellow papayas can be stored up to 2 weeks at 12–13 °C. The low temperature storage for particularly ripe fruits is recommended between 7 and 10 °C. The mature green papayas fruits are more susceptible to chilling injury than ripe fruits.

7.2 *Freezing*

Freezing is low temperature preservation technique where the product is frozen at −40 °C or more low temperature depending upon the type of freezing technique chosen and further stored at −20 °C. Freezing is cheaper than canning and frozen products are close to fresh products and of better quality. The metabolic activity and spoilage due to post-harvest chemical changes are retarded by freezing. Though the product preserved by freezing retains their quality appreciably, the major disadvantage of the process is that the low temperature has to be maintained during handling, transportation and storage before the product is finally consumed. Properly conducted freezing is effective for retaining the flavour, colour and nutritive value of food and is moderately effective for preservation of texture if quick freezing is practised. During freezing, there is a physical change within the food by conversion of water to ice crystals. This helps to capture the biological condition of the food at a point at which it is frozen and thereby upon defreezing the food, the some colour, flavour and taste is obtained.

The methods of food freezing include:

(a) **Slow freezing:** Technique involve freezing by air circulation either naturally or with forced circulation by fans. The temperature may vary from −15 to −29 °C and freezing may be accomplished in 3 to 72 h. The ice crystals formed are large and rupture the biological cells.
(b) **Quick freezing:** In this process the food attains the temperature of maximum ice crystal formation (0 to −40 °C) within 30 min or less. Quick freezing maintains the natural properties of the produce by reducing post-harvest changes and microbial deterioration to the barest minimum without any influence on the original qualities. The rate of freezing plays a great role in the quality of frozen fruits and vegetables. Faster freezing rate is required to obtain better quality. Liquid nitrogen is the most common cryogenic substance used in food freezing. Ultra quick freezing rate, minimum dehydration loss, freedom from oxidative changes, minimum freezing damage of the sensitive frozen products, maximum quality retention of texture, colour and flavour during freezing and the inert nature of refrigerant are the advantages of liquid nitrogen freezing. A systematic establishment of quick freezing can boost export trade. Methods have been standardized for the manufacture of cryogenically frozen, crack-free, peeled ripe mango slices having excellent retention of quality attributes, well comparable with those of fresh mangoes in ready to serve form with cent-percent

edible portion. Problems generally faced in the export of fresh mangoes, viz. short storage life, added bulk of stone, hidden disorders, live spongy tissue and weevils, etc., can be successfully overcome by producing the cryogenically frozen mango slices (Roy 2007).

The different methods of quick freezing are:

1. **Direct immersion freezing:** It is a very rapid method of freezing, in which the prepared fruits and vegetables are directly immersed in a refrigerated brine or sugar solution
2. **Indirect contact with refrigerants:** In this technique fruits and vegetables are frozen by placing them in contact with metal surface which is cooled by a refrigerant. This system can be batch type or continuous
3. **By freezing in air:** There are two types of air systems for fruit and vegetables freezing, namely still air and forced air. Still air freezing is accomplished by placing packaged or loose package in suitable freezing rooms. In the forced air method, refrigerated air at low temperature, −18 to −34 °C or below is blown across the material to be frozen.

(c) **Cryogenic freezing:** In this technique freezing is done at very low temperature (below −60 °C). The refrigerants used for cryogenic freezing are liquid nitrogen and liquid carbon-di-oxide. Cryogenic freezing is used for mushroom, sliced tomatoes, strawberry, etc.

(d) **Dehydro Freezing:** In this technique, freezing is preceded by partial dehydration. In case of some fruits and vegetables about 50 % moisture is removed by dehydration prior to freezing. This technique improves the quality of food.

7.3 Controlled Atmosphere Storage

The controlled atmosphere storage involves a system for holding fresh fruits and vegetables in an atmosphere that differs from normal air in respect to the proportion of nitrogen, oxygen and carbon-di-oxide. The composition of the atmosphere can be altered by restricted venting the storage room or the container, scrubbing the atmosphere of carbon-di-oxide and oxygen or by adding individual gases to the container. The controlled atmosphere storage, however, does not prevent deterioration but lengthens the storage life.

7.4 Modified Storage

Modified atmosphere storage refers to a storage system including a package in which air is removed and replaced with desired gases.

The main purpose of controlled and modified atmosphere storage is to reduce as much as possible the respiration rate of the stored produce.

7.5 *Sub-atmosphere Storage*

The produce in this storage system is stored under low atmospheric pressure at the required low temperature and optimal relative humidity conditions. It retains the fresh condition of the produce for much longer period than what is possible with conventional cold storage. The sub-atmosphere storage is suitable for a wide range of fruits, viz. apples, avocado, mango, papaya, pine apple, strawberry, etc.

8 Conclusion

Fruits and vegetables are an important nutritional requirement of human beings as these foods not only meet the quantitative needs to some extent but also supply vitamins and minerals which improve the qualities of diet and maintain the health. It is therefore necessary to make them available for consumption throughout the year in fresh or processed/preserved form. India is the second largest producer of fruits and vegetables after China. However it processes less than 2 % of the total production of the fruits and vegetables. Most of the processed products are consumed domestically. There is a need to develop for fresh preservation technologies to avoid the heavy post-harvest losses.

Acknowledgement Informations on recent research trends in post-harvest treatments of fruits have been collected from paper of Dr. R. Asrey and Mr. K. Barman presented in National Seminar held at CIPHET, Ludhiana during Dec. 19–20 which is duly acknowledged.

References

Asrey R, Berman R (2011) Recent research trend in post harvest treatment of fruits. Paper presented in national seminar on post harvest packaging, cold chain and instrumentation techniques for quality and safety of perishables. CIPHET, Ludhiana, Dec 19–20

Baumenn JW (1981) Application of enzymes in fruit juice technology. In: Birohan Brakebruogh HN, Parker KJ (eds) Enzymes and food processing. Applied Science, London, pp 29–34

Ben-Yehoshua S (1985) Individual seal packaging of fruits and vegetables in plastic film- a new post harvest technology. Hort Sci 20:32–37

Birch GG, Blaubrou HN, Parker KH (1981) Enzymes in the processing of the oilseeds. Curr Sci 55:775–776

Brar SS, Simmani SSA, Dhillon WS (1998) Effect of calcium nitrite and hydro-cooling on cold storage life of peach Cv. 'Shan-1-Rungeli'. J Food Sci Technol 35(3):223–225

Cao SF, Zhang YH, Tang SS, Jin P, Wang KT (2000) Effects of methyl jasmonate inhibition of *Collitotrichum* acute infection in loquat fruit and the possible mechanism. Post Harv Biol Technol 49:301–307

Castillo A, Mekenzie KS, Lucia LM, Acutt GR (2006) Ozone treatment for reduction of *Escherichia coli* and *Salmonella* genotype *Typhimurium* on beef carcass surfaces. J Food Prot 66(5):775–779

Chadha R, Kumbhar BK, Sarkar BC (2003) Enzymatic hydrolysis of carrot for increased juice recovery. J Food Sci Technol 40(1):35–39
Chanjirakul K, Wang SY, Wang CY, Siriphanich J (2006) Effects of natural volatile components on anti-oxidant capacity and anti-oxidant enzyme in raspberries. Post Harv Biol Technol 40:106–115
Cohen I, Lurie S, Shapin B, Ben-Yehoshua S, Shalim Y, Rosenbery I (1990) Prolonged storage of lemon. J Am Soc Hort Sci 115:251–255
Gupta OP, Jindal PC, Singh BP (1980) Effect of pre-harvest sprays of calcium nitrite on the storage behaviour of grapes. J Geophys Res 10(2):204–206
Kando S (2009) Fruit ripening and characteristics regulated by physiologically active substances. Fresh Produce J (Special Issue) 17:7–11
Kaushal BBL, Thakur S (1996) Influence of ambient and evaporative cool chamber storage condition on polyethylene protected packed kinnow fruit. J Hort Sci 10:179–184
Perez AG, Sang C, Rius J, Olias JM (1993) Effect of ozone treatment on post harvest quality of strawberry. J Agric Food Chem 47:1652–1656
Raj SK (2007) Recent trends in the processing of fruits and vegetables in India. In: Khetker (ed) Food science and technology, pp 212–220
Roy SK (2007) Recent trend in the processing of fruits and vegetables in India. In: Khetker BS (ed) Food science and technology, pp 212–220
Singh P, Kulshreshth K (2006) Studies on the effects of pre-treatment on the preparation of carrot powder. J Food Sci Technol 43(2):145–147
Smith SM, Stow JR (1984) The potential of sucrose ester coating material for improving the storage and shelf life qualities of Cox's orange Rippin apples. Annu Appl Biol 104:383–391
Smock RM (1940) Some additional effects of waxing apples. Am Soc Hort Sci 37:448–459
Thakur S, Kaushal BBL, Sharma RM (2002) Effect of different post harvest methods and storage condition on the fruit quality of kinnow. J Food Sci Technol 39(6):212–213
Xu X, Tian S (2008) Salicylic acid alleviated pathogen- induced oxidative stress in harvested sweet cherry fruit. Post Harv Biol Technol 49:579–593
Yao H, Tian S (2005) Effects of pre-harvest application of salicylic acid or methyl jasmonate on inducing disease resistance of sweet cherry-fruit in storage. Post Harv Biol Technol 35: 258–262
Zhang FS, Wang XQ, Ma SJ, Cao SI (2006) Effects of methyl Jasmonate in post harvest decay in straw-berry fruit and the possible mechanism involved. Acta Hort 712:693–698
Zhang WSUX, Wang XX, Wang GX, Zheng JT, Abey Singhe D (2007) Ethanol treatment alleviates post harvest decay and maintaining fruit quality in Chinese beans. Post Harv Biol Technol 46:195–198
Zhu L, Zhou J, Zhu S (2010) Effects of a combination of nitric oxide treatment and intermittent warming and prevention of chilling injury of peach fruit during storage. Post Harv Biol Technol 121:165–170

Development of Value-Added Products from Food Wastes

Canan Ece Tamer and Ömer Utku Çopur

1 Introduction

Food is lost or wasted throughout the supply chain, from initial agricultural production down to final household consumption. "Approximately one-third of the edible parts of food produced for human consumption gets lost or wasted globally, which is about 1.3 billion ton per year" is one of the main emphasizes described in the report "Global Food Losses and Food Waste-Extent, Causes and Prevention" presented by the Food and Agriculture Organization of the United Nations (FAO) (Gustavsson et al. 2011).

Food waste has large environmental impact across its lifecycle. According to studies from the European Commission the food sector is one of the three sectors (with housing and transport) with the greatest environmental impacts in the EU, representing 30 % of its global warming potential (Stenmarck et al. 2011). Since food industries produce a substantial amount of pollution, it is becoming more and more imperative to solve this problem. As the constraints related to environmental issues are becoming quite stringent, it is necessary to develop optimized systems for food waste treatment. There are several biological and chemical processes applied for food waste treatment such as composting, aerobic and anaerobic digestion, thermophilic anaerobic digestion, sequencing batch reactor, electrodialysis, wet oxidation, pyrolysis, incineration, solid state fermentation, and ozonation (Arvanitoyannis et al. 2008).

The causes of food losses and waste in low-income countries are mainly connected to financial, managerial, and technical limitations in harvesting techniques, storage and cooling facilities in difficult climatic conditions, infrastructure, packaging, and marketing systems. Given that many smallholder farmers in developing countries live on the margins of food insecurity, a reduction in food losses could have an immediate and significant influence on their livelihoods (Gustavsson et al. 2011).

C.E. Tamer (✉) • Ö.U. Çopur
Faculty of Agriculture, Department of Food Engineering, Uludag University, Bursa, Turkey
e-mail: etamer@uludag.edu.tr

A. Malik et al. (eds.), *Food Processing: Strategies for Quality Assessment*,
Food Engineering Series, DOI 10.1007/978-1-4939-1378-7_18

Overall, on a per-capita basis, much more food is wasted in the industrialized world than in developing countries. It is estimated that the per capita food waste by consumers in Europe and North-America is 95–115 kg/year, while this figure in Sub-Saharan Africa and South/Southeast Asia is only 6–11 kg/year (Gustavsson et al. 2011). Currently food wastage amounts in the EU 27–89 million, i.e., per annum (i.e., 179 kg per capita) and the projection for 2020—if no action is taken—is 126 million tonnes (i.e., a 40 % increase) (Anonymous 2013).

Across the world, waste management represents a growing developmental, environmental, and social challenge (Asam et al. 2011; Kim and Kim 2013). The management of the food wastes is becoming extremely difficult due to legislative restrictions on landfill. These are however an remarkable source of raw materials or added-value compounds and there is, therefore, the need to develop new recovery and reuse technologies, along with the development of sustainable ideas, technologies, and processes to avoid those disposals or, at least, to restrain the loss of added-value compounds attached to these wastes (Reis et al. 2012).

Studies suggest that food production could be increased to feed the burgeoning world population in 2050 but not without difficulties, such as sapping already constrained and finite resources like clean water, accelerating the overexploitation of fisheries, or impinging on land needed for urbanization, forests, and protected areas to the extent that compromises biodiversity and damages ecosystems (Godfray et al. 2010; Spencer and Butler 2010; The Royal Society 2010). Therefore, ensuring that a greater share of food that has already been produced fulfills its intended purpose of feeding people will become increasingly important (Buzby and Hyman 2012). The European Commission is considering making 2014 the "year against food waste". The parliament adopted a resolution aiming to halve food waste in the EU by 2025.

In this chapter, properties of byproducts and other wastes of various food industries and developed process techniques to obtain value-added products from these byproducts and the wastes are briefly discussed.

2 Definition of Food Losses, Food Waste, and Food Byproducts

Definitions of food loss and waste are not universal worldwide (Buzby and Hyman 2012). Food losses refer to the decrease in edible food mass throughout the part of the supply chain that specifically leads to edible food for human consumption. Food losses take place at production, postharvest, and processing stages in the food supply chain (Parfitt et al. 2010; Gustavsson et al. 2011). They can be qualitative, such as reduced nutrient value and undesirable changes to taste, texture, or color, or quantitative as measured by decreased weight or volume (Buzby and Hyman 2012).

Food losses occurring at the end of the food chain (retail and final consumption) are rather called food waste, which relates to retailers' and consumers' behavior (Parfitt et al. 2010; Gustavsson et al. 2011). Food waste, in general, can be said to be animal or vegetal waste from manufacture, distribution, sale, and consumption of food.

In the recent discussions about food waste, also the terms "unavoidable" and "avoidable" food waste have been used. The "unavoidable food waste" is animal or vegetal waste that origins from food but it is not likely that humans will eat it (bones, peelings etc.). Avoidable food waste often has the meaning disposed food that could have been consumed if managed differently (Stenmarck et al. 2011). According to Bloom (2010), food waste occurs when an edible item goes unconsumed as a result of human action or inaction and is often the result of a decision made farm-to-fork by businesses, governments, and individual consumers.

Food wastes are residues of high organic load, which are usually derived during raw materials processing to foodstuff and result in liquid or solid form. The fact that these substances are removed from the production process as undesirable materials defines them as "wastes" in most European legislations (Commission Regulations 442/1975/EEC; 689/1991/EEC). Nevertheless, discharging of wastes does not account the potentiality of re-utilizing them inside the food chain. For this reason, the term "food byproducts" is increasingly used among the related scientists in order to notify that "food wastes" are ultimate substrates for the recapture of functional compounds and the development of new products with a market value (Galanakis 2012).

2.1 *Extent of Food Losses and Waste*

Unlike most other commodity flows, food is biological material subject to degradation, and different food stuffs have different nutritional values. There are also moral and economic dimensions: the extent to which available food crops are used to meet global human needs directly, or diverted into feeding livestock, other "byproducts" and biofuels or biomaterials production (Parfitt et al. 2010).

In medium- and high-income countries food is to a great extent wasted, meaning that it is thrown away even if it is still suitable for human consumption. Significant food loss and waste do, however, also occur early in the food supply chain. In low-income countries food is mainly lost during the early and middle stages of the food supply chain; much less food is wasted at the consumer level. Food losses in industrialized countries are as high as in developing countries, but in developing countries more than 40 % of the food losses occur at postharvest and processing levels, while in industrialized countries, more than 40 % of the food losses occur at retail and consumer levels. Food waste at consumer level in industrialized countries (222 million ton) is almost as high as the total net food production in sub-Saharan Africa (230 million ton) (Gustavsson et al. 2011).

Buzby and Hyman (2012) reported that the estimated total value of food loss at the retail and consumer levels in the USA as purchased at retail prices was $165.6 billion in 2008. The top three food groups in terms of the value of food loss at these levels are meat, poultry, and fish (41 %); vegetables (17 %); and dairy products (14 %). Fresh fruits and vegetables are usually among the most-wasted items, followed by other perishables like bakery and dairy products, meat, and fish (Pekcan et al. 2006; WRAP 2008; Morgan 2009; Thönissen 2009; Parfitt et al. 2010).

Thönissen (2009) found an unusually high proportion of food waste consisted of dairy products, while in the Turkish data, wasted fresh fruits and vegetables accounted for the highest proportion (Pekcan et al. 2006). The extent to which such differences relate to consumption patterns or different wastage rates cannot be divined from these data alone, although the Turkish study noted the importance of fruit in the diets of households studied (Parfitt et al. 2010).

3 Properties of Food Waste

Some characteristics of food wastes that have been reported in the literature indicating moisture content of 74–90 %, volatile solids to total solids ratio of 80–97 %, and carbon to nitrogen ratio of 14.7–36.4. Due to relatively high moisture content of food waste, bioconversion technologies, such as anaerobic digestion, are more suitable compared to thermochemical conversion technologies, such as combustion and gasification (Zhang et al. 2007).

Although the best way of managing food waste should be to prevent and minimize its generation at the sources, the preventive approach is somewhat limited and sometimes impractical because of the nature of the waste stream (e.g., the biodegradable characteristics) and the increasing ratio of eat-out (Tsai 2008).

4 Development of Value-Added Products from Byproducts and Wastes

It is no longer practical to discard byproducts and wastes, especially when a significant amount of valuable raw materials have a strong economic potential like the production of new products and functional ingredients with a significant added-value (Zhang et al. 2010; Toldrá and Reig 2011). It is desirable to process all byproducts into valuable products, for human foods, pet foods, animal feeds, pharmaceuticals, or fertilizer and lately for biodiesel generation.

New aspects concerning the use of byproducts for further utilization on the production of food additives or supplements with high nutritional value and economically attractive have gained increasingly interest. Transformation of waste products with high value-added allows companies to reduce the global treatment costs, sometimes even to take some profits and thus improve their competitiveness. Moreover, the recovery process of byproducts is part of the current existing sustainable development and environmental protection (Sahraoui et al. 2011).

The overall strategy for the exploitation of food processing residues is generally built around fermentative or nonfermentative product development. In the case of fermentative utilization, the residues are used as microbial substrates for the development of a variety of value added products. In nonfermentative concepts, agro-industrial residues are processed for the extraction of bioactive molecules (Nawirska and Kwaúniewska 2005; Vendruscolo et al. 2008; Bhushan et al. 2008).

Nowadays, food wastes account as a source of valuable nutraceuticals and deal with the prospects of feeding fast growing population in twenty-first century. Perspectives originate from the enormous amounts of food related materials that are discharged worldwide and the existing technologies, which promise the recovery, recycling, and sustainability of high-added value ingredients inside food chain (Schieber et al. 2001; Sonja et al. 2009; Parfitt et al. 2010; Galanakis 2012).

The wastes originated by various branches of the food industry can be divided in two main groups (plant or animal origin) and subcategories such as cereals, root & tubers, fruits and vegetables, meat products, marine products, and dairy products (Galanakis 2012).

4.1 Cereals

With regard to the cereals, wheat is the dominant crop supply in medium and high-income countries, while in South and Southeast Asia rice is more prevalent (Gustavsson et al. 2011). Grinding of wheat and dehulling of rice generates byproducts like bran or straw, which are rich in high nutritional proteins (Prakash 1996), dietary fibers, and particularly glucuronoarabinoxylans (Sun and Tomkinson 2002; Hollmann and Lindhauer 2005).

For every 1 kg of grains harvested, 1–1.5 kg of straw, cobs, or other residues are generated (Blanch and Sciamanna 1980). Cereal straws are in use for microbial protein production, but still more can be used (especially if alkali treated) without any harm to the live weight gain of the animal fed with this (Palmer 1976a, b; Das and Singh 2004).

Cereals are rich in insoluble dietary fibers, which are mainly related to intestinal regulation. Apart from health effects, dietary fibers show some functional properties, such as water-holding and swelling capacity, increasing viscosity, or gel formation which are essential in formulating certain food products. Insoluble dietary fibers increase firmness of the products and provide higher fat absorption capacity. Formulated food products with high dietary fibers content are now commercially available. Dietary fibers incorporated in these products are obtained mainly by cereals (Oreopoulou and Tzia 2007).

Cereal bran, being particularly rich in different functional biopolymers, bioactive compounds and essential fatty acids, attracted the interest of pharmaceutical and food industry. Furthermore, the special properties of brans together with their particular physiological and nutritional aspects have led to a great interest in their incorporation as main or secondary components in different groups of food products including bakery and confectionery products, breakfast cereals and extruded foods, emulsions and functional dairy products, and pasta products (Soukoulis and Aprea 2012). Oat mill waste has been suggested for the extraction of β-glucan with advanced gelling properties (Patsioura et al. 2011; Galanakis 2012).

Brewers' spent grain (BSG) is a low-value byproduct of the brewing process consisting of the barley malt residue after mashing and lautering process. It is rich in cellulose and noncellulosic polysaccharides (mainly arabinoxylans) (Valverde 1994).

Due to the presence of polysaccharides and proteins, BSG has been used as a substitute to expensive carbon sources for industrial production of lactic acid (Mussatto et al. 2007). Recently, interest in the addition of BSG as a means to enhance the quality of food products for human consumption has increased due to its richness in oligosaccharides and phenolic compounds. BSG has been incorporated as a source of dietary fiber in bread, cookies and ready-to-eat products (Öztürk et al. 2002; Ainsworth et al. 2007). Gupta et al. (2013) investigated that the utilization of BSG for the development of a fermented liquid product rich in value-added phenolic compounds. The finished product had high antioxidant capacity and the phytochemical content remained unchanged during the storage period.

Bread is a major food waste in many countries around the world and in most European countries. Moreover, bread possesses the characteristics of an ideal substrate for solid state fermentations (Sakurai et al. 1985). Melikoglu et al. (2013) investigated production of a multi-enzyme solution, rich in glucoamylase and protease, from waste bread pieces using the fungus *Aspergillus awamori*. It was confirmed that waste bread could be successfully utilized as a primary raw material in cereal based biorefineries. Dorado et al. (2009) developed wheat-based bioprocess for the production of a nutrient-complete feedstock for the fermentative succinic acid production by *Actinobacillus succinogenes*. The researchers reported that the process could be potentially integrated into a wheat-milling process to upgrade the wheat flour milling byproducts into succinic acid. Leung et al. (2012) also selected succinic acid as the test case. However, they reported that thousands of other value-added products could be produced from waste bread. They suggested that the utilization of waste bread for the production of value-added products should be seriously considered by local governments as part of their strategy for tackling the municipal solid waste problem and for the environmentally friendly production of chemicals, materials and fuels.

Nowadays agro-industrial wastes for single cell oils (SCOs) production have received increasing attentions. As an oil source produced by microbial fermentation, SCOs can be used as substitutes for value-added lipids and as a feedstock for biodiesel production, having some advantages such as short producing period, little labor required and easy to scale up (Lin et al. 2010). There are several studies on SCOs production from hydrolysates prepared from various lignocellulosic biomass, such as rice straw (Huang et al. 2009), sugarcane bagasse (Tsigie et al. 2011), rice hulls (Economou et al. 2011), and corncob (Chen et al. 2012; Zhan et al. 2013).

4.2 Root and Tubers

Among the several roots and tubers, potato is the largest crop worldwide (FAO 2009). Processing of potatoes is conducted mainly for the production of chips or French fries and corresponding solid wastes consist of peels or cull potatoes. Fresh-cut processing generates huge amounts of natural byproducts as peels which could potentially be utilized as a source of valuable compounds such as fiber, minerals, antioxidants, and pigments. These could be further processed into functional food or even in nonfood areas such as pharmaceuticals and cosmetics with a high value added (Tarazona-Díaz and Aguayo 2013). It has been reported that peels and seeds from byproducts of fresh-cut fruit and vegetables contain high amounts of phenolic and flavonoid compounds with antioxidant and antimicrobial properties (Muthuswamy and Rupasinghe 2007). Tarazona-Díazand and Aguayo (2013) studied byproducts value from fresh-cut products as peels from apple, potato, cucumber, melon, and watermelon. Potato peel was high in Fe (190 mg/kg dry weight), a similar level to cooked liver beef (162.5 mg/kg dry weight equivalent; USDA 2010). However, iron present in vegetables, called non-heme iron, having a low level of absorbability, being less effectively absorbed than heme iron (about 1–8 % compared to 15–25 % of heme iron) (Lucarini et al. 2000).

Potato peels and processing wastewater investigated for the extraction of phenols. Microwave-assisted extraction has recently been applied for the extraction of phenols from potato byproducts (Oreopoulou and Tzia 2007). Rodriguez De Sotillo et al. (1998) tested the effectiveness of freeze-dried water extract of potato peels, containing chlorogenic, caffeic, gallic, and protocatechuic acids, as a natural antioxidant. The potato peels was examined for mutagenic activity using the in vitro *Salmonella typhimurium- Escherichia coli* microsome assay and found to be nonmutagenic. It was effective only at high concentration against gram negative and one gram positive bacteria but it was bacteriostatic.

Potato tubers represent an attractive feedstock for bioethanol production in some regions in Europe. Furthermore, considerable interest has been shown in using these agricultural crops and their products for ethanol production using solid-state fermentation (Kiran et al. 1999). Bioethanol production from potatoes is based on the utilization of rotten potatoes. Afifi et al. (2011) investigated the production of ethanol from industrial solid potato wastes and optimize the fermentation efficiency by *Saccharomyces cerevisiae*.

Beet pulp is a sugar-depleted and highly fibrous material that is produced after sugar is extracted from sugarbeet (*Beta vulgaris* L.). Mohdaly et al. (2010) evaluated the antioxidant effectiveness of potato peels and sugar beet pulp extracts during oxidation of sunflower and soybean oils by measuring both primary and secondary oxidation products and to compare its antioxidant activity with commercially antioxidants. It was determined that potato peels and sugar beet pulp could stabilize both sunflower and soybean oils very effectively. They inhibit thermal deterioration of oil by improving its hydrolytic stability, inhibiting double bond conjugation and reducing the losses of polyunsaturated fatty acids.

Cassava wastes can also be processed and converted into value-added components such as methane, ethanol, surfactant, fertilizer, etc. (Ubalua 2007).

4.3 *Fruits and Vegetables*

The growth of horticulture industries worldwide has generated huge quantities of wastes. In the fruit and vegetable industry, the preparation and processing procedures could lead to one third of the product being discarded. The residues are generally a good source of carbohydrates, especially cell wall polysaccharides and other functionally important bioactive molecules such as proteins, vitamins, minerals, and natural antioxidants (Bhushan et al. 2008; O'Shea et al. 2012). Recycling of fruit and vegetable wastes is one of the most important means of utilizing it in a number of innovative ways yielding new products and meeting the requirements of essential products required in human, animal, and plant nutrition as well as in the pharmaceutical industry (Ashok Kumar et al. 2011).

The production of orange juice on an industrial level leads to a considerable quantity of residue (around 8–20 million tons/year globally), which is still considered as waste or used as a complement in agriculture (Rezzadori et al. 2012). Once the juice has been extracted, the wastes that remain (pulp and molasses) are composed mainly of peel (flavedo and albedo), pulp, and seeds (Braddock 1999). The amount of residue obtained from citrus fruits account for 50 % of the original amount of whole fruit (Sahraoui et al. 2011). One alternative to improve the management of these residues is the implementation of new processes for their recovery, for instance, through the production of organic fertilizers, pectin, seed oil, bio-oil, essential oils, and antioxidant compounds, or as a substrate for the production of several compounds with high added value, such as microbial proteins, organic acids, ethanol, enzymes, and biologically active secondary metabolites and pollutant adsorbent materials (Siles et al. 2010).

The main advantage of dietary fiber from citrus fruits, when compared to other alternative sources, such as cereals, is its higher proportion of soluble dietary fiber. Moreover, citrus fruits have better quality than other sources of dietary fiber due to the presence of associated bioactive compounds (flavonoids and vitamin C) with antioxidant properties, which may provide additional health-promoting effects (Benavente-García et al. 1997; Marín et al. 2002, 2007).

The extracts rich in dietary fiber and natural antioxidants from the byproducts of the citrus processing industry is certainly viable and the extracts could be used as a functional ingredient in the meat (Aleson-Carbonell et al. 2005; Fernández-López et al. 2007), dairy product (García-Pérez et al. 2006; Sendra et al. 2010), bakery, jam, soup, and sauce industries (Grigelmo-Miguel and Martín-Belloso 1999; Grigelmo-Miguel et al. 1999; Larrea et al. 2005). The addition of such extract to meat products has the additional advantage of reducing the concentration of residual nitrite in the products, thus attenuating the effect of potentially harmful substances such as nitrosamines (Fernández-López et al. 2007; Viuda-Martos et al. 2011). Viuda-Martos et al. (2010) proved that orange fiber has a positive effect with regard to retarding oxidation and reducing the microbial growth of unwanted microorganisms, therefore increasing the shelf-life of the sausage. The levels of polyphenolic compounds in the sausage were successfully maintained, thereby conferring a health advantage to the consumer.

Citrus volatile oil is obtained during the industrial processing of citrus fruits. Apart from its versatile applications in flavor and aroma, this byproduct is also reported to have health beneficial properties such as antioxidant, antimicrobial, antiproliferative, and antifungal properties (Jayaprakasha et al. 2013).

Nighojkar et al. (2006) studied the use of orange peel as a substrate and inducer in the production of polygalacturonase by microorganisms and concluded that orange peel is a very good inducer. Mamma et al. (2008) concluded that it is possible to produce pectinolytic, cellulolytic, and xylanolytic enzymes from the fungal strains of the genera *Aspergillus*, *Fusarium*, *Neurospora*, and *Penicillium* and generate multi-enzyme activity using a simple growth medium consisting of a solid byproduct of the citrus processing industry.

The low-economic value citrus processing waste is the major waste generated by the fruit processing industry and has been suggested as a sustainable source of cellulosic biomass for biorefinery applications (Rivas-Cantu et al. 2013). A sustainable utilization potential of citrus processing wastes is their conversion to bioethanol to be used as biofuel and other value-added products by means of biorefinery processes that are dependent on their chemical composition (Wilkins et al. 2007).

The orange juice waste water can be suitable for applications on the food industry. It is an important source of phenolic compounds in which antioxidant properties could be very appreciated in a big number of food processing to avoid its oxidation during processing but also during storage period. Another important effect of orange juice waste water is the reduction of residual nitrite level in foods in which the nitrite is added as an additive, for example in meat products industry; the reduction of residual nitrite level could reduce the possibility to nitrosamine formation (Viuda-Martos et al. 2011).

Matthaus and Özcan (2012) examined the seed oils from Turkish and Vietnamese varieties of citrus fruits for their fatty acid composition, tocopherols, and sterol contents. As a result of the high oil content, the seeds of citrus species seem to be an interesting source for the production of vegetable oil. The oil contains linoleic acid as the major fatty acid accompanied by oleic acid. The content and composition of tocopherols are comparable to those of other sources such as sunflower, corn or rapeseed oil. The composition of sterols in citrus oil is dominated by β-sitosterol, which accounted for about 70 % of the total sterols in the oil. This is typical of many vegetable oils. The seed oils of citrus fruits might be used for edible applications as well as the production of potential value-added products.

Basavegowda and Lee (2013) determined the feasibility of using fruit wastes for the synthesis of silver nanoparticles. They described the use of satsuma mandarin fruit waste material for the consistent and rapid synthesis of silver nanoparticles. The UV-visible and transmission electron microscopy indicated that the nanoparticles produced were stable. Structural analysis by X-ray diffraction and elemental analysis by energy dispersive X-ray analysis proved the formation of elemental silver. The nanoparticles produced by this cost-effective, simple, fast, nontoxic, eco-friendly technique based on abundantly available fruit peel waste were characterized by a variety of standard analytical techniques. These silver nanoparticles have potential applications in the biomedical field and the straightforward procedure used has several advantages for a large-scale commercial production.

Fruit byproducts might also be utilized for their phenolic content and antioxidant properties (Djilas et al. 2009). Polyphenols or condensed tannins (proanthocyanidins) from fruit species have been used to enrich fruit extracts, cereal products, and dietary or nutraceutical agents, turning them into antioxidant functional products with a very positive image and a high acceptance by consumers (Sun-Waterhouse 2011). The antioxidant compounds may also be associated with dietary fiber and thus both may be approached jointly in nutrition and health as proposed in a recent publication which also describes that the transportation of antioxidants through the gastrointestinal tract may be an essential nutritional function of dietary fiber (Saura-Calixto 2011).

The potential use of various fruit pomaces as functional foods has been evaluated in various studies (Nawirska and Kwaúniewska 2005; Sun-Waterhouse 2011; O'Shea et al. 2012). Grape pomace consists of seeds, skins and stems, and in some cases this byproduct is used to extract grape seed oil. It is also used in the production of citric acid, methanol, ethanol, and xanthan gum as a result of fermentation (Deng et al. 2011). The phytochemicals in grape pomace have proven to confer many promising health attributes. Ruberto et al. (2007) and Maier et al. (2009) clearly demonstrated the availability of flavonoids and phenolic acids in grape pomace. Pedroza et al. (2013) added different dehydrated waste grape skins from the juice industry into aged and young red wines as an innovative way of compensating for color loss before bottling. Mixtures of dehydrated waste grape skins were found useful to improve the color and polyphenol profile of red wines, considering them a useful tool for correcting color loss before bottling.

Recent research trends reveal that there is an increase in the utilization of apple pomace as a food processing residue for the extraction of value added products such as dietary fiber, protein, natural antioxidants, biopolymers, pigments, and compounds with unique properties (Bhushan et al. 2008). The inclusion of apple pomace as an ingredient in food products could dramatically improve the nutritive properties of such products and perhaps the health of the consumer (O'Shea et al. 2012). These studies indicated that the polyphenols responsible for the antioxidant activity in apple are still present in the pomace and can easily be extracted for food fortification or nutraceutical product development. Apple pomace can therefore become an inexpensive and readily available source of dietary antioxidants (Bhushan et al. 2008). Moreover, addition of apple pomace conferred some favorable attributes such as a fruit aroma and taste, thus allowing the level of sugar added to be reduced (Masoodi et al. 2002; Sudha et al. 2007). Due to its functional characteristics (water holding, gelling, thickening and stabilizing abilities), nutrients (carbohydrates, proteins, vitamins A and C), minerals (P, K, Mn, Ca, Mg and Fe) and phytochemicals (catechins, procyanidins, caffeic acid, phloridzin, phloretin glycosides, quercetin glycosides, chlorogenic acid, etc.,) apple pomace has been used by researchers in a variety of food products such as sausages, jams, and baked goods (Bhushan et al. 2008; Henriquez et al. 2010; O'Shea et al. 2012). In addition to pectin, apple pomace is also a good source of cellulose and hemicelluloses-xyloglucan mostly fucogalactoxyloglucans (Caili et al. 2006). The conversion of xyloglucan into compounds such as thickening agents and texture modifiers or as a source of biologically active oligosaccharides having medicinal properties is possible. Therefore the extraction of xyloglucan from apple pomace can provide a new means for the development of high-value biomolecules (Bhushan et al. 2008).

Aguedo et al. (2012) analyzed the composition of the dried milled pomaces from cooked apples, pears, and dried dates. Apple and pear residues were composed mainly of cellulose, whereas lignin was the main fraction for dried date. The polyphenolic content and the antioxidant activity of the three products were also assessed and the values showed that their antioxidant characteristics were comparable to that of various raw fruits. No phenolic acids were detected, indicating that the cooking process resulted in their extraction.

Industrial processing of tomato generates a considerable amount of waste, consisting of peel, seeds, and a part of the pulp, which are known as tomato pomace. These residues contain valuable nutritional compounds, mainly fibers, proteins, and antioxidants (Lavelli and Torresani 2011). Lycopene, the carotenoid responsible for the red color of tomatoes, is a lipophilic compound with high antioxidant activity by scavenging oxygen radicals which reduce oxidative stress in the organism (Rao 2006; Maiani et al. 2009). This antioxidant activity results in a protective effect against cardiovascular disease, hypertension, atherosclerosis, cancer, and diabetes among others (Clinton et al. 1996; Kohlmeier et al. 1997; Etminan et al. 2004; Kong et al. 2010; Ried and Fakler 2011; Cámara et al. 2012).

Kalogeropoulos et al. (2012) investigated the presence of health-promoting phytochemicals like carotenoids, polyphenols, sterols, terpenes, and tocopherols in tomato processing byproducts, in order to evaluate the potential for their use as additives for the preparation of functional foods or for the shelf-life elongation of lipid-rich foods. Unprocessed industrial tomatoes were also analyzed for comparison. On a dry weight basis, tomato wastes contained significantly lower amounts of lycopene and increased amounts of β-carotene, tocopherols, sterols, and terpenes, while their fatty acid profile was similar to that of unprocessed tomatoes. Tomato byproducts contained similar amount of total polyphenols and exhibited similar DPPH radical scavenging activity and ferric reducing power with tomatoes. Among polyphenols determined, hydroxycinnamic acids predominated in whole tomatoes, while flavonoids predominated in tomato wastes with naringenin comprising 87 % of flavonoids. As most of the phytochemicals determined exert antioxidant activities, tomato processing byproducts could be successfully utilized as functional ingredient for the formulation of antioxidant rich functional foods.

Carrot is a good source of natural antioxidants, especially carotenoids and phenolic compounds. After processing, carrot residues, e.g., peels, pomace are usually discarded or used as animal feed. However, carrot byproducts still contain high contents of beneficial substances, especially bioactive compounds with antioxidant activities (Zhang and Hamauzu 2004). Carrot peels could be used as a good raw material to produce antioxidant dietary fiber powder. In addition to high dietary fiber fraction, the bioactive compounds with antioxidant activities associated with the material, made it more interesting. From industrial point of view, the peels which are the residues from processing and abundant could be further processed to add value to the product using conventional preparation methods. Unlike fruits which may contain kernels and seeds, the pomace received from carrots can easily be added to a product without introducing negative functional or flavor issues while still retaining a lot of its phytochemicals (Chantaro et al. 2008). Therefore, this byproduct can be used as an ideal ingredient for addition into food (O'Shea et al. 2012). Durrani et al. (2011) investigated the potential of carrot as the main ingredient in a honey based candy. The authors found that the product received positive sensory scores and acceptable physicochemical and microbiological results.

Broccoli byproducts have been proposed as a source of bioactive compounds (Domínguez-Perles et al. 2010). Domínguez-Perles et al. (2011) analyzed minimally-processed broccoli byproducts as a source of bioactive ingredients (glucosinolates, isothiocyanates, chlorogenic acid sinapic acid derivatives, flavonoids, minerals, vitamins) to design novel beverages, using organic green tea as a food matrix. Green tea enriched with broccoli concentrates showed improved physical quality, phytochemical composition, and antioxidant capacity.

Asparagus and its byproducts can be considered products of interest since a functional point of view due their contents of fiber and distinct phytochemicals, such as phenolics, mainly flavonoids, saponins, sterols, and fructans. Fuentes-Alventosa et al. (2013) developed a process for obtaining added-value compounds from asparagus byproducts by the hydrothermal treatment of the samples. Two fractions were separated after the hydrothermal treatment, consisting on an aqueous functional extract containing most soluble bioactive compounds from asparagus byproduct, and a fibrous residue that, after being dried, constitutes the asparagus bioactive fiber. The process includes a column purification step, using an adsorbent polymeric resin, which allows aqueous extracts, partially purified to be obtained and enriched in specific compounds (phenolics and/or saponins). The preliminary results showed that the distinct products obtained from asparagus byproducts are of interest for their biological activity and are suitable for being used as functional ingredients. Based on their antioxidant capacity, it could be proposed that their regular use could help the prevention of several diseases related to the oxidative damage.

Since oil press-cakes, byproducts obtained by mechanical pressing of oilseeds, contain substantial amounts of residual oil, are rich in proteins, nutrients, and minerals, and do not contain harmful organic solvents, their recovery and use in the food industry is significant. Oil press-cakes obtained by pressing oilseeds that do not contain antinutrients have strong potential to be used as sources of value-added, novel, and nutritious food products. Radócaj et al. (2012) developed a stable, oil-based spread rich in the omega-3 and omega-6 fatty acids using a hull-less pumpkin seed (*Cucurbita pepo* L.) oil press-cake, a byproduct of the pumpkin oil pressing process, along with cold-pressed hemp oil. Pumpkin seeds and seed oil have been implicated in providing many health benefits, which are attributed to their macro- and microconstituents composition. As indicated by recent research studies, pumpkin seeds (*Cucurbita pepo* L.) have components which have positive effects on the human body, including antidiabetic, antihypertensive, antitumor, antibacterial, anti-hypercholesterolemia, and anti-inflammatory actions (Makni et al. 2011). One of the most critical health benefits attributed to pumpkin seed oil is its activity against benign prostate hyperplasia (Xanthopoulou et al. 2009). Although nutritionally superior, pumpkin oil press-cake, as a byproduct of food processing, is currently not used for food production. This press-cake has a substantial amount of residual oil, which is rich in omega-6 fatty acids, γ -tocopherols, minerals and proteins, and as such, could have different applications in the development of functional food products (Radócaj et al. 2011).

Olive is a very popular oil crop in the Mediterranean region that can also be considered in the fruits' group. Olive mill pomace and wastewater have been widely valorized as a source of bioactive phenols and more recently pectin (Galanakis et al. 2010). Olive mill wastes represent an important environmental problem. Their high phenol, lipid and organic acid concentrations turn them into phytotoxic materials, but these wastes also contain valuable resources such as a large proportion of organic matter and a wide range of nutrients that could be recycled. Second oil extraction, combustion, gasification, anaerobic digestion, composting, and solid fermentation are some of the methods proposed for the valorization of olive mill wastes (Roig et al. 2006). Among the several fruits, olive is known to contain an appreciable amount of phenols with good antioxidant properties (Boskou et al. 2006) and dietary fibers with promising water holding capacity (Jiménez et al. 2000). Nevertheless, the majority of these compounds are lost in olive mill wastewater (OMW) during olive oil production. For example, OMW typically contains 98 % of the total phenols in the olive fruit (Obied et al. 2005; Suárez et al. 2009). OMW has also been referred to possess soluble dietary fibers and especially pectin material with satisfying gelling ability (Vierhuis et al. 2003; Cardoso et al. 2003). Galanakis et al. (2010) suggested that the clarification as well as the recovery of valuable compounds from pectin containing solution and phenol containing beverage is possible with the utilization of ultrafiltration and nanofiltration technologies. Lafka et al. (2011) investigated the extraction variables of conventional liquid solvent extraction for the recovery of phenolic compounds from olive oil mill wastes and evaluated and compared phenol content and the antioxidant activity of supercritical carbon dioxide extraction and the solvent phenol extracts. Ethanol was selected as the most appropriate solvent for the extraction of phenolic compounds from olive oil mill waste for the production of extracts with high phenol content and high antioxidant activity. Supercritical carbon dioxide was confirmed to be an efficient solvent for recovering phenolic compounds with relatively high antioxidant activity from olive oil mill waste. Therefore, the olive oil mill waste is a low-cost, renewable and abundant source of phenolic antioxidants. Phenolic extracts from olive oil mill waste can be used as alternatives to synthetic antioxidants in order to increase the stability of foods by preventing lipid peroxidation and protect living systems from oxidative damage by scavenging oxygen radicals. Aliakbarian et al. (2011) used a high pressure-high temperature reactor for extraction of bioactive compounds from olive pomace. Oleuropein and tyrosol were predominant phenolics in the extracts. Researchers were noted that methanolic extracts, can be used in food, cosmetic and pharmaceutical industries after additional purification processes. Solid residues, after phenolics extraction, are considered to have little environmental impact.

4.4 Meat Products

Slaughterhouses as well as the processing of meats generate a significant amount of solid and liquid byproducts (Toldrá et al. 2012). Slaughterhouse wastes consist of the portion of a slaughtered animal that cannot be sold as meat or used in meat-products. Such wastes include bones, tendons, skin, blood, and internal organs

(Russ and Meyer-Pittroff 2004). In the case of edible items, meat byproducts constitute an excellent source of nutrients like essential amino acids, minerals, and vitamins (Aristoy and Toldrá 2011; García-Llatas et al. 2011; Honikel 2011; Kim 2011). Added value products can be obtained in terms of shelf stability, improved technological functions (flavoring compounds, water bonding agents, emulsifiers), better organoleptic quality, or even more convenience (Toldrá et al. 2012). Another alternative is to produce functional ingredients like bioactive peptides and antioxidants.

Chicken processing plants all over the world generate large amount of solid byproducts in form of heads, legs, bones, viscera, and feather. These wastes are often processed into livestock feed, fertilizers, and pet foods or totally discarded. Inappropriate disposal of these wastes causes environmental pollution, diseases, and loss of useful biological resources like protein, enzymes, and lipids. Utilization methods that make use of these biological components for producing value added products rather than the direct use of the actual waste material might be another viable option for dealing with these wastes (Lasekan et al. 2013).

Since the outbreak of Bovine Spongiform Encephalopathy the use of animal byproducts for livestock feeding has come under strict regulation such that, only the low risk materials are permitted even in pet foods and as organic fertilizers (Cascarosa et al. 2012). Byproducts from different animal sources could be utilized as sources of protein hydrolysates, enzymes, and polyunsaturated fatty acids and also they could be converted to peptone for use in growth media or used directly as substrates for culturing microorganisms that can produce useful biotechnological products (Lasekan et al. 2013).

4.5 Marine Products

Byproducts and wastes generated from the fish processing industry are regarded as a source of highly valuable compounds, in some cases even higher in value than the starting material (Dumay et al. 2006; Ferraro et al. 2013). At present, the majority of these byproducts are sent to fish meal plants, where fish meal and fish oil are produced. Fish meal is by far the most valuable nonedible commodity produced from marine byproducts (Hardy and Tacon 2002). It is used either as animal feed or as plant fertilizer and is composed of protein (70 %), minerals (10 %), fat (10 %) and water (10 %), on a weight basis (Blanco et al. 2007).

Fish oil can have edible and nonedible applications depending on its composition. Production of soap, glycerol, varnishes, drying and hydraulic oils, fertilizers, and substrates for fermentation, are the most common uses, whilst its edible applications are essentially accounted for the production of margarine and shortenings (Ferraro et al. 2010).

Fish processing co-products (FPCP) protein hydrolysates have been found to possess desirable physicochemical properties (e.g., emulsifying, foaming, oil, and water-binding capacities) and many interesting bioactivities (antioxidative, antihypertensive, antimicrobial, and antianemia) with potential applications in food, nutritional and pharmaceutical products. Chemical hydrolysis has been the most common process for the production of crude FPCP protein hydrolysates, though with little ability to control product quality. The enzymatic hydrolysis process has emerged recently as the process of choice due to its mild reaction conditions, superior product quality, and functionality (He et al. 2013).

The main high added value compounds in marine byproducts and waste are polyunsaturated fatty acids (ω-3 and ω-6), free amino acids, chitin and chitosan, collagen and gelatin, hydroxyapatite, antifreeze proteins, astaxanthin, and enzymes (Ferraro et al. 2010). The use of fish processing byproducts as substrate provides a novel approach for the potential discovery of high-value bioactive products. Advances in bioprocess engineering technologies such as more efficient design and development of manufacturing equipment and processes such as bioreactors, together with novel use of nanotechnology methods such as using chitin, chitosan, and crab shell waste fermentation products to encapsulate functional food ingredients will mean that fish processing wastes in the future may serve as inexpensive raw materials in the generation of high-value bioactive compounds which may have a variety of applications (Kim and Mendis 2006; Srinivosa and Tharanatham 2007). Development of these technologies will ensure the exploitation of potential added-value products from this waste stream and will present unique challenges and opportunities for the seafood industry worldwide (Wilson et al. 2011).

4.6 Dairy Products

The dairy industry is based on the processing and manufacturing of raw milk into products such as yogurt, ice cream, butter, cheese, and various types of desserts by means of different processes, such as pasteurization, coagulation, filtration, centrifugation, chilling, etc. (Rivas et al. 2010). The characteristics of dairy effluents may vary significantly, depending on the final products, system type and operation methods used in the manufacturing plant (Gutiérrez et al. 1991). These wastewaters are mainly composed by different dilutions of milk (or transformed products), and washing water containing alkaline and acidic chemicals after the cleaning of bottles, tanks, and process equipment (Carvalho et al. 2013).

Treated dairy waste has found many applications among which the most important are for biogas/biodiesel production, fertilizer, animal feedstuff, the food industry, and miscellaneous uses (Arvanitoyannis and Kassaveti 2008). Milk or whey from dairy industry waste was concentrated with vacuum evaporator at 20 °C. Raw lactose was obtained on filter-centrifuge or vacuum filter from whey concentrate and processed by pre-crystallization and purified into lactose. Lactose remaining in the filtrate was purified from whey proteins by ultrafiltration and diafiltration. Permeate can be further separated from monovalent salts with nanofiltration to obtain minerals and vitamins. Whey concentrates or whey protein concentrates can be utilized both for sweet and salt foodstuffs, instant flours, and additives for soups.

On the other hand, lactose, minerals, and vitamins can be effectively used in dietetic foods for diabetics and hypertensic patients (Ostojic et al. 2005).

Cheese whey is the main by product from the dairy industry and is composed of approximately 93 % water, 5 % lactose, 0.9 % protein, 0.3 % fat, 0.2 % lactic acid, vitamins, and mineral salts (González-Siso 1996). Several industries recover a portion of the whey proteins via ultrafiltration for use in food supplements or in other milk products. However, cheese whey permeate resulting from this process still contains approximately 85–95 % of the whey lactose (Diniz et al. 2013). The lactose carbohydrate reservoir of whey and the presence of other nutrients essential for microbial growth, make whey a potential raw material for the production of various bioproducts through biotechnological means (Panesar et al. 2007). Several methods have been proposed for whey valorization. In this respect, lactose converting microorganisms have been evaluated for the production of potable and fuel-grade alcohol (Petsas et al. 2002), kefir-like whey drinks (Paraskevopoulou et al. 2003), lactic acid (Kourkoutas et al. 2005), baker's yeast (Plessas et al. 2004), single cell protein as livestock feed (Plessas et al. 2008), probiotic starter cultures for fermented milk products (Kourkoutas et al. 2005), and cheese ripening (Koutinas et al. 2009).

Whey proteins exhibit several functional properties that are essential for the formation of edible films. Viable edible films and coatings have been successfully produced from whey proteins; their ability to serve other functions, i.e., carrier of antimicrobials, antioxidants, or other nutraceuticals, without significantly compromising the desirable primary barrier and mechanical properties as packaging films, will add value for eventual commercial applications (Ramos et al. 2012).

References

Afifi MM, Abd El-Ghany TM, Mohamed AA, Taher MT, Khaled EG (2011) Biorefinery of industrial potato wastes to ethanol by solid state fermentation. Res J Agric Biol Sci 7(1): 126–134

Aguedo M, Kohnen S, Rabetafika N, Bossche SV, Sterckx J, Blecker C, Beauve C, Paquot M (2012) Composition of by-products from cooked fruit processing and potential use in food products. J Food Compos Anal 27:61–69

Ainsworth P, Ibanoglu S, Plunkett A, Ibanoglu E, Stojceska V (2007) Effect of brewers spent grain addition and screw speed on the selected physical and nutritional properties of an extruded snack. J Food Eng 81:702–709

Aleson-Carbonell L, Fernandez-Lopez J, Perez-Alvarez JA, Kuri V (2005) Functional and sensory effects of fibre-rich ingredients on breakfast fresh sausages manufacture. Food Sci Technol Int 11(2):89–97

Aliakbarian B, Casazza AA, Perego P (2011) Valorization of olive oil solid waste using high pressure-high temperature reactor. Food Chem 128:704–710

Anonymous (2013) Prevention and reduction of food waste. References: CESE 1917/2012 - NAT/570. http://www.eesc.europa.eu/?i=portal.en.nat-opinions.25955. Accessed 19 Oct 2013

Aristoy MC, Toldrá F (2011) Essential amino acids. In: Nollet LML, Toldrá F (eds) Handbook of analysis of edible animal by-products. CRC Press, Boca Raton

Arvanitoyannis IS, Kassaveti A (2008) Dairy waste management: treatment methods and potential uses of treated waste. In: Arvanitoyannis IS (ed) Waste management for the food industries. Academic, Amsterdam

Arvanitoyannis IS, Kassaveti A, Ladas D (2008) Food waste treatment methodologies. In: Arvanitoyannis IS (ed) Waste management for the food industries. Academic, Amsterdam

Asam ZZ, Poulsen TG, Nizami AS (2011) How can we improve biomethane production per unit of feedstock in biogas plants? Appl Energy 88:2013–2018

Ashok Kumar K, Narayani M, Subanthini A, Jayakumar M (2011) Antimicrobial activity and phytochemical analysis of citrus fruit peels -utilization of fruit waste. Int J Eng Sci Technol 3:5414–5421

Basavegowda N, Lee YR (2013) Synthesis of silver nanoparticles using Satsuma mandarin (*Citrus unshiu*) peel extract: a novel approach towards waste utilization. Mater Lett 109(15):31–33

Benavente-García O, Castillo J, Marín FR, Ortuño A, Del Río JA (1997) Uses and properties of citrus flavonoids. J Agric Food Chem 45:4505–4515

Bhushan S, Kalia K, Sharma M, Singh B, Ahuja PS (2008) Processing of apple pomace for bioactive molecules. Crit Rev Biotechnol 28:285–296

Blanch HW, Sciamanna A (1980) Composition of cellulosic raw materials for sugar production. Alcohol Fuel Process R/D Newsletter. Winter Quarter, USDOE Solar Energy Research Institute, p 69

Blanco M, Sotelo CG, Chapela MJ, Pérez-Martín RI (2007) Towards sustainable and efficient use of fishery resources: Present and future trends. Trends in Food Science and Technology. 18:29–36.

Bloom J (2010) American wasteland. Da Capo Press, Cambridge

Boskou G, Salta FN, Chrysostomou S, Mylona A, Chiou A, Andrikopoulos NK (2006) Antioxidant capacity and phenolic profile of table olives from the Greek market. Food Chem 94(4):558–564

Braddock RJ (1999) Handbook of citrus by-products and processing technology. Wiley, New York

Buzby JC, Hyman J (2012) Total and per capita value of food loss in the United States. Food Policy 37:561–570

Caili F, Haijun T, Quanhong L, Tongyi C, Wenjcian D (2006) Ultrasound assisted extraction of xyloglucan from apple pomace. Ultrason Sonochem 13:511–516

Cámara M, Fernández-Ruiz V, Redondo DF, Sánchez-Mata MC, Torrecilla JS (2012) Radial basis network analysis to estimate lycopene degradation kinetics in tomato-based products. Food Res Int 49:453–458

Cardoso SM, Coimbra MA, Lopez da Silva JA (2003) Calcium-mediated gelation of an olive pomace extract. Carbohydr Polym 52(2):125–133

Carvalho F, Prazeres AR, Rivas J (2013) Cheese whey wastewater: characterization and treatment. Sci Total Environ 445-446C:385–396

Cascarosa E, Gea G, Arauzo J (2012) Thermochemical processing of meat and bone meal: a review. Renew Sustain Energy Rev 16:942–957

Chantaro P, Devahastin S, Chiewchan N (2008) Production of antioxidant high dietary fiber powder from carrot peels. LWT food Sci Technol 41:1987–1994

Chen XF, Huang C, Xiong L, Chen XD, Ma LL (2012) Microbial oil production from corncob acid hydrolysate by *Trichosporon cutaneum*. Biotechnol Lett 34:1025–1028

Clinton SK, Emenhiser C, Schwartz SJ, Bostwick DG, Williams AW, Moore BJ, Erdman JW (1996) Cis-trans lycopene isomers, carotenoids, and retinol in the human prostate. Cancer Epidemiol Biomarkers Prev 5:823–833

Das H, Singh SK (2004) Useful byproducts from cellulosic wastes of agriculture and food industry-a critical appraisal. Crit Rev Food Sci Nutr 44:77–89

Deng Q, Penner MH, Zhao Y (2011) Chemical composition of dietary fiber and polyphenols of five different varieties of wine grape pomace skins. Food Res Int 44(9):2711–2719

Diniz RHS et al (2013) Optimizing and validating the production of ethanol from cheese whey permeate by *Kluyveromyces marxianus* UFV-3. Biocatal Agric Biotechnol. doi:10.1016/j.bcab.2013.09.002i

Djilas S, Cánadanović-Brunet J, Cétkovi G (2009) By-products of fruits processing as a source of phytochemicals. Chem Ind Chem Eng Q 15:191–202

Domínguez-Perles R, Martínez-Ballesta MC, Carvajal M, García-Viguera C, Moreno DA (2010) Broccoli-derived by-products - a promising source of bioactive ingredients. J Food Sci 75(4):383–392

Domínguez-Perles R, Moreno DA, Carvajal M, García-Viguera C (2011) Composition and antioxidant capacity of a novel beverage produced with green tea and minimally-processed byproducts of broccoli. Innov Food Sci Emerg Technol 12:361–368
Dorado MP, Lin SKC, Koutinas A, Du C, Wang R, Webb C (2009) Cereal-based biorefinery development: utilisation of wheat milling by-products for the production of succinic acid. J Biotechnol 143:51–59
Dumay J, Donnay-MorenoC BG, Jaouen P, Bergé JP (2006) Process Biochem 41(11):2327–2332
Durrani A, Srivastava P, Verma S (2011) Development and quality evaluation of honey based carrot candy. J Food Sci Technol 48(4):502–505
Economou CN, Aggelis G, Pavlou S, Vayenas DV (2011) Single cell oil production from rice hulls hydrolysate. Bioresour Technol 102:9737–9742
Etminan M, Takkouche B, Caamano-Isorna F (2004) The role of tomato products and lycopene in the prevention of prostate cancer: a meta-analysis of observational studies. Cancer Epidemiol Biomarkers Prev 13:340–345
FAO (2009) FAOSTAT. Data archives. http://faostat.fao.org/site/444/DesktopDefault.aspx?PageID¼444#ancor. Accessed 19 Oct 2013
Fernández-López J, Viuda-Martos M, Sendra E, Sayas-Barberá E, Navarro C, Pérez-Alvarez JA (2007) Orange fibre as potential functional ingredient for dry-cured sausages. Eur Food Res Technol 226(1–2):1–6
Ferraro V, CruzIB IB, Jorge RF, Malcata FX, Pintado ME, Castro PML (2010) Valorisation of natural extracts from marine source focused on marine by-products: a review. Food Res Int 43:2221–2233
Ferraro V, Carvalho AP, Piccirillo C, Santos MM, Castro PML, Pintado ME (2013) Extraction of high added value biological compounds from sardine, sardine-type fish and mackerel canning residues - a review. Mater Sci Eng 33:3111–3120
Fuentes-Alventosa JM, Jaramillo-Carmona S, Rodríguez-Gutiérrez G, Guillén-Bejarano R, Jiménez-Araujo A, Fernández-Bolanõs J, Rodríguez-Arcos R (2013) Preparation of bioactive extracts from asparagus by-product. Food Bioprod Process 91:74–82
Galanakis CM (2012) Recovery of high added-value components from food wastes: conventional, emerging technologies and commercialized applications. Trends Food Sci Technol 26:68–87
Galanakis CM, Tornberg E, Gekas V (2010) Clarification of high-added value products from olive mill wastewater. J Food Eng 99:190–197
García-Llatas G, Alegría A, Barberá R, Farré R (2011) Minerals and trace elements. In: Nollet LML, Toldrá F (eds) Handbook of analysis of edible animal by-products. CRC Press, Boca Raton
García-Pérez FJ, Sendra E, Lario Y, Fernández-López J, Sayas-Barbera E, Pérez-Alvarez JA (2006) Rheology of orange fiber enriched yogurt. Milchwissenschaft 61(1):5559
Godfray HJC, Beddington JR, Crute IR, Haddad L, Lawrence D, Muir JF, Pretty J, Robinson S, Thomas SM, Toulmin C (2010) Food security: the challenge of feeding 9 billion people. Science 327:812–818
González-Siso MI (1996) The biotechnological utilization of cheese whey: a review. Bioresour Technol 57(1):1–11
Grigelmo-Miguel N, Martín-Belloso O (1999) Characterization of dietary fibre from orange juice extraction. J Food Eng 41:13–21
Grigelmo-Miguel N, Carreras-Boladeras E, Martín-Belloso O (1999) Development of high-fruit-dietary-fibre muffins. Eur Food Res Technol 210:123–128
Gupta S, Jaiswal AK, Abu-Ghannam N (2013) Optimization of fermentation conditions for the utilization of brewing waste to develop a nutraceutical rich liquid product. Ind Crops Prod 44:272–282
Gustavsson J, Cederberg C, Sonesson U, van Otterdijk R, Meybeck A (2011) Global food losses and food waste - extent, causes and prevention. International Congress Save Food!. http://www.fao.org/fileadmin/user_upload/ags/publications/GFL_web.pdf. Accessed 19 Oct 2013
Gutiérrez JLR, Encina PAG, Polanco FF (1991) Anaerobic treatment of cheese-production wastewater using UASB reactor. Bioresour Technol 37(3):271–276

Hardy RW, Tacon AGJ (2002) Fish meal: historical uses, production trends and future outlook for sustainable supplies. In: Stickney RR, McVey JP (eds) Responsible marine aquaculture. doi:10.1079/9780851996042.0311
He S, Franco C, Zhang W (2013) Functions, applications and production of protein hydrolysates from fish processing co-products (FPCP). Food Res Int 50:289–297
Henriquez C, Speisky H, Chiffelle I, Valenzuela T, Araya M, Simpson R et al (2010) Development of an ingredient containing apple peel, as a source of polyphenols and dietary fiber. J Food Sci 75(6):H172–H181
Hollmann J, Lindhauer MG (2005) Pilot-scale isolation of glucuronoarabinoxylans from wheat bran. Carbohydr Polym 59:225–230
Honikel KO (2011) Composition and calorics. In: Nollet LML, Toldrá F (eds) Handbook of analysis of edible animal by-products. CRC Press, Boca Raton
Huang C, Zong MH, Wu H, Liu QP (2009) Microbial oil production from rice straw hydrolysate by *Trichosporon fermentans*. Bioresour Technol 4535–4538
Jayaprakasha GK, Chidambara Murthy KN, Uckooa RM, Patil BS (2013) Chemical composition of volatile oil from Citrus limettioides and their inhibition of colon cancer cell proliferation. Ind Crops Prod 45:200–207
Jiménez A, Rodríguez R, Fernández-Caro I, Guillén R, Fernández-Bolaños J, Heredia A (2000) Dietary fibre content of table olives processed under different European styles: of physico-chemical characteristics. J Sci Food Agric 80(13):1903–1908
Kalogeropoulos N, Chiou A, Pyriochou V, Peristeraki A, Karathanos VT (2012) Bioactive phytochemicals in industrial tomatoes and their processing byproducts. LWT food Sci Technol 49:213–216
Kim YN (2011) Vitamins. In: Nollet LML, Toldrá F (eds) Handbook of analysis of edible animal by-products. CRC Press, Boca Raton
Kim DH, Kim MS (2013) Development of a novel three-stage fermentation system converting food waste to hydrogen and methane. Bioresour Technol 127:267–274
Kim SK, Mendis E (2006) Bioactive compounds from marine processing byproducts - a review. Food Res Int 39(4):383–393
Kiran SN, Sridhar M, Rao LV, Pandey A (1999) Ethanol production in solid substrate fermentation using a thermotolerant yeast. Process Biochem 34:115–119
Kohlmeier L, Kark JD, Gomez-Gracia E, Martin BC, Steck SE, Kardinal AFM et al (1997) Lycopene and myocardial infarction risk in the EURAMIC study. Am J Epidemiol 146:618–626
Kong KW, Khoo HE, Prasad KN, Ismail A, Tan CP, Rajab NF (2010) Revealing the power of the natural red pigment lycopene. Molecules 15:959–987
Kourkoutas Y, Xolias V, Kallis M, Koutinas AA, Bezirtzoglou E, Kaliafas A, Kanellaki M (2005) *Lactobacillus casei* immobilization on fruit pieces for probiotic additive, fermented milk and lactic acid production. Process Biochem 40:411–416
Koutinas AA, Papapostolou H, Dimitrellou D, Kopsahelis N, Katechaki E, Bekatorou A, Bosnea LA (2009) Whey valorisation: a complete and novel technology development for dairy industry starter culture production. Bioresour Technol 100:3734–3739
Lafka TI, Lazou AE, Sinanoglou VJ, Lazos ES (2011) Phenolic and antioxidant potential of olive oil mill wastes. Food Chem 125:92–98
Larrea MA, Chang YK, Martinez-Bustos F (2005) Some functional properties of extruded orange pulp and its effect on the quality of cookies. LWT Food Sci Technol 38(3):213–220
Lasekan A, Bakar FA, Hashim D (2013) Potential of chicken by-products as sources of useful biological resources. Waste Manage 33:552–565
Lavelli V, Torresani MC (2011) Modelling the stability of lycopene-rich by-products of tomato processing. Food Chem 125:529–535
Leung CCJ, Cheung ASY, Zhang AYZ, Lam KF, Lin CSK (2012) Utilisation of waste bread for fermentative succinic acid production. Biochem Eng J 65:10–15
Lin H, Cheng W, Ding HT, Chen XJ, Zhou QF, Zhao YH (2010) Direct microbial conversion of wheat straw into lipid by a cellulolytic fungus of *Aspergillus oryzae* A-4 in solid-state fermentation. Bioresour Technol 101:7556–7562

Lucarini M, Di Lullo G, Cappelloni M, Lombardi-Boccia G (2000) In vitro estimation of iron and zinc dialysability from vegetables and composite dishes commonly consumed in Italy: effect of red wine. Food Chem 70:39–44

Maiani G, Periago Castón MJ, Catasta G, Toti E, Goñi Cambrodón I, Bysted A et al (2009) Actual knowledge on food sources, intakes, stability and bioavailability and their protective role in humans. Mol Nutr Food Res 53:S194–S218

Maier T, Schieber A, Kammerer DR, Carle R (2009) Residues of grape (*Vitis vinífera* L.) seed oil production as a valuable source of phenolic antioxidants. Food Chem 112:551–559

Makni M, Fetoui H, Gargouri NK, Garoui EM, Zeghal N (2011) Antidiabetic effect of flax and pumpkin seed mixture powder: effect on hyperlipidemia and antioxidant status in alloxan diabetic rats. J Diabetes Compl 25:339–345

Mamma D, Kourtoglou E, Christakopoulos P (2008) Fungal multienzyme production on industrial by-products of the citrus-processing industry. Bioresour Technol 99(7):2373–2383

Marín FR, Frutos MJ, Pérez-AlvarezJA M-SF, Del Río JA (2002) Flavonoids as nutraceuticals: structural related antioxidant properties and their role on ascorbic acid preservation. In: Atta-Ur-Rahman E (ed) Studies in natural products chemistry. Elsevier, New York

Marín FR, Soler-Rivas C, Benavente-García O, Castillo J, Pérez-Alvarez JA (2007) By-products from different citrus processes as a source of customized functional fibres. Food Chem 100:736–741

Masoodi FA, Sharma B, Chauhan GS (2002) Use of apple pomace as a source of dietary fiber in cakes. Plant Foods Hum Nutr 57(2):121–128

Matthaus B, Özcan MM (2012) Chemical evaluation of citrus seeds, an agro-industrial waste, as a new potential source of vegetable oils. Grasas Aceites 63(3):313–320

Melikoglu M, Lin CSK, Webb C (2013) Stepwise optimisation of enzyme production in solid state fermentation of waste bread pieces. Food Bioprod Process. doi:10.1016/j.fbp.2013.04.008

Mohdaly AAA, Sarhan MA, Mahmoud A, Ramadan MF, Smetanska I (2010) Antioxidant efficacy of potato peels and sugar beet pulp extracts in vegetable oils protection. Food Chem 123: 1019–1026

Morgan E (2009) Fruit and vegetable consumption and waste in Australia. Victoria, Australia: State Government of Victoria, Victorian Health Promotion Foundation.http://www.foodbankvictoria.org.au/what-we-do/uploadedFiles/1289866409126-8805.pdf. Accessed 19 Oct 2013

Mussatto SI, Fernandes M, Dragone G, Mancilha IM, Roberto IC (2007) Brewer's spent grain as raw material for lactic acid production by *Lactobacillus delbrueckii*. Biotechnol Lett 29:1973–1976

Muthuswamy S, Rupasinghe HPV (2007) Fruit phenolics as natural antimicrobial agents: selective antimicrobial activity of catechin, chlorogenic acid and phloridzin. J Food Agric Environ 5(3–4):81–85

Nawirska A, Kwaúniewska M (2005) Dietary fibre fractions from fruit and vegetable processing waste. Food Chem 91:221–225

Nighojkar S, Phanse Y, Sinha D, Nighojkar A, Kumar A (2006) Production of polygalacturonase by immobilized cells of *Aspergillus niger* using orange peel as inducer. Process Biochem 41:1136–1140

Obied HK, Allen MS, Bedgood DR, Prenzler PD, Robards K (2005) Investigation of Australian olive mill waste for recovery of biophenols. J Agric Food Chem 53(26):9911–9920

Oreopoulou V, Tzia C (2007) Utilization of plant by-products for the recovery of proteins, dietary fibers, antioxidants, and colorants. In: Oreopoulou V, Russ W (eds) Utilization of by-products and treatment of waste in the food industry. Springer, New York

O'Shea N, Arendt EK, Gallagher E (2012) Dietary fibre and phytochemical characteristics of fruit and vegetable by-products and their recent applications as novel ingredients in food products. Innov Food Sci Emerg Technol 16:1–10

Ostojic S, Pavlovic M, Zivic M et al (2005) Processing of whey from dairy industry waste. Environ Chem Lett 3:29–32

Öztürk S, Özboy O, Cavidoğlu I, Köksel H (2002) Effects of brewers' spent grain on the quality and dietary fiber content of cookies. J Inst Brew 108:23–27

Palmer FG (1976a) The feeding value of straw to ruminants. ADASQ Rev 21:220–234
Palmer FG (1976b) The feeding value and worthwhileness of chemically processed straw for ruminants. ADASQ Rev 22:247–266
Panesar PS, Kennedy JF, Gandhi DN, Bunko K (2007) Bio utilisation of whey for lactic acid production. Food Chem 105(1):1–14
Paraskevopoulou A, Athanasiadis I, Blekas G, Koutina AA, Kanellaki M, Kiosseoglou V (2003) Influence of polysaccharide addition on stability of a cheese whey kefir–milk mixture. Food Hydrocolloid 17:615–620
Parfitt J, Barthel M, Macnaughton S (2010) Food waste within food supply chains: quantification and potential for change to 2050. Philos Trans R Soc 365:3065–3081
Patsioura A, Galanakis CM, Gekas V (2011) Ultrafiltration optimization for the recovery of b-glucan from oat mill waste. J Membrane Sci 373:53–63
Pedroza MA, Carmona M, Alonso GL, Salinas MS, Zalacain (2013) Pre-bottling use of dehydrated waste grape skins to improve colour, phenolic and aroma composition of red wines. Food Chem 136:224–236
Pekcan G, Köksal E, Küçükerdönmez O, Ozel H (2006) Household food wastage in Turkey. FAO, Rome
Petsas I, Psarianos K, Bekatorou A, Koutinas AA, Banat IM, Marchant R (2002) Improvement of kefir yeast by mutation with N-methyl-N-nitrosoguanidine. Biotechnol Lett 24:557–560
Plessas S, Pherson L, Bekatorou A, Nigam P, Koutinas AA (2004) Bread making using kefir grains as baker's yeast. Food Chem 93:585–589
Plessas S, Koliopoulos D, Kourkoutas Y, Psarianos C, Alexopoulos A, Marchant R, Banat IM, Koutinas AA (2008) Upgrading of discarded oranges through fermentation using kefir in food industry. Food Chem 106:40–49
Prakash J (1996) Rice bran proteins: properties and food uses. Crit Rev Food Sci Nutr 36:537–552
Radócaj O, Dimíc E, Kakuda Y, Vujasinovíc V (2011) Chemical, nutritional and functional properties of a food by-product: hull-less pumpkin (*Cucurbita pepo* L.) seed oil press-cake. J Oil Soap Cosm 60:3–9
Radócaj O, Dimíc E, Vujasinovíc V (2012) Development of a hull-less pumpkin (*Cucurbita pepo* L.) seed oil press-cake spread. J Food Sci 77(9):1011–1017
Ramos OL, Fernandes JC, Silva SI, Pintado ME, Malcata FX (2012) Edible films and coatings from whey proteins: a review on formulation, and on mechanical and bioactive properties. Crit Rev Food Sci Nutr 52:533–552
Rao V (2006) Tomatoes, lycopene and human health. Caledonian Press, Barcelona
Reis IAO, Santos SB, Santos LA, Oliveira N, Freire MG, Pereira JFB, Ventura SPM, Coutinho JAP, Soares CMF, Lima ÁS (2012) Increased significance of food wastes: selective recovery of added-value compounds. Food Chem 135:2453–2461
Rezzadori K, Benedetti S, Amante ER (2012) Proposals for the residues recovery: orange waste as raw material for new products. Food Bioprod Process 90:606–614
Ried K, Fakler P (2011) Protective effect of lycopene on serum cholesterol and blood pressure: meta-analyses of intervention trials. Maturitas 68:299–310
Rivas J, Prazeres AR, Carvalho F, Beltrán F (2010) Treatment of cheese whey wastewater: combined coagulation–flocculation and aerobic biodegradation. J Agric Food Chem 58(13):7871–7877
Rivas-Cantu RC, Jones KD, Mills PL (2013) A citrus waste-based biorefinery as a source of renewable energy: technical advances and analysis of engineering challenges. Waste Manag Res. doi:10.1177/0734242X13479432
Rodriguez De Sotillo D, Hadley M, Wolf-Hall C (1998) Potato peel extract a nonmutagenic antioxidant with potential antimicrobial activity. J Food Sci 63(5):5417–5420
Roig A, Cayuela ML, Sánchez-Monedero MA (2006) An overview on olive mill wastes and their valorisation methods. Waste Manag 26:960–969
Ruberto G, Renda A, Daquino C, Amico V, Spatafora C, Tringali C et al (2007) Polyphenol constituents and antioxidant activity of grape pomace extracts from five Sicilian red grape cultivars. Food Chem 100(1):203–210

Russ W, Meyer-Pittroff R (2004) Utilizing waste products from the food production and processing industries. Crit Rev Food Sci Nutr 44:57–62
Sahraoui N, Vian MA, Maataoui ME, Boutekedjiret C, Chemat F (2011) Valorization of citrus by-products using microwave steam distillation (MSD). Innov Food Sci Emerg Technol 12:163–170
Sakurai Y, Misawa S, Shiota H (1985) Growth and respiratoryactivity of *Aspergillus oryzae* grown on solid state medium. Agric Biol Chem 49(3):745–750
Saura-Calixto F (2011) Dietary fiber as a carrier of dietary antioxidants: an essential physiological function. J Agric Food Chem 59:43–49
Schieber A, Stintzing FC, Carle R (2001) By-products of plant food processing as a source of functional compounds - recent developments. Trends Food Sci Technol 12:401–413
Sendra E, Kuri V, Fernández-López J, Sayas-Barberá E, Navarro C, Pérez-Alvarez JA (2010) Viscoelastic properties of orange fiber enriched yogurt as a function of fiber dose, size and thermal treatment. LWT Food Sci Technol 43(4):708–714
Siles JA, Li Q, Thompson I (2010) Biorefinery of waste orange peel. Crit Rev Biotechnol 30(1):63–69
Sonja D, Canadanovic-Brunet J, Cetkovic G (2009) By-products of fruits processing as a source of phytochemicals. Chem Ind Chem Eng Q 15:191–202
Soukoulis C, Aprea E (2012) Cereal bran fractionation: processing techniques for the recovery of functional components and their applications to the food industry. Recent Pat Food Nutr Agric 4(1):61–77
Spencer N, Butler D (2010) Food: the growing problem. Nature 466:546–547
Srinivosa PC, Tharanatham RN (2007) Chitin/chitosan-safe, ecofriendly packaging materials with multiple potential uses. Food Rev Int 23(1):53–72
Stenmarck Å, Hanssen OJ, Silvennoinen K, Katajajuuri JM, Werge M (2011) Initiatives on prevention of food waste in the retail and wholesale trades. Nordic Council of Ministers, Copenhagen, 87 pp
Suárez M, Romero MP, Ramo T, Macià A, Motilva MJ (2009) Methods for preparing phenolic extracts from olive cake for potential application as food antioxidants. J Agric Food Chem 57:1463–1472
Sudha ML, Baskaran V, Leelavathi K (2007) Apple pomace as a source of dietary fiber and polyphenols and its effect on the rheological characteristics and cake making. Food Chem 104(2):686–692
Sun RC, Tomkinson J (2002) Characterization of hemicelluloses obtained by classical and ultrasonically assisted extractions from wheat straw. Carbohydr Polym 50:263–271
Sun-Waterhouse D (2011) The development of fruit-based functional foods targeting the health and wellness market: a review. Int J Food Sci Technol 46:899–920
Tarazona-Díaz MP, Aguayo E (2013) Assessment of by-products from fresh-cut products for reuse as bioactive compounds. Food Sci Technol Int 19(5):439–446
The Royal Society (2010) Reaping the benefits: science and the sustainable intensification of global agriculture. The Royal Society, England
Thönissen R (2009) Food waste. The Netherlands. Presentation to the EU presidency climate smart food conference, Nov 2009, Lund, Sweden
Toldrá F, Reig M (2011) Innovations for healthier processed meats. Trends Food Sci Technol 22:517–522
Toldrá F, Aristoy MC, Mora L, Reig M (2012) Innovations in value-addition of edible meat by-products. Meat Sci 92:290–296
Tsai WT (2008) Management considerations and environmental benefit analysis for turning food garbage into agricultural resources. Bioresour Technol 99:5309–5316
Tsigie YA, Wang CY, Truong CT, Ju YH (2011) Lipid production from *Yarrowia lipolytica* Po1g grown in sugarcane bagasse hydrolysate. Bioresour Technol 102:9216–9222
Ubalua AO (2007) Cassava wastes: treatment options and value addition alternatives. Afr J Biotechnol 6(18):2065–2073

USDA (2010) United States Department of Agriculture, National nutrient database for standard reference [Online]. http://www.nal.usda.gov/fnic/foodcomp/search. Accessed 19 Oct 2013

Valverde P (1994) Barley spent grain and its future. Cerveza y Malta 122:7–26

Vendruscolo F, Albuquerque PM, Streit F, Esposito E, Ninow JL (2008) Apple pomace: a versatile substrate for biotechnological applications. Crit Rev Biotechnol 28:1–12

Vierhuis E, Korver M, Schols HA, Voragen AGJ (2003) Structural characteristics of pectic polysaccharides from olive fruit (*Olea europaea*cv moraiolo) in relation to processing for oil extraction. Carbohydr Polym 51(2):135–148

Viuda-Martos M, Ruiz-Navajas Y, Fernández-López J, Pérez-Álvarez JA (2010) Effect of orange dietary fibre, oregano essential oil and packaging conditions on shelf-life of bologna sausages. Food Control 21(4):436–443

Viuda-Martos M, Fernandez-Lopez J, Sayas-BarberaE SE, Perez-Alvarez JA (2011) Physicochemical characterization of the orange juice waste water of a citrus by-product. J Food Process Preserv 35:264–271

Wilkins MR, Widmer WW, Grohmann K, Cameron RG (2007) Hydrolysis of grapefruit peel waste with cellulase and pectinase enzymes. Bioresour Technol 98:1596–1601

Wilson J, Hayes M, Carney B (2011) Angiotensin-I-converting enzyme and prolyl endopeptidase inhibitory peptides from natural sources with a focus on marine processing by-products. Food Chem 129:235–244

WRAP (2008) The food we waste. WRAP, Banbury

Xanthopoulou MN, Nomikos T, Fragopoulou E, Anthnopoulou S (2009) Antioxidant and lipoxygenase inhibitory activities of pumpkin seed extracts. Food Res Int 42:642–646

Zhan J, Lin H, Shen Q, Zhou Q, Zhao Y (2013) Potential utilization of waste sweetpotato vines hydrolysate as a new source for single cell oils production by *Trichosporon fermentans*. Bioresour Technol 135:622–629

Zhang D, Hamauzu Y (2004) Phenolic compounds and their antioxidant properties in different tissues of carrots. Food Agric Environ 2:95–100

Zhang R, El-Mashad HM, Hartman K, Wang F, Liu G, Choate C, Gamble P (2007) Characterization of food waste as feedstock for anaerobic digestion. Bioresour Technol 98:929–935

Zhang W, Xiao S, Samaraweera H, Lee EJ, Ahn DU (2010) Improving functional value of meat products. Meat Sci 86:15–31

Nanotechnology in Food and Agriculture Industry

Emir Ayşe Özer, Melike Özcan, and Mustafa Didin

1 Introduction

Nanoscience and nanotechnology are new frontiers of this century. Nanotechnology is an enable technology that has the potential to revolutionize agriculture and food systems (Huang et al 2010). Nanotechnology is a multidisciplinary technological and scientific field undergoing a rapid development. Research in biology, chemistry, engineering, and physics has developed new materials using nanotechnology. Size reduction in materials to the nanometer scale (10^{-9} m) modifies the physical, chemical, and biological properties, resulting in new applications. The small particle size in combination with large surface area gives nanoparticles their unique features and then enormous potential for applications. These properties offer many unique and novel applications in various fields (Siegrist et al 2008; Bouwmeester et al 2009; Duncan 2011; Cushen et al 2012; Duran and Maezrcato 2013; Coles and Frewer 2013).

The nanoscale level of foods can affect the safety, efficiency, bioavailability, and nutritional value properties as well as the molecular synthesis of new products and ingredients (Rashidi and Darani 2011). In fact nanotechnology applications for food production include nanoformulated agrochemicals (e.g., fertilizers, pesticides, biocides, veterinary medicines) for improved efficacy, less use of farm chemicals,

E.A. Özer (✉) • M. Didin
Faculty of Agriculture, Department of Food Engineering, Mustafa Kemal University, Ürgen Paşa Mh., Hatay 31030, Turkey
e-mail: ayseozer@mku.edu.tr; mdidin@hotmail.com

M. Özcan
Technology, Research and Development Center, Mustafa Kemal University, Ürgen Paşa Mh., Hatay 31030, Turkey
e-mail: mtaskin80@hotmail.com

A. Malik et al. (eds.), *Food Processing: Strategies for Quality Assessment*, Food Engineering Series, DOI 10.1007/978-1-4939-1378-7_19

better control of applications (e.g., slow release pesticides), a safer and more nutritious animal feeds (e.g., fortified with nanosupplements, antimicrobial additives; detoxifying nanomaterials), and nanobiosensors for animal disease diagnostics. Example practices include nanodimension feed supplements and feed additives, such as nanoform of a biopolymer derived from yeast cell wall that can bind mycotoxins to protect animals against mycotoxicosis, and an aflatoxin-binding nanoadditive for animal feed derived from modified nano clay (Chaudhry and Castle 2011; Duncan 2011; Ezhilarasi et al 2013).

So the major links of nanotechnology to the food and agriculture industry are improving food security and processing, the ability of plants to absorb nutrients, flavor and nutrition, delivery methods, pathogen detection, functionality of foods, protection of the environment, and the cost-effectiveness of storage and distribution. The development of new functional materials and foods, microscale and nanoscale processing, and design of methods and instrumentation in food production can benefit from nanotechnology. The possible applications of nanotechnology in food production are shown in Table 1 (Rashidi and Darani 2011).

2 Nanotecnology in Food Industry

Nanotechnology has become increasingly important for the food sector. In the food industry, several novel applications of nanotechnologies have been became apparent, including the use of nanoparticles, such as micelles, liposomes, nanoemulsions, biopolymeric nanoparticles and cubosomes, as well as the development of nanosensors, which are been aimed at ensuring food safety (Sozer and Kokini 2009). Nanotechnology in food sectors such as agricultural production, food processing, food packaging and preservation, pathogen detection is important. Also direction involves nutraceuticals within nanocapsules, nanoencapsulated flavor enhancers, and nanoparticles that have the ability to selectively bind and remove chemicals from food (Sozer and Kokini 2009; Duran and Maezrcato 2013). The main target of new applications so far appears to be on food packaging and health food products, with only a few known examples in the mainstream food and beverage areas (Chaudhry and Castle 2011).

Nanotechnology in food industry is just beginning to be explored compared with several other areas, this technology has shown a great potential in this area. According to Lux Research, the nanotechnology industry is expected to grow to US$ 2.6 trillion in manufactured products by the year 2014. A prediction relating to the extension of nanotechnology in food and drink indicates that it might reach US$ 3.2 billions in 2015. The potential benefits of nanomaterials are large. In food industrial systems, nanotechnologies cover many aspects, such as food safety, packaging materials, nanosensors, nutrients delivery systems, bioavailability, and new materials for pathogen detection and others. It is possible to protect food against pathogens using packaging or edible films with antimicrobial proprieties. These several nanostructure applications have increased the food industry's investment in nanotechnology

Table 1 Application matrix of nanoscience and nanotechnology in main areas of food science and technology (Rashidi and Darani 2011)

Area of application	Purpose and fact	Approaches
Design of nanomaterial	Nanoparticles, Nanoemulsions, Nanocomposites, Nanobiocomposites (nanobiopolymeric starch) Nanolaminates	• Novel defined material, with self-assembling, self-healing, and manipulating properties
Nanosensors and nanobiosensors	Quality control and food safety	• Detection of very small amounts of chemical contaminants • Monitoring and tagging of food items • Electronic nose and tongue for sensor evaluation • Food born pathogen identification by measurement of nucleic acid, protein, or any other indicator metabolite of microorganism
Processing	Nanofiltration Nanoscale enzymatic reactor Heat and mass transfer Nanofabrication Nanocapsules for modification of absorption	• Selective passage of materials on the basis of shape and size • Improved understanding of process • Enhanced heat resistance of packages • Nanoceramic pan to reduce time of roasting and amount of consumed oil, reduction f trans fatty acids due to usage of plant oil instead of hydrogenated oil and finally resulted in safe nano-food development of nanocapsules that can be incorporated into food to deliver nutrients to enable increased absorption of nut
New products	Packaging Delivery Formulation Evaluation DNA recombinant technology	• Nanocomposites application as barriers, coating, release device, and novel packaging modifying the permeation behavior of foils, increasing barrier properties (mechanical, thermal, chemical, and microbial), improving mechanical and heat-resistance properties, developing active antimicrobial surfaces, sensing as well as signaling microbiological and biochemical changes, developing dirt repellent coatings for packages • Nanomycells for targeted delivery of nutrients (nutrition nanotherapy) • Nanocapsulation for controlled release of nutrients, proteins, antioxidants, and flavors • Production of nanoscale enzymatic reactor for development of new product. • Fortification of food by omega3 fatty acid, haem, lycopene, beta-carotene, phytosterols, DHA/EPA • Enzyme and protein evaluation as nanobiological system to development of new Products • Recombinant enzyme production in nanoporous media with special numerous application.

increasing the nanofood market from US$ 2.6 billion in 2005 to US$ 20.4 billion in 2010 (Dudo et al 2011; Duran and Maezrcato 2013)

Nanotechnology offers much promise to food science. "Food nanotechnology," simply, refers to the application of nanoscience to the food sector. More specifically, in this study we conceptualize food nanotechnology as novel breakthroughs and application developments made possible by nanolevel science and engineering applied to the "structure, texture, and quality of foodstuffs" and food-related products. Food nanotechnology includes a range of potential applications, including alterations to the properties of foods (e.g., nanoadditives and nanoingredients); improvements to the delivery, quality, and safety of food; and the development of enhanced food packaging (i.e., food contact materials). For example, scientists are creating food packages that contain nano-sized particles devised to warn consumers that a food product is unsafe to eat, and are inventing nanoencapsulated materials that can distribute nutrients to human cells (Dudo et al. 2011).

Advancements in nanosciences and nanotechnologies in recent years have opened up new prospects for food and agriculture industry. Applications of nanotechnologies in food industry offer a wide range of benefits to food industry and consumer. These include a possible reduction in the use of preservatives, salt, fat, and surfactants in food products; development of new or improved tastes textures and mouth sensations through nanoscale processing of foodstuffs. Antibacterial nanocoatings on food preparation surfaces can help maintain hygiene during food processing, whereas the use of "Smart" labels can help protect safety and authenticity of food products in the supply chain (Bouwmeester et al. 2009; Chaudhry and Castle 2011).

2.1 Food Processing

Nanotechnology has been bringing dramatic changes to food production, processing, and packaging. The word "nanofood" was recently developed. The concept of a nanofood is that "nanotechnology techniques or tools are used during the cultivation, production, processing, or packaging of the food; but not modified or produced food by nanotechnology machines" (Joseph and Morrison 2006; Bouwmeester et al 2009; Coles and Frewer 2013). Furthermore the application of nanotechnology includes smart packaging, on demand preservatives, and interactive foods (Gutiérrez et al 2012).

The use of nanostructures in food processed for improved/new tastes, textures, and mouth-feels. Nanostructuring of natural food materials may potentially enable the use of less fat but still produce tasteful food products. A typical product of this technology would be a nanostructured ice cream, mayonnaise, or spread, which is low-fat but is as "creamy" in texture as the full-fat equivalent. Such products would therefore offer a "healthy" option to the consumer (Chaudhry and Castle 2011).

Nano-sized or nanoencapsulated food additives and supplements can improve dispersability of fat-soluble additives in food products, improve food tastes, and enable hygienic food storage, reduction in the use of fat, salt, sugar and preservatives, and improvement in the uptake and bioavailability of nutrients and supplements. Currently available examples include vitamins, antioxidants, colors, flavors, and preservatives. Also developed for use in food products are nano-sized carrier systems for nutrients and supplements. These are based on nano-encapsulated substances in liposomes, micelles or protein based carriers. The nanocarrier systems are also used for taste masking of certain ingredients and additives or to protect them from degradation during processing. Examples include food additives, such as a synthetic form of the tomato carotenoid lycopene, benzoic acid, citric acid, ascorbic acid, and supplements such as vitamins A and E, isoflavones, ß-carotene, lutein, omega-3 fatty acids, coenzyme-Q10 (Chaudhry and Castle 2011).

Inorganic nanomaterials may also be potentially used in healthy food products. Example transition metals and metal oxides such as silver, iron, titanium dioxide, alkaline earth metals such as calcium, magnesium, and non-metals such as selenium, silicates.

Food packaging is currently the major area of application of metal and metal-oxide nanomaterials. Example nanomaterials finding use in packaging include plastic-polymer composites with nanoclay for gas barrier, nano-silver, and nano-zinc oxide for antimicrobial action, nano-titanium dioxide for UV protection, nano-titanium nitride for mechanical strength and as a processing aid, nano-silica for hydrophobic surface coating, etc. The use of nanosilver as an antimicrobial, antiodorant, and a (proclaimed) health supplement has already surpassed allother nanomaterials used in different sectors (Chaudhry and Castle 2011). The current use of nano-silver is mainly for health food and packaging applications, but its use as an additive in antibacterial wheat flour is the subject of a recent patent application (Park et al. 2006). Nano-silica is reported to be used in food contact surfaces and food packaging applications, and some reports suggest its use in clarifying beers and wines, and as a free-flowing agent in powdered soups. The conventional bulk forms of silica and titanium dioxide are permitted food additives (SiO2, E551, and TiO2, E171), but there is a concern that the conventional forms may also contain a nano-sized fraction due to natural size range variation (EFSA 2009).

2.2 Food Packaging

The application of nanotechnology in food packaging is to improve mechanical and barrier properties. Packaging applications includes use of nanosilver as a microbicide to extend the freshness of food and prevent contamination (Duncan 2011; Coles and Frewer 2013). Nanomaterials are used to improve the packaging barrier properties to regulate the passage of gases and moisture through the packaging to extend shelf life and maintain quality and freshness (Sozer and Kokini 2009; Coles and Frewer 2013).

Packaging of the future is likely to be more than just a physical container that provides food with protection from the surrounding environment. Further subdivision of nanopackaging is required; packaging from which migration into the food is purposeful and intended and packaging from which no nanoparticles migrate (in any significant amount). The former is likely to be subjected to greater safety assessments and negative consumer perceptions and for these reasons is less likely to advance as quickly as the latter. Using nanotechnologies to improve packaging materials is likely to be very costly and will not be introduced until methods are optimized; results are consistent and prove to weigh up favorably against costs (Cushen et al 2012).

The use of nanostructured materials, particularly nanocomposites, could considerably enhance the functional properties of packaging materials, and thus improve the shelf life of packaged foods. Nanocomposites are made out of nanoscale structures with unique morphology, increased modulus and strength, as well as good barrier properties (Sorrentino et al 2007; Rizvi et al 2010).

Most of the nanotechnology applications in food industry are in the packaging sector. It is expected that by 2015 this sector will be 19 % of nanotechnology food applications. This is mainly due to the big development of nanotechnology in this field, to a higher acceptance by consumers of the use of this technology in packaging than in food as ingredients and to the normative requirements, which are less restrictive than for the enforced current food legislation (Silvestre et al 2011; Gutiérrez et al. 2012). Adding nanocomposites or nanoparticles (e.g., silver, titanium dioxide, silicon dioxide, and nano-clay) into packaging materials to ensure better protection of foods by modifying the permeation behavior of foils, deodorizing, increasing barrier properties, blocking ultraviolet light, improving mechanical and heat-resistance properties, and developing antimicrobial and antifungal surfaces. Nylon nanocomposites providing barriers to oxygen and carbon dioxide flow have been used in food packaging (i.e., multi-layer PET bottles for beer and other alcoholic beverages) to keep freshness and block out smells (Chau et al 2007). All this contributes to the fact that there are many more applications in this area than in others of food sector, together with the necessity of a more sustainable, lighter and stronger at the same time, efficient and intelligent packages. These packages would be able to provide safer products with more quality and at the same time maintaining the products in the best possible conditions and with a longer shelf-life. During the last decade, the use of polymers as material for food packaging has incredibly increased due to its advantages over the use of traditional materials. In polymers world market, which has been increased of about 5 million tons in 1950 to almost 100 million tons at the present time, 42 % corresponds to packaging and containers (Silvestre et al 2011; Gutiérrez et al. 2012).

2.2.1 Nanocomposites

Nanoencapsulation is defined as a technology to pack substances in miniature making use of techniques such as nanocomposite, nanoemulsification, and nanoestructuration and provides final product functionality that includes controlled release of the core. The protection of bioactive compounds, such as vitamins, antioxidants, proteins, and lipids as well as carbohydrates may be achieved using this technique for the production of functional foods with enhanced functionality and stability (Sekhon 2010).

The use of nanocomposites for food packaging not only protects food but also increases shelf life of food products and solves environmental problems reducing the necessity of using plastics. Most of packaging materials are not degradable and current biodegradable films have poor barrier and mechanical properties, so these properties need to be considerably improved before these films can replaced traditional plastics and aid to manage worldwide waste problem (Sorrentino et al 2007).

2.2.2 Active Packaging

The active and intelligent food packaging category is a novel type of packaging than traditional methods. Active packaging is defined as an intelligent or smart system that involves interactions between package and package components. And food or internal gas atmosphere and complies with consumer demands for high quality, fresh-like, and safe products. In particular, active packaging changes the condition of packaged food to extend shelf life or improve food safety or sensory properties or improve the color and smell, while maintaining its quality (Imran et al 2010). Active packaging is thought to incorporate components that liberate or absorb substances in the package or in the air in contact to food. Up to now, active packaging has being mainly developed for antimicrobiological applications, nevertheless other promising applications include oxygen captation, ethylene elimination, CO2 absorption/emission, steam resistances and bad odors protection, liberation of antioxidants, preservatives addition, additives, or flavors. Nanoparticles more used in active packaging development are nanomaterials of metals and oxide of metals in antimicrobial packaging. Nanosilver use in packaging helps to maintain healthy conditions in the surface of food avoiding or reducing microbial growth. However, its action is not as a preservative even though, it is a biocide (Morones et al. 2005). Based on these properties, a big number of food contact materials, which inhibit microorganisms' growth have been created (i.e., plastic containers and bags to store food).

2.2.3 Intelligent Packaging

In general, the "intelligent" aspect of food packaging refers to the concept of monitoring information about the quality of the packed food. For instance, packaging materials incorporating nanosensors or nanocapsules based on nanotechnology will be able to detect food spoilage organisms and trigger a color change to alert the consumer that the shelf life is ending or has ended. This type of function can also incorporate a concept of "Release-on-Command," which will provide a basis for intelligent preservative packaging technology that will release a preservative if food begins to spoil.

Nanotechnology can be also applied in coatings or labels of packaging providing information about the traceability and tracking of outside as well as inside product conditions through the whole food chain. Some examples of these applications are leak detections for foodstuffs packed under vacuum or inert atmosphere (when inert atmosphere has been ruptured some compounds change of color warming consumers that air has come inside in where should be an inert atmosphere); temperature changes (freeze–thaw–refreezing, monitoring of cold chain by means of silicon with nanopores structure), humidity variations through the product shelf-life or foodstuffs being gone off (unusual microbial presence). Currently, sensors based on nanoparticles incrusted in a polymeric matrix isolated to detect and identify pathogens transmitted by food are being studied. These sensors work producing a specific pattern of answer against each microorganism (Yang et al 2007). Technology called "Electronic tongue" must be underlined, too. It is made up of sensor arrays to signal condition of the foodstuffs. The device consists of an array of nanosensors extremely sensitive to gases released by spoiling microorganisms, producing a color change which indicates whether the food is deteriorated. DNA-based biochips are also under development, which will be able to detect the presence of harmful bacteria in meat, fish, or fungi affecting fruit.

2.2.4 Antimicrobial Nanopackaging

The combination of food packaging materials and active substances is a new way to control surface microbial contamination of foods. Some nanomaterials exhibit antimicrobial effects. For such active packaging materials, sharing a common interface or physical contact with the food surface is essential. These active FCMs can extend the product shelf life, enhancing food quality and safety, and ultimately leading to less food waste (Cushen et al 2012).

2.3 Nutrient Delivery System

Nanostructures in foods can be designed for the targeted delivery of nutrients in the body for the most beneficial effects. By facilitating a precise control of properties and functionality at the molecular level, nanotechnology enabled the development of highly effective encapsulation and delivery systems. Examples include nanometer-sized association colloids such as surfactant micelles, vesicles, bilayers, reverse micelles, or liquid crystals. Such systems could be used in food applications as carrier or delivery systems for vitamins, antimicrobials, antioxidants, flavorings, colorants, or preservatives (Rizvi et al. 2010). Nanoformulations can be used to improve the uptake, absorption, and bioavailability of nutrients and supplements in the body (Chaudhry and Castle 2011).

Nanoparticles can, for instance, be used as bioactive compounds in functional foods. Bioactive compounds that can be found naturally in certain foods have physiological benefits and might help to reduce the risk of certain diseases, including cancer. By reducing particle size, nanotechnology can contribute to improve the properties of bioactive compounds, such as delivery properties, solubility, prolonged residence time in the gastrointestinal tract and efficient absorption through cells. Omega 3 and omega 6 fatty acids, probiotics, prebiotics, vitamins, and minerals have found their applications in food nanotechnology as bioactive compounds (Sozer and Kokini 2009; Kuan et al. 2012).

One important application of nanotechnology in food and nutrition is to design and development of novel functional food ingredients with improved water solubility, thermal stability, oral bioavailability, sensory attributes, and physiological performance (Huang et al 2010; Cushen et al. 2012). Nanoformulated food additives or supplements are available on the market, e.g., regulatory peptides from plants, nanodroplets/clusters, and nano-water. The applicants claim increased bioavailability but valid studies on relative bioavailability comparing these products with similar non-nanoformulated products are lacking (Garcia et al 2010).

Nanostructures in foods can be designed for the targeted delivery of nutrients in the body for the most beneficial effects. By facilitating a precise control of properties and functionality at the molecular level, nanotechnology enabled the development of highly effective encapsulation and delivery systems. Examples include nanometer sized association colloids such as surfactant micelles, vesicles, bilayers, reverse micelles, or liquid crystals. Such systems could be used in food applications as carrier or delivery systems for vitamins, antimicrobials, antioxidants, flavorings, colorants, or preservatives (Rizvi et al. 2010). Nanotechnology renders hydrophilic substances fat soluble and lipophilic ones water soluble, allowing nanoparticles of some functional ingredients (e.g., carotenoids, phytosterols, and antioxidants) to be dispersed in water or fruit drinks to improve their bioavailability Synthetic nanoparticles of lycopene are developed and accepted as GRAS-affirmed by the FDA for use in food in the USA Minute micelles (nanocapsules) are used as carriers for essential oils, flavor, antioxidant, coenzyme Q10, and vitamins, minerals, and phytochemicals to improve their bioavailability Encapsulating the nanoparticles of active ingredients (e.g., polyphenols, minerals, and micronutrients) to protect them from oxidation and getting to the taste receptor site, thus to reduce their undesirable off-tastes in the finished application Food industry application of liposomal nanovesicles for the encapsulation and delivery of nutrients and functional ingredients such as proteins, enzymes, flavors, and antimicrobial compounds Whey protein nanospheres (40 nm), which are internalized by cells and degraded therein to release the nutraceutical compounds, can be used as carriers for oral administration of nutraceutical agents to improve their bioavailability (Huang et al 2010; Kuan et al. 2012).

2.4 Safety and Sensing

Nanotechnology is being used to improve the quality and safety of food. Nanotechnology has benefited the area of food safety mostly through the development of highly sensitive biosensors for pathogen detection and the development of novel antimicrobial solutions (Rizvi et al. 2010)

Nano-derived innovations to food packaging are being designed to enhance food safety and help reduce food waste. Nanosensors are being developed that can detect and signal the presence of spoilage microorganisms, and potentially even differentiate the presence of pathogenic from benign microorganisms. Nanotechnology is also being used to create healthier foods that can deliver nutrients and medications to different parts of the human and can alleviate allergenic properties. And nanotechnology is also likely to enable substantial benefits to food manufacturing and agricultural production. Nanomaterials, for instance, might be developed to improve the delivery of nutrients and pesticides to crops, which some experts speculate could help developing countries (Dudo et al 2011).

Protein-coated nanocantilever, naturally vibrating at a specific frequency, is a new class of ultrasmall silicon sensors for the quick detection of viruses, bacteria, and other pathogens. When contaminants land on the devices, the slight mass changes can cause the nanocantilever to vibrate at a different frequency and be quickly detected. Development of synthetic tree-shaped DNA being tagged with color-coded probes, as a nanobarcode device, enables the identification of food pathogens, a miniature portable microbiodetector was developed using different nanowires, specific pathogen antibodies, and fluorescent antibodies for the simultaneous detection of toxins, pathogens, and chemicals in foodstuffs. Silver nanoparticles have been incorporated into different products from bandages to refrigerators for suppressing the spread of bacteria and other microbes.

2.5 *Potential Health Risks of Food Nanotechnology*

Whilst nano-sizing of materials offers lots of potential benefits, it can bring the prospect of consumer exposure to some insoluble and possibly biopersistent nanoparticles through consumption of food and drinks. Potential risks to both human health and the environment may exist. The concern is that, once in the body, nanoparticles with large reactive surfaces may cross biological barriers to reach those parts of the body which are otherwise protected from entry of (larger) particulate materials? All applications of this new technology must be assessed for safety of use (Chaudhry and Castle 2011; Cushen et al 2012). The profile of nanomaterials adsorption, distribution, metabolism and elimination and toxicological properties are not fully known at present. Some of the projected applications in the agricultural sector (e.g., nanopesticides) will also fall in this category. Another potential risk arising from nanotechnology-derived food contact materials will be dependent on the migration behavior of nanomaterials from packaging. The few experimental and modeling studies reported so far suggest that the likelihood of nanoparticle migration from polymer packaging to be either nil or very low.

Whilst nano-sizing of materials, their properties can change, which can have highly unpredictable impacts on humans, other animals and the environment. Some nanoparticles can cross biological barriers, including across the blood brain barrier, so potentially have the ability to enter cells and organs. As a consequence there is potential for these to interact with normal biological processes in an unpredictable manner. There is the possibility that nanomaterials such as nano-silver may bioaccumulate in food (from nano-packaging or from accumulation in plants and animals used in food production), or in the human body (Coles and Frewer 2013).

The migration of nanoparticles from packaging to packed food raised public concern, and had been corroborated by animal oral administration and in vitro cell experiments. In general, two types of mechanism can be adopted to explain the toxicity effects on humans. One is that the toxicity is independent of the nanoparticles and could be realized by generating the active oxygen species within the cells. Another is that the toxicity has a strong relationship with the chemical component of nanoparticles. For example, the crystallization and recrystallization of some metal or metal oxide nanoparticles will modify the secondary or tertiary conformation of the proteins. While other types of nanoparticles, such as the metal alloy, or single-wall.

But with these potential benefits come potential risks. Knowledge on the potential toxicity of nanoparticles is limited but rapidly growing. So far, very few studies focused on elucidating what happens when nanoscale materials, some designed to be biologically active, enter the human body or are dispersed in the environment (CDC/NIOSH 2009; Rizvi et al 2010). On one hand, while nanoscale components already occur naturally in many foods, food nanotechnologies may pose direct risks to human health. Recent research shows that inhaled nanoparticles can accumulate in the lungs and cause chronic diseases due to their small scale (Dudo et al 2011). Experimental studies in rats have shown that at equivalent mass doses, particles smaller than 100 nm are more potent than larger particles of similar composition in causing pulmonary inflammation and fibrosis, tissue damage, and lung tumors, and that toxicity increases with decreasing particle size/increasing surface area. In vitro studies performed on single- and multi-walled carbon nanotubes showed that they can enter cells and cause release of pro-inflammatory cytokines, oxidative stress, and decreased. According to the Centers for Disease Control/The National Institute for Occupational Safety and Health (CDC/NIOSH) report, epidemiological studies in workers exposed to aerosols of ultrafine particles have revealed depreciation of lung function, adverse respiratory symptoms, chronic obstructive pulmonary disease, and fibrosis, and even elevated levels of lung cancer (CDC/NIOSH 2009; Rizvi et al. 2010). Food-related nanotechnologies may also pose indirect threats to human health. For example, food could be contaminated by the use of nano-sized pesticides and nanoparticles could migrate into food from nano-packaging. There is also the possibility that nanoparticles could bioconcentrate in the environment and alter the food chain (Dudo et al 2011).

If nanoparticles accumulations occur in human body, concentrations might increase with time and ultimately affect health. Though it is broadly acknowledged that current quantities of nanomaterials produced today are relatively small and current impacts not large, the issue arises when nanomaterials become far more widely produced and used in many different products and processes. Some concerns stem from observations that the very same physical and chemical properties of nanomaterials that often lead to positive use and environmental benefits can also potentially cause health problems in human and animal communities.

2.6 *Regulation of Nanotechnologies in the Food Industry*

A number of regulatory gap studies have shown that developments in nanotechnologies are not taking place in a regulatory vacuum, as the potential risks will be controlled under the existing frameworks (Chaudhry and Castle 2011; Duval 2012). The current regulatory frameworks for food and food contact materials in different jurisdictions, such as the European Union, the United States, and Australia are broad enough to "capture" nanotechnology applications in the food sector. These include regulations relating to general food safety, food additives, novel foods, specific health claims, chemical safety, food contact materials, water quality, general product safety, and other specific regulations on the use of certain chemicals in food production/protection, such as biocides, pesticides, veterinary medicines, etc.

Despite rapid developments in food nanotechnology, little is known about the occurrence, fate, and toxicity of nanoparticles. Nanotechnology-derived food ingredients, food additives, and food contact materials have been reported in relation to potential implications for consumer safety and regulatory controls (Sekhon 2010). The current risk assessment approach used by FAO/WHO and Codex is available and appears suitable for ENMs in food.23 However; there is a need to consider the whole life cycle of engineered nanomaterials in agrifood applications.

3 Nanotecnology in Agriculture Industry

3.1 *Nanotechnologies in Plant-Based Agricultural Production and Products*

Plant-based agricultural production is the basis of broad agriculture systems providing food, feed, fiber, fire (thermal energy), and fuels through advancements in materials sciences, and biomass conversion technologies. While the demand for crop yield will rapidly increase in the future, the agriculture and natural resources such as land, water, and soil fertility are finite. Other production inputs including synthetic fertilizers and pesticides are predicted to be much more expensive due to the constraints of known petroleum reserve (Chen and Yada 2011).

3.1.1 Precision Farming

Precision farming has been a long-desired goal to maximize output (i.e., crop yields) while minimizing input (i.e., fertilizers, pesticides, herbicides, etc.) through monitoring environmental variables and applying targeted action. Precision farming makes use of computers, global satellite positioning systems, and remote sensing devices to measure highly localized environmental conditions, thus determining whether crops are growing at maximum efficiency or precisely identifying the nature and location of problems. Precision farming can also help to reduce agricultural waste and thus keep environmental pollution to a minimum. Although not fully implemented yet, tiny sensors and monitoring systems enabled by nanotechnology will have a large impact on future precision farming methodologies. Ultimately, precision farming, with the help of smart sensors, will allow enhanced productivity in agriculture by providing accurate information, thus helping farmers to make better decisions (Cioffi et al. 2004; Chen and Yada 2011).

3.1.2 Nanotechnology Make Possible Delivery of Agriculture Chemicals Like Fertilizers, Pesticides, Herbicides, Plant Growth Regulators

Many nanoscale carriers, including encapsulation and entrapment, polymers and dendrimers, surface ionic and weak bond attachments and other mechanisms may be used to store, protect, deliver, and release by control of intended payloads in crop production processes. One of the advantages of nanoscale delivery vehicles in agronomic applications is its improved stability of the payloads against degradation in the environment, thereby increasing its effectiveness while reducing the amount applied. This reduction helps address agricultural chemicals run-off and alleviate the environmental consequence. Controlled release mechanisms allow the active ingredients to be slowly taken up, hence, avoiding temporal overdose, reducing the amount of agricultural chemicals used, and minimizing the input and waste. Environmental consideration including precision farming can reduce pollution to a minimum (Chen and Yada 2011).

3.1.3 Environmental Stresses and Crop Condition Can Monitored by Field Sensing Systems

Nanotechnology may be developed and deployed for real time monitoring of the crop growth and field conditions including moisture level, soil fertility, temperature, crop nutrient status, insects, plant diseases, weeds, etc. Networks of wireless nanosensors positioned across cultivated fields provide essential data leading to best agronomic intelligence processes with the aim to minimize resource inputs and maximizing output and yield (Scott and Chen 2003). Such information and signals include the optimal times for planting and harvesting crops and the time and level of water, fertilizers, pesticides, herbicides, and other treatments that need to be administered given specific plant physiology, pathology, and environmental conditions (Chen and Yada 2011).

3.1.4 Plant Disease Mechanisms Can Studied by Nanotechnology

The advancement in nanofabrication and characterization tools have enabled studies of physical, chemical and biological interactions between plant cell organelles and various disease causing pathogens, i.e., plant pathology. A better understanding of plant pathogenic mechanisms such as flagella motility and biofilm formation will lead to improved treatment strategies to control the diseases and protect production (Cursino et al. 2009).

Antimicrobial activity of different metal nanoparticles particularly copper and silver nanoparticles has been investigated by some researchers against the plant pathogens. Cioffi et al (2004) reported the antifungal activity of polymer-based copper nanocomposite against plant pathogenic fungi. Park et al (2006) studied the efficacy of nanosized silica–silver (silica–silver nanoparticles) in the control of plant pathogenic fungi, viz. *Botrytis cinerea*, *Rhizoctonia solani*, *Colletotrichum gloeosporioides*, *Magnaporthe grisea*, and *Pythium ultimum*. They also demonstrated the effect of nano-based product prepared from these nanoparticles against the powdery mildew disease of pumpkin and found that the disease-causing pathogens were disappeared from the infected leaves within 3 days of spraying of this product. Nanopesticides, nanofungicides, and nanoherbicides are being used in agriculture (Owolade et al. 2008). Many companies made formulations which contain nanoparticles within the 100–250-nm size range that are able to dissolve in water more effectively than existing ones (thus increasing their activity). Some other companies employ suspensions of nanoscale particles (nano-emulsions), which can be either water or oil based and contain uniform suspensions of pesticidal or herbicidal nanoparticles in the range of 200–400 nm, which have multiple applications for preventative measures, treatment or preservation of the harvested product (Rickman et al. 1999; Goswami et al. 2010).

3.1.5 Plant Traits Can Be Improved Against Environmental Stresses and Diseases by Nanotechnology

Biotechnological research has been focusing on improving plant resilience against various environmental stresses such as drought, salinity, diseases, and others. Genomes of crop cultivars are currently being extensively studied. The advancement in nanotechnology-enabled gene sequencing is expected to introduce rapid and cost effective capability within a decade (Branton et al. 2008), hence leading to more effective identification and utilization of plant gene trait resources.

3.1.6 Lignocellulosic Nanomaterials

Recent studies have shown that nanoscale cellulosic nanomaterials can be obtained from crops and trees. It opens up a whole new market for novel and value-added nano-biomaterials and products of crops and forest. For example, cellulosic nanocrystals can be used as light weight reinforcement in polymeric matrix as nanocomposite (Laborie 2009; Mathew et al. 2009). Such applications may include food and other packaging, construction, and transportation vehicle body structures (Chen and Yada 2011).

3.2 *Nanotechnologies in Animal Production and Animal Health*

Agriculturally relevant animal production (livestock, poultry, and aquaculture) provides society with highly nutritious foods (meat, fish, egg, milk, and their processed products) which have been, and will continue to be, an important and integral part of human diets. There are a number of significant challenges in animal agricultural production, including production efficiency, animal health, and feed nutritional efficiency, diseases including zoonoses, product quality and value, by products and waste, and environmental footprints. Nanotechnologies may offer effective, sometimes novel, solutions to these challenges (Kuzma 2010).

3.2.1 Feeding Efficiency and Nutrition of Agricultural Animals Can Be Improved by Nanotechnology

For animal, human, and environmental health, and optimal farm-scale economics, the quality and safety of inputs, outputs, and byproducts of animal production are crucial. Nanotechnology can be used to enhance the safety of animal feed and waste (Kuzma 2010).

A critical element of sustainable agricultural production is the minimization of production input while maximizing output. One of the most significant inputs in animal production is feedstock. Low feeding efficiency results in high demand of feed, high discharges of waste, heavy environmental burden, high production cost, and competing with other uses of the grains, biomass, and other feed materials. Nanotechnology may significantly improve the nutrient profiles and efficiency of minor nutrient delivery of feeds. Most animal feeds are not nutritionally optimal, especially in developing countries (Chen and Yada 2011).

3.2.2 Nanotechnology Can Minimize Losses from Animal Diseases

Many animal diseases cause substantial losses in agricultural animal production. Some of the more significant diseases include bovine mastitis, tuberculosis, respiratory disease complex, Johne's disease, avian influenza, and porcine reproductive and respiratory syndrome (PRRS). Zoonotic diseases not only cause devastating economic losses to animal producers, but also impose serious threats to human health, e.g., Variant Creutzfeldt-Jakob Disease (vCJD). Nanotechnology has the potential to enable revolutionary changes in this area, and some specific technologies may be feasible in near future given the current state of research and development (Emerich and Thanos 2006; Scott 2007). Nanotechnology offers numerous advantages in detection and diagnostics including high specificity and sensitivity, simultaneous detection of multiple targets, rapid, robust, on-board signal processing, communication, automation, convenient to use, and low cost. The uses of portable, implantable or wearable devices are particularly welcome in agricultural field applications.

Early detection is imperative so that quick, simple and inexpensive treatment strategies can be taken to remedy the situation. Nanotechnology-based drugs and vaccines can be more effective in treating/preventing the diseases than current technologies, thus reducing cost. Precise delivery and controlled release of nanotechnology enabled drugs leave little footprint in the animal waste and the environment, which alleviate the increasing concern of antibiotic resistance, and decrease health and environmental risks associated with the use of antibiotics. The targeted delivery and active nanoparticles may enable new drug administrations that are convenient, fast, non-intrusive to animals, and cost effective (Chen and Yada 2011).

3.2.3 Nanotechnology Applications in Animal Reproduction and Fertility

Animal reproduction remains a challenge not only in developing countries but also in developed nations. Low fertility results in low production rate, increases in financial input, and reduced efficiency of livestock operations (Narducci 2007).

For centuries, animals have been bred for important characteristics such as disease resistance, improved performance and growth, and product quality. In the past decade, genetic engineering of several livestock species has been achieved. Genetic engineering can speed the process of introducing desirable traits into livestock and allows for the introduction of entirely new ones (NRC 2002). Genetic Engineering of livestock is faced with technical challenges that nanotechnology might be able to overcome (Kuzma 2010).

Also several technological fronts have been explored in order to improve animal reproduction. Microfluidic technology has matured over the last two decades, and has been integrated into many nanoscale processing and monitoring Technologies including food and water quality, animal health, and environmental contaminations. Nanoscale delivery vehicles are sought to substantially improve bioavailability and better control of release kinetics, reduce labor intensity, and minimize waste and discharge to the environment (Emerich and Thanos 2006; Narducci 2007). Another strategy that may be explored is to monitor animal hormone level using implanted nanotechnology-enabled sensing device with wireless transmission capability, thus the information of optimal fertility period can become available in real time to assist the livestock operators for reproduction decision making (Afrasiabi 2010).

3.2.4 Animal Product Quality, Value, and Safety

Modification of animal feeds has been effectively used to improve animal production as well as product quality and value. The regulation of nutrient utilization can be used to enhance the efficiency of animal production and to design animal derived foods consistent with health recommendations and consumer perceptions. For example, the concepts of nutrient regulation have been used to redesign foods, such as milk fatty acids, cis-9, trans-11 conjugated linoleic acid (CLA) and vaccinic acid (VA), that may have a potential role in the prevention of chronic human diseases such as cancer and atherogenesis (Bauman et al. 2008).

3.2.5 Nanotechnology Applications for Turning Animal by-Products and Waste and Environmental Matters into Value Added Products

Animal waste is a serious concern in the animal production industry. Stricter environmental policies will prevent irresponsible discharge of animal waste. The unpleasant odors that emanate from intensive animal production facilities adversely affect air quality, and in turn, living conditions and the real estate value of the adjacent area. However, bioconversion of animal waste into energy and electricity can result in new revenue, renewable energy, high quality organic fertilizer, and improved environmental quality while value added (Scott 2002). Nanotechnology-enabled catalysts will play a critical role in efficient and cost effective bioconversion and fuel cell for electricity production as well as enabling efficient energy storage which will greatly facilitate and benefit the development of distributed energy supplies, especially in rural communities where infrastructure is lacking (Davis et al. 2009; Soghomonian and Heremans 2009). Such an approach may result in the elimination of the need for system wide electricity grids, hence accelerate the rural development and improve productivity, and business and living environment and will be especially beneficial to developing countries (Chen and Yada 2011).

3.3 Nanotechnologies for Water Quality

Providing clean and abundant fresh water for human use and industry applications, including agricultural and farming uses, is one of the most daunting challenges facing the world (Vorosmarty et al. 2010). Agriculture requires considerable amount of fresh water, and in turn, often contributes substantially to pollution of ground water through the use of pesticides, fertilizers, and other agricultural chemicals. Effective technologies for remediation and purification will be needed to manage the volume of wastewater produced by farms on a continual basis, and be cost effective for all. Technical issues in the water challenges include water quality and quantity, treatment and reuse, safety due to chemical and biological hazards, monitoring and sensors. Nanotechnology R and D has shown great promises in providing novel and economically feasible solutions. Several aspects of nanotechnology solutions are briefly discussed below (Chen and Yada 2011).

3.3.1 Nanotechnology for Microbial Disinfection

In industrialized nations, chemical, and physical based (chlorine dioxide, ozone, and ultraviolet) microbial disinfection systems are commonly used. However, much of the world still does not have the industrial infra structure necessary to support chemical-based disinfection of water. Hence, alternative technologies that require less intensive infra structure and more cost effective approaches such as nanoscale oligodynamic metallic particles, those may exhibit a toxic effect on living cells even in relatively low concentrations, may be worthy of attention. Among the oligodynamic metallic nanoparticles, silver is considered the most promising nanomaterials with bactericidal and viricidal properties owing to its wide-range effectiveness, low toxicity, ease of use, its charge capacity, high surface to volume ratios, crystallographic structure, and adaptability to various substrates (Nangmenyi and Economy 2009). Its antimicrobial mechanism is due to the production of reactive oxygen species (ROS) that cleaves DNA. Another nanoscale technological development for microbial disinfection is visible light photocatalysts of transition metal oxides made into nanoparticles, nanoporous fibers, and nanoporous foams (Li et al 2009). In addition to its effectiveness in disinfecting microorganisms, it can also remove organic contaminants such as PPCPs and EDCs (Chen and Yada 2011).

3.3.2 Water Desalination and Nanotechnology

Given the limited fresh water supplies both above and underground, it is likely that the desalination of sea water will become a major source of fresh water. Conventional desalination technology is reverse osmosis (RO) membranes which generally require large amounts of energy. A number of nanotechnologies have been attempted to develop low energy alternatives. Among them, the 3 most promising technologies appear to be protein polymer biomimetic membranes, aligned-carbon nanotube membranes, and thin film nanocomposite membranes (Hoek and Ghosh 2009). Some of the prototypes have demonstrated up to 100 times better water permeability with nearly perfect salt rejection than RO. Carbon nanotube membranes, owing to its extremely high water permeability than other materials of similar size, have desalination efficiencies in the order of thousand times better than the current technology. Some of these membranes can also integrate other functionalities such as disinfection, de-odoring, de-fouling, and self-cleaning. Some of the above mentioned technologies are currently in commercial development stage, which may be introduced in the market place in near future (Chen and Yada 2011).

3.3.3 Heavy Metal Remove of Water and Nanotechnology

Functionalization of ligand-based nanocoating which is bonded to the surface of high surface and low cost filtration substrate can effectively absorb high concentrations of heavy metal contaminants. The system can be re-generated in situ by treatment with bifunctional self-assembling ligand of the previously used nanocoating media. A start-up company (Crystal Clear Technologies) has demonstrated that multiple layers of metal can be bonded to the same substrate (Farmen 2009). Such a water treatment unit should be available in near future for removal of various heavy metals in water. Another approach to remove heavy metals and ions is the use of dendrimer enhanced filtration (DEF) (Diallo 2009). Functionalized dendrimer scan bind cations and anions according to acidity.

References

Afrasiabi ZA (2010) New nanoscale biosensor for detection of luteinizing hormone in smallruminants to determine optimum breeding time. A USDS/NIFA Evans-Allenproject. http://cris.csrees.usda.gov/cgi-bin/starfinder/17118/crisassist.txt. As of 12-April-2011

Bauman DE, Perfield JW II, Harvatine KJ, Baumgard LH (2008) Regulation of fat synthesis by conjugated linoleic acid: lactation and the ruminant model. J Nutr 138:403–409

Bouwmeester H, Dekkers S, Noordam MY, Hagens WI, Bulder AS, Heer CD, Voorde ECG, Wijnhoven SWP, Marvin HJP, Sips AJAM (2009) Review of health safety aspects of nanotechnologies in food production. Regul Toxicol Pharmacol 53:52–62

Branton D, Deamer DW, Marziali A, Bayley H, Benner SA, Butler T (2008) The potential and challenges of nanopores sequencing. Nat Biotechnol 26(10):1146–1153

CDC/NIOSH (2009) Approaches to safe nanotechnology: an information exchange with NIOSH. Centers for disease control and prevention/National Institute for Occupational Safety and Health. Available at www.cdc.gov/niosh

Chau CF, Wu SH, Yen GC (2007) The development of regulations for food nanotechnology. Trends Food Sci Technol 18:269–280

Chaudhry Q, Castle L (2011) Food applications of nanotechnologies: an overview of opportunities and challenges for developing countries. Trends Food Sci Technol 22:595–603

Chen H, Yada R (2011) Nanotechnologies in Agriculture: New tools for sustainable developments. Trends FoodSci Technol 22:585–594

Cioffi N, Torsi L, Ditaranto N, Sabbatini L, Zambonin PG, Tantillo G, Ghibelli L, D'Alessio M, Bleve-Zacheo T, Traversa E (2004) Antifungal activity of polymer-based copper nanocomposite coatings. Appl Phys Lett 85:2417–2419

Coles D, Frewer LJ (2013) Nanotechnology applied to European food production—A review of ethical and regulatory issues. Trends Food Sci Technol 34:32–43

Cursino L, Li Y, Zaini PA, De La Fuente L, Hoch HC, Burr TJ (2009) Twitching motility and biofilm formation are associated with tonB1 in Xylellafastidiosa. FEMS Microbiol Lett 299:193–199

Cushen M, Morris JKM, Cruz-Romero M, Cummins E (2012) Nanotechnologies in the food industry recent developments, risks and regulation. Trends Food Sci Technol 24:30–46

Davis R, Guliants VV, Huber G, Lobo RF, Miller JT, Neurock M (2009) An international assessment of research in catalysis by nanostructure and materials. World Technology Evaluation Enter, Baltimore, MD, http://www.wtec.org/catalysis/WTEC-CatalysisReport-6Feb2009-color-hi-res.pdf

Diallo M (2009) Water treatment by dendrimer-enhanced filtration: principles and applications. In: Savage NF et al (eds) Nanotechnology applications for clean water. William Andrew, Norwich, NY, pp 143–155

Dudo A, Choi DH, Scheufele DA (2011) Food nanotechnology in the news Coverage patterns and thematic emphases during the last decade. Appetite 56:78–89

Duncan TV (2011) Applications of nanotechnology in food packaging and food safety: barrier materials, antimicrobials and sensors. J Colloid Interface Sci 363:1–24

Duran N, Maezrcato PD (2013) Nanobiotechnology perspectives Role of nanotechnology in the food industry: a review. Int J Food Sci Technol 48:1127–1134

Duval MN (2012) FDA Regulation of nanotechnology

EFSA (European Food Safety Authority) (2009) Scientific opinion on 'the potential risks arising from nanoscience and nanotechnologies on food and feed safety', scientific opinion of the scientific committee, adopted on 10 February 2009. EFSA J 958:1–39

Emerich DF, Thanos CG (2006) The pinpoint promise of nanoparticle-based drug delivery and molecular diagnosis. Biomol Eng 23:171–184

Ezhilarasi PN, Karthik P, Chhanwal N, Anandharamakrishnan C (2013) Nanoencapsulation techniques for food bioactive components: a review. Food Bioprocess Technol 6:628–647

Farmen L (2009) Commercialization of nanotechnology for removal of heavy metals in drinking water. In: Savage NF et al (eds) Nanotechnology applications for clean water. William Andrew, Norwich, NY, pp 115–130

Garcia M, Forbe T, Gonzales E (2010) Potential applications of nanotechnology in the agro-food sector. Ciênc Tecnol Aliment Campinas 30(3):573–581

Goswami A, Roy I, Sengupta S, Debnath N (2010) Novel applications of solid and liquid formulations of nanoparticles against insect pests and pathogens. Thin Solid Films 519:1252–1257

Gutiérrez FJ, Mussons ML, Gatón P, Rojo R (2012) Nanotechnology and food industry. In: Valdez B (ed) Scientific, health and social aspects of the food industry. Rijeka, InTech, p 488

Hoek EMV, Ghosh AK (2009) Nanotechnology-based membranes for water purification. In: Savage NF et al (eds) Nanotechnology applications for clean water. William Andrew, Norwich, NY, pp 47–58

Huang Q, Yu H, Ru Q (2010) Bioavailability and delivery of nutraceuticals using nanotechnology. J Food Sci 75(1):50–57

Imran M, Junelles AMR, Martyn A, Tehrany EA, Jacquot M, Linder M, Desobry S (2010) Active food packaging evolution: transformation from micro- to nanotechnology. Crit Rev Food Sci Nutr 50:799–821

Kuan CY, Fung WY, Yuen KH, Liong MT (2012) Nanotech: propensity in foods and bioactives. Crit Rev Food Sci Nutr 52:55–71

Kuzma J (2010) Nanotechnology in animal productione upstream assessment of applications. Livest Sci 130:14–24

Laborie MP (2009) Bacterialcellulose and its polymeric nanocomposites. In L. Lucia, O. Rojas (Eds.) The nanoscience and technology of renewable biomaterials. Blackwell:

Li Q, Wu P, Shang JK (2009) Nanostructured visible-lightphotocatalysts for water purification. In: Savage NF et al (eds) Nanotechnology applications for clean water. William Andrew, Norwich, NY, pp 17–37

Mathew AP, Laborie MP, Oksman K (2009) Cross-linked chitosan-chitin whiskers nanocomposites with improved permeation selectivity and pH stability. Biomacromolecules 10(6):1627–1632

Morones JR, elechiguerra JL, Camacho A (2005) The Bactericidal effect of Silver Nanoparticles. Nanotechnology, 16, 2346–2353

Nangmenyi G, Economy J (2009) Nonmetallic particles for oligodynamic microbial disinfection. In: Savage NF et al (eds) Nanotechnology applications for cleans water. William Andrew, Norwich, NY, pp 3–15

Narducci D (2007) An introduction to nanotechnologies: what's in it for us? Vet Res Commun 31(suppl 1):131–137

NRC (2002) National research council animal biotechnology: science-based concerns. National Academy Press, Washington DC

Owolade OF, Ogunleti DO, Adenekan MO (2008) Titanium dioxide affects disease development and yield of edible cowpea. Elect J Env Agric Food Chem 7(50):2942–2947

Park HJ, Kim SH, Kim HJ, Choi SH (2006) A new composition of nanosized silica-silver for control of various plant diseases. Plant Pathol J 22:295–302

Rashidi L, Darani KK (2011) The applications of nanotechnology in food industry. Crit Rev Food Sci Nutr 51:723–730

Rickman D, Luvall JC, Shaw J, Mask P, Kissel D, Sullivan D (1999) Precision agriculture: changing the face of farming. Geotimes feature article. www.ghcc.msfc.nasa.gove/precisionag/. Accessed 19 Nov 2011

Rizvi SSH, Moraru CI, Bouwmeester H, Kampers WH (2010) Nanotechnology and food safety. Ensuring Global Food Safety263–280. ISBN: 978-0-12-374845-4

Sekhon BS (2010) Food Nanotechnology- An overwiev. Journal of Nanotechnology Science and Applications, 3, 1–15

Scott NR (2002) Rethink, redesign, reengineer. Resource 9(9):8–10

Scott NR (2007) Nanoscience in veterinary medicine. Vet Res Commun 31(1):139–141

Scott NR, Chen H (2003) Nanoscale science and engineering or agriculture and food systems. In: Roadmap Report of National Planning Workshop. Washington DC, 18–19 Nov 2002. http://www.nseafs.cornell.edu/web.roadmap.pdf

Siegrist M, Stampfli N, Kastenholz H, Kelleri C (2008) Perceived risks and perceived benefits of different nanotechnology foods and nanotechnology food packaging. Appetite 51:283–290

Silvestre C, Duraccio D, Cimmino S (2011) Food packaging based on polymer nanomaterials. Prog Polym Sci 36:1766–1782

Soghomonian V, Heremans JJ (2009) Characterization of electrical conductivity in a zeolite like material. Appl Phys Lett 95(15):152112

Sorrentino A, Gorrasi G, Vittoria V (2007) Potential perspectives of bio-nanocomposites

Sozer N, Kokini JL (2009) Nanotechnology and its applications in the food sector. Trends Biotechnol 27(2):82–89

Vorosmarty CJ, Mc Intyre PB, Gessner MO, Dudgeon D, Prusevich A, Green P (2010) Global threats to human water security and river biodiversity. Nature 467:55–561

Yang C, Cheng C, Yiyun X, Xiuzhen Z, Yuxiang ZC (2007) Liposomes Encapsulating 10 Hydroxycamptothecin-Cyclodextrin Complexes and Their in Vitro Anti-Tumor Activities. Journal of Nanoscience and Nanotechnology, (19), 3986–3795

Index

A. Malik et al. (eds.), *Food Processing: Strategies for Quality Assessment*, Food Engineering Series, DOI 10.1007/978-1-4939-1378-7

D

H

I

J

K

L

Q

CPSIA information can be obtained at www.ICGtesting.com
Printed in the USA
LVOW02*0703091114

412690LV00001B/12/P

9 781493 913770